Ocean Engineering & Oceanography

Volume 8

Series editors

Manhar R. Dhanak, Atlantic University SeaTech, Dania Beach, USA
Nikolas I. Xiros, New Orleans, USA

More information about this series at http://www.springer.com/series/10524

Luiz Bruner de Miranda
Fernando Pinheiro Andutta
Björn Kjerfve · Belmiro Mendes de Castro Filho

Fundamentals of Estuarine Physical Oceanography

Springer

Luiz Bruner de Miranda
Oceanographic Institute
University of São Paulo (IOUSP)
São Paulo
Brazil

Fernando Pinheiro Andutta
Griffith Climate Change Response Program (GCCRP), Griffith Centre for Coastal Management (GCCM)
Griffith School of Engineering
Gold Coast, QLD
Australia

Björn Kjerfve
American University of Sharjah
Sharjah
United Arab Emirates

Belmiro Mendes de Castro Filho
Oceanographic Institute
University of Sao Paulo (IOUSP)
São Paulo
Brazil

ISSN 2194-6396 ISSN 2194-640X (electronic)
Ocean Engineering & Oceanography
ISBN 978-981-10-3040-6 ISBN 978-981-10-3041-3 (eBook)
DOI 10.1007/978-981-10-3041-3

Library of Congress Control Number: 2016958487

Printed on acid-free paper

This Springer imprint is published by Springer Nature
The registered company is Springer Nature Singapore Pte Ltd.
The registered company address is: 152 Beach Road, #22-06/08 Gateway East, Singapore 189721, Singapore

Preface

The first manuscript of this book was written in 1984, during the sabbatical year of Dr. Luiz Bruner de Miranda, as approved by the Oceanographic Institute, São Paulo University (IOUSP-USP), and financed by the Brazilian fellowship program of the National Council for the Development of Science and Technology (CNPq). During that sabbatical, Mr. Miranda worked with Dr. Björn Kjerfve, Belle W. Baruch Institute of South Carolina University (USA), where he deepened his knowledge on estuarine dynamics. Since 1986, Dr. Miranda has continuously improved the first manuscript of this book to incorporate his graduate-level teachings of *Kinematics and Estuarine Dynamics*. This book's first edition (2002) was published in Portuguese by the São Paulo University's Editorial Board (Editora da Universidade de São Paulo (EDUSP)). In 2003, this edition won the first place in the *Exact Science, Technology and Informatics* category of the biennial year's book edition of the Brazilian Book Chamber (Câmara Brasileira do Livro). Since then, this book has also been adopted for teaching undergraduate-level students at IOUSP-USP. The second edition of this book was published in 2012 and, and with Dr. Kjerfve's encouragement, an extensively revised and updated English version of this book was approved by EDUSP.

Estuaries are transitional environments between the continent and the ocean where rivers empty into the sea, resulting in a measurable dilution of the saltwater. In natural conditions, estuarine water masses are more productive than rivers and ocean waters due to its hydrodynamic characteristics which arrest nutrients, seaweeds, and plants, thus stimulating the primary and secondary productivities in these fascinating water bodies.

Investigations of Swedish rivers (Götaelf river) published by F.L. Ekman in 1876, in the Nova Acta Reg. Soc., based on hydrographic and current measurements, indicated that the outflow of river water in the estuary was accompanied by an inflow of seawater in the lower layers with the circulation continuity provided by *upwelling* motions from the bottom water. However, it was only by the middle of the last century that these ecosystems, very much vulnerable to the human influence, were more intensively investigated. Ekman's investigations resulted in valuable scientific knowledge that contributed to the understanding of how estuaries function, which is of fundamental importance to the management of this complex ecosystem. In physical oceanography, the estuaries investigations are based on (i) observational experiments in laboratory and in the field; (ii) data interpretation based on theoretical and semiempirical knowledge; (iii) analytical simulations; and (iv) numerical modeling.

About 60% of the great cities around the world were founded and have its development nearby estuaries. However, the estuarine water renewal and depuration of these environments are dependent on the interactions of physical, biologic, chemical, and geological processes not well known, and the direct and indirect introduction of substances and energy by the man may reach high concentration levels, causing the contamination of its waters with harmful influence on the living resources, danger to the human health, damage to the marine activities and fisheries, and reduction of its natural attraction.

In the Brazilian coastline with approximately 8500 km, there are thousands of estuaries, estuarine systems, and coastal lagoons, with extensions of a few kilometers up to hundreds of kilometers, bordered by swamp and mangroves. In the north of Brazil, we find one of the most spectacular deltaic estuarine systems formed by the Amazon river and in the south, the Patos lagoon, the biggest in the South America. The estuarine systems contributed to the development of great and small Brazilian cities, suffering modifications in its geometry and circulation and, consequently, in the processes of erosion, transport, and sedimentation, due to alteration in the hydrographic basin, the water's natural cycle, and its quality due to the human's activities during the last centuries.

Due to the great importance of these coastal environments in the Oceanographic Institute of the University of São Paulo (IOUSP), and in other Brazilian universities, there are disciplines, in the undergraduate and graduate levels, related to the estuarine studies. In this context, the objective of this book is to present twelve chapters an integrated coverage of the fundamental principles of the physical oceanography of estuaries based on scientific articles and classical and recent books written on estuarine physics and interrelated studies. In this book, we also try to permeate by our experience and expertise which was acquired in the interpretation of experimental results and publications in these fascinating transitional ecosystems in the past 60 years.

In Chap. 1, the reader will find the details of the importance of estuaries, its formation, and its recent geological age. In this chapter, it is also presented classical concepts and most recent definitions, as well as political actions and laws established to the estuaries preservations. In Chap. 2 are described the geomorphologic conditions and the forces imposed by the river discharge, tidal and density (salinity) pressure gradients and wind, and its influence on the circulation. In this chapter also have been introduced some characteristics of the tidal propagation in a uniform estuarine channel, as well as the velocity associated with this oscillating motion. Due to the great diversity of the estuaries and the possibility to forecast some general characteristics of the acting forces, circulation, and mixing processes, in Chap. 3 are presented the main criteria to estuaries classification taking into account its genesis, topographic characteristics, and vertical salinity stratification. The classical stratification–circulation diagram, the most recent diagrams based on the Ekman and Kelvin estuarine numbers, and the prognostic estuary classification based on non-dimensional tidal and freshwater velocities are also presented.

Planning and execution of an oceanographic research in an estuarine environment are presented in Chap. 4, with the description of the main methods of measurement of hydrographic properties and currents. The methodology applied to reduce, coordinate, edit, and analyze the experimental data are also presented in this chapter. The main concepts evolved in the reduction and edition of observational data to calculate the advective and diffusive flux and transport of properties, which may also be applied to the concentration of conservative and non-conservative properties, are presented in Chap. 5.

Studies on the mixing processes in an estuary have the main objective to calculate the classical residence and flushing times of salt and other properties of concentrations introduced into estuaries. As the freshwater volume may be used, in the first approximation and steady-state conditions to forecast the longitudinal salinity variation and flushing times, in Chap. 6 are presented the so-called simplified methods of mixing.

The main concepts related to the momentum, mass, and salt conservation equations, which are the starting points to formulate the hydrodynamic equations governing the physical processes in an estuary, are presented in Chaps. 7 and 8. This formulation initiated with tridimensional equations is reduced to simplified equations to formulate the two- and one-dimensional analytical methods. Some applications of integrated formulations of the equations of continuity and motion are also presented in these chapters.

In Chaps. 9 and 10 are presented steady-state analytical solutions of one- and two-dimensional of the classical salt wedge and well-mixed estuaries. At the end of these chapters, solutions of the u-velocity component and salinity vertical profiles of these estuaries types are presented.

In Chap. 11, steady-state analytical solutions for one- and two-dimensional estuaries of the partially mixed type are presented, as well as some classical solutions to calculate the velocity and salinity profiles and the fundamental concepts related to the stratification–circulation diagram used in the estuary classification.

As estuaries are three-dimensional and time-dependent, to overcome the simplifications of simple geometry and steady-state formulations to calculate the estuarine circulation and salinity distributions, numerical models have been developed. These models integrated numerically at selected grid points spatially distributed in the system domain and the algebraic expression of governing partial differential equations using methods of finite difference or finite element in a curvilinear coordinate system with sigma or non-dimensional vertical coordinates. The main principles used in the numerical model applications and some results are presented in Chap. 12.

Programs in MATLAB® computational environment of the main analytical equations developed in this book's chapters, using the Morgan's SEAWATER library, may be accessed in the web site of the Laboratory of Coastal and Estuaries Hydrodynamics, LHiCo (Oceanographic Institute of São Paulo University).

São Paulo, Brazil — Luiz Bruner de Miranda
Gold Coast, Australia — Fernando Pinheiro Andutta
Sharjah, United Arab Emirates — Björn Kjerfve
São Paulo, Brazil — Belmiro Mendes de Castro Filho

Acknowledgments

We are grateful to Drs. Alberto dos Santos Franco (in memoriam) and José Galicia Tundisi for the encouragement to write the first edition of this book, to Dr. Björn Kjerfve to check the language, to the Editorial Board of São Paulo University (EDUSP) and National Council for the Development Science and Technology (i.e., Conselho Nacional de Desenvolvimento Científico e Tecnológico (CNPq)), and for the visiting professor grant by the South Caroline University (USA) and the participation in the Research Productivity Fellowship Program, enabling the conclusion of this book. Thanks to our colleagues of the Library of the Oceanographic Institute of USP for their assistance and help and our families, especially wives, sons, and daughters for their comprehension and support.

We expect that this academic book, written with the greatest collaboration and the patient work of colleagues, administration and library staff, students, technicians, and research vessels crew of the Oceanographic Institute of the University of São Paulo, may help students and researchers of this extraordinary and fascinating science field. This text which is given now to its disposal has been due to the efforts of thousands of research hours.

Acknowledgements to Publishers and Material Reproduced by Permission

Academia Brasileira de Ciências © 1998, article Oceanographic and Ecological Aspects of the Itajaí-Açu River Plume During High Period by C.A.F. Schettini *et al.* [*An. Acad. Bras. Ci.,* 70(2) (Figure 6.16).

Academic Press Ltd., London (UK) © 1991, article On Estimating the Non-advective Tidal Exchanges and Advective Gravitacional Circulation Exchanges in an Estuary, by C.B. Officer & D.R. Kester [Estuarine, Coastal and Shelf Science, 1] (Figure 6.2).

Academic Press Ltd., London (UK) © 1973, article A simple Segmented Prism Model of Tidal Mixing in Well-Mixed Estuaries, by K.R. Dyer & P.A. Taylor [Estuarine, Coastal and Shelf Science, 20] (Figures 6.4 and 6.8).

Academic Press Ltd., London (UK) © 1985, article On Salinity Regimes and the Vertical Structure of Residual Flows in Narrow Tidal Estuaries, by D. Prandle [Estuarine, Coastal and Shelf Science, 20] (Figures 8.3 and 8.4).

Academic Press Inc., Florida (USA) © 1986, article Comparative Oceanography of Coastal Lagoons, by B. Kjerfve, Estuarine Variability (ed. D.A. Wolfe) (Figure 3.18).

Academic Press Inc., Florida (USA) © 1979, book Mixing Inland and Coastal Waters, by H.B. Fischer et al. (Figure 5.21).

American Association for the Advancement of Science © 1968, article Sea Level Changes During the Past 35,000 Years, by J.D. Milliman & K.O. Emery [Science, 162] (Figure 1.1).

American Geophysical Union © 1951, article A Mixing Length Theory of Tidal Flushing, by A.B. Arons & H. Stommel [*Trans. Amer. Geophys. Union, 32(3)*] (Figure 10.8).

American Society of Civil Engineers © 1953, article The Salt Wedge, by H.G. Farmer & G. W. Morgan. *Proceedings of the 3rd Conference on Coastal Engineering* (Figure 9.3).

Annual Review Fluid Mechanics © 1976, article Mixing Dispersion in Estuaries, by H.B. Fischer [*Annual Reviews* (8)] www.AnnualReviews.org, (Figure 3.13).

Cambridge University Press © 1985, article Estuarine Chemistry and General Survey Strategy. *Practical Estuarine Chemistry*: A Handbook (ed. P.C. Head, by A.W. Morris) (Figure 3.8).

Cambridge University Press © 2010, article Definition and classification of estuaries. Contemporary Issues in Estuarine Physics (ed. Arnoldo Valle-Levinson) (Figure 3.16).

Cambridge University Press © 2010, article Estuarines structure and circulation. Contemporary Issues in Estuarine Physics (ed. Arnoldo Valle-Levinson) (Figure 3.17).

Elsevier Science © 1997, article The Logarithmic Layer in a Tidal Channel, by R.G. Lueck & Y. Lu [*Continental Shelf Research*, 17(14)] (Figure 5.7).

John Wiley & Sons, Ltd © 1973 and 1997, book Estuaries: A Physical Introduction, 1st (1973) and 2nd (1997) ed. by K.R. D year, (reproduced and modified figures: 2.11, 2.12, 2.13, 3.1, 3.4 and 3.11).

John Wiley & Sons, Ltd © 1980, article The Estuary: Its Definition and Geodynamic Cycle, by R.W. Fairbridge. *Chemistry and Biochemistry of Estuaries* (eds. E. Olausson & I. Cato) (Figure 3.2).

Journal of Brazilian Association for Science Advancement © 1993, article Tidal Bores: First Ever Measurements, by B. Kjerfve & H. O. Ferreira [Ciência *e Cultura* 45(2)] (Figure 2.7).

Journal of Geophysical Research © 2000, article Convergence of lateral flow along coastal plain estuary, by Valle-Levinson; Li, C.; Wong, K-C & Lwiza, K.M.M., v. 105, NO C7 (Figure 11.10).

Journal of Marine Research © 1965, article Gravitational Circulation in Straits and Estuaries, by D.V. Hansen & M. Rattray jr., [J. Mar. Res. 23(1)] (Figures 11.5 and 11.6).

Louisiana State University Press © 1971, article Analysis of Major River Systems and their Deltas: Procedure and Rationale, with Two Examples, by M. Coleman & L. D. Wright (Figure 2.1).

Springer-Verlag GmbH & Co. Kg. © 1988, article Residual Circulation and Classification of Shallow Estuaries, by D.A. Jay & J.D. Smith © (1988). *Physical Processes in Estuaries* (eds. J. Dronkers & W. Leussen) (Figure 3.15).

The Humana Press Inc. © 1989, article Estuarine Classification — A Help or a Hindrance, by D. W. Pritchard © (1989), In: *Estuarine Circulation* (eds. B.J. Neilson, A. Kuo & J. Brubaka) (Figures 3.3, 3.5, 3.6 and 3.7).

The Oceanography Society © 1991, article The Physical Oceanography of the Amazon Outflow, by W.R. Geyer et al. [*Oceanography* (4)] (Figure 1.4).

The Open University Press © 1995 and Team Course 2001 *Ocean Circulation* (Figure 7.1).

University of South Caroline Press © 1978, article Some Simplified Tidal Mixing Flux Effects in Estuaries, by C. B. Officer, In: *Estuarine Transport Processes* (ed. B. Kjerfve) (Figures 6.12 and 6.13).

U. S. Geological Survey © 1972, article Measurement of Salt-wedge Excursion Distance in the Duwamish River Estuary, Seattle, Washington, by Means of Dissolved-oxygen Gradient, by W. A. Dawson & L. J. Tilley (Figure 9.1).

Water Environmental Federation © 1955, Alexandria, VA, article Distribution of Coliform Bacteria and Other Pollutants in Tidal Estuaries, [*Sewage Industl. Wastes* (27), by B. H. Ketchum (Figure 6.14).

Contents

About the Authors

Luiz Bruner de Miranda is an emeritus and senior professor of the Oceanographic Institute of São Paulo University. His career started in 1959 as a trainee at the very beginning of his graduate studies in Physics in the Faculty of Sciences and Letters of the São Paulo University. After his Lic. degree in Physics, he was contracted as an oceanographer and later as a head of the Marine Meteorology Section of the IOUSP. He was awarded a UNESCO fellowship in 1966 for theoretical studies in physical oceanography in the Geophysical Institute of Bergen (Norway) under the advice of Dr. Thor Kvinge and Haakon Mosby. During his studies in Bergen (Norway), he participated in several oceanographic cruises in fjords, Barents Sea and in the North Atlantic Ocean up to northern latitudes (70 °N). At the end of this program, the R/V Prof. W. Besnard was board in the Mjellem & Karlsen Shipyard (Norway, Bergen) and participated in the trial cruise to Shetland Islands and in the inaugural cruise of the Brazilian research vessel in a Joint Program (Brazil-Norway), to realize the investigations of Oceanographic Fisheries Research in the west coast of Africa (Cape Green/Canary Islands) and physical oceanographic work in Brazilian territorial waters. He started his carrier as a professor of physical oceanography in 1969 with Ph.D. from the Institute of Physics of the University of São Paulo. He underwent postdoctoral CNPq fellowship program with Dr. Björn Kjerfve in the South Caroline University (USA). He supervised 6 Ph.D. and 22 M.Sc. oceanography students (2 from Colombia and 1 from Taiwan) and 8 undergraduate students. He published 60 scientific articles, chapters, and books. Prêmio Jabuti (Câmara Brasileira do Livro) for the publication of *Princípos de Oceanografia Física de Estuários*, L.B. de Miranda, B.M. de Castro, and B. Kjerfve, 2003 (1st edition 2002 and 2nd edition 2012). Achievement awards are as follows: "Emeritus Professor (1989)," "Destaque em Pesquisa (USP-1992)," "Amigo da Marinha (2002)," and "Premio Jabuti" (CBL-2003). He was a visiting professor to the University of Conception (UNESCO-Bilateral Joint Course) and Escuela Naval de Cartagena (Colombia). He was the elected member of Academia de Letras, Ciências e Artes—AFPESP.

Fernando Pinheiro Andutta was a researcher fellow at the School of Engineering at Griffith University since August 2014; researcher at the Sino-Australian Research Centre for Coastal Management (SARCCM) since April 2012; associate lecturer at the University of New South Wales 2012–2014; researcher officer at James Cook University 2011–2012; and research assistant from 2008 to 2011. Dr. Andutta completed a technical course in electronics at Etec Aristóteles Ferreira in Santos (February 1996–November 1999). He then completed his B.Sc. (Honors) in mathematics at Unimonte in Brazil (February 2000–November 2003) and attended the Preparation Center of Officers R2 in Brazil (year 2000). Fernando did his M.Sc. (February 2004–June 2006) and his D.Sc. (January 2007–April 2011) in physical oceanography at the University of São Paulo, Brazil. Recently, he has completed his Ph.D. in applied physics at James Cook University in Australia (June 2008–March 2012). Dr. Andutta has been awarded scholarships for over 12 years to complete his B.Sc., M.Sc., Ph.D., and D.Sc. between 2000 and 2012.

Björn Kjerfve was a chancellor and professor of environmental sciences, American University of Sharjah, UAE, 2014; president of World Maritime University, Malmö, Sweden, from 2009 to 2014; professor/director of marine science, University of South Carolina, from 1973 to 2004; professor/dean, Texas A&M University; oversaw the Science Operation, IODP and *D/V Joides Resolution* Visiting Professor at the Federal Brazilian University (Universidade Federal Fluminense-UFF) during twenty years (1983–2003). He obtained B.A. in mathematics from Georgia Southern University in 1968; M.S. in oceanography from the University of Washington in 1970; and Ph.D. in marine sciences from Louisiana State University in 1973. He supervised 14 Ph.D. and 24 M.S. oceanography students—14 from Brazil; he taught 6000 oceanography students. He published 12 books, 235 scientific papers, chapters, and reports; he was elected as a corresponding member of Academia Brasileira de Ciências by the General Assembly on December 4, 2012, and inaugurated on May 7, 2013; he was awarded Coast Guard Legion of Honor (Degree of Maginoo), Manila, Philippines, on January 28, 2013; Prêmio Jabuti (Câmara Brasileira do Livro) for the publication of *Princípos de oceanografia física de estuários*, L.B. Miranda, B.M. Castro, and B. Kjerfve, 2003, second edition published 2012. Dubai Shiptek Lifetime Achievement Award 2011 for maritime academics; AOCEANO 2008 medal for 25 years of involvement with oceanography in Brazil; Scientific Council, Maria Tsakos Foundation, Greece 2012; International Executive Board, IAMU, Japan 2009–2014; Board of Governors IODP-MI 2008–2009, JOI 2004–2007, CORE 2004–2007 USA; Conselho Deliberativo, Instituto Acqua, Rio de Janeiro 1993–1996; Vice President, Coastal and Estuarine Research Federation, USA 1983–1985.

Belmiro Mendes de Castro Filho was a professor of physical oceanography at the Oceanographic Institute of the University of Sao Paulo and director of the same institute for the term 2001–2005. The author has a Ph.D. in physical oceanography and meteorology from the Rosenstiel School of Marine and Atmospheric Science (University of Miami, USA). Dr. Castro's main research interests are continental shelf dynamics, including the interactions between shelf and estuarine waters, and physical processes in marine pollution. In his work, the author uses mainly data collection and analysis, and numerical modeling as tools. In the 1990s, Castro was twice a visiting scientist to the Woods Hole Oceanographic Institute, Massachusetts, USA. He has published over 30 papers and 10 chapters in scientific journals and specialized books. Castro has also mentored and coadvised more than 20 graduate students. The author has been a coordinator of the Coastal Hydrodynamics Laboratory at the University of Sao Paulo for more than 25 years.

Symbols and Dimensions

Arabic Associated with Geometric Characteristics

$A(x, t)$ and $A(x)$	Longitudinal and time variation of a transverse section area and its steady state $[A] = [L^2]$
A_1, A_3 and A_2, A_4	Steady-state areas in the Knudsen theorem $[A_i] = [L^2]$
A_{su}	Surface area of the estuary $[A_{su}] = [L^2]$
A and A_T	Steady-state total areas $[A] = [A_T] = [L^2]$
$a(x, z)$	Right estuary margin coordinate $[a(x, z)] = [L]$
$b(x, z)$	Left estuary margin coordinate $[b(x, z)] = [L]$
B $[B = a(x, z) - b(x, z)]$	Estuary width $[B] = [L]$
CLC	Coastal boundary limit
CZ	Coastal zone
D	Estuary depth width $[D] = [L]$ or non-dimensional vertical scale
g	Gravity acceleration (≈ 9.8 m s^{-2}) $[g] = [L\ T^{-2}]$
$g' = g(\rho_2 - \rho_1)/\rho_2$	Reduce gravity $[g'] = [L\ T^{-2}]$
H_0	Water column height or depth $[H_0] = [h] = [L]$
H_1	Upper layer thickness $[H_1] = [L]$
$h(x, y, t) = H_0 + \eta(x, y, t)$	Local depth time variation $[h(x, y, t)] = [L]$
$H(x, y, t) = h(x, y) + \eta(x, y, t)$	Local depth with longitudinal and transverse bottom variations $[H(x, y, t)] = [L]$
$h(x, t)$	Local depth time variation $[h(x, t)] = [LT^{-1}]$
$h_1(x)$ and $h_2(x)$	Steady-state thicknesses of layers of salt wedge estuary $[h_{1,2}(x)] = [L]$
h_a, $\langle h \rangle$ or $h(x)$	Steady-state depth value $[h_a] = [L]$
$h_t(x, t) = h(x, t) - h_a$	Free surface time oscillation $[h_t] = [LT^{-1}]$
L	Estuary longitudinal length $[L] = [L]$
MZ	Mixing zone
NTZ	Nearshore turbidity zone

Oxyz	Cartesian reference system: Ox oriented positively seaward (or land ward), Oy in the transverse direction, and Oz upward (opposite to the gravity acceleration) or downward (in the gravity acceleration direction) with origin (z = 0) at the free surface or the bottom
TRZ	Tidal river zone
V	Geometric estuary volume [V] = $[L^3]$
x, y, z	Longitudinal, transverse, and vertical coordinates [x, y, z] = [L]
V	Geometric estuary volume [V] = $[L^3]$
X = x/L	Non-dimensional longitudinal distance
Y = y/L	Non-dimensional transverse distance
$Z = z/H_0)$	Non-dimensional depth (or sigma coordinate)

Arabic Associated with Physical Properties

a	Mean thermal expansion coefficient [a] = $[^oC^{-1}]$ or mixing parameter (dimensionless)
$a_1(X)$, $a_2(X)$, $a_3(X)$, $a_4(X)$	Non-dimensional integration variables
$A_{xx} = A_{xy} = A_x$	Turbulent dynamic viscosity coefficients $[A_{xx} = A_{xy} = A_x] = [M\ L^{-1}\ T^{-1}]$
$A_{yx} = A_{yy} = A_y$	Viscosity turbulent coefficients $[A_{yx} = A_{yy} = A_y] = [M\ L^{-1}\ T^{-1}]$
$A_{xz} = A_{yz} = A_z$	Viscosity turbulent coefficients $[A_{xz} = A_{yz} = A_z] = [M\ L^{-1}\ T^{-1}]$
$b_1(x)$, $b_2(x)$, $b_3(x)$, $b_4(x)$, b_5, b_6	Dimensionless integration variables
2B	Dimensionless proportional coefficient
B_c	Spectral width band $[B_c]$ = [L]
C	Any property concentration [C] = $[M\ M^{-1}]$ or [C] = $[M\ L^{-3}]$
c or c_o	Wave celerity or of phase velocity of the tidal wave [c] = $[c_o]$ = $[L\ T^{-1}]$
$c_o = W/Q_f$	Initial concentration of an effluent thrown into a river $[c_0] = [M\ L^{-3}]$
$c_o^* = W/Q_d$	Initial concentration of an effluent thrown in the MZ $\left[c_0^*\right] = \left[M\ L^{-3}\right]$
$(c_x)_d$ and $(c_x)_u$	Concentration of any property down and up the estuary, respectively $[c_x] = [M\ L^{-3}]$ or $[c_x]$ = $[M\ M^{-1}]$
$C_n^H = f_n^H$, $C_n^L = f_n^L$	Freshwater concentration in high and low tides (generic segment) [dimensionless]
$C_o(C_o = 1)$	Freshwater fraction in the estuary head [dimensionless]

C_1, C_2, C_3	Integration constants [dimensional or dimensionless]
C_D, C_E	Shear coefficient (dimensionless)
C_P	Circulation parameter (dimensionless)
CZ	Coastal zone
C_y	Chézy coefficient $[C_y] = [L^{1/2}\ T^{-1}]$
C_n^H	Freshwater concentration with freshwater fraction in a generic segment (dimensionless)
C_{100}	Dimensionless friction coefficient at 1 m from the bottom
db	Pressure unity in decibars (1 db = 10^{-1} bar = 10^5 dynes.cm^{-2}) [db] = $[M\ L^{-1}\ T^{-2}]$
D	Magnetic declination [D] = [°], mixing layer depth [D] = [L] or molecular diffusion coefficient $[M\ L^{-1}\ T^{-2}]$, of the non-dimensional vertical length scale
D_e	Salt-wedge thickness scaled by a constant value and non-dimensional relative salt-wedge intrusion scaled by the estuary length $[D_e] = [L]$
D_i	Pipe diameter $[D_i] = [L]$
D_m	Longitudinal displacement generated by the tide $[D_m] = [L]$
D_t	Thickness of the bottom turbulent layer $[D_t] = [L]$
dd	Current direction [dd] = [°]
dd_v	Wind direction $[dd_v]$ = [°]
E_k	Ekman number (dimensionless)
E, E_v	Evaporation, potential evapotranspiration [E] = [L], $[E_v] = [L\ T^{-1}]$
E_M	Tidal excursion $[E_M] = [L]$
$e = H/H_0$	Ratio of the tidal height by the mean depth (dimensionless)
$\exp(x) = e^x$	Exponential function (dimensionless)
F	Flushing rate, discharge number $[F] = [L^3\ T^{-1}]$, or the dimensionless flushing number
F_B	Internal Froude number (dimensionless)
F_1	Flushing rate due to the mixing process (dimensionless)
F_i	Interfacial Froude number (dimensionless)
F_m or γ	Densimetric Froude number (dimensionless)
F_R	Flux ratio (dimensionless)
F_T	Barotropic Froude number (dimensionless)
F_w	Mean freshwater concentration in a transversal section (dimensionless)

$f = f(i, j, k, t)$	A non-dimensional arbitrary discrete function in space and time
f_i^n, f_i^{n+1}	A mean time discrete value and a forward time step (dimensionless)
f, f_o, $f(x, y, z)$	Freshwater fraction (dimensionless)
$f(x, y, z, t)$	$[f]=[LT^{-1}]$
f_a	Sorting frequency $[T^{-1}]$
f	Mean freshwater fraction in the MZ (dimensionless)
f_n	Freshwater fraction in a generic segment (dimensionless)
f_0	Coriolis parameter $[T^{-1}]$
f_i	Interfacial stress $[M\ L^{-1}\ T^{-2}]$
∂$/dZ	Dimensionless vertical salinity gradient
f_x	Mean water concentration of freshwater in a transversal section (dimensionless)
$\vec{g}$, and g	Gravity acceleration vector and modulus $[L\ T^{-2}]$
$g\Delta\rho/\rho$	Reduced gravity $[LT^{-2}]$
G	Tidal kinetic energy dissipation tax (per mass) $[L^2\ T^{-3}]$ or internal Froude number (dimensionless)
G/J	Stratification parameter (dimensionless)
$H(x)$	Salt-wedge non-dimensional thickness $[H(x) = h_2(x)/H_0]$
H_o	Tidal height or mean water depth [L]
H_1	Upper layer depth $[H_1] = [L]$
H_m	Salt wedge dimensionless thickness at the estuary mouth
H_{max}	Maximum tidal height $[H_{max}] = [L]$
h_l	Logarithmic thickness height [L]
h_m	Salt-wedge depth at the estuary mouth [L]
I	Temperature correction index [°C]
J	Potential energy tax gain (per mass) $[L^2\ T^{-3}]$
K_e	Kelvin number (dimensionless)
K_1	Solar diurnal main tidal component $[K_1] = [L]$
K_2	Main semidiurnal tidal component [L]
K_T	Volumetric expansion coefficient $[^{o}C\ L^{-3}]$
K_x, K_y, K_z	Kinematic eddy diffusion coefficient $[L^2\ T^{-1}]$
K_{x0}	Longitudinal kinematic eddy diffusion coefficient (in $x = 0$) $[L^2\ T^{-1}]$
k	Bed friction coefficient (non-dimensional) $k = 2.5 \times 10^{-3}$) or $k = g/C_y^2$
l	Wire length [l] = [L]

L	Dimension of length
L_{length}	Theoretical salt-wedge length [L]
LS_i	Saline intrusion length [L]
ln	Neperian logarithmic
M_2, S_2, N_2	Main semidiurnal tidal components [L]
M_S	Instantaneous advective salt transport [M T^{-1}]
[M] and M	Mass dimension and mixing tidal parameter (dimensionless)
MZ	Mixing zone (dimensionless)
m	Mass or mass of dissociated salts [M]
n	Manning number [$L^{1/6}$]
N	True north [°]
NM	Magnetic north [°]
N_b	Buoyancy frequency [N_b] = [T^{-1}]
N_f	Tidal form number (dimensionless)
$N_x = A_x/\rho$ and $N_z = A_z/\rho$	Longitudinal and vertical kinematic eddy viscosity coefficient [L^2 T^{-1}]
NTZ	Nearshore turbid zone (dimensionless)
O_1	Solar diurnal main tidal component
P	Precipitation height or precipitation tax [L] or [L T^{-1}] or tidal period [T]
P(x, y, z, t)	Generic property in space and time dependent
P_m	Wet perimeter [L]
P_n	Tidal prism in a generic segment [L^3]
Pa (1kPa = 1db)	Pressure unity [M L^{-1} T^{-2}]
P(Z, t) or P(z, t)	Instantaneous vertical profile of generic property
$\langle P(Z_j) \rangle$ and $\overline{P}(t)$	Mean values of a generic property in time and space
p and p_a	Pressure and atmospheric pressure [M L^{-1} T^{-2}]
psu or symbol $^o/_{oo}$	Practical salinity unity [M M^{-1}]
Q	Volume transport [L^3 T^{-1}]
Q_d (or $\equiv$F)	Equivalent freshwater transport (or discharge tax) (non-dimensional)
Q_f [or $Q_t(t)$], Q^* and q_i	Freshwater discharge (or river discharge) [L^3 T^{-1}]
Q_b or Q_1	Volume transport per width [L^2 T^{-1}]
Q_1	Solar diurnal main tidal component [Q_1] = [L]
Q_2	Salt wedge volume transport [L^3 T^{-1}]
R ($R = TQ_f$)	River water volume discharged in a tidal cycle [L^3]
R/T_P	Flux ratio (dimensionless)
Ra	Estuarine Rayleigh number (dimensionless)
R_d	Deformation radius (dimensionless)
Re	Reynolds number (dimensionless)
Ri	Richardson number (dimensionless)

Ri_e	Estuarine Richardson number (dimensionless)
Ri_g	Gradient Richardson number (dimensionless)
Ri_L	Layer Richardson number (dimensionless)
R_t	Conductivity ratio (unity in mhos/cm)
R_H	Hydraulic radius [L]
r_n, r_n^*	Exchanged ratio in a generic segment (dimensionless)
S, S(x, y, z, t), S(x, y, t), S(x, t) and S(t)	Salinity (dimensionless)
S_A	Salinity in the absolute scale (dimensionless)
$S_a = \langle \overline{S} \rangle$	Steady-state salinity value (dimensionless)
S_n	Uniform salinity in a generic segment (dimensionless)
S_n^H, S_n^L	High- and low-tide salinity in generic segment (dimensionless)
S_N^H, S_N^L	High- and low-tide salinity at the mouth (dimensionless)
S_P	Stratification parameter or practical salinity (dimensionless)
S_s, S_f	Steady-state salinity values, at the surface and bottom (dimensionless)
S_0	Salinity in the coast region (not diluted) (dimensionless)
$S_s = \langle S \rangle - S_a$	Baroclinic salinity component (dimensionless)
$S_t = \overline{S} - S_a$	Barotropic salinity component (dimensionless)
S_x	Longitudinal salinity gradient ($\partial S/\partial x$) [L^{-1}]
S_y	Transversal salinity gradient ($\partial S/\partial y$) [L^{-1}]
S'	Salinity deviation (dimensionless)
$\langle \overline{S} \rangle = \overline{S}$	Steady-state salinity value in a water column (dimensionless)
$\overline{S}(t)$	Time mean salinity variation in the water column (dimensionless)
S/S_0	Dimensionless salinity
$\frac{S_f - S_s}{\overline{S}} = p_e$	Stratification parameter (dimensionless)
$\$ = \overline{S}/S_o$	Dimensionless mean salinity in the water column
$(S_n/S_0)^H, (S_n/S_0)^L$	Dimensionless relative salinity at high and low tides
S_ρ/F_u	Proportional ratio of the mixing parameter (dimensionless)
Sigma-t (or σ_t)	Density anomaly at atmospheric pressure (dimensionless)
T	Temperature (°C or °K), tidal period (T), or time dimension [T]

T'	Thermometric reading (°C)
T_E	Tidal excursion [L]
T_{PR} and V_e	Tidal prism [L^3]
T_V	Advective volume transport [$L^3\ T^{-1}$]
$(T_V)_L$	Mean advective volume transport per width unity [$L^2\ T^{-1}$]
T_M	Advective mass transport [$M\ T^{-1}$]
TM	Tidal exchange ratio (dimensionless)
T_C	Net advective transport of any property [Pr/T]
T_S	Mean advective salt mass transport [$M\ T^{-1}$]
t	Time [T]
t_a	Auxiliary thermometer reading [°C]
t_q	Flushing time—time interval in tidal period unit (dimensionless)
$(t_q)_n$	Flushing time in a generic segment [T]
t_q^H, t_q^L	Flushing time at high and low tides [T]
u(x, t)	Longitudinal velocity time variation or tidal induced velocity [$L\ T^{-1}$]
$\hat{u}$	Tidal amplitude [L]
$\vec{u} = [u, v, w]$	Velocity vector and components (Ox, Oy, and Oz, respectively) [$L\ T^{-1}$]
$\vec{u}' = [u', v', w']$	Eddy velocity and components (Ox, Oy, and Oz, respectively) [$L\ T^{-1}$]
$\bar{u}$	Mean velocity in the water column [$L\ T^{-1}$]
$<\bar{u}> = u_a \approx u_f$	Steady-state value of the u-velocity component [$L\ T^{-1}$]
u_B	Bottom velocity [$L\ T^{-1}$]
u_e	Entrainment velocity [$L\ T^{-1}$]
u_f	Freshwater velocity [$L\ T^{-1}$]
u_g	Gravitational circulation intensity [$L\ T^{-1}$]
$u_Q(z)$ and $u_W(z)$	Velocity profile generation of river discharge and wind [$L\ T^{-1}$]
u_{rms}	Root mean square of velocity (Ox component) [$L\ T^{-1}$]
$u_s = <u> - u_a$	Baroclinic velocity (Ox component) [$L\ T^{-1}$]
$u_t = \bar{u} - u_a$	Barotropic velocity (Ox component) [$L\ T^{-1}$]
$u_e(x, t)$	Stationary wave velocity (Ox component) [$L\ T^{-1}$]
u_f, $u_f(x, t)$	Freshwater velocity and its time variation [$L\ T^{-1}$]
u_s	Velocity at the surface (Ox component) [$L\ T^{-1}$]
u_s/u_f	Circulation parameter (dimensionless)
u_*	Friction velocity [$L\ T^{-1}$]
u_1 and u_2	Upper and lower layer velocity in the salt wedge [$L\ T^{-1}$]

u_{100}, u_{360}	Velocity values at 1.0 and 3.6 m above the bottom [L T^{-1}]
U	Wind intensity in the Ox direction or mean velocity in a pipe [L T^{-1}]
U_0 and U_T	Tidal velocity amplitude [L T^{-1}]
U_r	Reference velocity [LT^{-1}]
U_V and V_V	Ox and Oy wind velocity components, respectively [L T^{-1}]
U_s	Stokes drift velocity [L T^{-1}]
V	Volume [L^3]
V_M, T_P	Water volume at high tide (tidal prism) [L^3]
V_0	Mercury volume in the thermometric bulb at 0 ° C [L^3]
V_f	Freshwater volume in the mixing zone (MZ) [L^3]
$(V_f)_n$	Freshwater volume in a generic segment (n) [L^3]
V_{fn}^H, V_{fn}^L	Generic freshwater volume at high and low tides [L^3]
V_n	Volume at low tide in a generic segment [L^3]
V_N	Volume at low tide at the estuary mouth [L^3]
X_c	Salt-wedge length [L]
W	Effluent mass transport [M T^{-1}]
We	Wedderburn number
z_0	Roughness dynamic length (reflecting bottom roughness) [L]

Greek Associated with Physical Properties

α (α_0)	Specific volume (reference value) of seawater [L^3/M]
β	Coefficient of volume contraction (mean value $\approx 7.0 \times 10^{-4}$) (dimensionless)
γ	Rotation angle of the coordinate system [°] or dimensionless ratio associated with a salt wedge
δ	Factor of the reduced gravity (dimensionless)
δ_1, δ_2	Numeric values (dimensionless)
Δf	Surface runoff [L T^{-1}]
$\Delta f/P$	Runoff ratio (dimensionless)
Δp_a	Atmospheric pressure variation [M L^{-1} T^{-2}]
Δt	Time interval [T]
Δt_m	Minimum sampling time interval [T]
ΔT	Temperature correction (volumetric expansion) [°]
$\Delta \rho$	Density variation [M L^{-3}]

$\Delta\rho_H$	Density variation along the horizontal distance $[M\ L^{-3}]$
$\Delta\eta$	Surface elevation interval [L]
Δx	Interval of longitudinal distance [L]
Δz, ΔZ	Depth interval, non-dimensional depth interval [L]
Δf	Variation of freshwater volume $[L^3]$
$\delta S/<\overline{S}>$ or $\delta S/\overline{S}$	Stratification parameter ($\delta S = S_f - S_s$) (dimensionless)
$\delta f/\delta x$, $\delta f/\delta y$, $\delta f/\delta z$, $\delta f/\delta t$	Discrete differences in space and time of a function (dimensionless)
$\eta(t)$, $\eta(x, t)$, $\eta(x, y, t)$, $\eta(x)$	Tidal oscillation relative to a level surface and its steady state $[L\ T^{-1}]$ and [L]
$\partial\eta/\partial x \approx d\eta/dx = \eta_x$	Sea surface slope (dimensionless)
$\partial\rho/\partial x \approx d\rho/dx = \rho_x$	Longitudinal density gradient $[M\ L^{-4}]$
$\partial\rho/\partial y \approx d\rho/dy = \rho_y$	Transverse density gradient $[M\ L^{-4}]$
$\partial S/\partial x \approx dS/dx = S_x$	Longitudinal salinity gradient $[L^{-1}]$
$\partial S/\partial y \approx dS/dy = S_y$	Transverse salinity gradient $[L^{-1}]$
η_e	Sea surface elevation relative to a level surface (dimensionless)
η_o	Tidal wave amplitude [L]
η_1	Upper layer ordinate of surface slope of a salt wedge [L]
η_2	Upper layer ordinate of a salt wedge [L]
θ	Current velocity direction [°]
ϑ	Wind velocity direction [°]
ρ (ρ_0)	Density of estuarine water mass (reference density) $[M\ L^{-3}]$
ρ_{ar}	Air density $[M\ L^{-3}]$
ρ_η	Density at the surface $[M\ L^{-3}]$
$\overline{\rho}$	Mean density of water column $[M\ L^{-3}]$
$\Delta\rho_H$	Horizontal density difference in the MZ $[M\ L^{-3}]$
$\Delta\rho_V$	Vertical density difference in the MZ $[M\ L^{-3}]$
ΔS_V	Vertical salinity difference in the MZ (dimensionless)
α, $\alpha(S, T, p)$	Specific volume and *in situ* specific volume $[L^3\ M^{-1}]$
ξ ($\xi = u_f x/K_{x0}$)	Dimensionless horizontal distance
λ	Wave length [L]
κ	First-order linearized friction coefficient $[L\ T^{-1}]$
$\kappa = 0.4$, $\kappa = 2\pi/\lambda$	Von Kármán dimensionless and (wave number) $[L^{-1}]$
Φ_a, Φ_d	Advective and diffusive salt flux $[M\ L^{-2}\ T^{-1}]$
ϕ or Φ	Phase angle [°]
$\nu[\nu = \Phi_d/(\Phi_d + \Phi_a]$	Mixing parameter (dimensionless)

ν_c	Kinematic molecular viscosity coefficient $[L^2\ T^{-1}]$
ω ($\omega = 2\pi/T$)	Angular frequency $[T^{-1}]$
Ω	Angular earth velocity modulus $[T^{-1}]$
σ_t (or Sigma-t)	Density anomaly at atmospheric pressure $[M\ L^{-3}]$
τ	Shear stress $[M\ L^{-1}\ T^{-2}]$
τ_{xx}, τ_{xy}, τ_{xz}, τ_{yx}, τ_{yy}, τ_{yz}, τ_{zx}, τ_{zy}, τ_{zz}	Reynolds shear stress components $[M\ L^{-1}\ T^{-2}]$
τ_{Wx} or τ_W	Wind shear stress in the Ox component $[M\ L^{-1}\ T^{-2}]$
τ_{Bx} or τ_o	Bottom shear stress $[M\ L^{-1}\ T^{-2}]$
τ_{W0}	Reference value of the wind shear stress $[M\ L^{-1}\ T^{-2}]$
τ_{B0}	Reference value of the bottom shear stress $[M\ L^{-1}\ T^{-2}]$
$T_w = \tau_{Wx}/\tau_{W0}$	Dimensionless wind stress
$T_B = \tau_{Bx}/\tau_{B0}$	Dimensionless bottom shear stress
Φ	First positive arch value of an argument [°]
φ	Clinometer-measured angle [°]
φ_a, φ_d	Longitudinal advective and diffusive salt mass flux $[M\ L^{-2}\ T^{-1}]$
φ_V	Longitudinal advective volume flux $[L^1\ T^{-1}]$
χ	Anomaly of potential energy $[M\ L^2\ T^{-2}]$
φ_M	Advective flux of mass $[M\ L^{-2}\ T^{-1}]$
χ	Dissipative constant, proportional to the inverse of the wave number $[L^{-1}]$
Ψ	Dimensionless parameter proportional to the wind velocity
$\Psi = \Psi(x, z)$	Two-dimensional current function $[L^2\ T^{-1}]$
$\Psi = \psi/Q_f$	Dimensionless current stream function (ψ is a tridimensional current function $[\psi] = [L^3\ T^{-1}]$
$\psi = \psi(x, y)$	An arbitrary two-dimensional function $[L^2]$ or a dimensionless parameter
$\zeta = \zeta(x, t)$	Used in the place of the u-velocity component $[L\ T^{-1}]$
ε	Dimensionless normalized error value

Chapter 1
Introduction to Estuary Studies

1.1 Why to Study Estuaries?

The word estuary is derived from the Latin word *aestus* which means tide, also the adjective *aestuarium* has tidal or abrupt high wave as a meaning, an environment highly dynamic with changes due to natural forces. This word is generally used to indicate the place where the river meets the sea, characterizing a coastal river discharge. It is a transitional ecosystem between the continent and the ocean: complexity, variability and the human interference are the usual characteristic of all estuaries. In normal conditions they are biologically more productive than the rivers and the adjacent ocean, due to their high nutrient concentrations which stimulate primary production.

Fresh water discharge, incoming sea water from the continental shelf, associated suspended sediment, and organic and inorganic transport of substances are processes that are of great importance to the urban, social and economic development of the neighborhood of the estuarine regions. Some of these concentrations are used for the metabolism of marine organisms, and the pollutants which may also be carried and mixed with the natural substances, and are hazardous to the great variety of the marine biota and to the health of the populations which use these natural resources. According to Cronin (1967) it is, therefore, appropriate to identify and consider the past effects of man on the fundamental processes in estuaries and to contemplate future beneficial and detrimental influences on these fascinating, complex and important waters.

The influence of human activities on estuaries was not recognized as important until the second half of the XIX century. Previously, human impacts were limited to the outflow of domestic pollutants and, remotely, to the agriculture and farms due to the increasing discharges of the land sediments into the rivers. Thereafter, there was a huge expansion near the estuaries of the activities in industrial plants (manufacturers of a great diversity of material and substances), in agriculture (with the utilization of fertilizers and defensives), reservoir constructions (for power plants

L. Bruner de Miranda et al., *Fundamentals of Estuarine Physical Oceanography*,
Ocean Engineering & Oceanography 8, DOI 10.1007/978-981-10-3041-3_1

installation), as well as, natural harbors. As a consequence of expanding population in the cities, and the intensive water intake from river drainage basins, disturbing the natural equilibrium of estuarine and coastline water masses, there was a great increase in the human impact the sustainable development of these natural environments is put at risk.

The economic activity increase was ultimately associated with the estuaries due to the following reasons: ideal locations to place harbors and ports; high quantities of organic matter are produced by their fertility; a convenient inland navigation route as such, its water masses are periodically renewed under the river and tidal forcing. As an ecosystem, several vital functions are performed by estuaries, such as those presented by Ketchum (1983): a natural wildlife (birds, animals and fishes), a hatching place and nursery environment for several biological communities, and of fundamental importance as migratory routes for commercial fishes.

One motive for estuarine studies is to understand how this complex system functions. In the area of Physical Oceanography, experiments are of fundamental importance, they use these water bodies as a natural laboratory to interpret their circulation and mixing processes using theoretical and semi-theoretical knowledge. Small scale physical models may also be used together with analytical and numerical models. As in others areas of science, experiments and theory are activities mutually interactive and the results provide the basic knowledge for the development of the estuarine research.

Besides the scientific motivation, estuaries have a historic and continuing importance to human activities. Almost 60% of the most populated cities worldwide lie on or near estuaries, proportionally representing with its dimensions one of the richest regions of our planet. There is now a growing consciousness of the need to protect the estuarine environment. Until recently, little thought was given to ecological consequences (Geophysics Study Committee 1977). In Brazil, this proportional value of populated cities near estuaries is almost the same as globally and the scientific knowledge may be used to solve problems of practical nature, for instance: impacts related to changes in the hydrographic basin and in the estuary geometry due to dredging, identification of sedimentation places which represent an obstacle to the harbor navigation, residence times and effluent dispersal of substances in its interior, and physical, biological and chemical properties to the support projects of aquiculture.

As in foreign countries, the reason for the development of Brazilian cities in the estuaries proximities were: commercial and naval harbor facilities, a natural capacity for the renewal of the periodic and systematic water masses under tidal influence, natural communications with mangrove regions, plentiful biological community, and recreational and economic potentials.

Estuaries have been used as a receptacle of natural substances and wastes of industrial activities, which may degrade the water quality. An analysis of these products grouped according to their main sources, were described in detail according to their menace to human and biologic communities. None of the contaminants can be attributed entirely to any one class of activity, but each can be ascribed to be a major source (Schubel and Pritchard 1972).

- Pathogenic organisms, organic matter and nutrients: Municipal wastes.
- Pesticides and Herbicides: Agriculture.
- Heavy metal, oil, fresh-tainting, substances, toxic chemicals: Industry, shipping, from atmosphere, and highways and roads.
- Heat: Power plants.
- Sediments: Agriculture and construction in general

Natural suspended sediment transport in rivers due to the erosion processes in the hydrographic basin may be increased due to the activities already listed. The ultimate sedimentation process in an estuary may become a menace to its existence due to the possibility of an irreversible decrease in the water volume in its interior. Also, the input of contaminants associated with the sedimentation process may shorten its geological life and its natural biological health.

Although litter is an item not included in the above-list and is not an indicator of water quality, it represents a threat to the natural estuarine environment. However, being a behavioral problem it may be addressed only through environmental education, already applied in some countries. For instance, studies by Williams and Simmons (1997) in England, were carried out to investigate the problem extension of litter occurrence in beaches, and alarming results were obtained. Hundreds of plastic bottles, beverage cans and a diversity of variable materials were found on each kilometer of a beach during summer and winter seasons. Taking into account these problems, it was evident that further studies are needed at the interface river/estuary/beach to tackle this problem.

The water mass renewal in an estuary and its capacity to assimilate the by-products of man's activities that may lead to degradation of the estuarine environment, is dependent on the physical, chemical, biological and geological natural processes, which between them interact in a complex way and are vulnerable to the water quality. Although high-quality estuarine research programs do exist, these interactions were not comprehensively studied in an interdisciplinary way, because most investigations were developed in a specific line and consequently poses a threat to the human health. Therefore, a direct or indirect introduction of substances and energy by man may reach concentration levels so high that they cause estuary contamination with worse effects to the biota and human health. These substances may also affect the recreational activities and fisheries, and the water quality and reduction of its natural attractiveness, which may extend well beyond the estuary boundaries (Geophysics Study Committee 1977; GESAMEP 1995).

The physical processes common to estuaries are the circulation and the mixing of water masses with different origins: the fluvial freshwater and the seawater from the adjacent ocean. The resulting water mass is non-homogeneous and affected by a range of spatial and time scales: from microscopic to its geometric boundaries, as well as, time intervals from fractions of seconds to years and centuries.

The processes in estuaries acting on the distribution and variability of physical properties, the concentration of natural substances (salinity, nutrients, dissolved oxygen and suspended sediments) and the biological organisms, as well as, the local and remote inputs of wastes and pollutants, the research must be conducted on the basis of multidisciplinary investigations, including its drainage basin and the adjacent continental shelf, taking into account the interaction of the following main components (Geophysical Study Committee op. cit.):

- Circulation, transport and mixing of estuarine and coastal water masses.
- Environmental effects on the fauna and flora associated with basic biological phenomena.
- Erosion, transport and sedimentation.

These components are very interdependent on most estuarine processes. However, differences may be found between the estuaries or in specific regions of the estuarine system. The knowledge of these processes must be applied to the estuary management and decisions to be undertaken, for the best utilization and survival of local marine species as a natural and productive resource.

This book will focus on the first component which is ultimately related with the knowledge of the: hydrographic and hydrodynamic characteristic (salinity, temperature, density and circulation), mixing processes (advection and diffusion), residence times, as well as, the temporal and spatial scales which are important in focusing on the interdisciplinary characteristic of these environments.

The historic evolution of the ideas of the estuarine circulation related chronologically related with the main results of the Scandinavian and American research was presented in the article of Beardsley and Boicourt (1981). The first hypotheses about the bidirectional motions observed in the upper and lower estuarine layers, and compensation upward motions *from below* have been set in the pioneer work of F. L. Ekman in 1876 (quoted in Defant (1961), p. 539) in the analysis of experimental results on the longitudinal salinity distributions at the mouth of the Götaelf (Sweden) river flowing into a fjord. However, it was only after the Second World War that interest in the Physical Oceanography of estuaries increased significantly. This was firstly due to the military necessity and also to the growing concern of the need to protect the estuarine environment from activities of man.

In the last decades there has been much specialized literature on the investigation of estuaries. In preparing this book, not only classical books and articles were used, but also those published with the most recent achievements on the estuarine circulation and mixing processes.

1.2 Origin and Geological Age

With a few exceptions, estuaries were formed in relatively narrow regions between the sea and continent. They are transitional environments from a very recent geological epoch ($\approx$7 thousand years ago), formed during the last interglacial phenomena *eustatic* (volume variations of the oceans water), and *isostatic* (level changes in Earth crust), and also *tectonic* processes. Their locations, morphologies and lengths are dependent on the sea level, topographic features on the coastline, river discharges and ice melting. Major/minor alterations were suffered due to the natural processes of erosion, transport and deposition of sediments, and recently due to the exploitation of the drainage basins. Eustatic and isostatic sea level variations are mainly due to:

Eustatic—Mass variations due to the processes of freezing/defrost, and changes in the volume due to heating/cooling.
Isostatic—Earth's crust variation due to tectonic processes, geometric changes in the oceanic basins, and erosion and compaction of non-consolidated earth.

The glacial maximum occurred approximately fifteen or sixteen thousand years ago, when the sea level was approximately 130 m below the present level. The level evolution in the last 35 thousand years is shown in the eustatic curve of Millimann and Emery (1968) (Fig. 1.1). This result was complemented with samples of the Brazilian continental shelf, dated by the carbon-14 methodology by Kowsmann et al. (1977). The eustatic variation, affecting the water volume of all oceans, resulted in the deepest coastline of all continents, and was located approximately at the continental slope. Furthermore, the pleistocenic sediments transported during the last regressive event were reworked during a relatively short time period and transported in suspension by the continental drainage basin and deposited in the continental shelf and slope.

Due to global climatic changes, the increasing temperature of the Earth, the progressive defrost of the polar caps has increased the sea level with a ratio estimated at 1 m per century (Fig. 1.1). This last marine transgression (Flandrian Transgression) commenced about 15 or 17 thousand years ago and about 7 thousand years ago there was a rapid increase in the sea level which was interrupted by stabilizations around −110 and −60 m depth (Kowsmann et al. 1977). At the end of

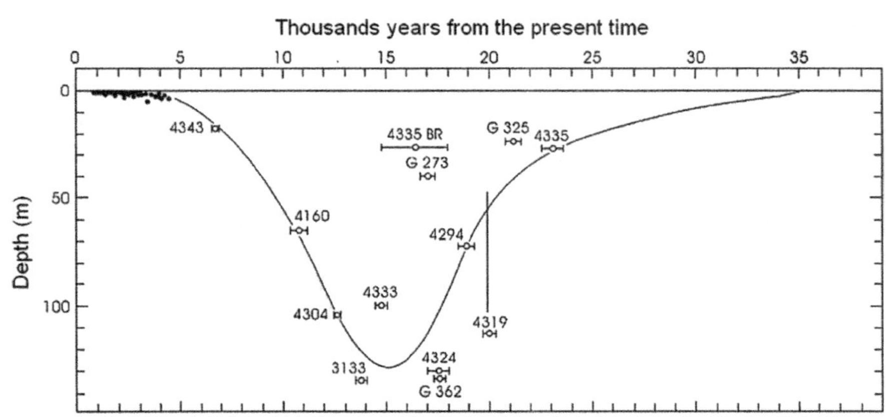

Fig. 1.1 Eustatic variation of the sea level in the USA continental shelf, inferred for the last glacial period (according to Milliman and Emery 1968), complemented with samples collected in the Brazilian continental shelf analysed by Kowsmann et al. (1977)

this transgressive event, between 7 and 2 thousand years ago, the sea reached the approximate current level and the coastal plains and the rivers valleys were slowly flooded, creating bays, estuaries, straits, inlets, coastal lagoons and inshore waters.

Recent research on sea level oscillations indicates that in the last millennia there were minor sea level fluctuations, shown in the more detailed eustatic curve of Fairbridge (1961) showing sea level oscillations of a few meters higher than the current sea level. This event has also been confirmed on the southeast and southern Brazilian coast in the investigations of Villwock (1972) and Suguio and Martin (1978).

Along the coastline of some countries there have been recent changes of sea level due to secular natural and man-made oscillations of the Earth crust (isostatic), which may have future implications on estuary life. The Scottish coastline, for example, is rising at a time rate of 3 mm/year (relative sea-level subsidence of 0.3 m/century), generated by ice melting due to climatic warming. However, the opposite occurs in the coast of Holland, where a subsidence with a time rate estimated in 2 mm/year (or 0.2 m/century). The southeast region of England is also being submitted to a serious overall coastline subsidence due to the simultaneous occurrence of sea level elevation and the coastline subsidence.

An extreme subsidence has been reported in the Bangkok coastline (Thailand), mainly due to the removal of subsurface spring water associated with a deltaic subsidence, representing a relative increase of the sea level at a time rate estimated as 4.5 m/century. Research on the annual fluctuations of the sea level in New York (USA) indicates an increase in the relative sea level of approximately 0.25 m/century between 1893 and 1991 (Leathermann et al. 1997).

The temperature of seawater and coastal environments may be regarded as a measure of the heat content. This physical property has daily, seasonal and secular variations; the latter is very important to the relative sea level and may be classified as eustatic because of its relationship with the heat budget of the ocean-atmospheric system and changes in the hydrologic cycle. In the last hundred years this property presented an increase of 0.5 °C, mainly due to the increase in the concentration of gas in the atmosphere (mainly carbon dioxide, methane and freon). These substances have the ability to absorb infrared radiation from the Earth's surface, forcing its reemission causing the warming of the planetary layer and the sea surface. For the next one hundred years a higher warming increase is expected (between 1.5 and 3.0 °C), then it will be a great influence on the hydrodynamic estuarine behavior and on other transitional environments and, as a consequence, on the distribution and productivity of the biological communities.

In the case of a catastrophic increase in the oceans temperature in a short time period, it is possible to predict great eustatic variations in the sea level. In the defrost hypothesis of the polar ice cap the sea level will be increased by 30 m (Dyer 1973). In this case, the actual estuaries will be inundated and new estuarine environments will be formed in the upper regions of the rivers. The sediment concentration of fluvial origin may be less, but due to the coastal erosion a great amount of sediments may be transported by waves and coastal currents. However, if the opposite occurs, a decrease of the present sea level may create shallow estuaries which, in a short time interval, may be submitted to sediments accretion. In both cases, there will be geomorphologic and depth changes and, consequently, in the currents generate by the tidal excursion. Hence, estuaries are coastal environments, well developed very recent, but in geological time scales their actual conditions may not last for several centuries from now.

1.3 Definition and Terminology

An estuary definition may be made in several ways and according to the immediate point of view. However, the definitions must take into account the characteristics and the essential processes, as well as the context into which the estuary is inserted, enabling the use of adequate classification criteria (Dyer 1973, 1997). For oceanographers, engineers, geographers and ecologists, the estuary word is used to indicate an inner coastal region where the fluvial waters meet the salt water transported by the tidal currents, extending up-river, as far as the tidal influence reaches. Due to the inclusion of the tidal river zone in definitions which will be presented we should mention the pioneer article by the naval official A. Rongel, published in the Anais Hidrográficos da Marinha do Brasil (Rongel 1943). In this article, focusing on the tidal wave propagation in estuaries, the superior range of the tidal influence was defined as the upper reaches of an estuary, later referred the *estuary head*. Further, some articles also have included this upper river region, and the continental shelf which receives the estuarine plume, as an integrated part of the estuary.

The first definition of an estuary was given by Ketchum (1951), focusing on the exchanges of fresh and salt water in tidal estuaries, and in this work we find the definition:

> Estuary is a region where river water mixes with, and measurably dilutes, sea water.

The most common definition for an estuary is that of Pritchard (1955); Cameron and Pritchard (1963) that state of:

> An estuary is a semi-enclosed coastal body of water having a free connection with the open sea and within which sea-water is measurably diluted with fresh water deriving from land drainage.

An analysis of this definition indicates that the circulation pattern in an estuary is influenced to a considerable degree by its lateral boundaries, and a fresh water volume of the river discharge which remains in the mixing zone of the estuary due to the tidal mixing. Further, being a semi-enclosed coastal feature it limits to some extent the size of the water bodies under consideration, and the salinity stratification will also be dependent on the geometry, and must allow an essentially continuous exchange of water between the estuary and the ocean (Pritchard 1967). In this way, a detailed understanding of the physical processes occurring in these environments is difficult to be achieved. As the biological, chemical and geological processes are strongly dependent on these forcing and physical variables, interdisciplinary studies are necessary for a better understanding in the context a coastal environment.

The fresh water discharge effect, which flows constantly from the river, gives rise to a downstream circulation component at the surface layer, while seawater dilution due to the mixing with fresh water generates an upstream baroclinic component which increases towards the bottom. The interaction of several properties and processes—river discharge, tidal currents, gradient pressure forces, advection and turbulent diffusion—generates, inside the geometric boundaries of the estuary, the salinity distribution which is a characteristic of each estuary (Officer 1983).

In the classical definition it is explicit from the *measurable dilution of the sea water by the fresh water from land drainage*, the occurrence of the mixing zone (MZ) and the water budged volume of the water arriving into the estuary due to the precipitation (P) and the one by the fresh water discharge (Q_f), must be higher than the water volume transferred to the atmosphere due to the evaporation (E_v). Then, according to the classical definition, the estuary holds the following condition $P + Q_f > E_v$, and it may be classified as a *positive estuary*, according to some authors (e.g. Ketchum 1953). In tropical estuaries the water budget may have the condition $P + Q_f < E_v$; this environment is usually referred as a *negative estuary*, but it may not be classified as an estuary according to the Pritchard's definition. As a transitional environment between an estuary and a negative one is $P + Q_f = E_v$, it is termed as *neutral estuary*.

The mean horizontal salinity distribution in an estuary is schematically shown in Fig. 1.2 by isolines with constant salinity values (isohalines), with variations between 1.0 and 36.0‰ (the salt concentration, in unities of g kg^{-1}, is denoted symbolically by ‰). This salinity variation shows that the sea water was measurably diluted by the river discharge and a volume parcel of this water was retained in the estuarine water body, generating the *estuarine water mass*.

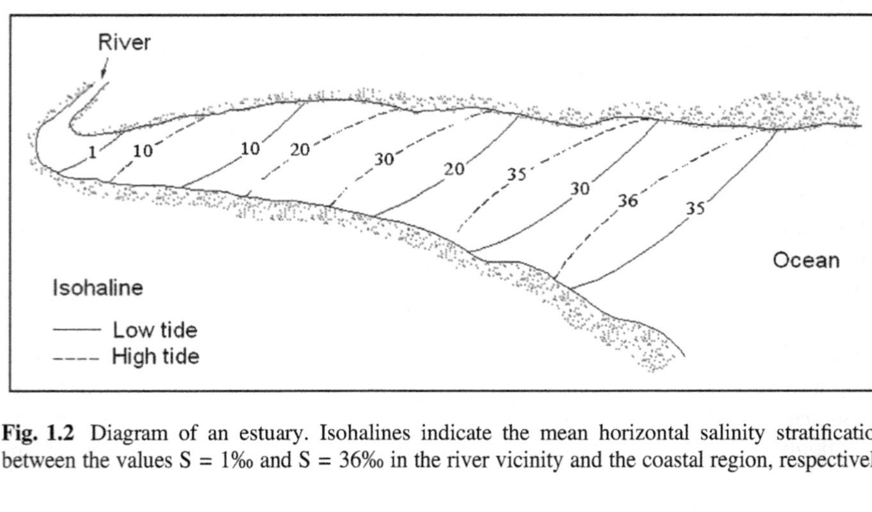

Fig. 1.2 Diagram of an estuary. Isohalines indicate the mean horizontal salinity stratification between the values S = 1‰ and S = 36‰ in the river vicinity and the coastal region, respectively

The estuary definition holds for environments with spatial scale much less than, for instance, the Baltic Sea between the Scandinavian countries and Western Europe. This sea is characterized by precipitation and fresh water discharge much higher than evaporation, so its water mass has low salt concentrations ($0 < S < 18‰$). However, this sea doesn't fit the estuary definition due to its high spatial scale and its geometric boundaries form the coast line rather than being a feature of the coastline (Pritchard 1967). The bays, as the Guanabara Bay in Rio de Janeiro (Brazil) may, in some seasons, show circulation and mixing processes which have estuarine characteristics (Bérgamo et al. 2006), however bays are too complex environments to be classified as estuaries (Kjerfve et al. 1997).

Pritchard's definition takes into account only the region influenced by the sea-water or *mixing zone*, Dionne (1963) suggested the following definition, to contemplate three explicitly sectors of zones along the estuarine domain:

> Estuary is an inner coastal region up to the river valley and the limit of the tidal influence, and generally split into three sectors: (a) Inferior or marine, with a free connection with the open sea; (b) Medium, subjected to the intense mixing of the sea water and the fresh water discharge; (c) Upper or fluvial, characterized by fresh water but influenced by tidal oscillation.

Thus, an estuary may be considered as a system composed of the region where the seawater dilution occurs (medium zone) up to the river portion subjected to the tidal oscillation (upper zone with density almost constant), and the inferior or marine zone near the mouth, and is an extended version of Pritchard's definition. The limits between these estuarine zones are variable and dependent of the river discharge intensity, tidal currents and the wind forcing. Afterwards, in the presentation of analytical and semi-analytical methods to obtain longitudinal dispersion coefficients in estuaries, Harleman (1971) devised the estuary as composed by two regions: one into which salt intrudes into the estuary (the classical MZ) of previous definitions, and the river tidal zone (upper and fluvial—the TRZ) as in the Dionne's definition.

To take into account aspects related to the estuarine processes of erosion and sedimentation by the tide, wave and fluvial sources Dalrymple et al. (1992) suggested the following definition:

> Estuary is the seaward portion of a drowned valley system which receives sediment from both fluvial and marine sources and which contains facies influenced by tides, wave and fluvial processes. The estuary is considered to extend from the landward limit of tidal facies at his head to the seaward limit of coastal facies at its mouth.

Scientific papers have increased in the last decades in the fields of Physical Oceanography, Ecology, Hydraulic and Environmental Engineering, focusing on the solution of pollution problems due to harbor construction, navigation, recreational, fresh water supply, discharges of domestic and industrial wastes, which may cause ecological impact in the marine environment. The term *estuarine zones* has also been used to mean an environmental system consisting of an estuary and those areas which are consistently influenced or affected by estuarine waters such as, but not limited to, salt marshes, coastal and intertidal areas, bays, harbors, lagoons, channels, inshore and offshore waters. Then, we may recognize that *estuarine zone* has been introduced to indicate all coastal transitional environments with higher or lesser influences of the river discharge and tide. As previously said, estuaries may have different definitions.

A definition by Kjerfve (1987) was proposed to be new, functional and helpful to those who work with the spectrum of estuaries types, and estuarine systems, including lagoons, river mouths and deltas.

An estuary is a coastal indentation that has a restrict connection to the ocean and remains open at least intermittently with the adjacent sea. The estuary can be subdivided into three regions:

(a) *The tidal river zone (TRZ) a fluvial zone characterized by lack of ocean salinity but subject to tidal rise and fall of sea level.*
(b) *The mixing zone (MZ) (the estuary proper) characterized by water mass mixing and existence of strong gradients of physical, chemical and, biotic quantities reaching from the river tidal zone to the seaward location of a river mouth bar or ebb-tidal delta.*
(c) *The nearshore turbid zone (NTZ) in the open ocean between the mixing zone and the seaward edge of the tidal plume at full ebb tide.*

The comparison of this definition with the previous (Dionne 1963) indicates the following correspondence: *zones TRZ* and MZ with the *upper and medium zone*, respectively. However, this definition differs considerably from those previously proposed in that it recognizes and includes a near-shore marine component, estuarine in character, which should be considered in the treatment of the physical or chemical dynamics, or ecology, of the estuarine system as a whole. In the interface of the coastal water and the estuarine water *frontal zones* and *fronts* are the lower and surface interfaces, respectively. The influence of the estuarine plumes in the biological production is not yet well known, however, there are several examples of the pollutant concentration increase in these convergent regions (Mann and Lazier 1991).

For a coastal environment as a river subjected to the tidal influence (also named estuarine channels), the river, mixing and the coastal zones are shown schematically in Fig. 1.3. This section may be representative of the longitudinal and vertical steady-state salinity distribution and the bidirectional velocity in the estuary shown in Fig. 1.2. In the TRZ, the salinity is zero and the motion is unidirectional down estuary as shown (or up estuary during the flood). The upper tidal limit of this zone is the region where the tidal current is zero, and is the estuary *head*. In the MZ there is a horizontal and vertical salinity changes, because in this region occurs the dilution of the seawater with the fresh water from the river discharge, and the isohalines configuration may have a wedge form, named *salt-wedge*. In the transitional region between the mixing zone (MZ) and the adjacent coastal region the estuary entrance or *mouth* is located; from this region towards the continental shelf it is possible to observe the near shore turbid zone (NTZ) formed by the *river plume* delimited from the oceanic coastal water by the estuary *front*. In this figure we may

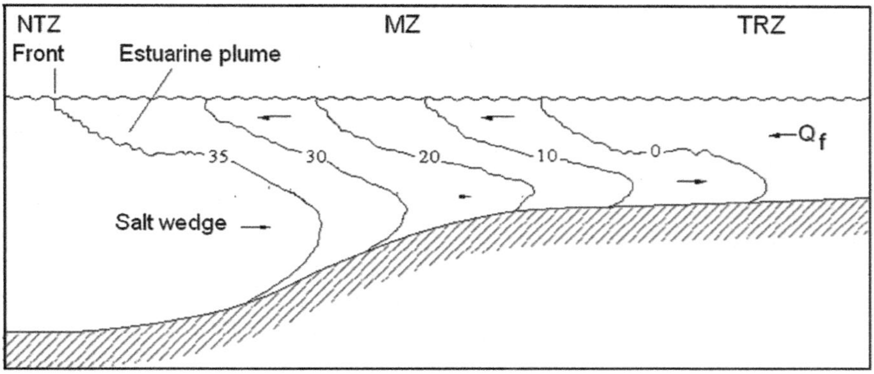

Fig. 1.3 Longitudinal section of an estuarine system showing: the tidal river (TRZ), the mixing (MZ) and near-shore turbidity (NTZ) zones. Also are shown characteristics of the vertical salinity stratification and the mean bidirectional circulation. Q_f is the river discharge (Adapted from Simpson 1997)

observe the salinity increase and decrease down and up-estuary on the layers located above and below the vertical salinity gradients, and the bidirectional mean circulation is seaward and landward in the surface and bottom layers, respectively.

The boundaries between the zones (TRZ, MZ and NTZ) indicated in Fig. 1.3 are dynamic, with spatial (longitudinal and lateral) positions and temporal variability with different time scales (tidal period, seasonal, annual and longer), in order to reach and equilibrium condition due to the estuary forcing: river discharge, tidal height, wind and coastal circulation.

Kjerfve (1987) definition takes into account all segments of the estuary, and the coastal zone during high river discharge may develop salinity stratification similar to the mixing zone (MZ). Thus, as the NTZ is part of the estuarine system can't be investigated separately when an oceanographic or ecologic study is being carried out. This zone is the oceanic region which is subjected to the highest impact from the humans and it has a singular effect due to the dynamic influence of the coastal currents and waves of several periods. The effects of vorticity, wind shear stress (whose intensity may be higher than the longitudinal gradient pressure force), and the inner and bottom frictions may also be important, contributing to a higher contrast of the NTZ physics and the continental shelf. Simple conceptual models of this zone were developed by Csanady (1977), Garvine (1974, 1975, 1977) just mentioning only the Garvine's first investigations on this subject. An up-to-date review on the dynamics of estuary plumes and fronts may be found in the recent article of O'Donnell (2010).

Property distributions in the MZ are due to the mixing process of two water masses with different characteristics: fresh and seawater. According to Garvine (1977) each one of these water masses must be considered as a *reservoir*, within which properties (heat and salt concentrations) vary slowly in time and space, in comparison with its variations in the estuarine zone (MZ). Off shore there will exist a body of seawater with fairly uniform salinity and temperature which vary somewhat seasonally which characterizes the NTZ formation. Upstream, at the estuary head (TRZ) it is also considered as a reservoir of zero salinity, but with temperature determined by a seasonal time scale related with its variation in the drainage basin. In the MZ we find only variations of temperature and salinity that are between the extreme values of these variables.

Due to the dynamical variability of the various estuarine zones (TRZ, MZ and NTZ) of an estuarine system these zones may not occur simultaneously in extreme climatic and whether conditions. In the estuaries located in equatorial and dry regions and forced by moderate tidal heights, the TRZ may not exist in some periods of the year when $P + Q_f < E_v$. Another extreme condition may occur during high river discharge, when the MZ may be displaced from the estuary to the coastal region and during this event the mixing processes of the river and seawater occurs in the continental shelf (Kjerfve 1987).

The horizontal extent of the influence of the river discharge in the ocean, named as estuarine or river plume (Garvine and Monk 1974), is dependent on the fresh water discharge and the coastal circulation. The most spectacular occurrence of this phenomenon is due to the fresh water discharge of the Amazon river located in the tropical Brazilian region, which is the highest fresh water source of all oceans. Its mean discharge, estimated at 1.8×10^5 m^3 s^{-1}, is almost 20% of all sources of continental water discharged in the oceans. The estuarine plume has a high percentage of river water, is displaced to the open sea in the NW direction along the coast and in the open sea, and it is the largest identifiable river plume in the ocean, with several millions of square kilometers in area within the Tropical Atlantic Ocean. During the Amazon river flood it is the largest estuarine plume and dominates the continental shelf hydrography of the Amazon Continental Shelf (Curtin 1986; Curtin and Legekis 1986; Geyer et al. 1991). Due to this high river discharge the MZ is displaced from the estuary to the continental shelf and the mixing of the river water with the high salinity ocean water occurs in the NTZ (Fig. 1.4a, b). Off shore of the estuary mouth the vertical salinity structure seems to be an arrested salt-wedge. In these figures it is possible to observe that the low salinity water (S < 10‰), usually found in the estuarine inner region (characterizing the MZ), is located in the inner continental shelf.

The importance of the wind stress of the nearly permanent Trade Winds, the tide, and the anisotropy of the bottom in the dynamic behavior of the estuarine plume, and the continental shelf circulation in the Amazon continental shelf was studied by Fontes (2000), using a tridimensional numerical model.

Interrelated estuarine studies, sampling simultaneously the hydrographic, biological, chemical and geological characteristics in the zones TRZ, MZ and NTZ, are not common due to the complicated logistic and the great experimental effort. However, we find in the Brazilian regional literature the paper by Schettini et al. (1998), on the analysis of almost simultaneous ecological and oceanographic data collected in the mixing zone, and the estuarine plume of the Itajaí-Açu river (Santa Catarina, Brazil). The experiment in these zones was conducted to investigate the estuarine plume influence in the coastal adjacent region. The results have shown

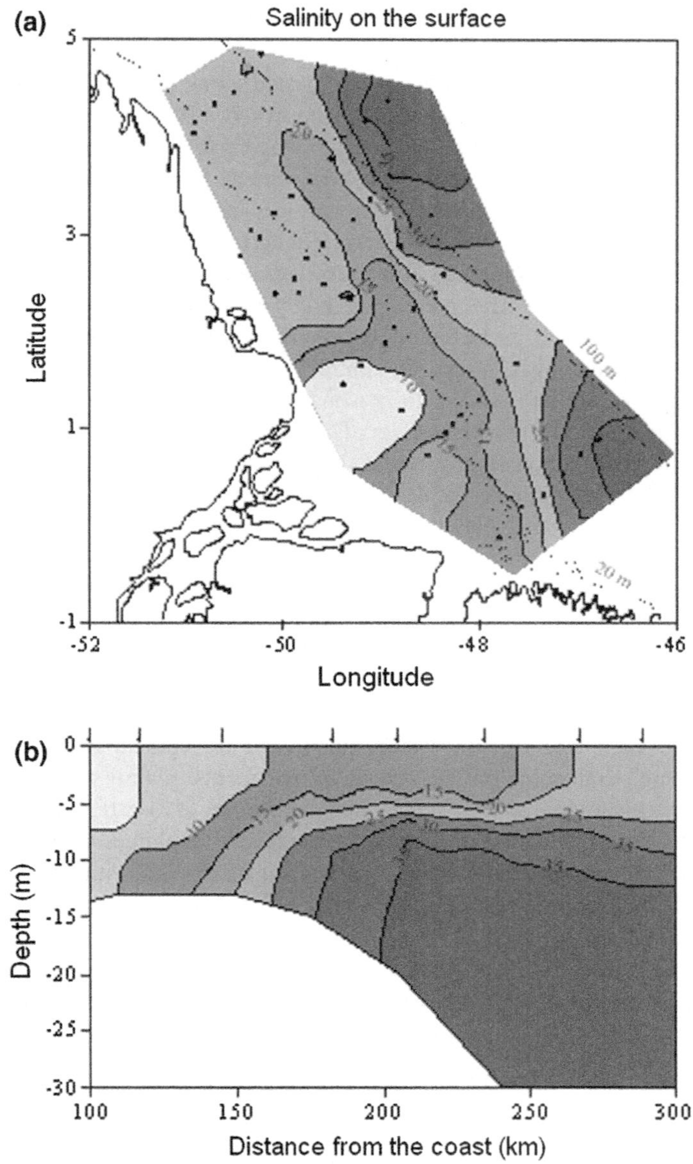

Fig. 1.4 Partial extent of the estuarine plume of the Amazon river shown by the salinity distribution on the surface (**a**) and the salt-wedge in the inner continental shelf (**b**). Experimental results obtained in February-March, 1990, during the joint project *A Multidisciplinary Amazon Shelf Sediment Study* (according to Geyer et al. 1991)

that the strong vertical stratification in the inner MZ inhibited the mixing of the fresh water discharge, with high nutrients concentration, with the salt-wedge water. Due to this behavior the biogeochemical processes occurred mainly in the continental shelf, after the river plume formation, extending its influence up to almost 10 km off shore of the estuary mouth.

Turbidity plumes of the main estuaries located along the NE/E Brazilian coast were analyzed for their dispersal patterns of Total Suspended Solids (TSS) concentration, using Landsat satellite imagery to map the concentrations of suspended solid particles and its seasonal movements by Oliveira et al. (2012). In situ sampling data were obtained from various oceanographic campaigns, conducted at the mouth of the rivers between 1994 and 2002. The behavior, dimension and degree of turbidity of the São Francisco river, during this time period indicated that estuarine plume have been greatly altered by environmental impacts in the river hydrological basin, and its turbidity has decreased due to the regulation of river flow in power plants installations. In contrast, the Doce and Paraíba do Sul rivers estuarine plumes are still subject to seasonal variations, and showed more turbid conditions than the São Francisco river. This behavior is because its dams are less numerous and the natural river flow has been maintained. The São Francisco and Doce estuarine plumes have the tendency to disperse obliquely along the coast, while the Paraíba do Sul plumes tend to disperse mainly parallel to the coast, enhancing near shore TSS retention.

The Caravelas and Peruípe estuarine system, also investigated by Oliveira et al. (op. cit.) is composed of several meandering channels which are connected to the ocean by a double inlet system in the north. The Caravelas estuary is tidally dominated and functions as a trap for inner shelf sediments. However, the Peruípe river at Nova Viçosa, in the south is forced by a higher river input, and its plume is restricted to nearshore shallow waters dominated by vertical mixing processes, producing high concentrations of suspended sediments, mainly in the spring tide (Schettini and Miranda 2010; Andutta 2011; Schettini et al 2013). During austral spring and summer, when NE-E winds system prevail, all plumes generally disperse southward, and northward reversals may occur in winter with the passage of atmospheric cold fronts. According to Herz (1992) the mangrove vegetation area in this system is 66.5 km^2.

Others definitions were presented, with emphasis in the ecological nature of these coastal environments, between those we may presente the one by Perillo (1995):

> An estuary is a semi-enclosed coastal body of water that extends to the effective limit of tidal influence, within which sea water entering from one or more free connections with the open sea, or any other saline coastal body of water, is significantly diluted with fresh water derived from land drainage, and can sustain euryhaline biological species from either part or the whole of their life cycle.

According to the Perillo's definition, rivers empting into bays or others near-shore water bodies may also be defined as estuaries. As examples, we may mention the James river estuary, empting its waters into the Cheasapeake Bay (Virginia, USA), the Cubatão, Bertioga, Mogi and Piaçaguera Brazilian rivers, discharging its waters in the upper reaches of the Santos Bay, which may also be classified as estuaries. Another example is the Paraguaçu river estuary empting into the northern region of the Todos os Santos Bay (Bahia, Brazil); the effects of the Pedra do Cavalo dam in the estuary circulation and stratifiacation was investigated by Genz (2006).

The definitions along with others by Dionne (1963), Kjerfve (1987), Dalrymple et al. (1992) and Perillo (1995) have the ability to include all basic marine disciplines which are necessary for studying this transitional water body as an ecosystem. There are interchanges between three estuarine domains: the tidal river zone (TRZ), the mixing zone (MZ) and the nearshore turbid zone (NTZ), which must be investigated together as a unique transitional system.

As a complement to the definitions already given follows the one by Dyer (1997) which is an extension of the classical Pritchard's definition and is possibly convenient for the physical oceanographers, this definition emphasizes the mixing and river zones:

> An estuary is a semi-enclosed coastal body of water which has a free connection with the open sea, extending upriver up to the tidal influence limit, and within which sea water is measurably diluted with fresh water derive from land drainage.

In a look on the map of a coastal region we observe that the geomorphologic configuration of an estuary may be more complex than the one presented in Fig. 1.2. These coastal environments usually were formed by a fluvial net of rivers discharging its water into several places around the semi-enclose region, including nearby its mouth, and the estuary is formed by several heads and eventually two or more mouths interacting with to the coastal ocean. The rivers forming these environments may have properties distributions and a hydrodynamic behavior like the one of an estuary, thus forming sub-estuaries systems. Those systems are referred in the literature as *estuarine system* in Pritchard's (1952a) pioneering paper on the oceanographic characteristics of the complex system of Chesapeake Bay (Virginia, USA), located in the east coast.

In the papers related to the regional oceanographic estuarine characteristics we find the terminology *system (or complex) estuarine-lagoon*, to indicate a coastal environment composed of several interlinked rivers and channels discharging through one or several mouths to the coastal region. This terminology has been widely used in studies of the oceanographic characteristics of the southern, and southeastern regions of the São Paulo State (Brazil), as the Estuarine-lagoon System of Cananéia-Iguape (Fig. 1.5), which is dominate by mangroves with an area of 23.5 km^2 (Herz 1992). Studies in this system of water properties related to physical, biological, chemical, ecological processes and material exchange between mangrove and the continental shelf, have been presented in the articles of Besnard (1950), Tundisi (1969), Miyao (1977), Tommasi (1979), Miyao et al. (1986), Tessler (1982), Schaeffer-Novelli et al. (1990) and Tessler and Souza (1998), and others. The Cananéia-Iguape system has two main mouths in the Cananéia and Iguape cities, and the Santos-São Vicente-Bertioga system interchanges its waters with the continental shelf through the Santos Bay and the Bertioga channel (Fig. 1.5).

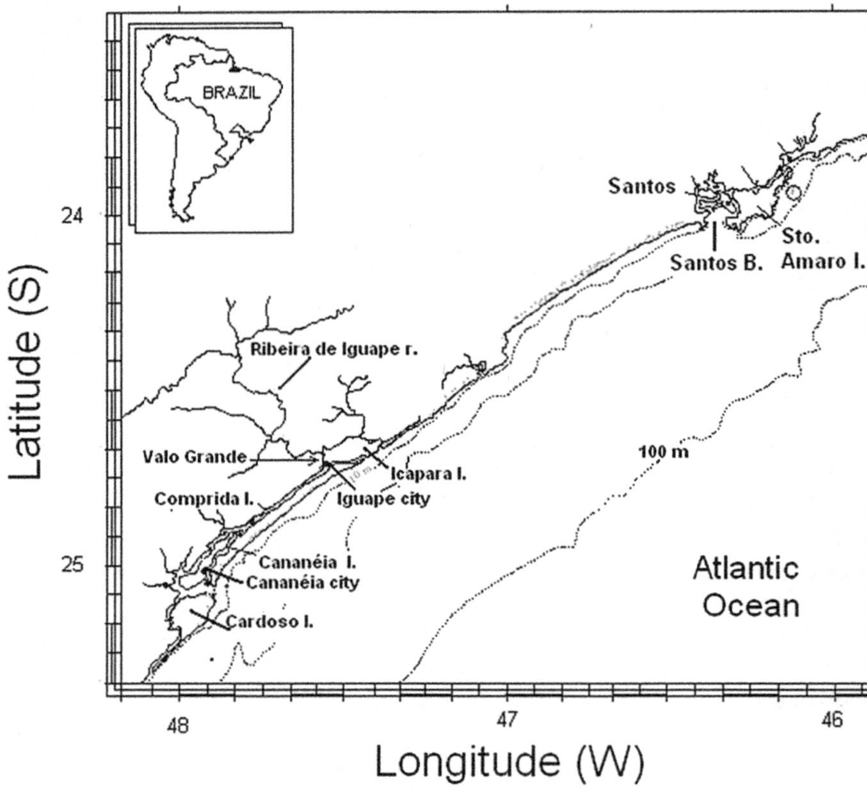

Fig. 1.5 The Estuarine-lagoon system of Cananéia-Iguape (Protection Environmental Area - PEA), the Valo Grande artificial channel, and the Estuarine System of Santos-São Vicente-Bertioga, in the southern and Southeastern regions of the São Paulo State (Brazil)

Interconnected to the Estuarine-lagoon System of Cananéia-Iguape is the estuarine system of Paranaguá Bay located in a region of least anthropogenic impact, and the present state of conservation being the result of the regional process of colonization. A synthesis on the physical, chemical and biotic components of this system may be found in Lana et al. (2001).

The nomenclature of estuaries and coastal environments will be expanded during the studies related with its classification (Chap. 3), according to its geomorphologic characteristics; a short description follows of the coastal and estuarine processes which exert influence on its dynamical behavior.

In coastal plain estuaries the motions in the *tidal river zone* (TRZ) are unidirectional up or down river during the flood or ebb tide, respectively, and it is filled with fluvial water. In the transition between the TRZ and MZ (mixing zone) there is a region with low or null velocity due to the bidirectional convergent motions. As the sediment concentration of fluvial or marine origin in this convergent region

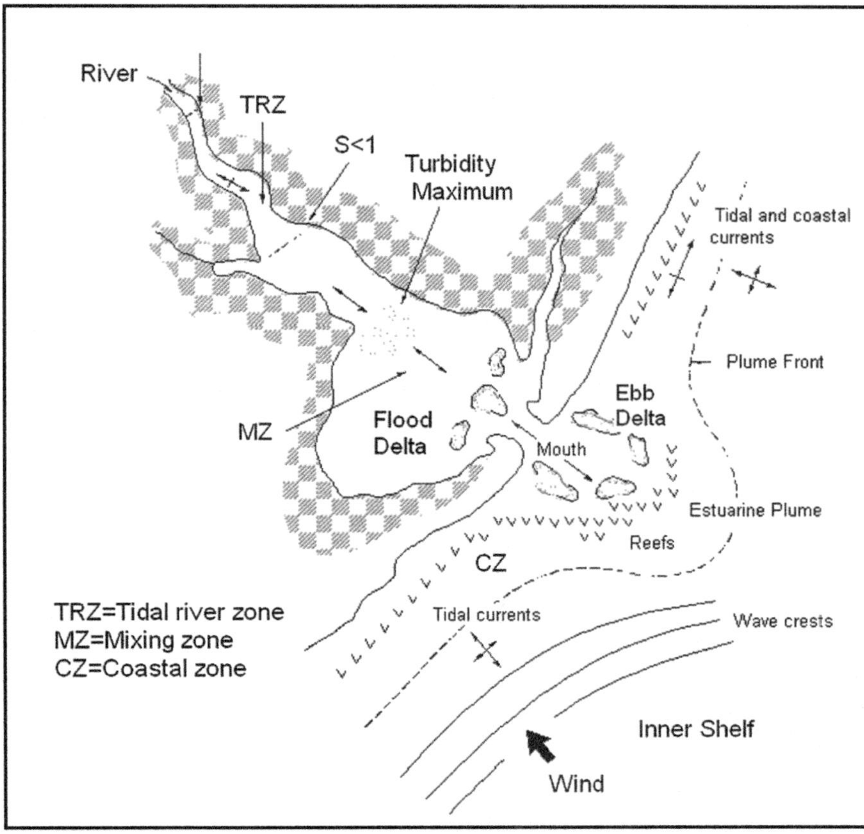

Fig. 1.6 Functional estuarine system delimitation. Geomorphologic characteristics and schematic representation of estuarine processes in the tidal river zone (TRZ), mixing zone (MZ), turbidity maximum, and nearshore turbidity zone (NTZ)

usually is high, it may sink slowly down and thus causing in the bottom zones of *turbidity maximums zones* (TM), schematically shown in Fig. 1.6. As the sediment concentration of fluvial or marine origin in this convergent region is high, it may sink slowly down and thus originating this zone. Due to the sedimentation process particles of mineral and organic origin accumulate on the estuary bottom and may reduce the estuarine channel depth and, in medium and long time scales may gradually form an obstacle to the navigation and is worst to the estuary environment. This process, accelerated by river erosion and sediment transport due to the estuarine circulation promote the sediment arrest in the estuary, prevent and retard the sediment outflow to the adjacent coastal region. Further details on the marine sedimentation processes may be found in Postma (1980).

The presence of reefs in the coastal zone off the northeastern Brazilian region may represent an obstacle to the water interchange between the estuaries and the continental shelf. The littoral drift generate by the breaking of waves may accumulate sediments in the estuary mouth and coastal lagoons, preventing the water interchanges between the MZ and the TRZ. The opposite effect, i.e., the erosive process may also occur at the estuary mouth and thus change its geomorphologic characteristics. The most energetic components to force these changes are the wind generated coastal currents and tides, mainly during the atmospheric frontal zones which may also generate storm surges; the tidal generated circulation has, in general, its main component of high frequency oriented orthogonally to the coast (Fig. 1.6).

1.4 Policy and Actions to Estuary Preservation

Until the second half of the XIX century the hydrographic basins feeding fresh water into the estuaries were subjected to relatively little alterations, and sewerage from cities being discharged into estuaries was diluted and renewed in these water bodies almost without serious impacts over these ecosystems. However, with the population increases in the cities, the agriculture expansion and the industrial revolution, the harbor workmanship and navigation channels, the quantity and the diversity of sediments and domestic and industrial sewerage began to increase the threat in the natural characteristics of the estuaries around the world. The direct and remote influences of human activities gave rise to the variations with different degrees of impact on the major environmental degradation in estuaries.

As in others countries, there is today in Brazil an increasing concern to introduce policy and actions to enable and support the sustainable development of the coastal regions in general and, particularly, the estuarine systems. The pioneering studies in the estuaries and coast protection were initiated in the last quarter of the XIX century, when the Swedish researcher F. L. Ekman performed experiments in the Götaelf River estuary. In the article by E. Goadoy published in the *Nature Magazin*, in 1870 (quoted in Tommasi (1994)), Ekman already foresaw that environmental impact studies were necessary to the constructions of great navigation channels (Tommasi 1994).

During decades, estuaries were and continue to be regions of conflict and policy debates, usually due to industrial and commercial interests. In the USA the legal question among municipal, state and federal government has always been a controversy question. Only in 1968 was the Protection Decree of Estuaries signed, with serious limitations in its use and control.

In the following year, 1969, environmental legislation was created in the USA, aiming to introduce the Environmental Impact Study (EIS) to be applied in January 1, 1970. Later, in 1972, subsequent appeals by the Interior State Department promulgated the Federal Law of Coastal Zones Control, establishing a planning forum to the Federal States and demanding that Federal actions in coastal areas were consistent with the approved State programs. This system has been approved to solve conflicts related to the maintenance of a health environment and the desired development. However, there is a general agreement of the political observers that the basic problem is that the public support for the system that controls the estuaries uses is missing.

Several estuarine characteristics that need to be protected don't affect immediately its aesthetic aspect, the marine organisms and the public well fair, but ecological studies must be conducted in order to identify in detail the environmental values to be protected. The most important is to be conscious that estuaries are environments that must be saved from the environmental degradation (Sewell 1978).

The Environmental Impact Study is considered as a political instrument aimed to minimize the environmental impact and to reduce its effects to tolerable levels. This study presents two distinguishable and complementary steps: the *diagnostic*, when all positive and negative aspects of the project are considered, and the *prognostic* when the structure of the project is made in order to generate the minor possible numbers of negative environmental and social effects. This system revealed important benefits to the natural resources uses and identify the social distribution of the project costs. Further details on the EIS objectives may be found in specific sources as for instance Tommasi (1994), among others.

The first Environmental Impact Study made in Brazil was the one related to the effluent toxic throw in the Bahia state elaborated in 1975, using float lines. A pioneer regulation of the use of the EIS was adopted in Rio de Janeiro State by means of the Norma Administrativa Ceca-NA-001, in compliment to the 1977 deliberation. However, it was only in August 31, 1981, that the Federal Law, No. 6.938, established the National Policy for the Environment indicating the instruments for the Environmental Impact Study (Art. 9°). This law was regulated by the Decree 88351/83, determining that the EIS be made with basic criteria established by the National Council for the Environment (NCE) in 1986, by means of the resolution 001/1986.

In July, 1983, a Sub-Commission of Coastal Administration of the Secretary of the Inter-Ministerial Commission for the Resources of the Sea (SECIRM) was designated, with the main attribution to elaborate a project of law for the Coastal Management. In this context the Coastal Management Program was prepared, under the coordination by the Sub-Commission, to promote the thematic recognition of the Brazilian coast, establishing in a great scale the coastal zones to coordinate the national coastal resources. Aiming to establish systematic work to orientate the action of the states that integrated the program the SECIRM funded, during the period of 1984 up to 1987, research activities of the Sectorial Plan for the Sea Resources of following universities, through its departments and institutes: State University of Rio de Janeiro (UERJ)—São Paulo State University (USP), Oceanographic Institute of São Paulo University (IOUSP) and the Geosciences Institute of the Federal University of Rio de Janeiro (FURJ) (Carvalho and Rizzo 1994).

The policy and actions to the preservation of coastal ecosystems in general and estuaries, in particular, were asserted by the law n° 7661, of May, 18, 1988, who created the National Plan for the Coastal Management (PNGC), as an integrating part of the National Politics for the Sea Resources and the National Policy for the Environment. Art. 2° of this law established, between other objectives, that its plan is aimed to orientate the rational utilization of the Coastal Zone, in order to contribute to increase the life quality of the populations and to protect its natural, historic, ethnic and cultural patrimony.

The activities which generated information on the scope of the PNGC, where discussed by researchers in specific meetings, including: oceanographic data sampling by remote sensing and simultaneous measurements of physical, biological, chemical and geological properties, in the water column, ocean bottom and estuarine systems (Covre and Calixto 1995; Agra Filho and Viegas 1995).

The Brazilian Federal Constitution, published in October, 1988, presents a chapter (Chap. 6—The Environment, Title VII—of the Social Order) on the environment protection, taking into account also the estuaries. Art. 225 (unique) has the following statement:

> All have the right to an environment ecologically equilibrated, as well as, with a common use by the people and essential to a health life, imposing to the Public Power and to the collectivity the duty to defend and preserve it to the present and future generations.

To assure this right, according to the insert IV of the first paragraph:

> It is a task of the Public Power to demand, according to the law, when the installation of a workmanship or activity which potentially may cause a meaningful degradation of the environment, a previous study of the environmental impact.

The importance of this Constitutional Act is related not only to the indication that the use of the coastal ecosystems must be accomplished in safe conditions to assure their preservation, but also recognizes their vulnerability and the need for their preservation for the future generations.

The demand of the Environmental Impact Study is also inserted in the Chap. 4 of the São Paulo State Constitution, which specifies the care that must be taken with the environment, natural resources and sanitation, including Brazil between the countries with an advanced environmental legislation.

References

Agra Filho, S. S. & Viegas, O. 1995. Planos de Gestão e Monitoramento Costeiro: Diretrizes de Elaboração. Brasília, Programa Nacional do Meio Ambiente. 85 p.

Andutta, F. P. 2011. O Sistema Estuarino dos Rios Caravelas e Peruipe (BA): Observações, Simulações, Tempo de Residência e Processos de Difusão e Advecção. Tese de Doutorado. São Paulo, Instituto Oceanográfico, Universidade de São Paulo. 121 p.

Beardsley, R. C. & Boicourt, W. C. 1981. "On Estuarine and Continental Shelf Circulation in the Middle Atlantic Bight". In: Warren B. A. & Wunsch C. (eds.). Evolution of Physical Oceanography, Cambridge, MIT Press, pp. 198–233.

Bérgamo, A.L.; Miranda, L.B. & Fontes, R.F.C. 2006. Current Measurements and Volume Transport in the Baía de Guanabara Entrance (Rio de Janeiro, RJ). (Org. Elisabete de Santis Braga). In: Oceanography and Global Changes, São Paulo, p. 357–369.

Besnard, W. 1950. "Considerações Gerais em torno da Região de Cananeia-Iguape I". Bolm Inst. Paul. Oceanogr., 1(1):9–26.

Cameron, W. M. & Pritchard, D. W. 1963. "Estuaries". In: Hill, M. N. (ed.). The Sea. Ideas and Observations on Progress in the Study of the Seas. New York, Interscience, pp. 306–324.

Carvalho, V. C. de & Rizzo, H. G. 1994. A Zona Costeira Brasileira—Subsídios para uma Avaliação Ambiental. Brasília, Ministério do Meio Ambiente e da Amazônia Legal/Secretaria de Coordenação de Assuntos de Meio Ambiente. 211 p.

Covre, M. & Calixto, R. J. 1995. O Sistema de Informações do Gerenciamento Costeiro no Âmbito do Plano Nacional de Gerenciamento Costeiro. Brasília, Programa Nacional do Meio Ambiente. 85 p.

Cronin, L.E. 1967. The role of Man in Estuarine Processes. In: Lauff G. H. (ed.). Estuaries. American Association for the Advancement of Science, Washington, D. C., pp. 667–689.

Csanady, G. T. 1977. The Coastal Boundary Layer. Estuaries, Geophysics and the Environment. National Academy of Sciences. Washington, D. C., pp. 57–68.

Curtin, T. B. 1986. Physical Observations in the Plume Region of the Amazon River during Peak Discharge-II. Water Masses. Cont. Shelf Res., 6(1/2):53–71.

Curtin, T. B. & Legeckis, R. V. 1986. Physical Observations in the Plume Region of the Amazon River during Peak Discharge-I. Surface Variability. Cont. Shelf Res., 6(1/2):31–51.

Dalrymple, R. W.; Zaitlin, B. B. & Boyd, R. 1992. "A Conceptual Model of Estuarine Sedimentation". J. Sedim. Petrology, 62:1130–1146.

Dionne, J-C. 1963. Vers une Définition plus adéquate de l'Éstuaire du Saint-Laurent. Zeitschrift für Geomorphologie, 7(1):36–44.

Dyer, K. R. 1973. Estuaries: A Physical Introduction. London, Wiley. 140 p.

Dyer, K. R. 1997. Estuaries: A Physical Introduction. 2. ed., Chichester, Wiley. 195 p.

Fairbridge, R. W. 1961. Eustatic Changes in Sea Level. Physics and Chemistry of the Earth. London, Pergamon Press, vol. 4, pp. 99–185.

Fontes, R. F. C. 2000. Estudo Numérico da Circulação na Plataforma Continental Amazônica. Tese de Doutorado. São Paulo, Instituto Oceanográfico, Universidade de São Paulo. 202 p.

Garvine R.W. 1974. Physical features of the Connecticut river flow during high discharge. *J. Geophsys. Res.* 79:841–846.

Garvine R.W. 1975. The distribution of salinity and temperature in the Connecticut river estuary. J. *Geophsys. Res.* 80:1176–1183.

Garvine R.W. 1977. River Plumes and Estuary Fronts. Estuaries, Geophysics and the Environment. Washington, D. C., National Academy of Sciences, pp. 30-35.

Garvine R.W. & Monk, J. D. 1974. Frontal structure of a river plume. *J. Geophys. Res.*, 79 (15):2251–2259.

Genz, F. 2006. Avaliação dos efeitos da Barragem Pedra do Cavalo sobre os efeitos da circulação estuarina do rio Paraguaçu e Baía de Iguape. Ph. D. Dissertation. Programa de Pós-Graduação em Geologia da Universidade Federal da Bahía. 266 p.

Geophysics Study Committee. 1977. Overview and Recommendations. Estuaries, Geophysics and the Environment. Washington, D. C., National Academy of Sciences, pp. 1–10.

GESAMEP-Joint Group of Experts on the Scientific Aspects of Marine Environmental Protection. 1995. Biological Indicators and their Use in the Measurement of the Condition of the Marine Environment. Repts Stud. Gesamp, n. 55. 56 p.

Geyer, W.R.; Beardsley, R. D.; Candela, J.; Castro Filho, B. M.; Legeckis, R. V.; Lentz, S. J.; Limeburner, R.; Miranda, L. B. & Trowbridge, J. H. 1991. The Physical Oceanography of the Amazo Outflow. Oceanography, vol. 4, pp. 8–14.

Harleman, D. R. F. 1971. One-Dimensional Models. In: Ward Jr., G. H. & Espey Jr., W. H. (eds.). Estuarine Modelling: An Assessment Capabilities and Limitations for Resource Management and Pollution Control. Austin, Tracor, pp. 34–89.

Herz, R. 1992. Manguezais do Brasil. Instituto Oceanográfico da Universidade de São Paulo, Publ. Especial, 237 p.

Ketchum, B. H. 1951. The Exchanges of Fresh and Salt Waters in Tidal Estuaries. J. Mar. Res., 10 (1):18–38.

Ketchum, B. H. 1953. Circulation in Estuaries. In: Johnson, J. W. (ed.). Proceedings of Third Conference on Coastal Engineering, Council on Wave Research. Cambridge, The Engineering Foundation, pp. 65–76.

Ketchum, B.H. 1983. Estuaries and Enclosed Seas (Ecosystems of the World). Elsevier, Amsterdam. 500 p.

Kjerfve, B. 1987. Estuarine Geomorfology and Physical Oceanography. In: Day Jr., J.W., C.H.A. S. Hall, Kemp W.M. & Yáñez-Arancibia, A. (eds.) Estuarine Ecology. New York, Wiley, pp. 47–78.

Kjerfve, B.; Ribeiro, C. H. A.; Dias, G. T. M.; Fillippo, A. M. & Quaresma, V. S. 1997. Oceanographic Characteristics of an Impacted Coastal Bay: Baía de Guanabara, Rio de Janeiro, Brazil. Continent. Shelf Res., 17(13):1609–1643.

Kowsmann, R. O.; Costa, M. P. A.; Vicalvi, M. A.; Coutinho, M. G. N. & Gambôa, L. A. P. 1977. Modelo de Sedimentação Holocênica na Plataforma Continental Sul Brasileira. Rio de Janeiro, Projeto Remac. Petrobras, Cenpes, Dintep, pp. 8–26.

Lana, P.C.; Marone, E.; Lopes, R.M & Machado, E.C. 2001. The Subtropical Estuarine Complex of Paranaguá Bay, Brazil. In: U. Seeliger and B. Kjerfve (eds.) of Coastal Marine Ecosystems of Latin America. Springer-Verlag, Berlim, Heildelberg, pp. 131–143.

Leatherman, S. P.; Douglas, B. C. & Crowell, M. 1997. Beach Erosion Trends and Shoreline Forecasting. J. Coast. Res., 13(4):3–4.

Mann, K. H. & Lazier, J. R. N. 1991. Dynamics of the Marine Ecosystems. Biological-Physical Interactions in the Oceans. Cambridge, Blackwell/ Cambridge University Press. 466 p.

Millimann, J. D. & Emery, K. O. 1968. Sea Level Changes during the Past 35 000 years Science, 162(3858):1121–1123.

Miyao, S. Y. 1977. Contribuição ao Estudo da Oceanografia Física da Região de Cananeia (25°S–048°W). Dissertação de Mestrado. São Paulo, Instituto Oceanográfico, Universidade de São Paulo. 87 p.

Miyao, S. Y.; Nishihara, L. & Sarti, C. C. 1986. Características Físicas e Químicas do Sistema Estuarino-Lagunar de Cananeia-Iguape. Bolm. Inst. oceanogr., São Paulo, 34:23–36.

O'Donnell, J. 2010. The dynamics of estuary plumes and Fronts. In: ed. Valle-Levinson A. Contemporary Issues in Estuarine Physics. Cambridge University Press, pp. 186–246.

Officer, C. B. 1983. Physics of Estuarine Circulation. In: Ketchum, B. H. (ed.). Estuaries and Enclosed Seas. Amsterdam, Elsevier, 26, pp. 15–41, (Ecosystems of the World).

Oliveira, E.N.; Knoppers, B.A.; Lorenzzetti, J.A.; Medeiros, P.R.P.; Carneiro, M.E. & Souza, W. F.L. 2012. A satellite view of riverine turbidity plumes on the NE-E Brazilian coastal zone. Braz. j. oceanogr., vol. 60, n.3, pp. 283–298.

Perillo, G. M. E. 1995. Definition and Geomorfologic Classification of Estuaries. In: Perillo, G. M. E; Picollo, M. C. & Pino-Quivira (eds.). Geomorfology and Sedimentology of Estuaries. Berlin, Springer-Verlag, pp. 17–49 (Environmental Science).

Postma, H. 1980. Sediment Transport and Sedimentation. In: Olausson E. & Cato, I. (eds.). Chemistry and Biogeochemistry of Estuaries. Chichester, Wiley, pp. 153–186.

Pritchard, D. W. 1952a. Salinity Distribution and Circulation in the Chesapeake Bay Estuarine System. J. Mar. Res., 11(1):106–123.

Pritchard, D. W. 1955. Estuarine Circulation Patterns. Proc. Am. Soc. Civ. Eng., 81:717:1–11.

Pritchard, D. W. 1967. What is an Estuary: Physical View Point. In: Lauff G. H. (ed.). Estuaries. Washington, D. C., American Association for the Advance of Science, pp. 3–5.

Rongel, A. 1943. Marés Fluviais. Anais Hidrográficos. Marinha do Brasil-Hidrografia. Tomo X, pp. 31:45.

Schaeffer-Novelli, Y.; Cintrón-Molero, G.; Adaime, R. R. & Camargo, T. M. 1990. Variability of Mangrove Ecosystems along the Brazilian Coast. Estuaries, 13(2):204–218.

Schettini, C. A. F. & Miranda, L. B. 2010. Circulation and Suspended Particulate Matter Transport in a Tidally Dominate Estuary: Caravelas Estuary, Bahia, Brazil. Braz. J. Oceanogr. 58(1):1–11.

Schettini, C. A. F.; Kuroshima, K. N.; Pereira, J. Fo.; Rörig, L. R. & Resgalla Jr., C. 1998. Oceanographic and Ecological Aspects of the Itajaí-açu River Plume During High Discharge Period. Anais. Acad. Bras. Ci., 70(2):335–351.

Schettini, C. A. F. Pereira M.D.; Siegle, E.; Miranda, L.B.; & Silva, M.P. 2013. Residual fluxes of suspended sediment in a tidally dominated tropical estuary. Continental Shelf Research, 70:27–35.

Schubel, J. R. & Pritchard, D. W. 1972. The Estuarine Environment, Part 2. J. Geol., 20(4):179–188.

Sewell, G. H. 1978. Administração e Controle de Qualidade Ambiental. Trad. e adapt. Gildo M. S. Filho. São Paulo, E.P.U./Edusp/Cetesb. 295 p.

Simpson, J. H., Gong. W.K. & Ong J.E. 1997.The determination of net fluxes from a mangrove estuary system. Estuaries, 20, 103-109.

Suguio, K. & Martin, L. 1978. Formações Quaternárias Marinhas do Litoral Paulista e Sul Fluminense. In: 1978 International Symposium on Coastal Evolution in the Quarternary. São Paulo. The Brazilian National Working Group for the IGCP. pp. 1–55 (Special Publication, 1).

Tessler, M. G. 1982. Sedimentação Atual na Região Lagunar de Cananeia-Iguape, Estado de São Paulo. Dissertação de Mestrado. São Paulo, Instituto de Geociências, Universidade de São Paulo, vol. 1. 110 p.

Tessler, M. G. & Souza, L. A. P. 1998. Dinâmica Sedimentar e Feições Sedimentares Identificadas na Superfície de Fundo do Sistema Cananeia-Iguape. Rev. Bras. Oceanogr., São Paulo, 46 (1):69–83.

Tommasi, L. R. 1979. Considerações Ecológicas sobre o Sistema Estuarino de Santos (SP). Tese de Livre-docência. São Paulo, Instituto Oceanográfico, Universidade de São Paulo, 2 vols. 489 p.

Tommasi, L. R. 1994. Estudo de Impacto Ambiental. São Paulo, Cetesb: Terragraph Artes e Informática. 354 p.

Tundisi, J. G. 1969. Produção Primária, Standing Stock e Fracionamento do Fitoplancton na Região Lagunar de Cananeia. Tese de Doutorado. São Paulo, Instituto de Biociências, Universidade de São Paulo. 130 p.

Villwock, J. A. 1972. Contribuição à Geologia do Holoceno da Província Costeira do Rio Grande do Sul, Brasil. Dissertação de Mestrado. Porto Alegre, Universidade Federal do Rio Grande do Sul. 133 p.

Williams, A. T. & Simmons, S. L. 1997. Estuarine Litter at the River/Beach Interface in the Bristol Channel, United Kingdom. J. Coast. Res., 13(4):1159–1165.

Quoted Reference

Defant, A. 1961. Physical Oceanography. Oxford, Pergamon Press, vol. 1. 729 p.

Chapter 2
Circulation and Mixing in Estuaries

2.1 Hydrologic Processes: Ocean-Drainage Basin-Estuary

An estuary is a coastal transitional environment between the continental land masses and the ocean, where the sea water is diluted by fresh water from continental drainage basin. This environment is forced by local and remote processes generated by climatic, oceanographic, geological, hydrographic, biological and chemical events, occurring in the ocean and drainage basin at distances, as far as, thousands of kilometers away.

The drainage basin is the catchment origin of the river system, which empties into the estuary, supplying not only water but also sediments, organic and inorganic substances and eventually pollutants. The fresh water budget of a drainage basin is dependent on the climatic conditions, soil characteristics, vegetation area and uses (urban, agriculture and industrial), evaporation and transpiration from surface water, and its interactions are illustrated in Fig. 2.1, according to Coleman and Wright (1971). This fresh water budget of a hydrographic basin is only a small fraction of the global hydrologic cycle, which is defined as the movement and endless recycling of water between the atmosphere, the land surface, and the ground.

With human development and occupation of the land adjacent to estuaries occurring after the second half of the XIX century, the geometry and hydrologic conditions of the drainage basins of rivers were gradually and drastically altered. These alterations resulted from the construction of dams, water barriers and channels (such as the Valo Grande channel in the estuarine- lagoon system of Cananéia-Iguape—Fig. 1.5), forest destruction, and road and pavement construction, which interfered deeply in the natural ecological characteristics of these coastal environments.

Solar radiation is the main energy source in our planet. Its distribution on the earth's surface varies according to the geographic latitude and the seasons of the year. The cloud cover, aerosols and particle concentrations in the atmosphere, also

L. Bruner de Miranda et al., *Fundamentals of Estuarine Physical Oceanography*,
Ocean Engineering & Oceanography 8, DOI 10.1007/978-981-10-3041-3_2

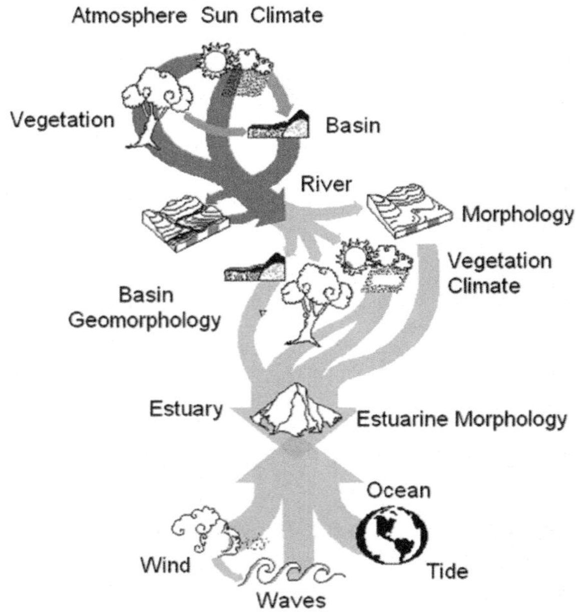

Fig. 2.1 Local and remote processes and forces in the catchments and in the ocean, which contributes to the estuaries characteristics and dynamics (according to Coleman and Wright 1971)

influence the solar energy reaching the earth's and ocean's surfaces. This energy flux controls the heat concentration in the oceans and atmosphere. It is the only energy source that drives the processes of evaporation and transpiration in the estuary catchment region, and primary production through the photosynthesis of microscopic organisms (phytoplankton) and zooplankton, which supports the estuaries feeding chain in the estuaries.

The wind stress forcing promotes the aeration of the estuary surface layers and the mixing of the estuarine and coastal water masses, and may erode the vertical salinity gradients, mainly in the estuary's mouth sheltered from local weather. This forcing may also generate currents, waves, intensifying the vertical mixing, mainly in estuaries with large surface area. Characteristics on the induced circulation in the Potomac river estuary (Virginia, USA), has been investigated in the paper of Elliott (1978), based on the analysis of meteorological data and current measurements at three depths. It was shown that the non-tidal currents were responding to two distinct forcing mechanisms of almost equal importance: the local and non-local wind effects propagating into the estuary from the Chesapeake Bay.

The precipitation, river discharge and evaporation-transpiration in the estuary catchment is always positive, i.e., the sum of the fresh-water source (precipitation and river discharge) is always higher than the sink (evaporation-transpiration). This is in agreement with the Pritchard (1955) definition, stating that in the estuary the seawater is measurably diluted by the fresh water discharge. The main factors that may influence this balance are the temperature and relative air humidity, the wind direction and intensity, the geomorphology of the catchment region, the soil characteristics and vegetation (Fig. 2.1).

The river discharge into the estuary gradually dilutes the sea water due to the advection and turbulent diffusion, generating the longitudinal salinity (density) gradient, which are of fundamental importance to the estuary dynamics. The river discharge (Q_f) is physically defined as a volume transport (volume per time unity). The ratio of the volume transport by any transversal section of the estuary is the intensity of the mean velocity u_f, (distance per time) across the section forced by the river discharge; these quantities have dimensions of $[Q_f] = [L^3T^{-1}]$ and $[u_f] = [LT^{-1}]$, respectively. As will be seen later, this quantity is one of the low frequency components of the estuary seaward motion, and it is the result of the application of the mass (volume) conservation equation integrated over the estuary domain.

In most nations, there are governmental entities which are responsible for monitoring the river discharges of small and large rivers. In medium and high latitude regions, the snow height accumulated during the winter time is used to forecast the availability of fresh water; as the snow melts at the end of winter and during spring, an important volume of fresh water is added to the catchment regions. In general, the fresh water discharge is monitored with measurements of the relative height of the river surface (once a day or with continuous sampling), which are converted to discharge estimates with a calibration equation of the equipment. In the rivers flowing into the sea, these measurements are made far from the estuary head or tidal river zone (TRZ), where the motion is unidirectional. These data are stored in official water and energy management institutions, and may be easily accessed.

It is desirable that the river discharge (Q_f) is considered as known data when focusing on problems related to estuarine physics, because its long term monitoring and determination is from the dominion of Hydrology. Because, in general, this physical quantity is measured for the purpose of controlling the availability of the fresh water supply for urban communities, industrial plantations and agriculture, and not specifically for estuarine research.

In Fig. 2.2 is presented schematically a catchment area with river discharge towards the coastal region occurring through an estuary. Let us assume that this catchment is composed of several rivers and tributaries with its discharges being measured at the monitoring stations, indicated by black circles. In this figure, A_1 is the total monitored catchment area and, lets us indicate by q_i a generic river discharge of the station i (in this case i = 1, 2, 3 and 4). Then, the river discharge contribution (Q^*) of the monitored area (A_1) is calculated by:

$$Q^* = \sum_i q_i, \quad i = 1,\ldots 4, \tag{2.1}$$

which doesn't take into account the partial catchment area (A_2), localized seaward of the monitoring stations, thus $Q^* < Q_f$.

However, as the total river discharge to be calculated is Q_f, it is necessary for a correction to be applied to the partial discharge Q^*. In the hypothesis that the precipitation and evapo-transpiration processes have the same values in both areas

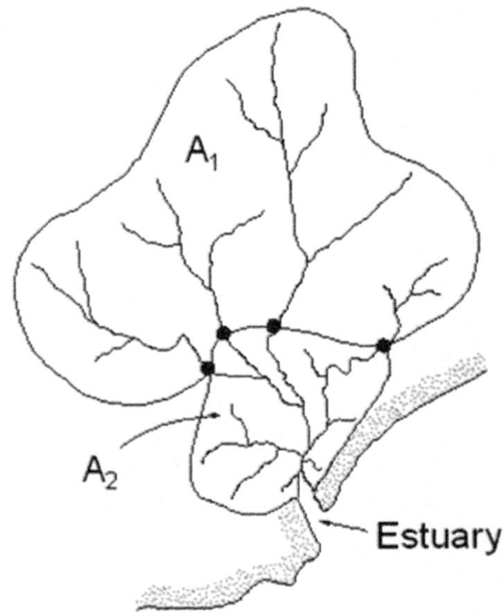

Fig. 2.2 Schematic diagram of an estuary catchment area ($A_1 + A_2$) of main rivers and its tributaries. A_1 is the partial area of the monitored rivers discharges at the stations indicated by *black dots*

and are uniform in the total area ($A_1 + A_2$) and that the soil characteristics are also the same, we may estimate a dimensionless correction (c), which is given by:

$$c = \frac{(A_1 + A_2)}{A_1}, \qquad (2.2)$$

with $c \geq 1$; and $c = 1$ when $A_2 \to 0$, and the river discharge monitoring stations are located at the estuary mouth, which is not the case. Then, with these results it follows that the best value of the river discharge at the estuary head is given by:

$$Q_f = cQ^*, \qquad (2.3)$$

where c is a correction factor. From this equation, it follows that if $A_1 \gg A_2$, then $c \approx 1$ and $Q^* \approx Q_f$.

The river discharge input at the estuary head occurs with a delay in relationship its measurements at the up-river gauging stations. Because gauging stations are often located further upstream and away from the estuarine zone. In turn, these measurements also have a delay in relationship to the precipitations at the catchment area, where the processes that determine the water volume of the surface flow in the river and its tributaries occur; however, in general, these time intervals are very difficult to calculate and are usually not taken into account. Besides the seasonal time scale dominance in the river discharge (Q_f), daily time variations may also occur due to the abnormal short time precipitations. Then, in order to have representative values of river discharges for a given experiment, it is advisable to use daily mean values of a time series measurements.

The problem of estimating the river discharge in an estuarine system is illustrated for the river system of the catchment area emptying into the Winyah Bay (33° 15′N), localized in the South Caroline State (USA). Based on monitored river discharges from the main river and some tributaries, this estuarine system has an annual mean fresh water discharge of 557 $m^3 s^{-1}$. The fresh water input and low contributions. Taking into account the monitored main catchment area, (A_1), and the catchment area localized seaward of the monitoring stations (A_2), Eq. (2.2) was used to calculate the correction factor c (c = 2.1). This factor was applied to the daily discharge values of the tributaries in the area A_1 (Eq. 2.3) in order to obtain a more representative time series of the river discharge into the Winyah Bay (Fig. 2.3). In this figure it may be observed that the $Q_f = Q_f(t)$ time series is a non-stationary quantity, presenting daily and seasonal variability. The following river discharge regimes were also found in this data analysis:

(i) A moderate river discharge from April to June, varying from 200 to 1000 $m^3 s^{-1}$;
(ii) In the transitional period from June to July, a secondary maximum of 1600 $m^3 s^{-1}$ occurred, which decreased in a few days to reach the lowest values between August and November (200 $m^3 s^{-1}$);
(iii) As the result of the ice melting at the end of the winter season, the highest river discharges occurred during February ($Q_f > 4000$ $m^3 s^{-1}$).

River discharges have been monitored for many rivers around the world from the early 1900s. In Brazil, the National Electricity Agency (ANEEL), trough the National Water Agency (NWA), is responsible for discharges and water quality measurements, for most Brazilian states, with exception of the São Paulo state, whose river discharge measurements are made by the Department of Water and Electrical Energy (DAEE). In the Jaguaribe river in NE Brazil (Ceará), for example,

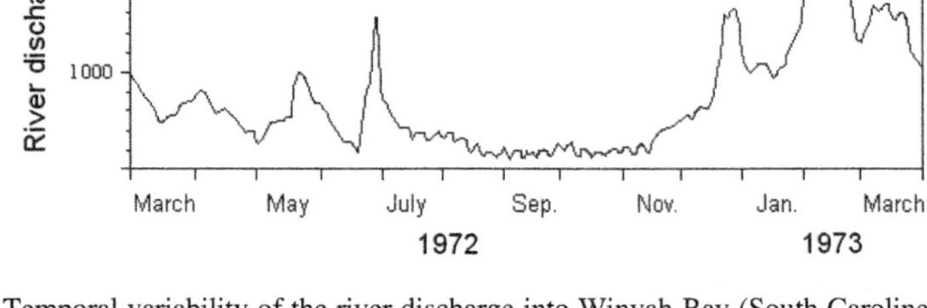

Fig. 2.3 Temporal variability of the river discharge into Winyah Bay (South Caroline, USA). Q_f was estimated by the sum of the daily discharges in the catchments of the main tributaries, extrapolated to the estuary head with the application of the correction factor (Eq. 2.2)

there are 11 pluviometric stations with 80 years of monitoring, plus 11 river discharges monitoring stations, performing 25 years of continuous data (Lacerda et al. 2002), and the Itajai-açu river (Santa Catarina) has been monitored since 1934 (Schettini 2002). However, it should be noted that major rivers affecting the coastal zone may have cause environmental problems due to river diversion, damming, and industrial effluent releases, which affect the river discharge, addition to eutrophication, pathogen contamination, toxic chemicals, loss of habitat, and declines in fish and wildlife.

For estuary catchment areas, or any coastal environment (bay, inlet, coastal lagoon) where rivers discharges have not been monitored, this quantity may be indirectly estimated using semi-empirical methods. First of all, lets introduce the definition of *surface runoff tax*, Δf, as the precipitation rate (P) generating a runoff which discharges into the river; both Δf and P are expressed as height per time $[LT^{-1}]$. The ratio of these quantities defines the dimensionless *surface runoff ratio* $(\Delta f/P)$ as a dimensionless quantity, which must depend on the rainfall itself; when rainfall is so heavy that evaporation can only remove a small fraction of the annual water accumulation in the catchment area, the runoff ratio must approach unity $(\Delta f/P \rightarrow 1)$. Conversely, in extreme cases when rainfall is so sparse that evaporation can easily remove the annual water accumulation, the runoff ratio approaches zero $(\Delta f/P \rightarrow 0)$.

The maximum annual evaporation is, of course, strongly dependent on the intensity of solar radiation. Various functions have been proposed that relate the runoff ratio $(\Delta f/P)$ to the rainfall, P. The simplest of these functions was introduced by the German hydrologist P. Schreiber, in 1904 (quoted in Holland 1978), who found that the runoff ratio for central European rivers was related to the rainfall (P) by the expression, already adapted for the effect of latitude on the runoff ratio (Holland, op.cit.):

$$\frac{\Delta f}{P} = \exp\left(-\frac{E_v}{P}\right), \tag{2.4}$$

which states the relationship of the surface runoff ratio and the exponential ratio of E_v/P (the annual mean potential evapotranspiration rate and the corresponding precipitation).

The dependence of the first term in Eq. (2.4), on the evaporation rainfall tax, has already been discussed, and is in agreement with this equation because it decreases exponentially between the extreme values 0 and 1 when $E_v \gg P$ and $E_v \ll P$, respectively, and represents the rainfall tax which will effectively take part in the surface runoff.

Schreiber's equation (2.4) predicts that the runoff ratio should be latitude dependent, because the evaporation decreases rapidly due to the low temperatures at high latitudes, which has been confirmed by Kosoun et al. (1974). Thus, the exponential decrease of E_v with increasing latitude between the extreme values 0 and 1, for $E_v \gg P$ and $E_v \ll P$, respectively, confirms the physical basis of the Schreiber's equation.

The evapo-transpiration potential (E_v) has an accentuated decrease with temperature towards high latitudes. The runoff ratio's ($\Delta f/P$) dependence on latitude was experimentally obtained with data of representative catchments of Europe, Asia, Africa, Australia and North and South America, including the Amazon and São Francisco rivers. Although this dependence indicates a greater scatter at low latitudes, it was possible to confirm the physical basis of the P. Schreiber equation (Eq. 2.4) and to establish, with curve fitting procedure, the following equation for determining E_v as function of the annual mean absolute surface air temperature (T), according to Holland (1978):

$$E_v = 1.2 \times 10^9 \exp(-\frac{4.62 \times 10^3}{T}). \tag{2.5}$$

In this equation the temperature is expressed as K, and E_v, $[E_v] = [LT^{-1}]$ is calculated in cm year^{-1}, with satisfactory can be applied in the range from the Equator to about 70° of latitude (north and south), and thus excluding high latitude areas. This method only gives representative values when calculated using monthly or annual mean temperature and precipitation values for long time periods (Kjerfve 1990).

Thus, to estimate the runoff ratio ($\Delta f/P$), it is necessary to know the mean values of the surface air temperature (T), the rainfall tax (P), the catchment area (A_T), and the time interval (Δt) used to calculate the mean values of E_v and P. Then, the river discharge may be obtained with the following equation:

$$Q_f = \frac{\Delta f}{P}(PA_T) = \Delta f A_T. \tag{2.6}$$

We can observe in this equation that the river discharge, calculated by the product of the precipitation tax by the catchment area (PA_T), is corrected by the runoff ratio ($\Delta f/P$). The result this equation to calculate the river discharge Q_f is in unities of cm^3 year^{-1}, according to the units indicated above. However, using a convenient conversion factor, it is possible to calculate the river discharge in the International System of Units (m^3 s^{-1}).

In order to exemplify this method, the following data of the catchment area and meteorological data of the Santee river (South Caroline, USA) is used: $A_T = 41 \times 10^3$ km^2 $= 41 \times 10^9$ m^2, T = 20 °C = 293 K and P = 1.25 m year^{-1}. Applying Eqs. (2.5) and (2.6), it follows: E_v = 170 cm ano^{-1} and $\Delta f/P = 0.25$, and in Eq. (2.6) the final result is: $Q_f = 1.28 \times 10^{10}$ m^3 ano^{-1} ≈ 406 m^3 s^{-1}; this value is about 25% lower than the mean river discharge based on river measurements previously indicated (557 m^3 s^{-1}).

As already indicated, the factor 1.2×10^9 of the exponential function in Eq. (2.5) was adjusted for the runoff ratio $\Delta f/P$ to be calculated with Eq. (2.4) in units of cm^3 year^{-1}. However, with monthly mean values of a long time series of air temperature, it is possible to obtain the monthly and the seasonal variation of the river discharge in an estuary (or in any other coastal environment). In this case, it is easily to shown that to calculate E_v in mm monthly^{-1}, the factor must be altered to 1.0×10^9.

This method has been applied by Medeiros and Kjerfve (1993) to calculate the seasonal variation of the river discharge into the tropical estuarine system of the Itamaracá (Pernambuco, Brazil—07° 50′S; 034° 50′W) during the peak rainy (57.7 $m^3 s^{-1}$) and dry seasons (0.2 $m^3 s^{-1}$). With these results, it was possible to conclude that the freshwater input variation dominates the system behavior. During the rainy season the system is partially mixed, and gravitational circulation prevails; however, in the dry season the estuary becomes well-mixed due to the low freshwater discharge. Normalized data of air temperature and precipitation time series of several years, from meteorological stations representative to the Saquarema and Araruama coastal lagoons, Guanabara bay (Rio de Janeiro, Brazil) and for the estuarine-lagoon system of Cananéia-Iguape (Fig. 1.5), were also used to estimate the freshwater discharges in these systems (Schettini 1994; Kjerfve et al. 1996, 1997; and Bonetti and Miranda 1997).

When the catchment area extends over large areas, the results of the fresh water discharge will be more representative, if mean data values of air temperature and rainfall in the subareas of the catchment are known, enabling calculation of partial results. Applying the procedure presented above, the sum of these partial results will be the total freshwater discharge into the system.

Equilibrium conditions between the freshwater input in the estuary head and the outflow to the adjacent coastal region remains, in general, under nearly steady-state conditions. In this condition, the resulting time mean volume transport during several tidal cycles, across transversal sections located in the estuary head and mouth, will have almost the same value. The evaporation, precipitation, spring water and percolation at the bottom processes, usually have a small contribution to the freshwater balance in the estuarine system.

Besides the secular influence of the air temperature over the hydrologic cycle in the last one hundred years, abnormal events of dry weather and flooding related to the anomalies of the sea surface temperature (SST) have also occurred, with intervals varying from one to seven years.

The more catastrophic dry climate that remained for 83 days occurred in the USA in 1986. This was correlated with high sea surface temperatures (>32 °C) in the North Atlantic, whose influence on the hydrologic cycle caused a reduction by almost 50% on the rivers discharges. Besides the social and economic problem, this dry climate also produced an increase in the salinity of the estuarine water masses with the following consequences to the biological community:

- Great mortality of fishes and crustacean community.
- Reduction of larvae recruitment.
- Turbidity increase in the water mass during the recovery of the normal hydrologic cycle, with great impact on biological and algae production.

The SST increase also creates favorable conditions to the occurrence of violent storms and hurricanes in tropical and subtropical regions. These transient events, reaching coastal environments, are usually strong enough to produce impacts with alterations in the sea level, coastal and estuarine circulation, biological, chemical and erosive processes. Then, we may propose the following questions of fundamental ecological importance:

- What will be the biological and physical alterations in the estuaries, due to the climatic transient changes, and due to long time periods (decades, centuries and millennia)?
- When will there be an occurrence of changes in the coastal region?
- How to grant and maintain the sustainability of the estuary development due to the natural and human alterations?

2.2 Temporal and Spatial Scales of Sea-Level Variations

The formation history and localization of estuaries were dependent on the secular variations of the relative sea level which occurred during the Holocene sea level transgression (Fig. 1.1, Chap. 1). After reaching the current sea level, its variation on time scales of seconds (wind waves), hours (astronomic tidal waves), days (waves generated by meteorological forcing), months (vortices and meanders of oceanic currents), annual (seasonal variations in the ocean-atmosphere processes and steric level variations), inter-annual and decadal (climatic interactions generated by global processes as El Niño), started to exert influences with different intensities on the hydrodynamic behavior of estuaries.

The time scales, characterizing the estuarine variability can be described as being either *intratidal* or *subtidal* (Jay 2010):

(a) Intratidal—variability which occurs at semidiurnal or diurnal tidal frequencies >1 cycle/day (periods of 12–25 h), or shorter time periods (Fig. 2.4a, b), and their *overtides* driven by non-linear processes occurring at sums or multiples of the basic astronomical frequencies, are the most obvious examples of intratidal variability. Also classified in this category are (i) the variations of scalar properties (e.g., salinity, temperature and density) driven directly by tidal currents; (ii) the effects of daily sea breeze, harbor seiches with periods of minutes to hours; (iii) internal waves; and (iv) in large estuaries inertial motion at the local pendulum frequency (periods of 12–20 h at mid-latitudes).

(b) Subtidal—variability at frequency of <1 cycle/day. In this category are included low-frequency tidal motions as the fortnightly and monthly tidal modulation of the M_2 and S_2 tidal components—periods of 13–15 days, and M_2 and N_2—periods of 28–30 days, and variability related with weather systems with typical periods 3–10 days (Fig. 2.4c), or even at longer time periods, such as seasonal in response to the changes in the river discharge and snow melting events (Fig. 2.4d).

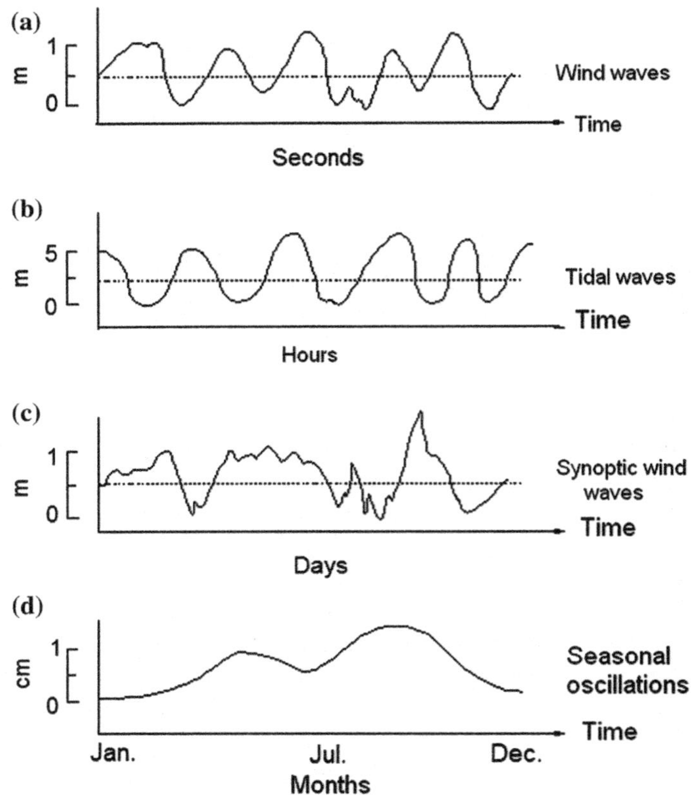

Fig. 2.4 Time and height variations of the sea level. Intratidal (**a**, **b**); wind waves (or gravity waves) and tidal waves and, subtidal (**c**, **d**); synoptic wind waves and seasonal oscillations

(c) Motions that do not vary in time and space are classified as *steady-state* and *uniform*, respectively.

Wind waves generally have a greater influence over the mixing processes in shallow estuaries with large surface area. In some estuarine systems, this influence is more accentuated at the estuary mouth, which is the region more intensively forced by these waves remotely generated on the continental shelf and the open ocean.

The astronomic tide forcing is dominant in generating motions (turbulent and circulation, in low scale large scale, respectively), producing mixing of the freshwater and salt water, thus evolving the processes of advection and diffusion, which vary spatially and are also influenced by the estuary geometry. In relation to this forcing, it is necessary to distinguish between the tidal oscillation and tidal co-oscillation. It is known that the tidal phenomenon is generated by the gravitational attraction of the Moon and the Sun, associated with centripetal acceleration, acting directly in the great ocean water masses. The action of this gravitational phenomenon on the estuarine water masses is negligible. However, the global tide generation in the ocean propagates towards the continental shelf in different types of waves (Kelvin, Poincaré and gravity long waves), being one of the main generating motions and mixing processes in the estuaries; their influence propagates up and down estuary as a long wave of gravity. The tidal forcing in the estuary, which has the same period as the oceanic tide is denominated as *co-oscillation tide* (Defant 1960).

The origins of the sub-tidal fortnightly modulation are not as simple as the equilibrium tidal theory implies, and to understand them requires knowledge of both equilibrium and dynamic tidal theory, and the inter-relationships with the tidal constituents: synoptically driven neap-spring cycles occurs, when the M_2 and S_2 tidal constituents dominate the tides, and tropically driven neap-spring cycles occurs when the tidal constituents O_1 and K_1 are prevalent (Kvale 2006). In this low frequency band there is also a distinctive separation between its spectral characteristics. With periods of days and weeks we may observe the random occurrence of continental shelf waves. These topographic waves are generated by the oscillation in the synoptic scale of the wind component parallel to the coast (Gill and Schumann 1979). Also, within this time scale the sea level may respond in phase to the wind components (mainly parallel to the coast) and to the atmospheric pressure oscillations generating wind storms. Since the sea surface responds as an inverted barometer to the atmospheric pressure, these oscillations are also able to rise up or lower the sea surface level by about one centimeter for each decrease or increase of one millibar in the atmospheric pressure, respectively.

Meanders and vortices of oceanic currents may propagate through the continental shelf, reaching the coastline. The typical temporal scale of this process is about one month, having the same signature of the up and down elevation of the sea level, considering these to be anti-cyclonic and cyclonic, respectively. However, the possibility of influence of these phenomena in the estuaries is restricted to narrow continental shelves with abrupt topography. Seasonal meteorological processes, such as the predominant wind direction or atmospheric pressure fields over the continental shelf, may also be the forcing of the annual oscillations of the sea level at the estuaries mouth.

At larger time scales, such as several years, climatic global processes such as the El-Niño-Southern Oscillation may influence the sea surface height along the coastline where its intensity may be stronger and with longer duration. The typical El-Niño influence, for example, is characterized by the upwelling interruption usually present along a great extension of the Equatorial South America west coast, mainly Peru. It is known that the upwelling occurrence is only possible when the sea level at the coastline remains lower than the mean sea level (from approximately 0.01 to 0.1 m). Thus, during the time occurrence of this global phenomenon which may last for several months the sea level in the Peru coast and, consequently, at the estuaries mouth will remain lower or higher than in normal conditions, with influences on the physical and ecological characteristics of these transitional water bodies.

However, although this large temporal time scale of the sea level oscillation and with simultaneous high frequency oscillations, the hydrodynamics characteristics of estuaries are mainly controlled by the semi-diurnal and diurnal tidal oscillations. The differences in magnitude and duration between ebb and flood currents, produced by the distortion of the tide wave propagating on the costal shelf and entering bays and estuaries, are named *tidal asymmetry*.

The oceanic tide was one of the first oceanographic phenomena theoretically studied. Using Newtonian concepts of Mechanics, the equilibrium tidal theory was developed in the pioneering scientific work of Daniel Bernoulli, in 1740. The main components of the tidal oscillation were calculated by P.S. Laplace, in 1775. In subsequent work, the tide effect was subdivided into oceanic and shallow water tides, with contributions from several notable researchers, for example, those mentioned in the von Arx (1962) book. However, even with the work of these researchers and attempts to find the tidal heights in real time using the solutions of the Laplace equations, the problem of how to compute the theoretical tide forecast from its basic principles remained to be solved. The previson of tidal heights needs experimental data from sea level height measurements in coastal and open ocean tidal stations.

Over the continental shelf, the tide wave is transported by a complex set of waves influenced by the Earth's rotation. This transport is highly dependent on the width and bottom topography on the continental shelf. In continental shelves with wide width (order of 10^5 m), the tide co-oscillates with the adjacent deep ocean, which may present amplifications during its propagation from the continental shelf break up to the coastline. Poincaré waves predominate in the energy transport across the continental shelf. However, in narrow continental shelves the tidal wave is usually a propagate wave; along the coastline it propagates as Kelvin and Poincaré waves. In general, however, the tidal wave propagation on the continental shelf is a combination of co-oscillation and propagation and the influence of these characteristics on estuarine processes are not well understood. Upon entering an estuary, the water depth becomes shallow and the tidal wave is observed to become more asymmetric as it travels upstream. The explanation for this increasing asymmetry lies in the fact that friction causes the wave to travel at a speed (celerity —Eq. 2.20) governed by the water depth.

The height of the tidal wave (H_o) is defined as the difference in the elevation between the highest level (crest) and lowest level (trough), indicated by HW and LW (Fig. 2.5). The distance between two consecutive crests and lows is the wave length (λ) and the time interval of the propagation of these events (tidal cycle) one or two times per day is the tidal period (T_P). The tidal height varies periodically according to the gravitational intensity, and the time interval of these cycles is approximately equal to 12.4 and 25.0 h (semi-diurnal or diurnal, respectively). The highest (higher HW) and lowest (lower LW) tidal wave amplitude occurs when the Earth, Sun and the Moon are aligned or in quarter, respectively (Fig. 2.5).

The tidal amplitude (η_0) is the difference between the crest level (or high water) and the mean sea level, equal to the half of the height ($\eta_0 = H_o/2$). The tidal height in the open sea is usually less than one meter. However, in crossing the continental shelf, or in semi-closed coastal regions, such as bays, inlets and estuaries, there will be amplification of the tidal height which may reach heights of several meters.

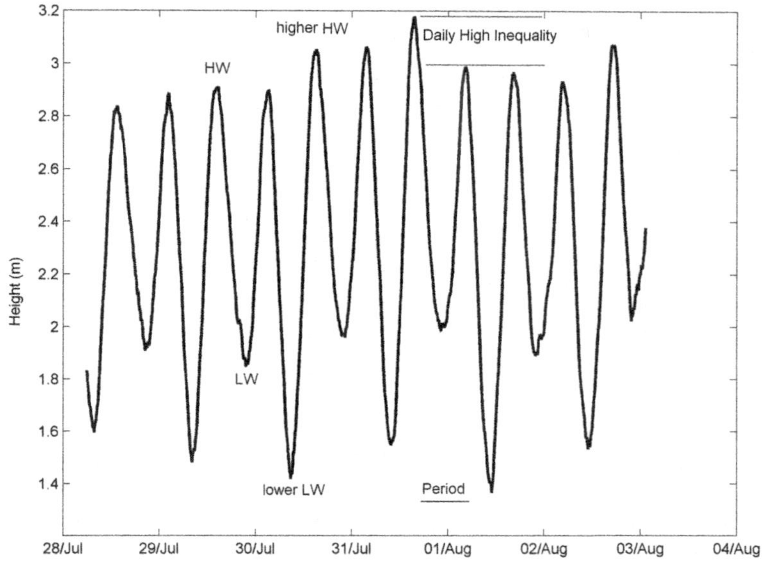

Fig. 2.5 Periodic oscillations of the tide in the Estuarine-lagoon system of Cananéia (Fig. 1.5) showing the high (HW) and low (LW) water oscillations, the corresponding higher HW and lower LW heights, the daily high inequality and the tidal period

Thus, according to the maximum height (H_{MAX}), estuaries are forced by tides which may be classified as (Davies 1964):

Microtide: $H_{MAX} < 2$ m.
Mesotide: $2 < H_{MAX} < 4$ m.
Macrotide: $4 < H_{MAX} < 6$ m.
Hypertide: $H_{MAX} > 6$ m.

The tidal height importance is related to the fact that the periodic tidal inundated areas that may have favorable conditions to the development of vegetation, such as *mangroves* and *salt marshes*. The importance of tidal range lies in the fact that these intertidal habitats will occur only in areas which can be submerged by the tide. Because the highest elevations are only covered for a short time period around high tide, they receive less sediment and nutrients than other areas lower down. Hence, the higher the vegetation, the less nutrients it receives and the slower will be the rate at which it accretes; this has been confirmed by studies performed in the Severn estuary (England) by French (1997).

The tidal prism (T_{PR}) is defined as the sea water volume input into the estuary during the flood tide (from low to high tide), and is related to the tidal height. It may be approximately calculated as $T_{PR} = H_oA_s$, where A_s is the surface area of the estuary. In microtidal regions, the capacity of the water renewal during the tidal cycle is small when compared with estuaries forced by hypertides.

To establish the relative importance of the tidal forcing and the river discharge, a dimensionless number was defined as the ratio of the volume of the fresh water discharge into the estuary during a tidal cycle ($R = Q_fT_P$), and the tidal prism (T_{PR}). This dimensionless number, $F_R = R/T_{PR}$, is named *flux ratio*. Then, according to this definition, when $F_R > 1$ or $F_R < 1$ the estuary is dominated by the river discharge or the tide, respectively. An alternative method to calculate the flux ratio will be presented at the end of this chapter, and as will be seen in Chap. 3, this number has also been used as a criterion for estuary classification.

Besides the dominant semi-diurnal and diurnal tidal oscillations, the tide has components of higher periods: fortnightly and seasonally are the main oscillations. The fortnightly period is the time between successive spring tides and is modulated by Moon phases; this period is 14.8 days, which is the period of maximum constructive interference between the main semi-diurnal tidal constituents M_2 and S_2, which reinforce at spring tides and are in opposition at neap tides (Leblond and Mysak 1978). Its basic modulation features in shallow rivers were explained through scaling arguments, which show that this wave is generated by the fortnightly modulation of frictional forces due to the variation in tidal velocities (Leblond 1979). There is also a fourth level of height fluctuations, which occurs over longer periods (18.6 years). This oscillation, with frequencies lower than the more energetic ones (semidiurnal and diurnal), are in the sub-tidal domain.

The time variability of the tidal height during events of spring and neap tide, sampled in the estuarine channel of Cananéia (São Paulo State, Brazil), is presented in Fig. 2.6. This figure clearly shows the fortnightly tidal modulation, and the occurrence of the highest and lowest tidal heights during spring and neap tide ($H_o \approx 2.0$ and 0.8 m, respectively), and the sub-tidal oscillation. The high frequency oscillations of temperature simultaneously recorded are coherent with the tidal oscillation, indicating the advective influence of the tidal current in the redistribution of the heat concentration. The low frequency variations of tide and temperature, calculated with a low frequency filter to eliminate the periodic and high energetic variations, are also shown in Fig. 2.6.

The physics of the tidal oscillations decomposition in its components is very complex, being related to the relative motions of the Earth-Sun-Moon system, and may be found in specialized books (Defant 1960; von Arx 1962; Franco 1988, 2009, and others). There is another important aspect related to the tidal wave propagation into the estuary: in this environment, the tidal wave may be composed of short period harmonic oscillations generated by the local topography, the tidal excursion is blocked up by the river discharge and, simultaneously, the tidal wave may be subjected to an energy loss due to the frictional drag in the estuary bottom and margins. Consequently, the tidal wave propagation up and down the estuary undergoes significant modification and substantially deforms, as for instance, in generating the tidal bore.

The interaction between the up estuary tidal wave propagation and the estuary morphology is responsible for the variations in the tidal height and current intensities. Convergent margins of the estuary forces the tidal wave to be laterally compressed and, in the absence of friction, there will be an increase in its height due

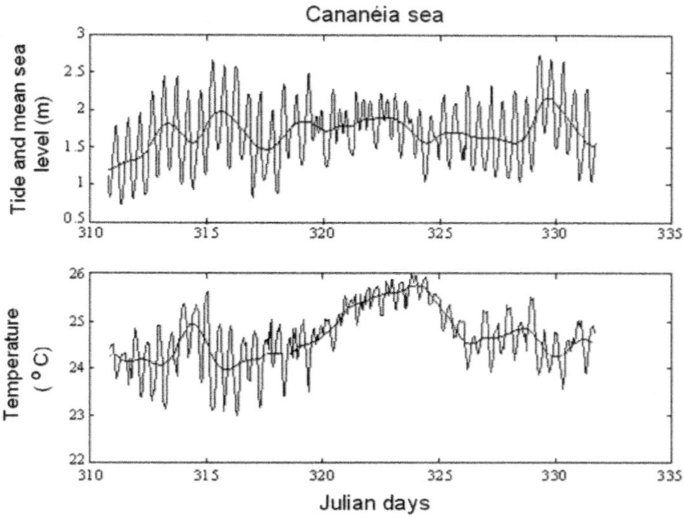

Fig. 2.6 Tidal record in the Lagoon System of Cananéia (southern São Paulo State) showing high and low frequency time variations modulated during spring and neap tides, and the corresponding temperature variations. Time scale in Julian days

to the energy conservation. Inversely, due to the frictional influence, there will be a decrease in the tidal height (Dyer 1997). According to the relative influence of the geometric characteristics of the estuarine channel, the following conditions may be found (Nichols and Biggs 1985; Dyer 1997): (a) hypersynchronous, (b) synchronous and (c) hyposynchronous estuaries.

(a) Hypersynchronous: convergence exceeds friction and, consequently, the tidal range and tidal currents increase up-estuary until the tidal river zone (TRZ), where convergence diminishes, friction becomes the larger effect and the tide height reduces. These estuaries are generally funnel shaped.
(b) Synchronous: friction and convergence have equal and opposite effects on the tidal height, and range is constant along the estuary until the river zone (TRZ) is reached.
(c) Hyposynchronous: friction exceeds the effects of convergence, and the tide range diminishes throughout the estuary. These estuaries tend to have restricted mouths, with the water entering through the mouth effectively spreading out within the estuary. The highest velocities occur at the mouth.

In the northern Brazilian coast hypersynchronous conditions may be observed in shallow estuarine channels in the spring tide and autumn-winter transitional period, with an increase in the tidal height sufficiently large to produce a tidal bore. This occurrence was observed in the Mearin river located in the São Marcos Bay (Maranhão, Brazil), and analyzed in the article of Kjerfve and Ferreirra (1993). In this semi-diurnal tidal region the time intervals between flood and ebb have an

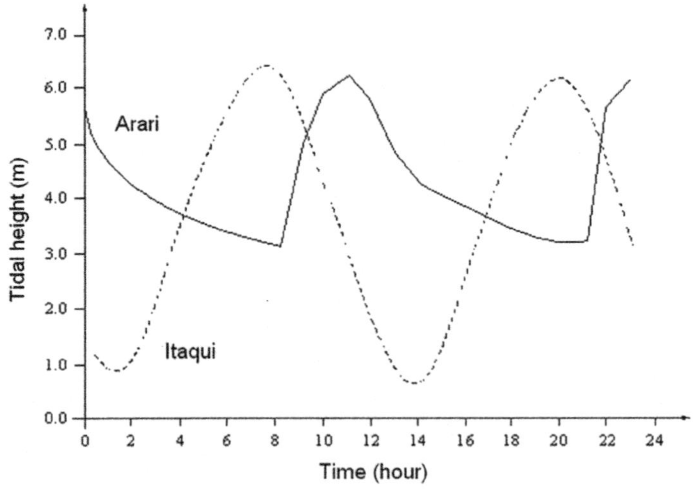

Fig. 2.7 Comparison of the tide forecast in the Mearin River (Maranhão State, Brazil) in Itaqui, and the tide record in Arari, on August 21, 1990. The sinuous sea level macro oscillation in Itaqui, contrasting with the amplitude damping and asymmetric oscillation in Arari (*bold line*) (according to Kjerfve & Ferreira 1993)

asymmetrical behavior, as observed in the simultaneous records of the estuary mouth and up-estuary in the Arari community (Fig. 2.7). The comparative analysis of these records indicates that the sinuous variation changes to a very asymmetric oscillation, indicating a hyposynchronous condition, with time intervals of flood and ebb being 4.0 and 9.0 h, respectively. This indicates, as we will see later, the *flood dominance* phenomenon.

The phenomenon of diurnal inequalities has also been observed in the Santana harbor (Pará, Brazil) located in the left margin of the North Channel of the estuarine delta of the Amazon river, were the time intervals of flood and ebb were 4.0 and 8.0 h, respectively, and similar to the observations in the Mearin River. A simple physical interpretation of this phenomenon, taking into account the propagation of a progressive wave in a shallow channel, will be presented in Chap. 8.

In the shallow water of estuaries, two processes affect the tidal wave propagation. In the first process, even in a frictionless estuary, when the tidal variation of the water depth is large, the wave crest will move more quickly than the through. The crest of the tide may partially overtake the trough, resulting in a shorter flood and longer ebb; the highest velocities thereby occur in the flood tide. The second is the effect of the bottom friction. This is a non-linear process which depends on the square of the current velocity and its effect is to produce greater friction in shallow water. This also slows down the water movement more at low water levels relative to high water levels. Thus the time delay between low water at the mouth and that at the head is greater than the time delay of the high water. The combined effects of these two processes produces a short duration of the flood phase of the tide and the fast flood currents, creating what is known as *flood dominance*, which has hypersynchronous characteristics. *Ebb dominance* can also be produced within estuaries, essentially through interactions between deep channels and shallow water areas, and the varying distribution of friction during the tide (Dyer 1997).

The knowledge of tide asymmetric behavior is very important to estuary management, particularly in relationship to the sediment erosion, transport and deposition, as indicated in the analysis of French (1997):

> The tide moves a set amount of water into and out of an estuary during each tidal cycle. If the flood and ebb tides are of unequal length, then in the shorter tide, the fixed volume of water has to move faster than in the longer tide. That is, if the flood tide takes 4.0 h and the ebb takes 8.0 h, then the estuary has, in effect, 4.0 h to fill and 8.0 h to empty. The only way that this can be achieved is by variations in velocity and, therefore, energy. If, as in the situation above, flood periods are shorter, and energy greater, more sediment can be carried in on the flood tide than can be moved out on the ebb tide (assuming adequate sediment supply). Hence, the estuary will experience a net input of sediment. In contrast, if the situations were reversed, and the flood tide took longer than the ebb, there would be greater velocities and greater energy during the ebb, and, therefore, more sediment moved and a net sediment loss to the system. This net movement of sediment is also modified by other factors, such as residual currents, the mechanism of sediment transport (bedload or suspension), sediment stability, and sediment entrapment.

With the development of the classical method of Doodson (1928) to the analysis, the astronomic tidal height forecast and the techniques using spectral analysis (Franco 1988, 2009), the hourly time series values of the tidal height measurements are able to be decomposed into a set of harmonic components by a computational procedure using the Pacmaré (Franco 2000); this program performs analyses for long periods (18.6 years) and shorter periods (1 year or less). The theoretical foundation of the classical tidal harmonic analysis has also been presented by Pawlowicz et al. (2002), and complemented with a set of programs written in MATLAB® to perform analyses for periods of one year or shorter. The main components are semidiurnal and diurnal, which take into account about 83% of the total tidal amplitude. Some of these components are presented in Table 2.1, with its amplitudes and phase values determined from a 30 day time series of hourly tidal height records in the Bertioga estuarine channel (São Paulo, Brazil).

Table 2.1 Harmonic constants (amplitude and phase) of the main tidal components in the Bertioga estuarine channel, computed according to Franco (2000), using a 30 days of hourly measurements

Diurnal components	Amplitude (cm)	Phase (°)
O_1—main-lunar	11.2	27.5
K_1—lunar-solar	6.9	189.5
P_1—main-solar	6.3	109.0
Semi-diurnal components	Amplitude (cm)	Phase (°)
M_2—main lunar	32.6	174.9
S_2—main solar	23.5	184.3
N_2—lunar-elliptic	5.6	240.5

The relative importance of diurnal and semidiurnal tidal components may be obtained with the *form number* (N_f) defined by A. Courtier in 1938 (quoted by Defant 1960), by the following ratio:

$$N_f = \frac{K_1 + O_1}{M_2 + S_2}, \tag{2.7}$$

where K_1, O_1, M_2 and S_2 are the main diurnal (index 1) and semidiurnal (index 2) tidal components. According to the variation of this dimensionless number the tide may be classified as:

$0 < N_f < 0.25$—semidiurnal;
$0.25 < N_f < 1.5$—mixed, with semidiurnal predominance;
$1.5 < N_f < 3.0$—mixed, with diurnal predominance;
$N_f > 3.0$—diurnal.

Using this classification criteria the tidal oscillation in the Bertioga estuarine channel is $N_f = 0.32$, classified as mixed, with semidiurnal predominance (Miranda et al. 1998).

The co-oscillating tide at the estuary mouth propagates up-estuary as a shallow water wave, because its length is too great in comparison to the estuarine depth. The horizontal motions associated with the tidal oscillations are named *tidal currents*. These motions undergo significant modification due to the frictional drag, morphology, relative vorticity and also by the Earth's rotation due to the Coriolis acceleration; this last influence is more evident in estuaries with great dimensions and forced by diurnal tides. When the natural oscillation period of the water body of the estuarine system is equal or close to the main tidal component, the resonance phenomenon may be observed, and the height of the stationary wave may reach several meters. Classic examples are its occurrence in the Bay of Fundy (Golf of Maine, USA) and the Igarapé do Inferno in the Amazon continental shelf (Brazil). In this condition, the tidal prism (T_{PR}) is very large compared with the estuary volume in low tide, and the water mass renewal in the estuary is efficient in removing unwanted wastes.

In theoretical problems when the tide variation must be taken into account, it may be convenient to simplify the theory by considering this forcing to be analytically represented by only one tidal component as:

$$\eta(x,t) = \eta_o \cos(\kappa x - \omega t), \tag{2.8}$$

where $\omega = 2\pi/T$ and $\kappa = 2\pi/\lambda$ denote the angular frequency and the wave number, respectively. This equation may be, for instance, used as a boundary condition to simulate the tidal forcing at the estuary mouth.

2.3 Dimensional Analysis Applied to Equations and Processes

Although simple uses of this technique have already been given in this chapter, the application of *dimensional analysis* is very useful in theoretical development of complex and non-complex processes, and to check the correctness of its intermediate and final results. As an introduction to this approach, let us consider some definitions of non-dimensional parameters already given and a physical problem related to the main forces which drives the estuarine processes.

In dimensional analysis, the dimension of a quantity is denoted by square brackets [ ] and is expressed in reference to the fundamental quantities of Physics: mass (M), length (L) and time (T). As simple examples of applying this analysis let us take as first the already defined tidal prism $T_{PR} = H_oA_S$, which is a volume $[L^3]$ which depends on the product of tidal height (H_o) and the estuary surface area (A_S); then the dimension of T_{PR} are: $[T_{PR}] = [LL^2]$, which may also be abbreviated by $[T_{PR}] = [L^3]$, and the water volume R discharged by the river discharge (Q_f) into the estuary during the time interval of one tidal period (T_P) is calculated by $R = Q_fT_P$ and $[R] = [L^3T^{-1}T] = [L^3]$. The ratio of these quantities (T_{PR}/R) was defined as the *flux ratio* (F_R) which, is a non-dimensional quantity, as well as, the form number N_f (Eq. 2.7).

In the next topic of this chapter we are going to deal with the *gradient pressure force* in terms of acceleration, that is, force per unit of mass, which has two main components: the barotropic and the baroclinic, whose dimensions are,

$$[g\frac{\partial\eta}{\partial x}] = [LT^{-2}\frac{L}{L}] = [LT^{-2}],$$

and,

$$[\frac{g}{\rho}\int_{z}^{\eta}\frac{\partial\rho}{\partial x}dz] = [\frac{LT^{-2}}{ML^{-3}}\frac{ML^{-3}}{L}L] = [LT^{-2}].$$

Another example of using dimensional analysis will follow from the steady-state equation of motion of a well-mixed and laterally homogeneous estuary, forced by the barotropic and baroclinic components of the gradient pressure force. Under the assumption of a constant kinematic eddy viscosity coefficient (N_z), this equation may be written as (Eq. 10.4, Chap. 10),

$$\frac{\partial^2u}{\partial z^2} - \frac{g}{\rho N_z}\frac{\partial\rho}{\partial x}z + \frac{g}{N_z}\frac{\partial\eta}{\partial x} = 0.$$

The general solution of this second order differential equation is dependent on two integrations constants (C_1 and C_2), which may be calculated under specified upper and lower boundary conditions,

$$u(x, z) = \frac{1}{6}\frac{g}{\rho N_z}\frac{\partial \rho}{\partial x}z^3 - \frac{1}{2}\frac{g}{N_z}\frac{\partial \eta}{\partial x}z^2 + C_1 z + C_2.$$

The dimension of the first member of this equation is $[LT^{-1}]$, and for first and second terms on the right-hand-side, the analytical expression and the corresponding dimensions are:

$$[\frac{1}{6}\frac{g}{\rho N_z}\frac{\partial \rho}{\partial x}z^3] = [\frac{LT^{-2}}{L^2T^{-1}}\frac{L^3}{L}] = [LT^{-1}],$$

and

$$[\frac{1}{2}\frac{g}{N_z}\frac{\partial \eta}{\partial x}z^2] = [\frac{LT^{-2}}{L^2T^{-1}}L^2] = [LT^{-1}],$$

which indicates that the analytical expressions of these terms, as would be expected, have dimension of velocity.

The analytical expressions of the constants C_1 and C_2, obtained by applying the boundary conditions, will have a correct analytic expressions if $[C_1] = [T^{-1}]$ and $[C_2] = [LT^{-1}]$, which may be easily checked. Further examples on the application of dimensional analysis to physical and biological problems may be found in the book edited by Platt et al. (1981).

2.4 What Generates the Circulation and Mixing in the Estuary?

To answer this question, which we judge important in this chapter, it is necessary to use some fundamental concepts and the hydrodynamic equations of sea water, which are studied in others areas of Physical Oceanography, and will be presented in detail under the optics of estuarine physics in Chaps. 7 and 8.

The estuary dimension vary between small and medium spatial scales, in relation to the adjacent ocean, and for most estuaries their lengths (L) is much greater than their width (B). Between the estuary head and mouth, the salinity varies between the fluvial water (salinity practically zero), and that of the coastal region (S_o), generating a mean longitudinal salinity gradient ($\Delta S/\Delta x$) equal to S_o/L.

The geometric estuarine characteristic has a great influence on its motions and on the distribution of properties, which are often studied using a right-handed rectangular Cartesian co-ordinates system Oxyz (Fig. 2.8a); the origin being in the mean level of the free surface. Oxy plane is located on the free surface and

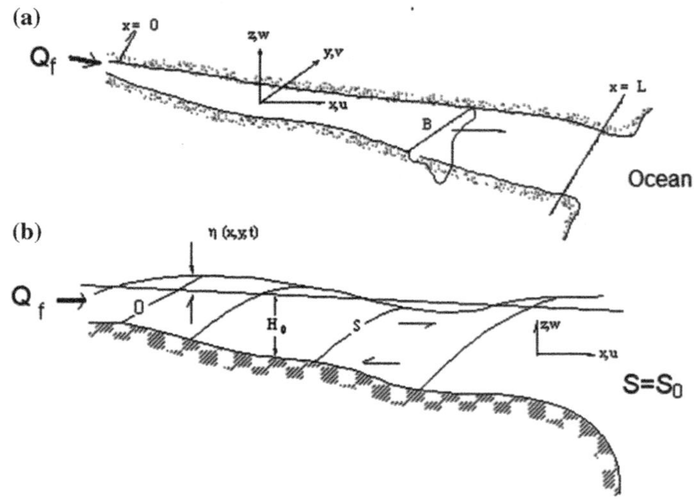

Fig. 2.8 **a** Scheme of a coastal plain estuary indicating the Oxyz reference system, the u, v, w velocity components, the cross section with width B and the river discharge Q_f. **b** The longitudinal section with the isohalines, S, the co-oscillating tide, η(x, y, t), and the depth $H_0(x, y)$. The Oz axis is oriented against the acceleration of gravity

orthogonal to the gravity acceleration, the Ox axis is usually oriented longitudinally in the positive sense seaward (or eventually landward), the Oy axis, indicating the transverse estuary dimension, is perpendicular to the Oxz plane and the Oz axis vertically upwards will be oriented against, or in the same sense of the gravity acceleration. The velocity components, in relationship to the coordinate axis, are denoted by u, v, and w, respectively; the components u and v are denominated longitudinal and transversal (or secondary), respectively.

In the estuary longitudinal section (Fig. 2.8b), the Oxz axes indicates the longitudinal and depth characteristics of the salinity field by its isohalines distribution (S = const.). This figure also indicates that the estuarine water mass is also forced by the tidal co-oscillation in the longitudinal direction, and its height variation is denoted by $\eta = \eta(x, y, t)$.

The motions in this transitional coastal environment are generated by the following forcings: sea level variations, freshwater discharge, gradient pressure force (due to the thermohaline influence on the density), continental shelf circulation and wind stress acting on the free surface. These forcings are functions of time and space and simultaneously act on the estuarine water body. The circulation on the continental shelf is mainly driven by the wind stress; its influence, generating the circulation and mixing processes may be observed on synoptic and seasonal time and space scales. Our study is restricted to the conditions found in the majority of coastal plain estuaries, and we will examine the motions generated by: the co-oscillating tide, $\eta = \eta(x, y, t)$, the fresh water discharge, Q_f; and the longitudinal salinity ($\partial S/\partial x$) or density ($\partial \rho/\partial x$) gradients.

The time scale of the estuarine water body to the impulsive forces is large, and usually separated in motions in the dominium of high and low frequency. The dilution of the sea water by the freshwater of the river discharge will also produce variations in the longitudinal salinity gradient that will contribute to the estuary's dynamic structure.

The various dynamic physical mechanisms that generate the estuarine circulation and control the salinity distribution and the chemical, biological and geological properties are liable to a hydrodynamic formulation. This formulation will be presented later using tools of the geophysical fluid dynamics, aiming to forecast its motions, the water interchanges with the adjacent ocean and the physical characterization of the estuary.

As the oscillation periods of the co-oscillating tide are well known, it is possible to separate the effects of the low frequency forces using a statistic procedure of experimental data. Theoretically, this separation is made with the integration (from the generic depth z, at a pressure p), up to the free surface η (submitted to the atmospheric pressure $p = p_a$) of the fundamental hydrostatic equation ($\partial p/\partial z = -\rho g$ or $\partial p = -\rho g dz$) with the Oz axis oriented according to Fig. 2.8,

$$p_a - p = -g \int_z^{\eta(x,t)} \rho dz. \tag{2.9a}$$

To calculate the longitudinal pressure gradient, the expression (2.9a) must be derived in relation to x, and applying to the term in the right-hand-side the Leibnitz rule of the derivation of an integral, because its upper integration limit $z = \eta(x, y, t)$ is function of x, and the result is:

$$\frac{\partial p_a}{\partial x} - \frac{\partial p}{\partial x} = -g(\rho_\eta \frac{\partial \eta}{\partial x}) + (\int_z^{\eta(x,t)} \frac{\partial \rho}{\partial x} dz). \tag{2.9b}$$

Then, calculating from this expression the longitudinal components of the pressure gradient force per mass unity, which generate motions oriented from high to low pressure regions, is obtained by multiplying by the factor $(-1/\rho)$ all terms of Eq. (2.9b),

$$-\frac{1}{\rho}\frac{\partial p}{\partial x} = -\frac{1}{\rho}\frac{\partial p_a}{\partial x} - g\frac{\partial \eta(x, t)}{\partial x} - \frac{g}{\rho}\int_z^{\eta(x,t)} \frac{\partial \rho}{\partial x} dz, \tag{2.9c}$$

With the approximation $\rho_\eta/\rho = \rho_o/\rho \approx 1$. All terms of Eq. (2.9c) have dimension of acceleration (force per mass unity) and, in the right-hand-side we have the following longitudinal components of the pressure gradient force: the barometric, the barotropic and the baroclinic, respectively.

Lets us consider some simple solutions of the Eq. (2.9c), under steady-state condition and at the sea surface, $z = \eta(x)$, then, it is reduced to

$$0 = -\frac{1}{\rho}\frac{\partial p_a}{\partial x} - g\frac{\partial \eta}{\partial x}. \tag{2.9d}$$

Solving this equation by finite increments to the estuary free surface elevation $\Delta\eta$, it follows that:

$$\Delta\eta = -\frac{1}{\rho g}\Delta p_a. \tag{2.9e}$$

This result indicates that for an increase of the atmospheric pressure ($\Delta p_a > 0$), the sea surface responds as an inverted barometer, because $\Delta\eta < 0$, and the following order of magnitude may be estimate: for an increase (decrease) of the atmospheric pressure of 1.0 mbar = 10^2 N m^{-2}, the sea surface is depressed (elevated) by 0.01 m. However, in nature, the static influence of low atmospheric pressure centers are usually related to the low frontal zones propagations, that may reaches the coastal regions with strong winds, and the association of these influences (static and dynamic) may generate dangerous storm-surges at the coastline.

As the salinity influences on the density is predominant over the pressure (p) and temperature (T), the longitudinal density gradient in Eq. (2.9c) may be substituted by the corresponding longitudinal salinity gradient ($\partial S/\partial x$), calculated using a linear simplified equation of state, $\rho(S) = \rho_o(1 + \beta S)$ (Eq. 4.12, Chap. 4), where ρ_o is a constant density value, and β is the mean saline or haline contraction coefficient, define by $\beta = (1/\rho)(\partial\rho/\partial S)$. However, this coefficient varies with the S, T and p properties, but for estuarine water mass this coefficient may be approximate by a constant value ($\beta \approx 7.5 \times 10^{-4}$ (‰)$^{-1}$). With these simplifications, and disregarding the barometric pressure gradient, the longitudinal component of the pressure gradient force, per mass unity has the following expression:

$$-\frac{1}{\rho}\frac{\partial p}{\partial x} = -g\frac{\partial \eta}{\partial x} - \frac{g}{\rho}\int_z^\eta \frac{\partial \rho}{\partial x}dz, \tag{2.10a}$$

or

$$-\frac{1}{\rho}\frac{\partial p}{\partial x} = -g\frac{\partial \eta}{\partial x} - g\beta\int_z^\eta \frac{\partial S}{\partial x}dz. \tag{2.10b}$$

It should be noted that in the article published in 1952 (Pritchard 1952), D.V. Pritchard became the first researcher to link estuarine circulation to the forcing by horizontal density gradient, using observations from the James river estuary (Virginia, USA) to demonstrate this mechanism (quoted in Geyer 2010, p. 13).

The first parcel on the right-hand-side of Eq. (2.10a) indicates that the dynamic influence generated by the tidal forcing is independent of the depth (barotropic), and its intensity is proportional to the negative value of the sea surface slope. In estuaries, the time variations in the velocity intensity sequentially assume the following trend: when $\partial\eta/\partial x > 0$, reach maximum landward intensity at the mid-flood tide, the velocity decreases to zero at high and low tide ($\partial\eta/\partial x = 0$), before reversing ($\partial\eta/\partial x < 0$) to maximum ebb velocity midway through the ebb, where almost zero intensity is reached closing the tidal cycle.

For an estuary with a length of 10^4 m (10 km), forced by meso-tidal oscillation ($2 < H_o < 4$ m), the maximum intensity, per mass unity may be estimated as varying between -10^{-3} and 10^{-3} m s^{-2}, in the flood and ebb tide, respectively. Then, this gradient force component varies in the interval -10^{-3} to 10^{-3} m s^{-2}, with the null intensity occurring at high and low tide.

The dynamical influence of the distribution of the longitudinal density (salinity) gradient is proportional to the longitudinal density gradient, integrated in the interval from the depth z up to the free surface level, $\eta = \eta(x, y, t)$, (Eqs. 2.10a). Its intensity is zero at the free surface ($z = \eta$) and increases with depth, but its value is always negative because the density (salinity) increases seaward $\partial\rho/\partial x > 0$ ($\partial S/\partial x > 0$). The dynamical influence of this component of the gradient force (baroclinic), per mass unity, always generates landward (flood) motions; its intensity increases with depth, and its maximum value is on the bottom. For an estuary with a length of 10^4 m (10 km), a mean depth of 10 m, and a longitudinal salinity variation from zero to 30‰ at the head and mouth, respectively, its maximum intensity may be estimated as a relative value of -10^{-4} m s^{-2}. Hence, the intensity of this component varies in the interval -10^{-4} m s^{-2} to 0, with zero at the surface. Then, according to these estimated intensities of the baroclinic pressure gradient component, which always generates landward or flood motions, its maximum value is 10 times less than the modulus of the barotropic component. However, its intensity will be higher than the barotropic component near the slacken water.

The river discharge (Q_f), besides to diluting the sea water and generating a longitudinal density gradient, is also responsible for another low frequency component that generates velocity in the estuarine water mass. However, in opposition to the baroclinic component, the mean steady-state value of this component (u_f) always generates seaward motion. By the volume integrated mass conservation principle, the intensity of this component is given by the following ratio:

$$u_f(x) = \frac{Q_f(t)}{A(x)}, \tag{2.11}$$

where $A = A(x)$ is the mean area of a transversal section during the tidal cycles. The time variation of this velocity component (u_f) is dependent on the low frequency variations (usually seasonal) of the river discharge.

Under steady-state conditions the tidal oscillation vanish and so the barotropic pressure gradient, and the resulting low frequency motion forced by the

Table 2.2 Vertical profiles of the intensity and direction of the current (velocity vector) measured in the Cananéia estuarine channel, at two different times, t_1 and t_2 of ebb tide

Depth (m)	Vel. (m s^{-1}) (t_1)	Direction (°) (t_1)	Vel. (m s^{-1}) (t_2)	Direction (°) (t_2)
0.0	1.09	201	0.22	198
1.0	1.09	188	0.21	187
2.0	0.94	187	0.10	199
3.0	0.85	191	0.02	168
4.0	0.74	197	0.19	011
5.0	0.59	198	0.30	024
6.0	0.44	175	0.33	010
7.0	0.30	172	0.47	017

longitudinal baroclinic pressure gradient of density and by the river discharge in landward and seaward direction, respectively, forces to generate a two layer bidirectional motion. This motion is named *gravitational circulation*, *gravity current* or the *classical estuarine circulation*. It is essentially a two dimensional advective motion which drives the volume and salt transports seaward and landward in the upper and lower layers, respectively.

To illustrate the dynamics effects of the barotropic and baroclinic pressure gradients in the velocity components, experimental data are presented in Table 2.2. These vertical velocity profiles were measured in the estuarine channel of Cananéia sea (Fig. 1.5, Chap. 1), in two times (t_1 and t_2) of the ebb tide, in the position where the channel is oriented in the north-south direction.

The experimental data indicate that in the vertical profile at time t_1, the estuarine circulation is southwards (seaward) with direction between 172° and 201°, and with relatively high speeds from 0.30 to 1.09 m s^{-1}. This physical characteristic indicate that during this ebbing event the barotropic gradient component of the pressure gradient force prevailed over the baroclinic component. In the following instant (t_2), the velocity minimum is 0.02 m s^{-1} at 3 m depth; in this layer between the surface and 3 m depth, the motion is also southward (direction between 168° and 199°), but with very low intensities (<0.22 m s^{-1}) compared to the preceding profile (t_1). At depths bellow 3 m, the intensity increases, but the circulation is in the opposite direction (landward) to the upper layer, as indicated by its direction from 11° to 24°. These results indicate that the water column, with only seven meters depth, has bidirectional circulation during the ebbing tide, with the non-motion depth at the middle of the water column (≈3 m). Then, with the decrease in the intensity of the barotropic component of the pressure gradient force, there was an increase in the relative importance of the baroclinic pressure gradient in the deepest layers of the estuarine channel. As demonstrated in this experiment, the baroclinic component of the pressure gradient force always generates landward circulation, and its intensity increases with depth (second term on the right-hand-side in Eq. 2.10).

The motions described in the preceding paragraphs are macroscopic, usually named as advective, to distinguish them from the microscopic or small-scale motions that generate turbulent diffusion. In the following paragraphs, the influence of the microscopic motions in the redistributions of properties in the estuarine water mass by *mixing process* (advection and diffusion) will be examined. Focusing our attention on one volume element of control, the local salinity variation ($\partial S/\partial t < 0$ or $\partial S/\partial t > 0$) indicates the occurrence of salt dilution (or the salt concentrate)—dilution by the river water and concentration by the seaward input. This process is a simple description of the mixing between water masses. Salinity was used as an indicator, but this process is generated by the salt transport due to the macroscopic motion (advective) and the simultaneous microscopic flux of properties due to: interchanges of kinetic energy, heat, salt and contaminants introduced in the estuary. These fluxes of properties are due to the following motions:

(a) Macroscopic or medium motion (generated by the tide, river discharge and density gradients); this process contributes to the mixing and is named *advective*;
(b) Microscopic and random motions, whose effects on the mixing are denominated as *molecular* or *turbulent diffusion* in laminar and turbulent motions, respectively.

The advective influence changing the salinity concentration during the tidal cycle is evidenced in the comparison of the u-velocity component and vertical salinity profiles (Fig. 2.9a, b), which were simultaneously sampled at half-hourly time intervals during a semi-diurnal tidal cycle. The turbulent diffusion generates the erosion of the vertical salinity variation shown in the salinity profiles from 31.0 and ≈36.0‰ (Fig. 2.9a), which migrate in the water column due to the velocity intensity variation. This process is due to the turbulent mixing generated by internal friction water layer with different intensities, and the estuary friction at the bottom. The layer with the higher vertical salinity gradient is called *halocline*, although in oceanic water this term is used to indicate a salinity decrease with depth.

The time variability of the u-velocity profiles during the ebb ($u > 0$) and flood ($u < 0$) is shown in Fig. 2.9b. The ebbing intensities (≈ 0.6 m s^{-1}) are higher than in the flood (≈ -0.3 m s^{-1}), due to the influence of the barotropic gradient pressure force and the river discharge. This asymmetry, when analysed in terms of steady-state mean values, indicates the presence of the river discharge and the influence of the barotropic pressure gradient. It should be also pointed out, that this figure also indicates the influence of the baroclinic pressure gradient in the bi-direction characteristic of the vertical velocity profiles in the neighborhood of the HW and LW tides when the barotropic forcing tends to zero.

By analogy with the tide, the circulation variability in estuaries is classified as *intertidal* when it occurs at semi-diurnal or diurnal tidal frequencies (>1 cycle/day), or *subtidal* at lower frequencies (<1 cycle/day). The main subtidal frequency is the fortnightly, and is due to the time period modulation between successive spring tides (≈15 days).

The longitudinal salinity advective flux (ϕ_{adv}) and the diffusive flux (ϕ_{dif}) fluxes, associated with the macroscopic (mean) and the turbulent (microscopic) motions,

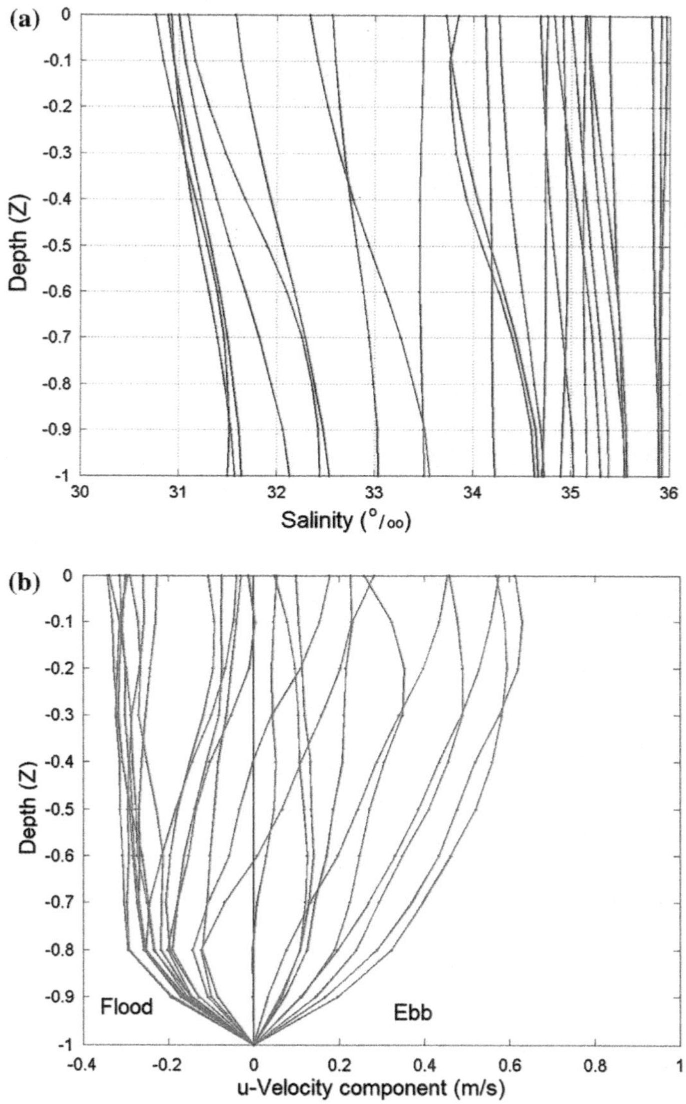

Fig. 2.9 Time variability of half-hour vertical profiles of salinity (**a**) and the u-velocity component (**b**) during a complete semidiurnal tidal cycle. Positive and negative velocity values indicate flood and ebb, respectively, and Z is the non-dimensional depth (after Andutta 2011)

are defined by: $\phi_{adv} = \rho Su$ and $\phi_{dif} = -(\rho K_{xS})\partial S/\partial x$, respectively; these fluxes have both dimensions of $[ML^{-2}T^{-1}]$. The coefficients K_{xS} and ρK_{xS} are the kinematic and dynamic eddy diffusion coefficients of salinity, with dimensions $[K_{xS}] = [L^2T^{-1}]$ and $[\rho K_{xS}] = [ML^{-1}T^{-1}]$. The formulation of the diffusive flux is parameterized with the Fickian law, and is oriented from the regions of high concentration to lower concentration. In the SI system of unit these fluxes are expressed in kg m^{-2} s^{-1}. The equivalent expressions may be written for concentration of any property, using the corresponding diffusion coefficients.

2.5 Tidal Wave Propagation in an Estuary

The tidal wave propagation in estuaries involves a relatively intense advective process, usually in a region with complex topography. Some of the inter-connected features are the phase and the related phase differences of the wave propagation, tidal current and salinity variations, the tidal excursion and length of the saline intrusion. The most elaborated theory of tidal wave propagation in a channel, taking into account the most important characteristics, was developed in the pioneering paper of H. Poincaré, in 1910. The dynamics of the tidal currents were studied by Fjeldstad (1929), taking into account the bottom frictional forces (quoted in Defant 1961).

It follows a classical and simplified solution for the wave propagation in estuaries, according to Defant (1960), Ippen and Harleman (1961), Dyer (1973) and others, that is useful to illustrate the physical processes occurring in estuarine channels, although of limited interest because it uses a simple geometry (long channel and rectangular transverse section), and doesn't take to account the non-linear dissipative frictional forces, it is a useful approach on the wave propagation in channels.

As by hypothesis the estuary depth is small when compared to the wave length of the tidal co-oscillation, the nature of this oscillatory motion is that of an oscillatory wave in shallow water, which differs in physical behavior from that of a short wave propagation. As the ratio of the water depth (H_0) to the wave length (λ) is much less than one [$(H_0/\lambda) \ll 1$], the solution is reduced to that of a non-dimensional shallow water problem (Pedloski 1979).

The hypotheses applied to the theoretical development are:

(a) The salinity (density) field is steady and uniform.
(b) The channel length (L) is less than the wave length ($L < \lambda$).
(c) Channel long and narrow ($L \gg B$), one-dimensional motion ($v = w = 0$), uniform transversal section, and there is no deflection due to the Earth's rotation (Coriolis acceleration is disregarded).
(d) With the tidal co-oscillating motion in the channel entrance there is the water storage in the channel during the flood (tidal prism), and the subsequent exit of this water in the ebb.

In order to develop the theory of the oscillating tidal motion, a coordinate system indicated in Fig. 2.10 will be used. The vertical axis (Oz) is oriented positively against the gravity acceleration ($\vec{g}$) with its origin on a uniform bottom with the

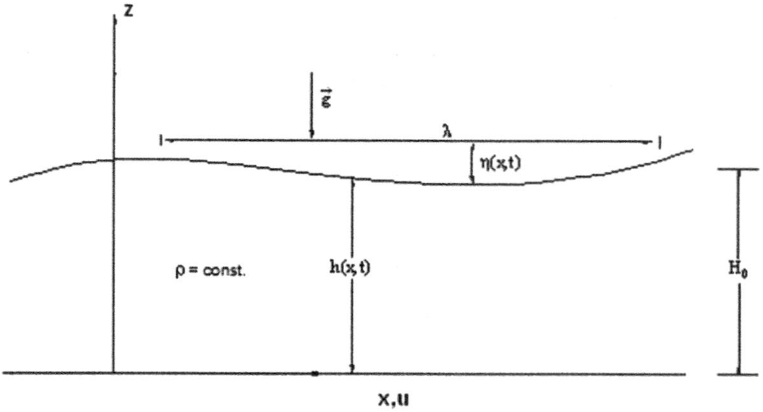

Fig. 2.10 Reference system of a wave oscillation, $\eta(x, t) = h(x, t) - H_0$, in a long shallow water channel of rectangular cross-section. The longitudinal axis Ox is oriented positively in the up-channel direction. The salinity (density) field is steady and uniform

water layer undisturbed, indicated by H_0. The origin of the horizontal axis ($x = 0$) is located at the estuary mouth and oriented positively in the up-channel direction.

In this development, the simplified equation of continuity and the Ox component of the equation of motion will be used, which are obtained from its tri-dimensional equations with a mathematical procedure which will be presented in Chap. 7 and 8. We will be considering the shallow water approximation. As $\eta = \eta(x, t)$ is the oscillation of the free surface related to the depth, H_0, and the salinity is constant, the longitudinal component of the gradient pressure force, per mass unity (Eq. 2.10a), is simplified, to the barotropic component,

$$\frac{1}{\rho}\frac{\partial p}{\partial x} = g\frac{\partial \eta}{\partial x}. \tag{2.12}$$

With the indicated simplifications, the equation of motion is reduce to the following one-dimensional expression (Harleman 1971; Pedloski 1979):

$$\frac{\partial u}{\partial t} + u\frac{\partial u}{\partial x} = -\frac{1}{\rho}\frac{\partial p}{\partial x} = -g\frac{\partial \eta}{\partial x}. \tag{2.13}$$

Under the hypothesis that $\eta(x, t) \ll H_0$, this equation may be transformed in a linear equation disregarding the advective acceleration ($u\frac{\partial u}{\partial x} = 0$). Then, it is reduced to:

$$\frac{\partial u}{\partial t} = -g\frac{\partial \eta}{\partial x}, \tag{2.14}$$

which is the simplified expression of the linear shallow water equation, without friction and with the assumption that the Coriolis acceleration is negligible The unknowns of this equation are the u-velocity component, $u = u(x, t)$, and the sea surface elevation, $\eta = \eta(x, t)$.

Taking into account that the motion is one-dimensional, it is necessary to use the corresponding continuity equation to transform it into a closed hydrodynamic system; for this to be accomplished, the continuity equation has the following analytic expression (Pritchard 1958):

$$\frac{\partial(uA)}{\partial x} + \frac{\partial A}{\partial t} = 0, \tag{2.15}$$

where A is the cross section area, $A = B(H_0 + \eta)$. Combining this area with Eq. (2.15), and performing the derivations in relation to x, it follows:

$$(H_0 + \eta)\frac{\partial u}{\partial x} + u\frac{\partial(H_0 + \eta)}{\partial x} = -\frac{\partial \eta}{\partial t}. \tag{2.16}$$

Taking into account the approximation $\eta \ll H_0$ the equation reduces to:

$$H_0\frac{\partial u}{\partial x} + \frac{\partial \eta}{\partial t} = 0, \quad \text{or} \quad \frac{\partial(H_0 u)}{\partial x} + \frac{\partial \eta}{\partial t} = 0, \tag{2.17}$$

which is the simplified form of the continuity equation of shallow water. Equations (2.14) and (2.17) form a system of two equations of partial derivatives and two unknowns: the free surface oscillation $\eta = \eta(x, t)$ and the velocity $u = u(x, t)$. The solution of this system of equations, assuming a linear frictional dependence proportional to the u-velocity in the right-hand-side of Eq. (2.14), which dissipates the amplitude of the tidal wave with an exponential decrease, $e^{-\mu x}$, with $(\mu = (\tau/2)\sqrt{gH_0})$, is presented by Blumberg (1975).

Eliminating by cross derivation from Eqs. (2.14) and (2.17) the variables $\eta = \eta(x, t)$ and $u = u(x, t)$, respectively, we have:

$$\frac{\partial^2 \eta}{\partial t^2} = gH_0\frac{\partial^2 \eta}{\partial x^2}, \tag{2.18}$$

and

$$\frac{\partial^2 u}{\partial t^2} = gH_0\frac{\partial^2 u}{\partial x^2}. \tag{2.19}$$

This equation set is the classical equations of progressive waves (the wave profile propagates horizontally) in a channel with a uniform transverse section, with a phase propagation velocity (celerity) c_0 expressed by:

$$c_0 = \sqrt{gH_0}. \quad (2.20)$$

A possible solution of Eq. (2.18) is the harmonic function,

$$\eta(x, t) = \eta_0 \cos(\kappa x - \omega t + \Phi), \quad (2.21)$$

simulating the wave propagation in the x > 0 direction (landward). Φ is an arch of positive first determination, that is, $0 \leq \Phi \leq 2\pi$, which may be determined by the initial and boundary conditions. If, for example, the surface elevation in the estuary mouth at time t = 0 is equal to η_0, then $\eta(0, 0) = \eta_0$ and, consequently, $\cos(\Phi) = 1$ and $\Phi = 0$. In this case, at the estuary mouth the higher elevation values of the oscillation, $\eta = \eta(x, t)$, will occur at times t = 0, T, 2T, … For $x = \lambda$ and any of multiples of this length, the wave will occur in phase with the position at x = 0. The simplifying hypothesis generated a symmetric tidal wave. However, in nature, the wave propagation may be greatly distorted, as exemplified for the Mearin river (Fig. 2.7).

As estuaries have a finite length, which is generally small compared to one quarter of the tidal wave length ($L < \lambda/4$), the oscillation of the free surface is usually uniform along the estuarine channel. Due to this behavior, this oscillation may be simulated by $\eta(t) = \eta_0 \cos(\omega t)$, as found in the classical work of Arons and Stommel (1951), when the stationary salinity distribution in an estuary was studied with these characteristics (one dimension, simple geometry and frictionless).

As it is a liner problem, velocity solution (Eq. 2.19) is also analytically represented by the same harmonic dependence:

$$u(x, t) = U_0 \cos(\kappa x - \omega t + \Phi). \quad (2.22)$$

In this solution U_0 is the velocity amplitude, which may be determined by calculating the first derivatives of solutions (2.21) and (2.22) relative to the variables t and x, respectively, and combining this result with the continuity Eq. (2.17), resulting the following relationships: $U_0 = \eta_0\omega/H_0$. $\kappa = \eta_0\lambda/H_0T = \eta_0 c_0/H_0$, where $c_0 = \lambda/T$ is the phase velocity of the wave. Then, the solution of Eq. (2.22) may be re-written as,

$$u(x, t) = \frac{\eta_0 c_0}{H_0} \cos(\kappa x + \omega t + \Phi), \quad (2.23)$$

or, combining with Eq. (2.21),

$$u(x, t) = \frac{c_0}{H_0} \eta(x, t), \quad (2.24a)$$

or

$$u(x, t) = \eta_0 \sqrt{\frac{g}{H_0}} \cos(\kappa x + \omega t). \quad (2.24b)$$

From this result it follow that the amplitude U_0 of the tidal velocity may be approximate by $U_0 = \eta_o\sqrt{\frac{g}{H_0}}$; for a tidal wave with an amplitude $\eta_0 = 1$ m, and a channel depth of $H_0 = 10$ m $\rightarrow U_0 \approx 1$ m s^{-1}. As will be seen later (Chap. 8) the product of a non-dimensional coefficient k by U_0H_0, may be used to simulate the vertical kinematic eddy viscosity coefficient ($N_z = kU_0H_0$).

Due to the initial hypothesis this solution is valid for $\eta \ll H_0$ and, from this inequality, it follows that for $\eta_0 \ll H_0$, the amplitude U_0 will always be less than c_0 ($U_0/c_0 \ll 1$). Despite the approximations made to reach the simplified solution it is useful to show a fundamental characteristic of the flow: (i) the u-velocity due to the wave propagation, does not exceed the phase velocity of the proper wave (Mello 1998), and (ii) It may be used to estimate the tidal wave velocity, for example: for a tidal amplitude $\eta_0 = 1$ m propagating in an estuary with 10 m depth the maximum tide velocity will be 1 m s^{-1}.

Equation (2.24a) indicates that the free surface elevation (η) and the velocity (u) are in phase, which is a characteristic of the gravity progressive wave. As for the hypothesis that the velocity is uniform in the transversal section, let's assume that the tidal wave is forcing a vertically homogeneous estuarine channel. In the case that the salinity redistribution is only due to advection, and during the flood the salinity increases gradually towards the mixing zone (MZ). After this flooding event, there will be a salinity decrease due to the influence of the river discharge dammed during the flood. Then, the salinity time variation in a determined longitudinal position (x) will be out of phase by approximately $\pi/2$ or a time interval of 3 h, for a semi-diurnal tidal oscillation in relation to the tidal current as shown in Fig. 2.11.

In this hypothetical estuary, besides the oscillation of the barotropic velocity with the same frequency as the tide, there will also be a low frequency uniform velocity at

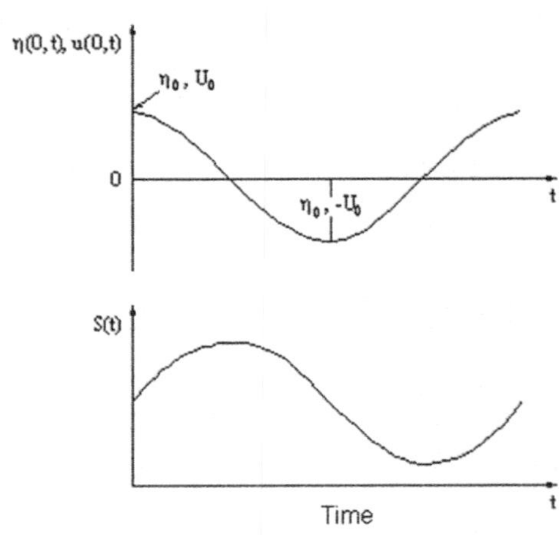

Fig. 2.11 Tidal co-oscillation at the mouth of an estuary with infinite length, generating a wave propagation, $\eta = \eta(0, t)$, a tidal current, $u = u(0, t)$ and salinity temporal variation $S = S(t)$ (adapted from Dyer 1973)

the cross section area (A). This component is generated by the river discharge (Q_f), and its intensity (u_f) is given by the ratio Q_f/A. This motion is always oriented seaward ($u_f > 0$), and the resulting velocity in the estuarine channel is:

$$u(x,t) = u_f + U_0 \cos(\kappa x - \omega t + \Phi). \tag{2.25}$$

The tidal excursion (T_E) is defined as the distance travelled by a water particle along the mixing zone, MZ, starting from the rest at the estuarine mouth, during the elapsed time interval between low and high tide. According to this definition, this motion occurs during a time interval of half a tidal period ($T_P/2$). In the assumption that the horizontal axis, Ox, has its origin (x = 0) at the estuary mouth, the landward velocity of this volume element is, according to Eq. (2.22), $u(0, t) = U_0 \cos(-\omega t + \Phi)$. To satisfy the initial condition, $u(0, 0) = 0$, it is necessary that, $\Phi = \pi/2$. Then, the tidal excursion, T_E, may be calculated by the mean velocity at the origin during the time interval Δt of half a tidal period ($\Delta t = T_P/2$), times Δt (Harleman 1971) multiplied by the time interval of:

$$T_E = [\frac{2}{T_P}\int_0^{\frac{T_P}{2}} u(0,\ t)dt]\frac{T_P}{2} = U_0 \int_0^{\frac{T_P}{2}} \cos(-\omega t + \frac{\pi}{2})dt = (\frac{2U_0}{\pi}\frac{T_P}{2}) = \frac{U_0 T_P}{\pi}. \tag{2.26}$$

In Eq. (2.26) the quantity $2U_0/\pi$ is the theoretical mean velocity at the estuary mouth during the time interval of half a tidal period ($T_P/2$). From this equation it is possible to verify that the tidal excursion is proportional to the velocity amplitude (U_0) and the tidal period (T_P); for a semi-diurnal tide $T_P = 43082.0$ s, and, $T_E = 13{,}713.0 U_0$. Then, if the estuary is forced by semidiurnal tides and the amplitude of the tidal velocity is 1.4 m s^{-1}, using Eq. (2.26) the following value for the tidal excursion is calculated: $T_E = 19{,}200.0$ m ($\approx$19 km). This value is only a first approximation of the tidal excursion and the mixing zone MZ length, because the energy dissipation and the channel geometry have not been taken into account. From the relationship between U_0 and η_0 ($\eta_0 c_0/H_0$), the tidal excursion may be calculated as $T_E = c_0 \eta_0 T_P/\pi H_0$, and is directly proportional to the tidal amplitude η_0, to the wave celerity and the tidal period T_P, and is inversely proportional to the depth of the estuarine channel, H_0.

It should be noted that the tidal excursion and the length of saline intrusion are different physical quantities and, for the determination of the second property, extensive theoretical and experimental studies were carried out and presented by Prandle (2009). Its determination in salt wedge and partially mixed estuaries will be studied in Chaps. 9 and 11.

Another quantity associated with the velocity, u = u(x, t) (Eq. 2.22) is the mean root square (u_{rms}). This statistical quantity is equal to the positive value of the root mean square of the mean velocity during one or more complete tidal cycles, that is,

$$u_{rms} = \sqrt{\frac{1}{T_P}\int_0^{T_P} u^2(x, t)\,dt} = \sqrt{\frac{1}{2}}U_0 = 0.7U_0. \tag{2.27}$$

This final has been obtained using trigonometric identities and reducing to the simplest expression.

Let's consider an estuarine channel with finite length L and a tidal wave travelling landward which is reflected at the estuary head, originating a wave in the opposite direction to the original wave. According to the superposition principle, there will be a superposition of these waves. If they have the same frequency, phase velocity and amplitude, and are propagating in opposite directions, with the free surface elevation of the incident and reflected waves denoted by $\eta_1 = \eta_1(x, t)$ and $\eta_2 = \eta_2(x, t)$, respectively, and under the assumption that the phase difference is zero ($\Phi = 0$), it follows the expression for the composite wave:

$$\eta_e(x,t) = \eta_1(x,t) + \eta_2(x,t) = \eta_0[\cos(\kappa x - \omega t) + \cos(\kappa x + \omega t)], \tag{2.28}$$

where ($\eta_e(x, t)$ is the resulting oscillation. By applying the cosine of the addition and subtraction rules to the angles ($\kappa x \pm \omega t$), this equation may be simplified to:

$$\eta_e(x,t) = 2\eta_0 \cos(\kappa x)\cos(\omega t). \tag{2.29}$$

This solution indicates that in a x longitudinal position, the motion of all particles oscillate as a simple harmonic motion. However, the wave amplitude $2\eta_o\cos(\kappa x)$ varies along the estuary with the following characteristics: it is zero in determined and fixed points and has extreme positive and negative values, resulting in an oscillation of a *stationary wave*. In the above solution (2.29), it is observed that the motion amplitude has the value of zero when $\cos(\kappa x) = 0$, which occurs at the longitudinal points $x = \lambda/4, 3\lambda/4, 5\lambda/4\ldots$, named *nodal* points In turn, the maximum and minimum values of the amplitude are $2\eta_0$ and $-2\eta_0$, respectively, occurring when $\cos(\kappa x) = \pm 1$, that is, in the points $x = 0, \lambda/2, \lambda, 3\lambda/2, \ldots$, named *anti-nodal* points, which are separated by half a wave length ($\lambda/2$). The nodal points are permanently at rest and the wave energy remains steady.

To calculate the longitudinal velocity component of the stationary wave $u_e = u_e(x, t)$, it is necessary to use the continuity Eq. (2.17), resulting the following expression:

$$u_e(x,t) = 2\frac{\eta_0\omega}{H_0}\sin(\omega t)\int \cos(\kappa x)dx = 2U_0\sin(\kappa x)\sin(\omega t), \tag{2.30}$$

which satisfies the boundary condition $u_e(0, t) = 0$. The amplitude of the longitudinal velocity of the water driven by the stationary wave is also dependent of its longitudinal position. When $\sin(\kappa x) = 0$, corresponding to the positions $x = 0, \lambda/2, \lambda, 3\lambda/2, \ldots$(*anti-nodal* points), the velocity is zero. When $\sin(\kappa x) = \pm 1$, that is, in the nodal points, x = λ/4, 3λ/4, 5λ/4, … the amplitude of the velocity is equal to two times those observed in a progressive gravity wave ($2U_0$).

For a fixed longitudinal position (x), Eqs. (2.29) and (2.30) indicate that in the time domain the nodal and anti-nodal points occur at t = T/4, 3T/4, 5T/4 and t = 0, T/2, T, 3T/2 …, respectively, and the free surface oscillation and the velocity are out of phase by a time angle π/2. By analogy in the case of a progressive wave, considering only the advection of the tidal current in the salinity redistribution of a vertically homogeneous estuary, and under the assumption that the channel length is an integer multiple of λ/2 (L = λ, for example), a stationary oscillations will be generate in the mixing zone (MZ). In the anti-nodal times the velocity is zero and the advection in the salinity redistribution is null. However, in the nodal times, flood and ebb tides will occur, driven by wave advection. In a real estuary, the salinity will increase during the flood and decrease during the ebb, due to the fresh water discharge mixing with the coastal saline waters, and the higher salinity values are out of phase in the stationary wave of a π/2 (Fig. 2.12).

Estuaries don't have a uniform geometry, and in general they become more narrow and shallow towards the estuary head. As such, the tidal amplitude has a

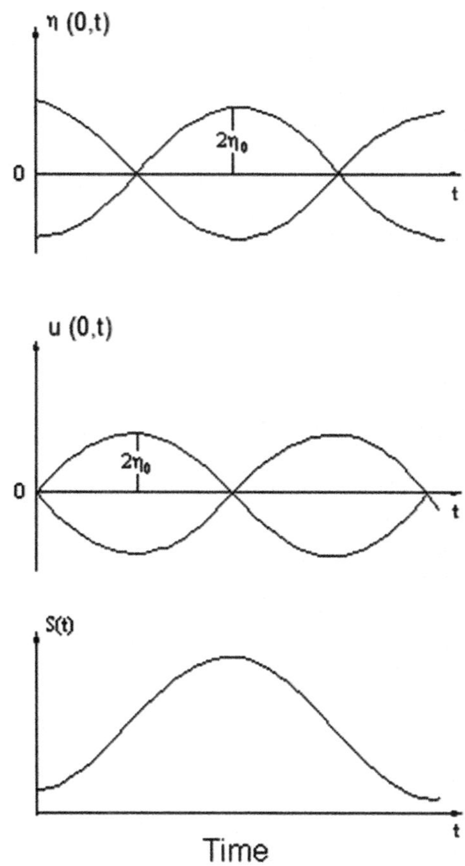

Fig. 2.12 Tidal co-oscillation at the mouth of an estuarine channel with finite length generating a stationary wave $\eta_e(0, t)$, forced by the tidal current $u_e(0, t)$ and salinity S(t) local variations (adapted from Dyer 1973)

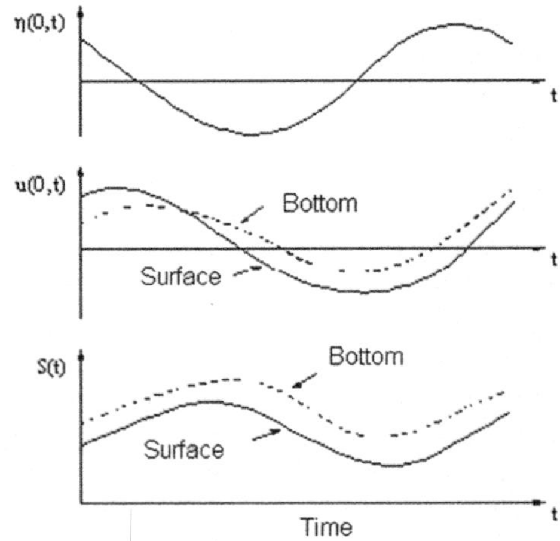

Fig. 2.13 Complex oscillations due to the composition of progressive and stationary waves at the estuary mouth η(0, t), due to the tidal u(0, t) and salinity S (0, t) possible scenarios (adapted from Dyer 1973)

tendency to increase up-estuary, but may simultaneously decrease due to the energy dissipation resulting from friction drag, which has been neglected in the theoretical development. As the estuary head usually isn't a close extremity and there will also be energy dissipation in the reflected wave, in most estuaries the wave oscillation is a complex composition of progressive and stationary waves. According to the relative importance of these interacting waves the resulting surface oscillation, the tidal currents and the salinity vary from one estuary to another. A possible scenario is presented in Fig. 2.13.

Applying Eq. (2.27), which defines the root mean square velocity (u_{rms}) to the velocity of the stationary wave (Eq. 2.30), results in:

$$u_{rms} = \sqrt{\frac{1}{T}\int_{0}^{T} u_e^2(x,t)dt} = \sqrt{2U_0}\sin(\kappa x) = 1.4U_0\sin(\kappa x). \tag{2.31}$$

Then, the root mean square of a stationary wave depends on the longitudinal distance (x). As $\kappa = 2\pi/\lambda$, it follows that the $u_{rms} = 0$ or $u_{rms} = 1.4U_0$ at the nodal and anti-nodal points, respectively.

If a non-linear friction drag or a simple linear dependence with a friction coefficient is introduced into the equation of motion (2.14), the solutions for the wave amplitude $\eta = \eta(x, t)$ and the velocity $u = u(x, t)$ will, respectively, show exponential decline and be out of phase with the longitudinal distance (x). Classical solutions may be found in Defant (1960) and, more recently, in the article of Melo and Jorden (1999); in the latter development the channel declivity was also take into account.

2.6 Non-dimensional Numbers

The English physicist and engineer Osborne Reynolds, was able to demonstrate in a classical paper published in 1883, based in laboratory experiments of fluid flow in pipes, that the transition of a laminar (well behaved) motion to a turbulent motion may be determined by the multiplying the product of its mean velocity U by the pipe diameter (Di), and then dividing by the coefficient of viscosity ν_c, $[\nu_c] = [L^2T^{-1}]$. Then, this dimensionless number is calculated by the ratio UDi/ν_c, and now is known as the *Reynolds number*; physically, this number compares the relative importance of inertial and viscous forces to determine the characteristics of the flow.

Further investigations of fluid flow in channels with free surface, demonstrated that the pipe diameter may be substituted by the channel depth (H_0) to forecast the regime of motion of liquids with uniform density; the *Reynolds number* is thus defined as:

$$Re = \frac{UH_0}{\nu_c}. \tag{2.32}$$

In general, when $Re < 2.0 \times 10^3$, the flow regime is laminar; however, when Re is of the order of 1.0×10^5 or higher, the motion is fully turbulent, and in a non-homogeneous fluid the mixing is intensified; between the orders of magnitude 2.0×10^3 and 1.0×10^5 the fluid flow regime is transitional.

In estuaries, the mixing of the fresh and salt water is also dependent on the vertical stability of the layers in motion. In experimental investigations (Sternberg 1968) in which the Reynolds number was calculated for estuaries forced by tides and with different bed roughness characteristics, it was observed that Re varied between 1.5×10^5 and 3.6×10^5, for turbulent motions and the flow over geometrically simple beds became fully turbulent at lower Re values than for beds of complex roughness. In natural water bodies, such as rivers and estuaries, the $Re = 1.5 \times 10^5$ always indicates a transitional change from laminar to turbulent motion.

There are several mechanisms that produce turbulent motions in an estuary: the bed roughness, the velocity vertical shear, the wind stress on the surface, the surface gravity waves and the internal waves. The intensity of the turbulent motion controls the vertical distribution of the physical, chemical, biological and geological properties of the estuarine water mass.

The competition between the vertical stratification and the mixing forms an important process in estuarine dynamics; if the vertical density (salinity) gradient opposes the exchanges of motion by turbulence, an extra velocity shear is necessary to cause mixing (Dyer 1977).

The analysis of the regime of motion in estuaries may also be investigate by the dimensionless *gradient Richardson number* (Ri), defined by the English Meteorologist Lewis Fry Richardson, in 1920, related to studies to forecast of turbulent motions in the atmosphere. This number compares the stabilizing capacity of the vertical density gradient ($\partial\rho/\partial z$) with the destabilizing capacity produced by the vertical velocity shear (∂u/∂z), and is very important for indicating the turbulence occurrence in the fluid flow. Its analytical differential expression is:

$$\mathrm{Ri} = -\frac{\frac{g}{\rho}\frac{\partial \rho}{\partial z}}{\left(\frac{\partial u}{\partial z}\right)^2} \approx -\frac{g\beta\frac{\partial S}{\partial z}}{\left(\frac{\partial u}{\partial z}\right)^2}, \tag{2.33}$$

where the negative signal depends on the orientation of the vertical axis (Oz); in this expression, it is positively oriented in the direction contrary to the gravity acceleration. In the last term on right-hand-side of this definition, it was used the linear equation of state of seawater, [$\rho(S) = \rho_0(1 + \beta S)$, see Chap. 4)], and the simplifications $\rho/\rho_0 = 1$, and β = const. ($\approx 7.0 \times 10^{-4}$ is the approximate value for *salinity contraction coefficient*). This number also provides the quantitative information on the relationship between the buoyancy stabilizing force of the vertical density (salinity) gradient $\partial\rho/\partial z$ ($\partial S/\partial z$), to the destabilizing capacity produced by the vertical velocity shear $\partial u/\partial z$.

The Ri number also indicates the tendency of the water column to be either: very stable and stratified, or weak stratified, indicating the possibility of a turbulent fluid flow. From Eq. (2.33), it follows that Ri > 0 correspond to a vertical stability ($\partial\rho/\partial z < 0$ or $\partial S/\partial z < 0$), or alternatively for Ri < 0 vertical instability ($\partial\rho/\partial z > 0$ or $\partial S/\partial z > 0$). It should be pointed out that usually the estuarine water mass is in a non-steady-state and the Richardson number is also time dependent Ri = Ri (x, y, z, t).

When the vertical salinity gradient in a given depth is negative ($\partial S/\partial z < 0$), the turbulent flow may be attenuated and the flow regime may develop into a laminar flow. However, above and below this depth the flow regime may be turbulent. Several scientific theoretical and experimental investigations were conducted to understand the mechanisms of the formation and increases of instabilities in stratified fluid interfaces. From these studies it has been found that in uniform motions the transition of the laminar to the turbulent regimes usually occurs for Ri = 0.25. When Ri < 0.25, the turbulence surpasses the vertical density stratification generating vertical mixing.

The vertical salinity (density) stratification and the velocity shear during a tidal cycle generate variations in the Richardson number at different positions.

For instance, in a given position in the MZ, due to these variations Ri is usually named the local *Richardson number*. An analysis on the variation of this number was made by Blumberg (1975), with hourly measurements of velocity and salinity in the Potomac River (Virginia, USA), during tidal cycles in a water column of 10 m depth. From this analysis, the Ri values were found to vary almost randomly with depth and time, with the maximum positive value occurring at the depth of the maximum vertical salinity gradient. The vertical distribution of Ri presented high vertical stability in some depths, separated by layers of low stability. However, using time mean values of salinity and velocity during the tidal cycle, the highest Ri values occurred at 6 m depth, associated with the highest vertical salinity gradients.

The Richardson number was also applied in studies of the internal structure of the Frazer river estuary (Vancouver, Canada), during length variations of the saline intrusion forced by the tidal oscillations (Geyer and Farmer 1989). The analysis of the calculated values at 0.5 m depth intervals were compared to the vertical profiles of the longitudinal velocity component and the salinity. In the obtained results the Ri value remained very close to 0.25 in the layers above and below the halocline. This number presented high values (Ri > 25.0) in the halocline layer with the maximum density vertical gradient and small velocity shear.

Although the variables needed to calculate the Ri number may be accurately measured with the equipment now available to perform profiling of physical properties, it is also useful to use global properties of the water column, and the *layer Richardson number* (Ri_L) has been defined as:

$$Ri_L(t) = \frac{gh(t)\Delta h_V}{\rho \overline{u}^2} \approx \frac{gh(t)\Delta h_V}{\overline{u}^2}, \tag{2.34}$$

where h = h(t) is the local depth, $\Delta\rho_v$ (ΔS_v) is the difference between the bottom density (salinity) and its value at the surface, and $\overline{u} = \overline{u}(t)$ is the mean u-velocity component in the water column which also varies during the tidal cycle Bowden (1978). The time dependence of this dimensionless number was investigated by Dyer and New (1986), showing that; (i) $Ri_L = 20$ is the upper limit for which turbulent mixing occurs near the halocline in partially mixed estuaries; (ii) below this critical number, $Ri_L < 20$, the bottom turbulence becomes effective to the vertical mixing process in the water column; (iii) below $Ri_L = 2$, turbulent mixing is completely isotropic and the water column is unstable; and (iv) for $Ri_L > 20$, the water column is stable with low vertical mixing.

The criteria of Dyer and New (op.cit.) were applied by Miranda et al. (2012) in the analysis of hourly current and hydrographic properties measurements during spring and neap tide in the Piaçaguera estuarine channel, located in the upper reaches of the Santos-São Vicente estuary (Chap. 1, Fig. 1.5). The sampling was made in the austral winter, when the channel was partially mixed and weakly stratified. In the spring tidal cycle (Fig. 2.14a), the calculated values were lower than the critical value, i.e. $Ri_L < 20$ with one exception and it is expected that bottom turbulent mixing became effective, making easier the interchange of the water mass of the bottom and upper layers. During the neap tidal cycle, with a few exceptions, the Ri_L is higher than 20 ($Ri_L > 20$), as shown in Fig. 2.14b, indicating a stable vertical water column, preventing the water masses vertical interchange.

The river discharge may be thought of as a source of deficit of potential energy due to the seawater dilution in the MZ, and, due to the density increase, and the tide as source of kinetic energy to overcome the deficit. Otherwise, the river discharge may be interpreted as a buoyancy source which may be quantified by the product, $g\Delta\rho_H Q_f$, $[g\Delta\rho_H Q_f] = [MLT^{-3}]$, where $\Delta\rho_H$ is the density difference between seawater and freshwater, and is proportional to the barotropic velocity. These physical arguments, associated with laboratory experimental results of circulation in open

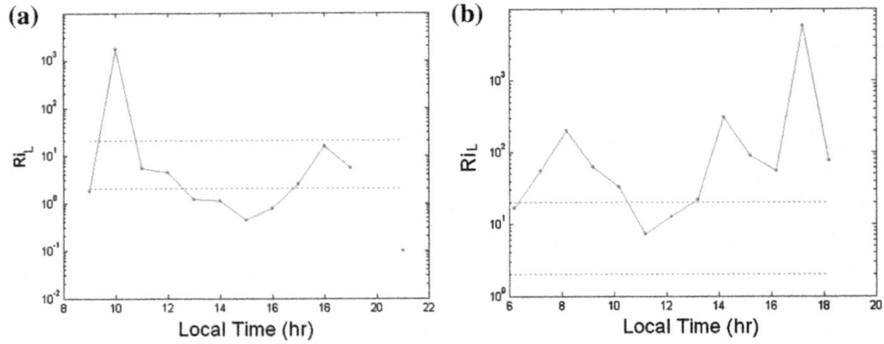

Fig. 2.14 Hourly variation of the layer Richardson number in the Piaçaguera estuarine channel (São Paulo State, Brazil) during spring (**a**) and neap (**b**) tidal measurements (according to Miranda et al. 2012)

channels, were used by Fischer (1972) to define a dimensionless number (Ri_e) named *Richardson estuarine*:

$$Ri_e = \frac{g\frac{\Delta\rho_H}{\overline{\rho}}Q_f}{Bu_{rms}^3}. \qquad (2.35)$$

In this equation, B is the mean (or local) estuary width, $\overline{\rho}$ is the mean density of the estuarine water mass, and u_{rms}^3 is the third power of the root mean square of the u-velocity component. When the tidal wave is approximate to a progressive wave, we have already shown that $u_{rms} = 0.7U_0$ (Eq. 2.27, Chap. 2). As the horizontal stratification of the estuarine water mass is associated with the low frequency variation of the river, mean density values during tidal cycles must be used to calculate this dimensionless number. This estuarine number is physically interpreted as: when Ri_e is relatively large, the estuary is highly stratified and dominated by the river discharge. Alternatively, if Ri_e has a low value, the estuary is weakly stratified and forced by the tide. In a first approximation, the transitional regime occurs when $0.08 < Ri_e < 0.8$ (Fischer et al. 1979).

To forecast the regime of the flow in highly stratified estuaries an alternative dimensionless number was defined by the comparison of the relative velocity on the upper layer, u_1, with the velocity of the progressive internal wave propagating in the interface of a highly stratified halocline. Considering a shallow upper layer with depth, H_1, and density, ρ_1, in motion with velocity u_1, over a deep layer with density, ρ_2 ($\rho_2 > \rho_1$), the internal wave celerity (c_0) at the interface, is given by (Defant 1960):

$$c_0 = \sqrt{H_1 g\frac{\Delta\rho}{\rho_2}} = \sqrt{H_1 g'}. \qquad (2.36)$$

In this definition $\Delta\rho$ is the density difference of the lower and upper layers ($\rho_2 - \rho_1$) and $g\Delta\rho/\rho_2$ is the *reduced gravity* (g'). An alternative number, which has been called *interfacial Froude number* [1] (Fi), was defined based on a simple theory and experimental results (Stommel and Farmer 1952):

$$F_i = \frac{u_1^2}{H_1 g \frac{\Delta\rho}{\rho_2}} = \frac{u_1^2}{c_0}, \tag{2.37}$$

where u_1 is the mean velocity in the upper layer. Thus, the interfacial Froude number compares the fluid speed to the wave speed propagation at the fluid interface. According to the Dyer (1997) in *subcritical* condition Fi < 1 the fluid speed is lower than the internal wave celerity. However, when u_1 approaches the internal wave celerity, a wave or perturbation can only travel upstream very slowly and, as the wave energy accumulates, the wave amplitude grows until critical conditions for Fi = 1, and the wave will break with energetic mixing. The thickness of the flowing layer will abruptly increase generating an increase in fluid velocity, and the flow changes to a supercritical condition Fi > 1. This situation is known as the *internal hydraulic jump*, in analogy to the *hydraulic jump* phenomenon generated when a stratified fluid passes over an obstacle located at the bottom.

The interfacial Froude number may be written as a function of the river discharge and geometric dimensions of the estuary, such as the width (B) and depth, H_0. Writing the river discharge as $Q_f = u_f BH_0$, the *densimetric Froude number* may be defined as (Hansen and Rattray 1966):

$$F_m = \frac{u_f}{\sqrt{H_1 g \frac{\Delta\rho}{\rho_2}}} = \frac{Q_f}{BH_0\sqrt{H_1 g \frac{\Delta\rho}{\rho_2}}} = \frac{Q_f}{BH_o\sqrt{g'H_1}}. \tag{2.38}$$

This number is also known as the *estuarine or internal Froude number*, and a critical number can be achieved by a constriction at the mouth or by a shallow sill depth (Dyer 1997). When the river discharge is weak, $F_m \to 0$, and the phenomenon of *over-mixing* can occur, and may be observed when the fluid interface (H_1), between the upper and lower layers, is equal to half of the water depth ($1/2H_0$). This phenomenon has been investigated in laboratory experiments and theoretically by Stommel and Farmer (1952), to explain why, for a stratified estuary, its mouth is a limiting condition to the exchange of salt with the continental shelf.

Equation (2.38) is equivalent to another dimensionless number obtained by the ratio, $u_f/(gH_0\rho^{-1}\Delta\rho)^{1/2}$, named *densimetric Froude number*, which is the ratio the velocity generated by the river discharge (u_f) divided by the internal circulation

[1] The designation of the Froude number was given in honor of William Froude (1810–1879), engineer and naval architect who established the main similarity of reduced ship models and the natural dimensions, introducing the dimensionless number $V/(gL)^{1/2}$ where V and L are the velocity and the ship length at the water level, respectively, and g is the gravity acceleration.

potentially generated by the density gradient $(gH_0\rho^{-1}\Delta\rho)^{1/2}$ (*densimetric velocity*), thus defining the *densimetric Froude number*. This dimensionless number may be applied to estuaries that are highly stratified, where the velocity above the halocline is mainly driven by the river discharge. There is also the *internal Froude number* (Fi), expressed by:

$$F_i = \frac{u}{\sqrt{hg\frac{\Delta\rho}{\rho_2}}} = \frac{u}{\sqrt{hg'}}, \tag{2.39}$$

where the velocity u is a mean velocity value and h is the water layer depth.

The *estuary number* (N_e) was defined by Dyer (1997) by the following ratio,

$$N_e = \frac{T_{PR}F_m^2}{T_PQ_f} = \frac{T_PF_m^2}{R}. \tag{2.40}$$

where T_{PR} is the tidal prism, F_m is the densimetric Froude number (Eq. 2.38), and T_P is the tidal period and Q_f is the river discharge. This definition involves the ratio of the tidal prism to the fresh water volume discharged during a tidal cycle $R = TQ_f$, and some experiments indicate that when $N_e > 0.1$ or $N_e < 0.1$, the estuary is well-mixed or stratified, respectively.

When the wind stress on the surface layer is an important driving force its influence in a highly stratified estuary may be estimated by a dimensionless number named *Wedderburn*. It has been used to investigate the influence of the wind stress in lakes and reservoirs with high vertical temperature stratification. This number, defined by Thompson and Imberger (1980), is calculate by the ratio of the Richardson number and the estuary aspect ratio δ:

$$W = \frac{Ri}{\delta}, \tag{2.41}$$

where, $\delta = L/H_0$, is the ratio of the longitudinal dimension of the water body, L, to its depth H_0.

Considering an estuary that is vertically stratified and forced predominantly by the wind, the numerator of Eq. (2.41) may be substituted by the following modified Richardson number:

$$Ri = \frac{g\Delta\rho_H H_1}{\rho u_1^2}, \tag{2.42}$$

In this definition $\Delta\rho_H$, u_1 and H_1 are the density variation along the estuary, the velocity, and the depth of the layer over the halocline, respectively. Combining

Eqs. (2.41) and (2.42), it follows the analytical expression of the *estuarine Wedderburn number*, W_e:

$$W_e = \frac{g\Delta\rho_H H_1^2}{\rho u_1^2 L} \approx \frac{g\Delta\rho_H H_1^2}{\tau_w L}, \tag{2.43}$$

in the last expression the quantity ρu_1^2 was approximated by the wind stress shear, τ_w.

Equation (2.43) indicates that W_e is directly proportional to H_1^2, the horizontal density stratification, and is inversely proportional to the wind stress multiplied by the estuary length. When the wind stress forcing is predominant, $W_e < 1$, and it has the main contribution to the dynamic behavior and mixing of the layer above the halocline. Alternatively, when $W_e \geq 1$ the longitudinal density stratification and the buoyancy due to river input prevails over mixing.

An example on the use of the Wedderburn number to establish the importance of wind stress on the dynamics of two small and shallow (mid-channel depth 1–2 m) estuaries, the Childs and Quashnet, located in the Waquoit Bay (Cape Cod, USA) is presented in the article of Geyer (1997). This study demonstrated the strong influence of wind forcing on salinity structure and flushing characteristics of these estuaries. For most of the observations $W_e < 1$, indicating the important role of wind stress in these estuaries. The sensitive dependence of this number on depth explained why the gentle winds observed during the surveys had such a profound influence on the estuarine dynamics; onshore winds inhibit the estuarine circulation, increasing the longitudinal estuary salinity gradient, and reducing the flushing rate. However, offshore winds enhanced the surface outflow, flushed out the freshwater and reduced the longitudinal salinity gradient.

An alternative expression to the dimensionless flux rate, F_R, may be obtained with theoretical results previously defined in this chapter. This number is determined by the ratio of the volume of river water discharged during a tidal cycle, R, and the tidal prism, T_{PR}. If, A, is the mean transversal section of the estuary mouth, the fresh water volume may be calculated by the product $R = Q_f T = Au_f T$. Thus, under the assumption of a progressive tidal wave propagating landward, $T_{PR} = A(2U_0/\pi)$ and $2U_0/\pi$, are the tidal prism and mean velocity during half tidal cycle, respectively. Then, an alternative expression for the flux ratio is:

$$F_R = \pi\frac{u_f}{U_0} = 3.14\frac{u_f}{U_0}, \tag{2.44}$$

which is directly and inversely proportional to the velocity of the river discharge and the amplitude of the velocity generated by the tide, respectively. As such, this number will be large or small or when the estuary is dominated by the river discharge or the tide, respectively.

Others dimensionless numbers will be defined in the following chapters related to the estuary classification, according to the acting processes: river discharge, salinity stratification, and mixing processes. Among these there are the dimensionless numbers: (i) Ekman number, $Ek = \nu_k/f_C D^2$, where ν_k, f_C and D are the kinematic viscosity coefficient, the Coriolis parameter and the vertical length scale, respectively; and; (ii) Kelvin number, $Ke = B/R_D$, where B is the estuary width and R_D is the deformation radius, $R_D = (N_{BV})/f_C$, with N_{BV} indicating the Brunt-Väisälä frequency which arises in the presence of a background stratification, and the fluid parcel perturbed vertically from its starting position experiences a vertical acceleration.

2.7 Mixing and Entrainment

The term mixing is applied to multitude of physical processes occurring internally in the seawater, which tend to produce uniformities in concentrations of physical, chemical, biological and geological properties. There are two processes that contribute to mixing: (a) *advection* (or advective), characterized by regular patterns of water movement on a macroscopic scale; and (b) diffusion (or diffusive), characterized by microscopic or small-scale irregular movements called turbulence, which together with molecular diffusion give rise to the local exchange of property, without any net transport of water (Okubo 1970; Bowden 1975, 1977). The advective and diffusive tidal forcing are the main processes which produces to the mixing of the river and salt water, contributing to the salinity stratification characteristics.

The mixing of salt and fresh water is carried out by a combination of turbulence generated by the current shear mainly due to tidal currents in the water column, and at the sea bed. These two effects vary in their magnitudes and timing during the tide, as well as in different estuarine types, as the salinity stratification and tidal velocities changes. Competition between stratification and mixing play a crucial part in estuarine dynamics because when the fluid is stratified, the density gradient resists to the exchanges of momentum by the turbulence and an extra velocity shear is necessary to cause mixing, and thus influence the distribution of natural water properties and those discharged into the estuaries. The role of internal mixing processes (mixing, turbulent diffusion and internal waves) and its main characteristics are presented in the Dyer's (1997) book and a brief outline is given below.

In an estuary with small tidal amplitude that may be neglected and a moderate or high river discharge, the water discharged into the estuary head moves seaward flowing persistently on the surface layer due to its relatively low density. During this flow, and at $Ri > 0.25$, instabilities at the stratified interface take the form of *cusps* or progressive interfacial *Holmboe waves*, which grow in height and became sharper crested. Eventually they break, and wip-like elements of denser water are ejected from the crests into the upper lighter layer. This upward transference of salt water is unidirectional, and this phenomenon is known as *entrainment*. Its importance to the estuarine dynamics has been presented in the pioneering article of Keulegan (1949) based on laboratory experiments. A vivid description of mixing in a natural estuary has been given by Farrel (1970—quoted in Dyer 1997, p. 45), showing that entrainment is a one-way process in which a less turbulent water mass become drawn into a turbulent layer; it is an internal mechanism, besides the advection and turbulent diffusion to the estuarine mixing, occasioning salinity increase in the upper layer. The balance of the volume transport down and up-estuary in the upper and lower layers, as well as the transport associated with the entrainment, were calculated using the continuity of fresh water by Tully (1958).

The above process may be superimposed to the turbulent diffusion which may be subdivided into three types (Bowden 1977): (1) On or near the bottom due to the frictional shear, which propagates into the water column; (2) Generated inside the water layer due to the turbulent diffusion, which may be dumped by the vertical stratification; and (3) Turbulent diffusion on the free surface due to the wind stress, generating gravity waves and drift currents. Generally, when the river discharge is intense and the tidal amplitude is small the entrainment is the mechanism which predominates; however, the higher the tidal amplitude, the higher is its influence to generate mixing through the processes of advection and diffusion.

The estuarine flow usually occurs in a transitional regime from laminar to turbulent, generated by microscopic and macroscopic scales of motion, respectively. Mainly due to tidal oscillations the entrainment, turbulent diffusion and advection are the processes responsible for the mixing of the fresh and salty water masses, the local and spatial salinity, and temperature, as well as, the concentrations of natural properties and pollutants launched into the estuaries. Unfortunately the fresh water discharged into estuaries and its geometry has been altered and this external interference will alter the circulation, the assimilation and mixing processes, and water mass renewal in these very important coastal ecosystems.

References

Andutta, F. P. 2011. O Sistema Estuarino dos Rios Caravelas e Peruipe (BA): Observações, Simulações, Tempo de Residência e Processos de Difusão e Advecção. Tese de Doutorado. São Paulo, Instituto Oceanográfico, Universidade de São Paulo. 121 p.

Arons, A. B. & Stommel, H. 1951. "A Mixing-Lenght Theory of Tidal Flushing". Trans. Am. Geophys.Un., 32(3):419–421.

Blumberg, A. F. 1975. A Numerical Investigation into the Dynamics of Estuarine Circulation. Tech. Rept. Chesapeake Bay Institute, The Jonhs Hopkins University. n. 91. 110 p. + Apêndices.

Bonetti, J. F. & Miranda, L. B. de. 1997. "Estimativa da Descarga Fluvial no Sistema Estuarino-Lagunar de Cananeia-Iguape". Rev. Bras. Oceanogr., São Paulo, 45(1/2):89–94.

Bowden, K. F. 1975. Oceanic and Estuarine Mixing Processes. In Chemical Oceanography, Academic Press, London, Chapter 1, pp. 1–41.

Bowden, K. F. 1977. Turbulent Processes in Estuaries. Estuaries, Geophysics, and the Environment. National Academy of Sciences, Washington, D. C., pp. 46–56.

Bowden, K. F. 1978. "Mixing Processes in Estuaries". In: Kjerfve, B. (ed.). Estuarine Transport Processes. University of South Carolina Press, Columbia, pp. 11–36. (Belle W Baruch Library in Marine. Science, 7).

Coleman, J. M. & Wright, L. D. 1971. Analysis of Major River Systems and Their Deltas: Procedures and Rationale, with two Examples. Baton Rouge, Lousiana State University Press. 125 p.

Davies, J. H. 1964. "A Morphogenic Approach of World Shorelines". Z. Geomorphology, 8:127–142.

Defant, A. 1960. Physical Oceanography. Oxford, Pergamon Press, vol. 2. 598 p.

Doodson, A. T. 1928. The Analysis of Tidal Observations. Phyl. Trans. R. Soc. Lond., Series A, 227:223–279.

Dyer, K. R. 1973. Estuaries: A Physical Introduction. London, Wiley. 140 p.

Dyer, K. R. 1977. Lateral Circulation Effects in Estuaries. Estuaries, Geophysics and the Environment. Washington, D. C., National Academy of Sciences, pp. 22–29.

Elliott, A.J. 1978. Observations of the Meteorologically Induced Circulation in the Potomac Estuary. *Estuarine, Costal Marine Science,* 6, 285–299.

Farrel, S.C. 1970. Sediment Distribution and Hydrodynamics Saco River and Scarboro Estuaries, Maine. Cont. 6-CRG, Dept. Geol. Univ. Mass. (quoted in Dyer, 1997), p. 45).

Fischer, H. B. 1972. Mass Transport Mechanisms in Partially Stratified Estuaries. J. Fluid Mech., 53:672–687.

Fischer, H. B.; List, E. J.; Koh, R. C. Y.; Imberger, J. & Brooks, N. H. 1979. Mixing in Inland and Coastal Waters. New York, Academic Press. 483 p.

Franco, A. S. 1988. Tides: Fundamentals, Analysis and Prediction. São Paulo, Fundação Centro Tecnológico de Hidráulica. 249 p.

Franco, A. S. 2000. MARÉS: Programa para Previsão e Análise. In: Manual, BSP, São Paulo. 36 p.

Franco, A. S. 2009. Marés – Fundamentos, Análise e Previsão. Diretoria de Hidrografia e Navegação. Rio de Janeiro, 2nd ed., 344 p.

French, P. W. 1997. Coastal and Estuarine Management. London, Routledge. 251 p. (Environmental Management Series).

Geyer, W. R. 1997. Influence of Wind on Dynamics and Flushing of Shallow Estuaries. Estuar. Coast. Shelf Sci., 44:713–722.

Geyer, W. R. & Farmer, D. M. 1989. Tide-Induced Variations of the Dynamics of a Salt Wedge Estuary. J. Phys. Oceanogr., v. 19, pp. 1060–1072.

Gill, A. E. & Schumann, E. H. 1979. The Generation of Long Shelf Waves by the Wind. J. Phys. Oceanogr., 4:83–90.

Hansen, D. V. & Rattray Jr., M. 1966. New Dimensions in Estuary Classification. Limnol. Oceanogr., 11(3):319–325.

Harleman, D. R. F. 1971. One-Dimensional Models. In: Ward Jr., G. H. & Espey Jr., W. H. (eds.). Estuarine Modelling: An Assessment Capabilities and Limitations for Resource Management and Pollution Control. Austin, Tracor, pp. 34–89.

Holland, H. D. 1978. The Chemistry of the Atmosphere and Oceans. New York, Willey. 351 p.

Ippen, A. T. & Harleman, D. R. F. 1961. One-Dimensional Analysis of Salinity Intrusion in Estuaries. Committee on Tidal Hydraulics. Tech. Bull. Corps of Engineers U. S. Army, n. 5. 120 p.

Jay, D. A. 2010. Estuarine Variability. In: ed. Valle-Levinson A. Contemporary Issues in Estuarine Physics. Cambridge University Press, pp. 62–99.

Keulegan, G. H. 1949. Interfacial Instability and Mixing in Stratified Flows. J. Res. U. S. Geol. Surv., 43:487–500.

Kjerfve, B. 1990. Manual for Investigation of Hydrological Processes in Mangrove Ecosystems. New Delhi, Unesco/UNDP. 79 p.

Kjerfve, B. & Ferreira, H. O. 1993. Tidal Bores: First Ever Measurements. Ciência Cult., São Paulo, 45(2):135–137.

Kjerfve, B.; Schettini, C. A. F.; Knoppers, B.; Lessa, G. & Ferreira, H. O. 1996. Hydrology and Salt Balance in a Large, Hypersaline Coastal Lagoon: Lagoa de Araruama, Brazil. Estuar. Coast. Shelf Sci., 42:701–725.

Kjerfve, B.; Ribeiro, C. H. A.; Dias, G. T. M.; Fillippo, A. M. & Quaresma, V. S. 1997. Oceanographic Characteristics of an Impacted Coastal Bay: Baía de Guanabara, Rio de Janeiro, Brazil. Continent. Shelf Res., 17(13):1609–1643.

Kosoun, V.I.; Sokolov, A.A.; Budyko. M.I.; Voskresensky, K.P.; Kalinin, G.P.; Konoplyantsev, A.A.; Korotkevich, E.S.; Kuzin, P.S. & Lvovitch, M.I. 1974. World Water Balance and Water Resources of the Earth, prepared by the U.S.S.R. National Committee for the International Hydrological Decade, V.I. Kosoun, Editor-in-Chief, Leningrad.

Kvale, E.P. 2006. The origin of neap-spring tidal cycles. Marine Geology, 235, pp. 5–18.

Lacerda, L.D., Kjerfve, B., Salomons, W. and Kremer, H.H. 2002. Regional assessment and synthesis: South America. In (Eds. Lacerda, L.D; Kremer, H.H.; Kjerfve, B.; Salomons, W.; Marshall Crossland J.I.

Leblond, P.H. 1979. Forced Fortnightly Tides in Shallow Rivers. Atmosphere. Ocean. 17(3):253–264.

Leblond, P.H. & Mysak, L.A. 1978. Waves in the Ocean. Elsevier. Amsterdam, 602 p.

Miranda, L. B.; Castro, B. M. de. & Kjerfve, B. 1998. Circulation and Mixing in the Bertioga Channel (SP, Brazil) Due to Tidal Forcing. Estuaries, 21(2):204–214.

Miranda, L. B.; Dalle Olle, E.; Bérgamo, A.L.; Silva, L.S. & Andutta, F.P. 2012. Circulation and salt intrusion in the Piaçaguera Channel, Santos (SP). Braz. J. Oceanography, 60(1):11–23.

Medeiros, C. & Kjerfve, B. 1993. Hydrology of a Tropical Estuarine System: Itamaracá, Brazil. Estuar. Coast. Shelf Sci., 36:495–515.

Mello, E. 1998. Considerações sobre a Hidráulica de Canais Fluviais e Canais de Maré. Rev. Bras. Rec. Hid. 3(2):95–107.

Melo, E. & Jorden, V. 1999. Tide Penetration in Coastal Waters. Proceedings of the Fifth International Conference on Coastal and Port Engineering in Developing Countries. Cape Town, South Africa, pp. 1771–1781.

Nichols, M. M. & Biggs, R. B. 1985. Estuaries. In: Davis R. A. (ed.). Coastal Sedimentary Environments. Berlin, Springer-Verlag, pp. 77–186.

Okubo, A. 1970. Oceanic Mixing. The Johns Hopkins University. Chesapeake Bay Institute. Tech. Rept., n. 62. 140 p.

Pawlowicz, R.; Beardsley, B. & Lentz, S. 2002. Classical tidal harmonic analysis including error estimates in MATLAB® using T-TIDE. Computers & Geosciences, v. 28:939–937.

Pedloski, J. 1979. Geophysical Fluid Dynamics. 2. ed., New York, Springer-Verlag. 624 p.

Prandle, D. 2009. Estuaries: Dynamics, Mixing, Sedimentation and Morphology. Cambridge University Press, N.Y., 236 p.

Pritchard, D. W. 1952. Salinity Distribution and Circulation in the Chesapeake Bay Estuarine System. J. Mar. Res., 11(1):106–123.

Pritchard, D. W. 1955. Estuarine Circulation Patterns. Proc. Am. Soc. Civ. Eng., 81:717:1–11.

Pritchard, D. W. 1958. The Equations of Mass Continuity and Salt Continuity in Estuaries. J. Mar. Res., 17:412–423.

Platt, T.; Mann, K.H. & Ulanowicz, R.E. (Eds.) 1981. Mathematical Models in Biological Oceanography. The Unesco Press, 156 p.

Schettini, C. A. F. 1994. Determinantes Hidrológicos na Manutenção da Condição Hipersalina da Lagoa de Saquarema. Dissertação de Mestrado, Niterói, Universidade Federal Fluminense. 75 p.

Schettini, C. A. F. 2002. Caracterização Física do Estuário do Rio Itajaí-açu, SC. Revista Brasileira Recursos Hídricos, 7(1):123–142.

Sternberg, R. W. 1968. Friction Factors in Tidal Channels with Differing Bed Roughness. J. Mar. Geol., 6:243–260.

Stommel, H. & Farmer, H. G. 1952. Abrupt Change in Width in Two-layer Open Channel Flow. J. Mar. Res., 11:203–214.

Thompson, R. O. R. Y. & Imberger, J. 1980. Response of a Numerical Model of a Stratified Lake to Wind Stress. Proc. 2nd Intl Symp. Stratified Flows, Trondheim, pp. 562–570.

Tully, J. P. 1958. On Structure, Entrainment and Transport in Estuarine Embayments. J. Mar. Res., 17:523–535.

von Arx, W. S. 1962. An Introduction to Physical Oceanography. Addison-Wesley, Massachusetts. 422 p.

Quoted References

Defant, A. 1961. Physical Oceanography. Oxford, Pergamon Press, vol. 1. 729 p.
Dyer, K. R. 1997. Estuaries: A Physical Introduction. 2. ed., Chichester, Wiley. 195 p.
Geyer, W.R. 2010. Estuarine salinity structure and circulation. In: ed. Valle-Levinson A. Contemporary Issues in Estuarine Physics. Cambridge University Press, pp. 12–26.

Chapter 3
Estuary Classification

The hydrographic and morphologic characteristics of estuaries are very diverse, varying with changes in response to natural and non-natural phenomena as climate changes and human interference, and often show conflicting situations that make oceanographic generalizations impossible. Until sixty years ago, studies of these environments interpreted such variations as local phenomena.

Estuary classification criteria were developed with the aim of comparing different estuaries by categorizing them, using a database of their main characteristics. The known estuary classification schemes are based on salinity-stratification, river discharge, circulation and mixing and they allow the main characteristics an estuary's circulation and mixing processes to be forecast.

The first estuary classification was suggested by Stommel (1951), taking into account the main forces such as tides, fresh water discharge and wind. The estuaries in the USA, which are formed by the Raritan, Pamlico Sound and Mississippi rivers in the states of New Jersey, North Caroline and Lousianna have tide, wind and river discharge, as primary forcing, respectively. The remarkable characteristic of those estuaries is the vertical salinity stratification. The Raritan estuary, which is the shallowest, is forced by tides of moderate height and is almost vertically homogeneous. The Mississippi estuary, which is the deepest, is forced by micro-tides and is the most stratified due to the huge river discharge of the Mississippi river into the Gulf of Mexico characterized by microtides. However, in most estuaries, it is difficult to identify a single force that predominates the circulation and mixing processes.

Estuaries forced by tides were the most studied type of estuary in the 1950s. They were characterised by tidal generate velocities which were more intense than velocities generated by river discharge. The most noticeable difference between the

L. Bruner de Miranda et al., *Fundamentals of Estuarine Physical Oceanography*,
Ocean Engineering & Oceanography 8, DOI 10.1007/978-981-10-3041-3_3

three previously mentioned estuarine types is the vertical salinity stratification, the most extreme case being the vertically homogeneous Raritan river estuary, with a mean depth of only 3 m.

One of the classical classification criteria was developed taking into account geomorphologic characteristics (Pritchard 1952a), which grouped estuaries into the following types: coastal plain (formed due to the river valleys flooding), fjord and bar built.

The density (salinity) variations in estuaries that generate currents due to the mixing of the fresh and salt water, motivated Stommel (1953) to examine the physical characteristics of estuaries using the following examples: estuaries with negligible mixing (salt wedge), deep estuaries (such as fjords), and estuaries with moderate and intense mixing. These classification systems received improvements through the work of Pritchard (1955) and Cameron and Pritchard (1963), taking into account characteristics related to vertical salinity stratification, salt budget and the steady-state circulation. For classification criteria, Simmons (1955) used the *flux ratio* (F_R) as a parameter defined by the ratio of the fresh water volume discharged into the estuary in the tidal period (Q_fT_P) by the tidal prism (T_{PR}), $F_R = Q_fT_P/T_{PR}$, as a parameter to indicate the vertical stratification changes in the estuary. Ippen and Harleman (1961) and the contemporary papers of Harlemann and Abraham (1966) and Hansen and Rattray (1966) contributed towards a quantitative estuary classification system; initially, the *stratification number* was introduced, followed by the theoretical development of a classification method with the Stratification-circulation Diagram, which uses the *stratification* and *circulation* parameters.

The Stratification-circulation Diagram was theoretically deduced with the most complete and comprehensive theoretical development of a steady-state bi-dimensional analytical model with a simple geometry, whose dynamical formulation will be presented later. In this diagram it is also possible to obtain the relative contributions of the salt transport due to advection and mixing, which control the concentration of dissolved substances in the estuary. This diagram, representing the state of the art of estuary classification, was further complemented by several researchers, confirming the originality of this classical theoretical development (Fischer 1972; Hamilton and Rattray 1978; Rattray and Uncles 1983; Prandle 1985, among others).

It is useful to add the dynamics of the ecology aspects of the estuarine systems to these classification systems, enabling biologists and ecologists to use comparative methods more adequately. The main elements of the ecological classification were chronologically described by Yãnes-Arancibia (1987), emphasizing that the hydrodynamic and geomorphologic processes are connected to the ecology of marine ecosystems. In addition, processes such as erosion, deposition and sediment transport are natural features of coastal environment, and thus also affecting the ecology of the marine ecosystem. Yãnes-Arancibia (op. cit.) provided the ecological classification taking into account interdisciplinary criteria related to the energy sources and the ecological diagnostic. Among the studies with ecological scope of Brazilian estuaries we must mention the works of Tommasi (1979) in the Santos-São Vicente Estuary (São Paulo), Knoppers et al. (1987) and Brandini et al. (1988) in the estuarine region of Paranaguá Bay (Paraná), Sankarankutty et al. (1995) in the Potengi river Estuary (Rio Grande do Norte), and Tundisi and Matsumura-Tundisi (2001) in the Estuarine-lagoon System of Cananéia (São Paulo).

3.1 Geomorphologic Types of Estuaries

Geomorphologic classification criteria was presented in Pritchard (1952a, 1967), grouping the estuaries according to their geomorphological structure into the four types already mentioned: coastal plain estuaries, fjord, bar built (or coastal lagoons), and tectonically formed estuaries. Each type exhibits a somewhat similar dynamic behavior in terms of water circulation and mixing.

3.1.1 Coastal Plain

These estuaries are typical of coastal plain regions and were formed during the Holocenic sea level transgression, when former river valleys became increasing more flooded by glaciers melting. The flooding process was more intense than the sedimentation, and the geomorphic characteristics of the estuaries remained similar to those of the rivers. These estuaries are relatively shallow, with very few deeper than 30 m. They are bordered by broad shallow flats, and a typical cross-section sometimes increases in size down estuary and consists of a V-shaped channel. The width to depth ratio is usually large, but it depends on the rock type where the river valley has been formed. Due to recent erosion processes in the river, the bottom in the upper estuary is filled with mud and fine sediments, but the estuary bed in its lower reaches may be filled with coarse sediments due to the bottom erosion and sediment transport.

In general, coastal plain estuaries are localized in tropical and subtropical regions and are common along the east coast of the USA and South America. These estuaries vary greatly in size. For example, the Chesapeake Bay with nearly 300 km of length, and 25 km wide, this is the largest coastal plain estuary in the USA. Delaware Bay, South Caroline, Charleston Harbor, and the Hudson River are other examples of coastal plain estuaries. Estuaries of this type are also common in Brazil and include the Potengi, São Francisco, Doce, Contas, Caravelas and Peruípe rivers in the east and northeastern Brazilian coast. Others have been described by Officer (1976), Dyer (1973, 1997) and in South America by Kjerfve (1989).

In these estuaries the width to depth ratio is large, and the flux ratio is dependent on the river discharge intensity and the tidal height. The suspended sediment transport in estuaries formed by rivers with high discharge usually is also large.

3.1.2 Fjord

Fjords also owed their origin to the last glaciation cycle during the Pleistocene. Due to the advance of glaciers, the high pressure of these ice caps over the continental blocks and their erosive power during melting, tongues of the leading ice edge scoured out many river valleys in latitudes above 45° (Fairbridge 1980; Dyer 1997). As a consequence of the decreasing power of this erosive process towards the sea, a steep rock bar (sill) usually formed seaward of the leading ice edge. This sill is caused by the scoured material from the basin was pushed forward by the advancing glacier and deposited at the leading edge. After the retreat of the glacier the sill remains. Whereas the water column over the sill generally varies between 10 and 90 m, the depth inside the fjord basin often exceeds 800 m, and extends to several hundred kilometers inland. Fjords are common in both hemispheres in high latitude regions, where there has been glacial activity (e.g. in South of Chile, New Zealand, Norway and other).

Because fjords are too deep and the water exchanges with the adjacent ocean is limited by the sill depth, the fresh water flow and the circulation are confined to a shallow upper layer; the motion towards the sea, due to the entrainment in the fresh and salt water interface, generates a salt transport into the upper layer. Below this and down to the bottom, the water mass is almost isohaline and the fresh water transport during the spring and summer is dominant in the tidal prism. The upper layer depth has little variation along the fjord and its volume transport increases towards the sea.

The water depth above the halocline is almost the same as the sill depth, and during events of high fresh water discharge it is an almost homogeneous layer; the entrainment process intensifies generating a moderately or highly stratified salinity in the upper layer. However, when the river discharge decreases, the upper layer characteristics change to high stratification near the sill, and high stratification up to the surface. Due to the large depth, the temperature decreases towards the bottom, but due to fresh water inflow from ice melting, temperature inversions in subsurface layers may occurs (Pickard 1961, 1971).

Sill depths restrict the water interchange between the fjord basin and the ocean, and may cause anoxic conditions which are detrimental to the biological community especially during the summer when there is high vertical stability of the water column. Favorable conditions to the biological community may be re-established in the autumn and winter, with water renewal occurring by deep convective

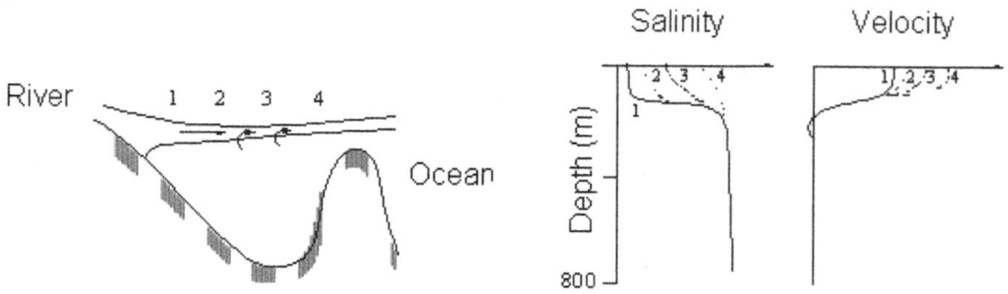

Fig. 3.1 Schematic vertical profiles of salinity and velocity in fjords (according to Dyer 1973)

overturning. Examples of deep and very long fjords in Norway are the Sögnne and Hardanger fjords, which have depths of almost 1200 m Saelen (1967).

As fjords are deep estuaries, their width to depth ratio is relatively small compared with coastal plain estuaries, and they are characterized by an almost rectangular vertical section. Their geologic formation justifies both these characteristics: a rock bottom and recent sedimentation process occurring mainly in the river mouths. In the winter their fluvial discharge may be very low or even absent. Due to the higher velocities in the sill, the circulation in the mouth may be very different from the circulation observed in the shallow mixing zone. Schematic vertical salinity and velocity profiles are presented in Fig. 3.1.

3.1.3 Bar-Built (or Coastal Lagoons)

The bar-built estuary is the third largest group of estuaries, according to Pritchard (1952b). They were also formed as the coastal plain estuaries (former river valleys) became increasingly more flooded by glaciers melting, but the recent coastal sedimentation processes caused the development of an offshore bar on a shore line of shallow water. In general, bar-built estuaries are associated with coastal regions subjected to erosive processes which can easily produce great quantities of sediments that are reworked by waves and transported by coastal currents. These estuaries are usually shallow (less than 20–30 m deep) and may be composed of a channel and lagoons that may exists between the open sea and the estuary. The discharging river (or river system) into the estuary, besides its variable discharge depending on the season, may transport a great quantity of suspended sediments, generating seasonal changes in the mouth or bar entrance. During the seasons of high river discharge, bar-built estuaries may be partially or completely eroded, but they are reestablished in periods of lower river discharge. This estuary type is common in tropical regions and is referred to in the Brazilian terminology as *estuarine-lagunar*, such as in the Cananéia-Iguape region in the Southern São Paulo State (Fig. 1.5, Chap. 1).

Lagoons exhibit a larger fractional area of open surface area oriented parallel to the coast, whereas coastal plain estuaries are most often oriented normal to the coast (Fairbridge 1980). Lagoons also have a less well-drained subaqueous drainage channel and are uniformly shallow over large areas. The physical processes are mostly wind-dominated, and tidal forcing having a minor influence. Further details on coastal lagoons will be described later.

3.1.4 Tectonic, Deltas and Rias

Estuaries that do not fit in the preceding geomorphologic classification are categorized as tectonic estuaries, formed due to: tectonic fault, earth shakes and slides, and volcanic eruption. This category also includes estuaries affected by morphodynamic processes of sedimentation with emphasis on coastal environments like *deltas* and *rias*.

The best example and extensively studied estuary in this group is San Francisco Bay (see Officer 1976; Conomos 1979). Tectonically caused estuaries exhibit much variability, and some may behave oceanographically similarly to coastal plain estuaries, fjords or lagoons, depending on the local constraints (Kjerfve 1989).

In the age and geological estuary formation presented in Chap. 1, it was stated that estuary formation was largely due to the extensive glaciations of about 10,000 years ago. From this epoch, these coastal environments were progressively eroded and filled by sediments furnished by rivers and littoral transports, and re-worked by the occurrence of eustatic oscillations in this time period (Fairbridge 1980). In the regions forced by large tidal oscillations (macrotides) with moderate or high waves and fluvial transport with high sediment concentrations, the recent sedimentary processes resulted in the generation of islands inside the estuary. This type of estuary is called an *estuarine delta* or *flood delta*. One of the most spectacular is the tropical delta of the Amazon River in the Brazilian littoral north, with a funneled geometry. Again, in the case of high sediment concentrations, but in regions forced by tides with low height (microtides) with moderate wave energy, the sedimentation process will be localized in the inner continental shelf and in the vicinity of the estuary mouth, generating sand banks and islands, forming an *ebb delta* or a simple *delta*. The delta of the Mississippi river, in the Mexican Golf is a classic example (Wright 1970). These deltaic formations, *estuarine deltas* or *flood and ebb deltas* are dominated by tidal height and river discharge, respectively.

Ria estuaries are typical of high latitude mountain regions, formerly occupied by glaciers. They originated tectonically due to the *increase in e*levation of the continental region where former river valleys were located, as a result of the weight decrease of the glacial melting. The river valleys were flooded by the sea level rise, generating the estuaries. In general a ria estuaries has an irregular morphology and

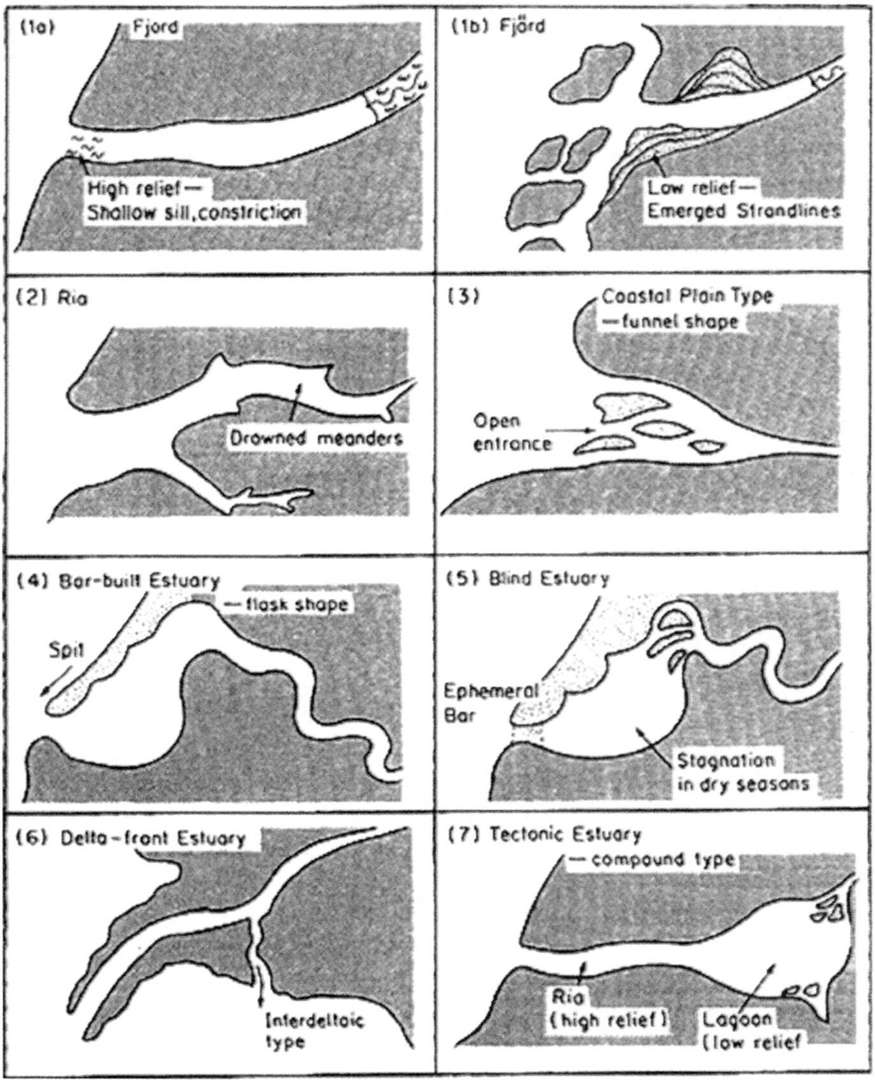

Fig. 3.2 Basic estuarine physiographic types (according to Fairbridge 1980)

are fed by rivers basins located in the adjacent region. Their geometry may be a channel in between mountains or a funneled geometry with depth increasing seaward which may amplify the co-oscillating tide.

Figure 3.2 schematically shows some described estuarine types, according to their physiographic classification of Fairbridge (1980).

3.2 Salinity Stratification

Estuaries classified according to geomorphologic type, as previously described, have great differences in their circulation patterns, salinity stratification and mixing processes. The classification criteria according to the salinity stratification may enable the main circulation characteristics in the MZ to be quantitatively established, taking into account the great majority of coastal plain estuaries. This classification considers the dominant terms of the steady-state equation of salt conservation, establishing the balance of the advection and diffusion terms, which will be studied in detail in Chap. 7. According to the vertical salinity stratification the following estuary types were defined: *salt wedge* (type A), *moderately or partially mixed* (type B), *vertically well-mixed and laterally stratified* (type C) and *well-mixed* (type D). The transition between these estuarine types is dependent upon the river discharge, tidal amplitude, the baroclinic component of the gradient pressure force and the geometric characteristics as the ratio of width/depth (Pritchard 1955; Cameron and Pritchard 1963).

3.2.1 Salt Wedge Estuary (Type A)

Salt wedge estuaries are typical in regions that experience micro and meso tides and high river discharge. They are mainly dominated by river discharge, and the entrainment process causes a seaward increase in salinity in the upper layer, while the mixing by turbulent diffusion is neglected, and a slow landward circulation in the bottom layers is observed. Due to variations in the river discharge intensity and the tidal forcing, the salt water wedge doesn't remain stationary but has slow displacements (Stommel 1953; Geyer 1986).

Due to the continuous seaward motion in the surface layer, which may be deflected by the Coriolis acceleration, the velocity shear in the interface (fresh and salt water) may led to an entrapment of some salt water from the wedge into the upper fresh water layer. In this situation there is little or no mixture of fresh water into the salt wedge. The upper layer now increases its salt water volume as it moves seaward, and there must be a slow upstream movement of water in the salt water wedge to compensate for the loss upward into the fresh water. According to Pritchard (1955) laboratory studies of salt wedge estuaries performed by Keulegan (1949), utilizing flumes of various sizes, described this process well. When the relative velocity between these interfaces becomes very intense, the interfacial Froude number (F_i—Eq. 2.38, Chap. 2) reaches the critical value ($F_i = 1.0$); in this condition, internal breaking waves occur in the interface generating turbulent motion which produces the mixing of the upper and lower layers. In such an estuarine system, the two dominant terms in the salt balance would be the horizontal and vertical advections. For more details see Pritchard (op. cit.).

This estuarine type is schematically shown in Fig. 3.3 and, in agreement with the experimental results of Keulegan (1949) it follows that: the mass continuity is

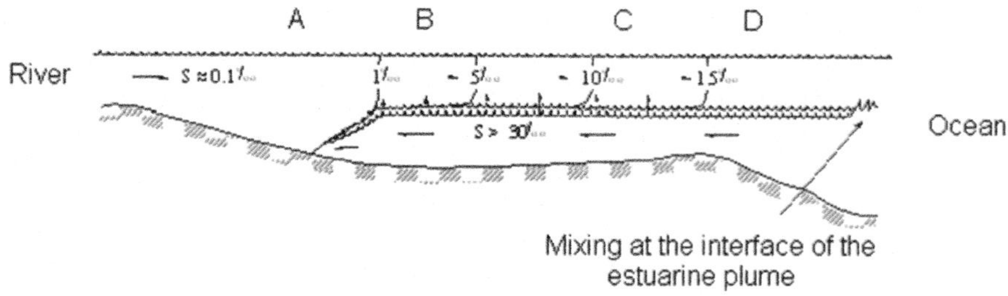

Fig. 3.3 Salt wedge estuary diagram. The *vertical arrows* in the interface between the bidirectional motions, indicate the entrainment process occurrence (according to Pritchard 1989)

preserved by the low intensity motion of the salt-wedge up-river, to replace the sea water parcel that is advected seaward in the upper surface. This process adds salt water into the estuary's upper layer separated by sharp halocline (and hence picnocline) and the volume transport and salinity increase towards its mouth. The depth of null velocity due to the bidirectional motion is located in the halocline, and the thickness of the salt water layer is controlled by the critical interfacial Froude number (Stommel and Farmer 1952).

Longitudinal and vertical distributions of salinity and velocity vertical profiles, in the hypotheses of absence and presence of shear stress due to the interfacial friction are presented in Fig. 3.4a, b. In this type of estuary, in which the width/depth ratio is usually great, the flux ratio (the fresh water volume discharged during a tidal cycle is greater than the input of sea water volume during the low and high water) is great, showing that the circulation is dominated by the river discharge.

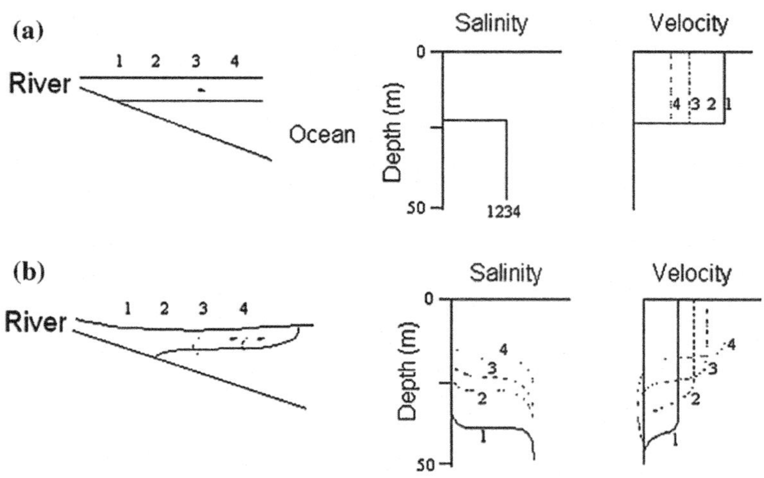

Fig. 3.4 Schematic diagrams of the salinity stratification and velocity profiles in a salt wedge estuary. Conditions (**a**) and (**b**) indicate without and with interfacial friction, respectively (according to Dyer 1973)

The interfacial Froude number, which is an indicator of the vertical stability, was calculated for the South Channel of the delta of Mississippi river (Mississippi, USA) in different positions in the salt-wedge, and the values obtained varied from 0.41 (in the upper region of the salt-wedge) to 0.69 and 0.80 in the mouth, showing a stable vertical stratification at the wedge interface (Wright 1970).

Analytical uni and bi-dimensional models of salt wedge estuary will be studied in Chap. 9. Specifically, the circulation and the upland distance of the salt-wedge penetration will be determined by considering the longitudinal pressure gradient force and intensity of the fresh water input, which is controlled by the friction at the river/salt water interface. Classical examples of salt-wedge structures have been observed in several estuaries, such as: Velar estuary in India (Dyer and Ramamoorthy 1969), the delta of the Mississippi river (Wright 1970), Duwamish river estuary (Dawson and Tilley 1972) in the USA, and the estuary of the Fraser river in British Columbia (Canada) which has high fresh water discharge and is forced by mesotides (Geyer 1986; Geyer and Farmer 1989).

As a synthesis of what has been described above, the salt wedge estuary is characterized by high river discharge, low tidal currents and entrainment, generating a highly vertical stratified estuary. Further knowledge on the steady-state dynamical characteristics of this estuary and the salt-wedge intrusion length will be studied in Chap. 9.

3.2.2 Moderately or Partially Mixed (Type B)

With the co-oscillating tidal forcing, the entire water mass in the estuary will be periodically agitated. This occurs even at small tidal heights; however, the agitation will only be intense enough to bring about an accentuate erosion of the halocline when the flux ratio is small (<1). Estuaries with moderate vertical salinity gradients are classified as partially mixed (or type B), and tidal mixing plays an important role in the circulation in this estuary type (Pritchard 1955).

The tidal kinetic energy involved in this oscillating process must be enough to not only work against the stabilizing action of the buoyancy, but also to generate turbulent diffusion at the fluid interface and the friction at the geometric limits of the estuary. The energy of the turbulent vortices generated at the interface will be dissipated, and work against the stabilizing buoyancy forces, eroding the vertical salinity gradients and producing the mixing of the fresh and salt water. Then, there will be an increase in the potential energy of the water due to the increase in the salinity (density) of the upper layer, and, consequently, an increase in the seaward volume transport above the halocline. Simultaneously, due to the mass continuity, the opposite volume transport of salt water will also be increased below the halocline, developing a bidirectional two layer flow up and down the estuary (Fig. 3.5). This phenomena has been observed by Pritchard (1955) who estimated that in steady-state conditions, the volume transport towards the sea and through the estuary mouth of the James river estuary (USA), in the upper layer above the

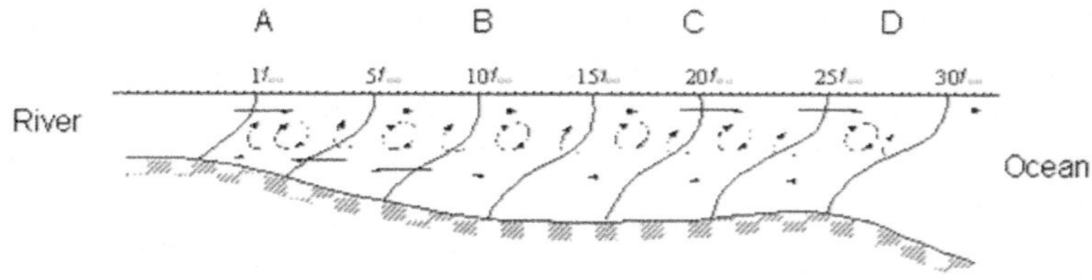

Fig. 3.5 Longitudinal salinity distribution and circulation in a partially mixed estuary (according to Pritchard 1989)

halocline is twenty times larger than the river discharge ($20Q_f$), and a compensating motion up-estuary below the halocline with a volume transport equivalent to nineteen times the river discharge ($19Q_f$) has been estimated.

The bidirectional circulation that occurs in the water column is much less intense than the oscillatory tidal currents. As these circulation motions are superimposed, the first motions can only be detected by calculating the steady-state temporal averages during one or more complete tidal cycles.

The longitudinal salinity distribution shows similar vertical profiles; however, the vertical salinity gradients in the halocline are greater in the seaward direction (Fig. 3.5). The bottom geometry and the occurrence of transversal (or secondary) circulation may alter the typical configuration. Due to the increase and decrease of the potential energy in the layers above and below the halocline, respectively, the isobaric surfaces have ascending configurations in the upper layer and in the opposite direction in the bottom layer, forced by the estuarine circulation. These motions may also be influenced by the Coriolis acceleration, mainly in wide estuaries and in regions where the diurnal tidal component is predominant.

This estuary type will be studied with uni and bi-dimensional analytical models. These models have the objective to determine the circulation and mixing processes, as well as to obtain steady-state vertical profiles of velocity and salinity forced by gradient pressure forces, fresh water discharge and wind stress.

As the result of the more efficient water exchange of the river and the adjacent sea, due to the turbulent diffusion, the salinity stratification mainly changes in the following frequencies: (i) *intratidal*, due to the diurnal or semi-diurnal periodic tidal variation, and *subtidal* (fortnightly) which is due to the time period of the modulation between successive spring tides ($\approx$15 days), caused by the constructive interference of the main semi-diurnal tidal components M_2 and S_2, and; (ii) on a seasonal time scale (*subtidal variability*) due to the variations in the river discharge.

3.2.3 Vertically Well-Mixed (Types C and D)

These estuaries are formed in former shallow river valleys. They have low fresh water discharge and are forced by meso or macro tides, generating a great bottom turbulent shear stress vigorous enough to vertically mix the water column and produce the halocline erosion. In natural conditions these estuaries have low vertical salinity stratification; the upstream flux of salt necessary to balance the downstream advective flux, due to the river flow through the estuary, must result from the non-advective flux (Pritchard 1955; Dyer 1973). The intensity of the longitudinal salinity (density) gradient is less than in partially mixed estuaries, and its low intensity is unable to generate gravitational circulation. Thus, the steady-state circulation in a well-mixed estuary is mainly unidirectional and seaward direction, and because the intensity of the longitudinal density gradient force is obtained by vertical integration (Eq. 2.10a, b), a vertically homogeneous estuary that is relatively deep may also develop a weak gravitational circulation.

Well-mixed estuaries may be separated into the following sub-types:

(a) Laterally Stratified (type C)

When the estuary has a relatively large width/depth ratio, the Coriolis acceleration deflection may generate a lateral salt stratification. The resulting seaward cyclic circulation and the up-estuary circulation is intensified to the left in the Southern Hemisphere, generating lateral salinity variations, although the water column may remain almost vertically homogeneous (Fig. 3.6).

(b) Non-Laterally Stratified (type D)

In narrow tidal rivers, the vertical and lateral shear stress in their boundaries may be strong enough to generate laterally homogeneous salinity stratification. In this condition, the salinity increases gradually down-estuary and the steady-state motion at all are oriented in this direction at all depths. Although the tendency of this advective motion is to generate a down estuary salt transport, it is in balance with the up-estuary turbulent diffusion (or salt dispersion) associated with topographic irregularities and the bottom shear. Some salt water may be trapped in the inner embayment during the flood and return to the main channel during the ebbing current.

In this estuarine type, the current shear, which is homogeneous in the transversal section, generates and intense turbulent diffusion that is higher than the previously presented estuarine types. Although the tidal wave has a predominant progressive component, the flooding and ebbing maxima currents' intensities are usually out of phase from the high and low tide, respectively.

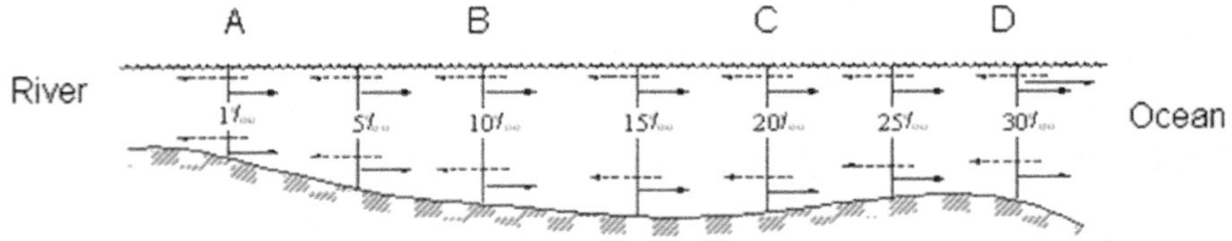

Fig. 3.6 Salinity distribution and circulation in a well-mixed estuary with small lateral salinity stratification (according to Pritchard 1989)

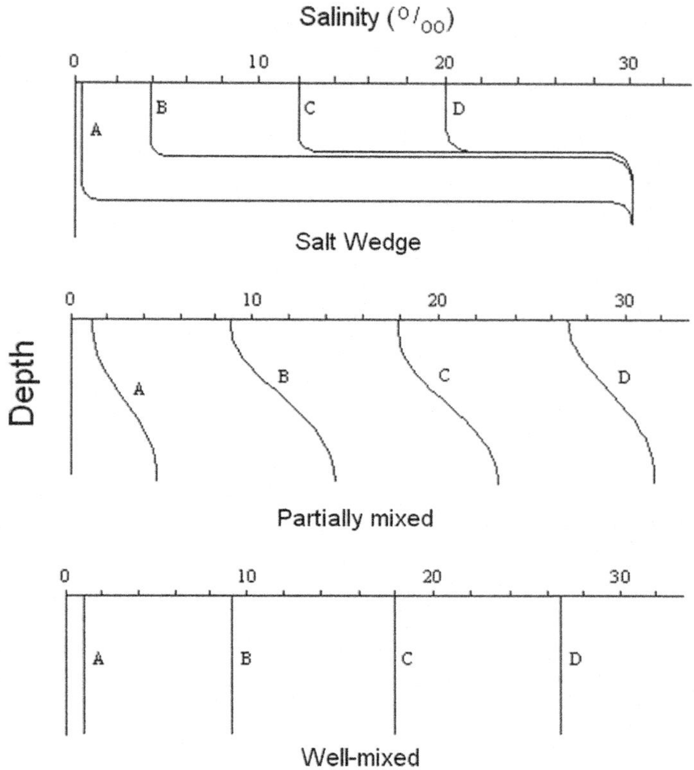

Fig. 3.7 Vertical salinity profiles of salt wedge, partially mixed and well-mixed estuaries. A, B, C and D are longitudinal positions from the head down to the mouth (according to Pritchard 1989)

The important physical parameters that control the sequence of estuarine types are river flow, tide, and mean depth, width and length of the estuary. Tidal modulation of these parameters and the influence in the estuarine classification have been evaluated by Pritchard (1955).

The longitudinal variations in vertical salinity profiles from the head to the mouth (at positions A, B, C and D) for different estuary types are presented in Fig. 3.7. This figure shows the vertical salinity gradient attenuations in the following estuary types: salt wedge (dominated by the river discharge), partially mixed (dominated by the fresh water discharge and the tide), and the vertically homogeneous (dominated by the tide). Of course, between the salt wedge and well-mixed types, there will be a sequence of estuarine types characterized by vertical salinity gradients.

The classification criteria according to the geomorphology and salinity stratification, although separately studied, have some agreement between them. For instance, *coastal plain estuaries* are generally *partially mixed*, but may have a tendency to be well-mixed if the fresh water discharge decreases and the tidal input changes to macro or hyper tide. In some regions of high river input and micro or meso-tides, salt wedge and highly stratified estuary types prevail.

Among the first quantitative criteria for estuary classification was that of Simmons (1955)—contemporary to the Pritchard's paper—using the flux ratio number, $F_R = Q_f T_P / T_{PR} = R/T_{PR}$, the ratio of the fresh water volume discharged into the estuary during the tidal cycle to the tidal prism. The hypothesis was that the estuarine types form a continuous sequence, which should be classified by the flux ratio. This value, defined in Chap. 2, was introduced as a control number to indicate changes in the vertical salinity stratification due to the predominance of the river discharge or the tide forcing. Using this ratio and taking into account experimental data of several estuaries, the following sequence of estuarine types was established:

- *Highly stratified*, when the flux ratio is equal to or higher than one ($F_R \geq 1.0$).
- *Partially mixed* and *well-mixed*, when the flux ratio values are almost 0.25 and less than 0.1, respectively.

The Simmons (1955) results to classify the estuarine type sequence are too general, because the geometric characteristics of the estuaries (width, bottom topography and nature) also have influence in generating the turbulent diffusion that controls the salinity stratification that a certain tidal height may produced. For example, according to Dyer (1973), the flux ratio of the Mersey river estuary (England) is close to 0.01 and 0.02, suggesting a well-mixed estuary, using the Simmons's classification criteria. However, by applying the vertical salinity stratification criteria, it is classified as a partially stratified estuary due to the measurable differences of salinity between the bottom and surface.

It should be pointed out that the estuary classification may be time variable, and dependent on the position along the estuary. Considering the estuary head as the transitional limit of the MZ and TRZ, the tidal amplitude reach this region very attenuated and the river discharge is the main forcing, the entrainment is predominant and vertical stratification will remain relatively high. Downstream regions in the estuary, tidal amplitude often increases, as well as the vertical turbulent diffusion, and the estuary may be classified as partially mixed; however, at the estuary mouth the high tidal amplitude intensify the vertical mixing.

The variability of the vertical salinity stratification is a complex process on a large temporal scale. In order to classify an estuary, using classification criteria, we must use steady-state data. However, the intensity variation of the tidal currents between the neap and spring tidal cycles (fortnightly tidal modulation) usually have a significant influence on the vertical salinity stratification, and hence, on the estuary classification. On a larger temporal scale, the seasonal variations in the fresh water discharge also influence the vertical salinity stratification and the criteria to be applied to the estuary classification.

As mentioned already, geometric changes related to the width, depth, length and bottom characteristics may alter the circulation intensity and the mixing of the fresh and the salt water, according to the following sequence: when the river discharge and the tidal amplitude remain constant, but the estuary width increases, the *flux ratio* will decrease because of the tidal prism increase, and this result looks similar to a decrease in the river discharge. Due to these changes, the vertical salinity stratification decreases, and the estuary tends to be less stratified or eventually well-mixed. In the hypothesis of an increase in the estuary depth, the *flux ratio* remains unmodified, but as the river induced velocity decreases, the vertical mixing (in relation to the former case) and the vertical salinity stratification may evolve to a partially mixed estuary. These effects may occur along the length of the estuary, and when a given cross section has its area reduced, the currents (tidal and river discharge) increase due to the mass (volume) continuity, and the vertical shear may promote halocline erosion. Consequently, the vertical salinity stratification tends to be less intense than in cross sections with higher area.

The problem with the salinity stratification classification criteria is that there is no an agreement concerning the salinity gradient in the halocline which can be used to determine the following transitions: highly stratified/partially mixed and partially mixed/well-mixed. Therefore, it is convenient to introduce the following suggestion to classify estuaries, according to Officer (1977):

Salt Wedge: there is a distinct interface separating the river discharge from the salt water, with a low water flux into the upper layer, indicating that the entrainment process is predominant.
Highly Stratified: the salinity profile has a strong vertical stratification, and the difference between the bottom and the surface salinities has several unities.
Partially Mixed: the salinity profile has a moderate vertical stratification, and the salinity difference between the bottom and the surface salinities has a few unities.
Well-mixed: there is practically no difference between the bottom and surface salinities.

The longitudinal salinity structure in steady-state conditions and its circulation in the salt wedge, partially mixed and well-mixed estuaries are presented in Fig. 3.8.

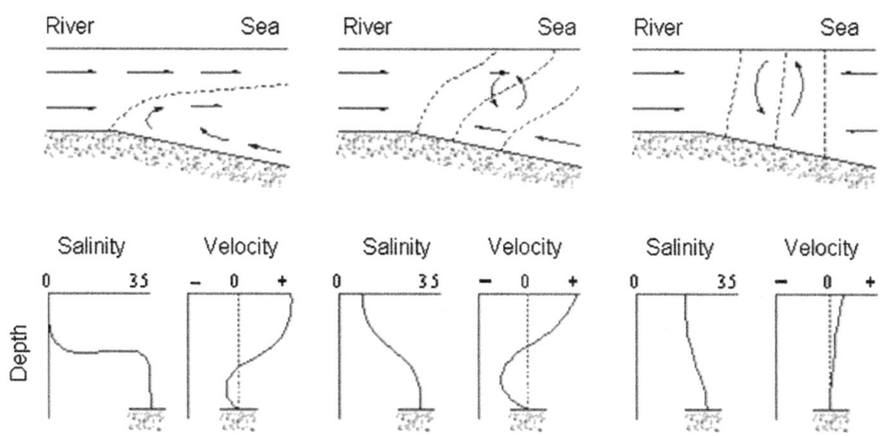

Fig. 3.8 Characteristics of the circulation, salinity stratification and typical mean vertical salinity and longitudinal velocity profiles of the main estuary types classified according to the salinity stratification, in the following sequence from left to right: salt wedge, partially mixed and well-mixed (according to Morris 1985)

In the salt wedge estuary the vertical salinity profile has the halocline with strong gradient, the circulation is dominated by the river discharge, the entrainment generates the small salinity increase in the surface layer, and the slow motion of the salt-wedge is up-estuary. The partially mixed is forced by both the river discharge and the tidal currents, and its steady-state circulation occurs in two layers. In well-mixed estuaries the vertical salinity gradient is almost zero, an indication that this estuary is mainly forced by the tide, and its mean circulation is unidirectional and seaward and with low intensity.

3.3 Classification Diagrams

The first theoretical classification studies using uni-dimensional parameters were published in the 1960 decade, and further advances in estuary classification have more recently been added to the oceanographic literature. Included among these pioneering studies, whose results were compared with experimental data from natural estuaries, and physical models with different characteristics of salinity stratification, circulation and river discharges, were the works of Ippen and Harleman (1961), Harleman and Abraham (1966) and Hansen and Rattray (1966). The methods used share a common characteristic of the salinity stratification, adopting steady-state conditions and a simple geometry. However, in addition to the salinity structure, the classification diagrams take into account the following processes contributing to the vertical mixing: energy dissipation, potential energy gain by the water column, stationary velocity on the surface, velocity generated by the input of river discharge, and the gravitational circulation.

The first analytical method to classify estuaries using a stratification parameter was developed by Ippen and Harleman (1961) taking into account theoretical and semi-empirical results, obtained in idealized estuaries and with physical models simulated in channels. These researchers demonstrated experimentally that the vertical mixing is related to the energy lost by the tidal progressive wave propagation (G) and the corresponding increase in the potential energy gain of the water column (J) due to the density (salinity) increase of the estuarine water mass. These energy fluxes, per width and mass unity along the estuary length, are given by:

$$G = \frac{1}{2}c_0\left(\frac{\eta_0}{H_0}\right)^2\frac{gH_0}{L}\frac{\sinh(2\mu L)}{\cos(2\kappa L)}, \tag{3.1}$$

$$J = \left(\frac{\Delta\rho}{\rho}\right)gH_0\frac{u_f}{L}, \tag{3.2}$$

and its non-dimensional ratio, named *stratification number*, is expressed by,

$$\frac{G}{J} = (\frac{\rho}{\Delta\rho})\frac{1}{2}\frac{c_0}{u_f}(\frac{\eta_0}{H_0})^2\frac{\sinh(2\mu L)}{\cos(2\kappa L)}, \quad (3.3)$$

where $\Delta\rho$ is the density difference between the river and seawater and $\Delta\rho/\rho$ is a measure of the channel stratification, u_f is the river discharge, H_0 and L are the estuary depth and length, c_0 and η_0 are the phase velocity and the amplitude of the simulated tidal wave, μ and κ are related by $\mu = (\chi/2\pi)\kappa$, with χ is a constant denoting a energy dissipation.

The physical quantities, G and J in the above expressions, are given in units of the energy per mass and time unities, [G] = [J] = $[L^2T^{-3}]$, and the ratio G/J was named *stratification number*, which is physically analogous to the estuarine Richardson number (defined in Chap. 2) and used as a criterion to indicate the stability of fluids with density stratification. When the relative stratification $(\Delta\rho/\rho)$ decreases, the ratio G/J increases, and in the limit-case $\Delta\rho \to 0$ it tends to the infinite (G/J $\to \infty$), characterizing an estuary that is vertically homogeneous; however, when this ratio decreases (G/J $\to$ 0), there is an increase in the potential energy of the water column and the estuary becomes highly stratified. This has been confirmed by Ippen and Harleman (1961) results: for G/J ratio values from 20 and 200 the estuary is highly stratified and well-mixed, respectively. In the comparison of the flux ratio $F_R = (T_PQ_f/T_{PR})$ (Eq. 2.43, Chap. 2), with the G/J, we may observe that both are dependent on the river velocity (u_f), but with opposite behavior. In turn, experiments in physical models, forced by distinct tidal amplitudes, indicated variations in the stratification number for the same flux ratio. Therefore, the flux ratio isn't a criterion to indicate similarities in relationship to the vertical diffusion (stratification) characteristics; rather the non-dimensional (G/J) number is more appropriate.

Without going into the theoretical details, which will be presented in Chap. 11, the classical methodology which uses the orthogonal system of Cartesian coordinates, named Stratification-Circulation Diagram, deduced by Hansen and Rattray (1966) will be now presented. Various authorities in the field of Physical Oceanography of estuaries—Dyer (1973, 1986, 1997), Officer (1976), and Pritchard (1989), among others, consider this to be one of the best methods for estuary classification. Its theoretical formulation was confirmed with the introduction of alternative parameters by Fischer (1972), Prandle (1985) and Scott (1993).

The coordinate axes of this diagram are non-dimensional parameters: the *stratification parameter* $(S_b - S_s)/\bar{S} = \delta S/\bar{S} = S_p$, where S_b, S_s and $\bar{S}$ are the steady-state salinity values at the bottom, surface, and the mean-depth value, respectively, measuring the vertical stratification of the water column, and the *circulation parameter* $(u_s/u_f) = C_P$, which is defined as the ratio of the steady-state surface velocity, u_s, by the velocity generate by the river discharge, u_f. When the river discharge isn't adequately known, instead of u_f the mean value (in time and depth) of the velocity in the water column, known as residual velocity it may be

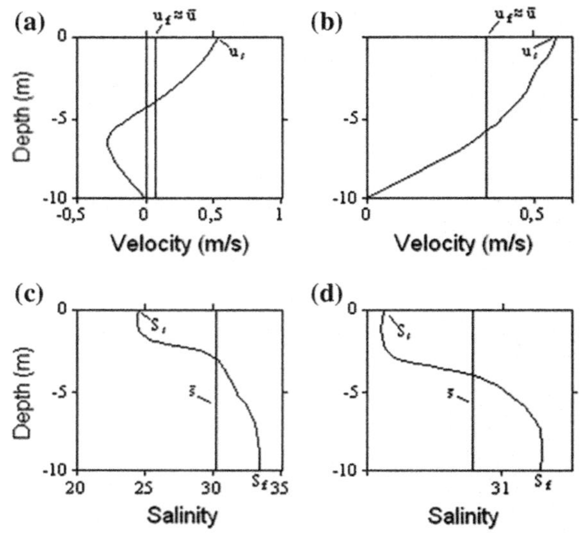

Fig. 3.9 Time-mean vertical velocity and salinity profiles and the definition of the quantities $u_f \lesssim \bar{u} \geq u_a$ and u_s, and S_s, S_b, which are necessary to calculate the circulation (u_s/u_f) and the stratification ($\delta S/\bar{S}$) parameters. Partially mixed (**a**–**c**) and well-mixed (**b**–**d**) estuaries

substituted by the steady-state mean-depth value of the velocity measured in the water column, known as residual velocity $\langle \bar{u} \rangle = u_a$ is used. This velocity is formally defined as $\langle \bar{u} \rangle = u_a = (1/\Delta T)[\int_0^{\Delta T}[\frac{1}{H}\int_0^H u(z,t)dz]dt$, where ΔT is a time interval of one or more tidal cycles, and H is the local depth; the usual practice to determine these parameters is illustrated for partially stratified estuaries and well-mixed estuaries in Fig. 3.9a, c and b, d, respectively.

The circulation parameter (C_P) is a measure of the gravitational circulation, and its variation is from the theoretical value of 1.5, typical for well-mixed estuaries, up to 10^3 (for partially stratified estuaries), and reaching higher values as 10^5 for fjords. The inequality $u_s > u_f$ is due to the fact that numerator of this non-dimensional number is the sum of the current velocity generated by the river discharge (u_f) plus the velocity generated by the gravitational circulation due to the entrainment and mixing processes that elevate salt water into the upper layer. Typical values for the stratification parameter S_P varies between 1.0×10^{-4} and a few unities, for low and high stratified estuaries.

In deep estuaries, such as fjords, the gravitational circulation is confined in a relatively shallow surface layer, and, as the velocity generated by the fresh water discharge usually has very low values, the circulation parameter may reach values up to 10^5. In the opposite situation, there is the well-mixed estuary dominated by tidal forcing; in this estuary, the residual velocity on the surface layer tends to the velocity generated by the fresh water discharge ($u_s \rightarrow u_f$) and, consequently this parameter approaches unity (Fig. 3.9b) almost confirming the theoretical value $u_s/u_f = 1.5$.

The deduction of the Stratification-circulation diagram was applied for a laterally homogeneous (narrow) estuary under steady-state conditions. Thus, the quantities, S_b, S_s, $\bar{S}$, u_s and u_f (or $\langle \bar{u} \rangle \approx u_a$), used in the definition of the stratification and circulation parameters are for narrows estuaries, must be time independent, and, in practice, they are calculated using time mean values for one or more tidal cycles (Fig. 3.9a–d).

The Hansen-Rattray solution includes another non-dimensional parameter, ν, and shows that this parameter is linked to the salt flux $[ML^{-2}T^{-1}]$ processes by the equation $\rho K_{x0}(\partial S/\partial x) = \nu\rho u_f\bar{S}$ (Eq. 11.96, Chap. 11). The first term of this equation represents the landward salt flux by all processes other than gravitational circulation and the seaward salt flux due to the river discharge fraction (named by Fischer (1972) as non-gravitational). Thus, the difference (1 − ν) is the salt flux due to the gravitational circulation.

The stratification $\delta S/\bar{S}$ and circulation u_s/u_f parameters are theoretically linked to the mixing parameter (ν) which represents the relative proportion of the up estuary salt transport due to advection and turbulent diffusion, which is formally expressed by: $\nu = \nu(\frac{\delta S}{\bar{S}}, \frac{u_s}{u_f})$, or, $\nu = \nu(S_P, C_P)$]. As will be demonstrated in the Chap. 11, this correlation is given by the following second order algebraic equation (Hansen and Rattray 1966):

$$\left(\frac{\delta S}{\bar{S}}\right)^{-1}[210+252(\frac{u_s}{u_f}-\frac{3}{2})]\nu^2+[32-(\frac{\delta S}{\bar{S}})^{-1}(210+252(\frac{u_s}{u_f}-\frac{3}{2}))$$
$$+76(\frac{u_s}{u_f}-\frac{3}{2})+\frac{152}{3}(\frac{u_s}{u_f}-\frac{3}{2})^2]\nu=0. \quad (3.4)$$

The parameter ν, which is the unknown of this equation, vary in the interval from zero (0) to one (1), that is, $0 < \nu \leq 1$, indicating, according its definition, that the salt transport is generated exclusively by the advective process or due to the turbulent diffusion (dispersion); for ν = 1, the salt transport up estuary is generated exclusively by turbulent diffusion and, for ν = 0, the Eq. (3.4) has no physical meaning; but in the case of the salt transport being generated only by the turbulent diffusion (ν = 1), the equation is reduced to:

$$32+76(\frac{u_s}{u_f}-\frac{3}{2})+\frac{152}{3}(\frac{u_s}{u_f}-\frac{3}{2})^2=0. \quad (3.5)$$

This second order equation in the variable (u_s/u_f − 3/2) only has a real solution if the numeric term 32 is neglected. If this is the case, the equation has two solutions: $u_s/u_f = 0$ and (u_s/u_f = 3/2 = 1.5). The first solution has no physical meaning, and the second solution indicates that if the up-estuary salt transport is only generated by turbulent diffusion (ν = 1) the solution is independent of the stratification parameter. Then, disregarding the numeric constant, 32, from Eqs. (3.4) and (3.5), it is possible to draw a set of parametric curves with the parameter ν = const. For this, we shall calculate the corresponding values of the stratification parameter using fixed values of the circulation parameter ($C_P = u_s/u_f$) with ν varying from 0.01 to 1.0. Thus, it is possible to generate graphically the correlation $\nu = (S_P, C_P)$ in a orthogonal reference system with the parameters S_P and C_P as ordinate and abscissa axis, respectively, which, according to Hansen and Rattray (1966), is called Stratification-circulation diagram. The solution of Eq. (3.5), determines the relative contribution of the tidal diffusive (ν) and the advective ($1 - \nu$) processes, which are responsible for the up-estuary salt transport. From these results it follows:

(i) When $\nu = 1$, there is no gravitational circulation and the up-estuary salt transport is only due to the tidal turbulent diffusion;
(ii) When $\nu \rightarrow 0$, the tidal turbulent diffusion is negligible and the advective process alone is responsible for the up-estuary salt transport.

The Stratification-circulation diagram with its coordinate ($\delta S/\bar{S}$) and ordinate (u_s/u_f) axes and a set of isolines of the parameter ν (ν = 1.0, 0.90, 0.70 and 0.01) is presented in Fig. 3.10. The S and Q points represent these parameters obtained with experimental data from the estuarine Channel of Bertioga (Santos-São Vicente Estuary) and indicate theoretically the relative up-estuary salt transport due to the advection and turbulent diffusion. These points were obtained from observations during two complete tidal cycles of neap (Q) and spring (S) conditions, and represent nearly steady-state conditions. In the neap tide the, ν parameter is equal to 0.7 and indicates that the processes of advection (0.3) and turbulent diffusion (0.7) were important to the up-estuary salt transport. However, in the spring tide, the tidal currents caused the more mixing between fresh and salt water, this is due to an

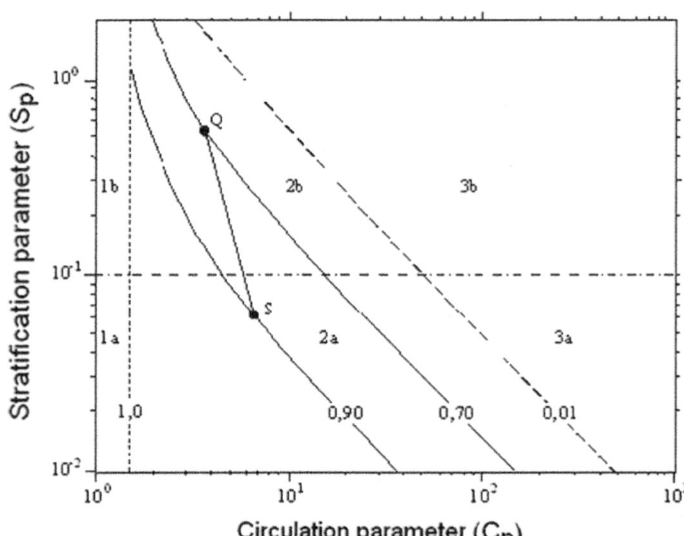

Fig. 3.10 Theoretical parametric curves of the mixing parameter ($0 < \nu \leq 1$) due to diffusive and advective processes obtained using Eq. 3.6, with the correlation of the stratification ($\delta S/\bar{S}$) and circulation (u_s/u_f) parameters. Q and S are the images of these parameters in neap and spring tidal cycles, respectively, in the Bertioga estuary (Santos-São Vicente Estuary, SP). Diagram axis in log-log scales (adapted from Miranda et al. 1998)

increase in the turbulent diffusion, with the parameter increased to ν = 0.9. Consequently, the tidal generated mixing process dominated the salt transport and the advective contribution decreased to 0.1.

A simple procedure for estimating the parameter (ν) was presented (Officer and Kester 1991). In this estimate were used monthly mean salinity, ocean salinity and freshwater input from observations at different times of the Narragansett Bay (Rhode Island, USA). The analysis of these data indicated variations from low values, in the winter $\nu \approx 0.3$, up to higher values in the summer $0.6 < \nu < 0.8$ indicating a salt turbulent tidal diffusion more efficient.

The physical interpretation of the circulation and mixing processes related with the theoretical results presented in Fig. 3.11, were further used by Hansen and Rattray (1966) to improve this diagram interpretation, aiming to establish the Stratification-circulation Diagram as a quantitative estuary classification. As the stratification parameter is a measure of the vertical stratification of the water column, the value 10^{-1} was adopted to indicate the transition between high (type b) and a low (type a) stratification of estuaries. Hansen and Rattray (op. cit.) confirmed this theory by analyzing an experimental data set from several estuaries, from which four previously classified estuarine types emerged.

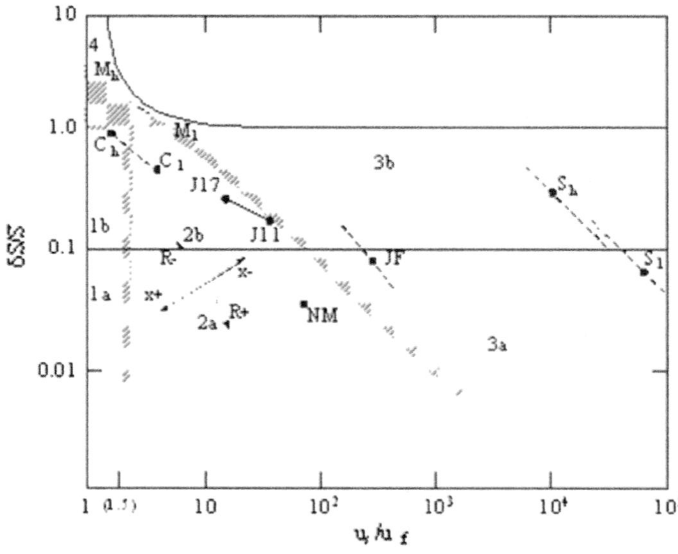

Fig. 3.11 Stratification-circulation diagram of Hansen and Rattray (1966). X+ and X− indicate longitudinal variations towards the mouth, respectively, R+ and R− indicate increase and decrease in the river discharge. Experimental data from various estuaries in conditions of high (h) and low (l) river discharges, and figures also indicate the longitudinal distance from the mouth. $M_{h,l}$, $C_{h,l}$, $J_{17,11}$, NM, JF and $S_{h,l}$ are from the following estuaries: Mississippi river mouth, Columbia river, James river, Narrows of the Mersey river, Strait of San Juan de Fuca and Silver Bay (Diagram axis in log-log scales according to Dyer 1997)

According to Hansen and Rattray (1966) estuaries may be classified as the following types (Fig. 3.11):

Type 1—Well-mixed estuary. The residual flow is seaward at all depths (unidirectional circulation), and the upstream salt transfer is effected by turbulent diffusion with no upstream salt transport by gravitational circulation.
Type 2—Partially mixed estuary. The residual flow reverses at depth (bidirectional circulation) and both advection and diffusion processes contribute to the upstream salt flux.
Type 3—Fjord. It is distinguished from Type 2 primary due to the dominance of gravitational circulation (advection), accounting for over 99% of the upstream salt transfer.
Type 4—Salty wedge estuary. The stratification is still greater than other estuary types, and the flow undergoes a gradual change towards the mouth from a thick upper layer flowing over a thin lower layer near the head, and changes to a thin upper layer flowing over a thick lower layer near the mouth.

Of course, as pointed out by Hansen and Rattray (1966), a certain arbitrariness was necessary in the separation of classes. Subdivisions (a) and (b) for types 1, 2 and 3 mean low and high stratification when $S_P < 0.1$ and $S_P > 0.1$, respectively. Types 2a and 2b, previously classified as partially mixed (low and highly stratified), include the majority of the coastal plain estuaries, and the inclusion of types 3a and 3b (fjord) and 4 (salt wedge) in the Stratification-circulation Diagram, were based on experimental results because the theoretical analysis would not apply to this type of estuary. Further studies on gravitational circulation in fjords may be found in Hamilton and Rattray (1978).

As we have seen, the theoretical foundation of the Stratification-circulation Diagram (Fig. 3.11) has been confirmed by experimental data processed according to the theoretical hypothesis. The circulation and stratification parameters represented in the diagram were based on experimental results of several estuaries, confirming its great potential for estuary classification. In this figure it is possible to observe that, under different intensities of the river discharge and at different longitudinal positions along the estuary length, the parameters $(\delta S/\bar{S})$ and (u_s/u_f) migrate on the diagram. This conclusively shows that straight line segments, rather than points, better define the estuary classification, enabling to the main characteristics of its driving forces and geometry to be taken into account.

Similarly it is also expected that a straight line segment will better classify estuaries as a result of the variations in tidal current intensity which result from the fortnightly tidal modulation generating spring and neap tidal cycles. This is illustrated in Fig. 3.12 with experimental results from the Bertioga estuarine channel, where appreciable fortnightly tidal modulation was observed (Miranda et al. 1998). In this figure, the channel is classified as type 2 (partially mixed), but it alternates between type 2a (moderately stratified) and type (2b) (highly stratified) for the spring (S) and neap (Q), respectively. As seen in the figure, in the first event, turbulent diffusion and advection were active in the upstream salt flux, with relative

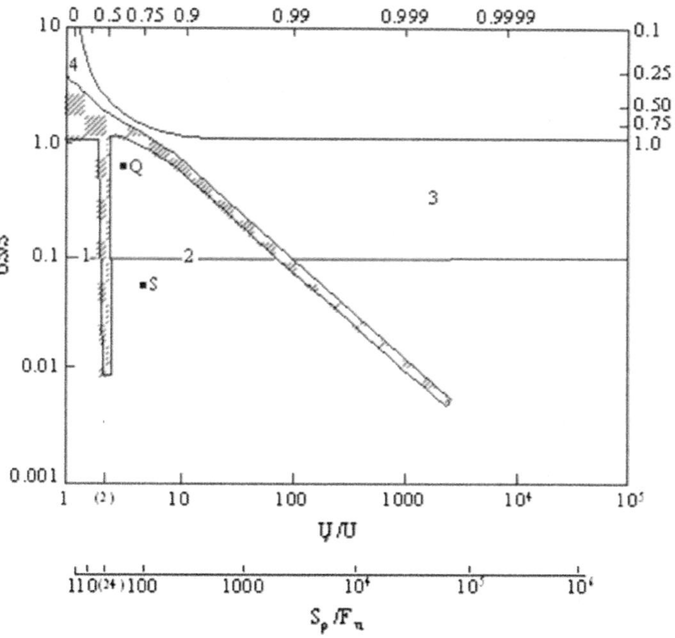

Fig. 3.12 Stratification-circulation Diagram with the classification of the Bertioga channel (Santos, SP, Brazil), for the spring (S) and neap (Q) tidal cycles. The alternative axis for the circulation parameter (S_p/F_u), as proposed by Prandle (1985) was included. Diagram axis in log-log scales

contributions of 70% (ν = 0.7) and 90% (ν = 0.9) in the neap and spring tide, respectively. Others examples of this tidally modulate process may found in Andutta (2011) and Andutta et al. (2013), for the Caravelas-Peruípe estuarine system (Southern Bahia State, Brazil).

This classification diagram has also been used to classify the estuarine channel which links the Patos Lagoon (South of the Rio Grande do Sul, Brazil) to the sea. A data series of experiments from 1988 was used, the channel was classified as type 1b (well-mixed), with the salt balance dominated by turbulent diffusion. However, two exceptions were observed due to the abnormal forcing of cold meteorological fronts, when the estuarine channel classification alternated from being type 2b (partially mixed) to type 4 (salt wedge), dominated by the increase in the river discharge due to the post-frontal condition (Möller 1996; Möller and Casting 1999).

In the upper reaches of the Piaçaguera estuarine channel (Santos, Brazil), little change in estuary classification was observed during the fortnightly tidal modulation in the winter which remaining as type 2a, and the associated estuarine Richardson number was ≈1.6 (Miranda et al. 2012).

As another examples of this diagram application, to classify a more complex estuarine environment as the Paranaguá Bay (Paraná, Brazil), we should mention the preliminary investigation of Knoppers et al. (1987), which classified the bay as type 2a (partially mixed and low stratified), and the more elaborated investigation of Mantovanelli (1999) and Mantovanelli et al. (2004), which used intensive measurements of eight semidiurnal tidal cycles during the 1997 winter and in the 1997/1998 summer, comprising two spring and two neap cycles in each season. The sampled transversal section was localized at 25° 30′S 030° 42′W, and the results indicate that the salinity stratification and the mixing processes were controlled primarily by tidal currents and secondarily by seasonal freshwater discharge and the fortnightly tidal modulation. As the results related to this topic, the estuarine characteristics of the investigate region changed in the summer from being type 2b (partially mixed and high stratified) to type 2a (partially mixed and low stratified) in the spring and the neap tidal cycle, respectively. In turn, in the winter observation period it changed to type 1 (well-mixed) in the spring tidal cycle.

The circulation and the salinity stratification for well-mixed and partially mixed estuaries are dependent on the river discharge, the tidal oscillation and the estuarine geometry. These global characteristics may be related to the stratification-circulation diagram using the following dimensionless bulk parameters: the Froude densimetric number defined as $F_m = u_f/(gH_0\rho^{-1}\Delta\rho)^{1/2}$, which expresses the ratio of the river flow to the potential density-induced internal circulation, and the parameter calculated by the ratio $P = u_f/u_{rms}$, where, u_{rms}, is the root mean square of the longitudinal velocity. Using experimental data from coastal plain estuaries and the analysis of theoretical parameters of the Hansen and Rattray (1965) analytical model, a reasonable correlation between these parameters were obtained for six estuaries where observations have been interpreted in terms of the Stratification-circulation diagram. The indicated straight-line relationships, not necessarily the best fit, were according to Hansen and Rattray (1966):

$$\frac{u_s}{u_f} = 1.5 + \frac{1}{3}F_m^{-3/4}, \tag{3.6}$$

and

$$\frac{\delta S}{\bar{S}}[0.125 + 0.15(\frac{u_s}{u_f} - 1.5)] = 0.05(P)^{-\frac{7}{5}}, \tag{3.7}$$

These equations indicate that the circulation parameter (u_s/u_f), depends on the value of F_m, but the stratification parameter ($\delta S/\bar{S}$) depends on both F_m and P. Knowing these functional dependencies, the isolines of P and F_m were included in the Stratification-circulation Diagram, only approximately as a test, because as yet there is no reliable set of observations to confirm those theoretical results. A few years later, Dyer and Ramamoorthy (1969) applied the Stratification-circulation diagram to classify the Vellar estuary (India), to complement the experimental data set analysis of salinity and circulation measurements. Their results indicated that this bar-built estuary was type 4 (salt wedge) and type 1 (well-mixed) at high and low river discharge, respectively. The bulk parameters, F_m and P were calculated from the velocity data, fresh water discharge and hydrographic data for the time

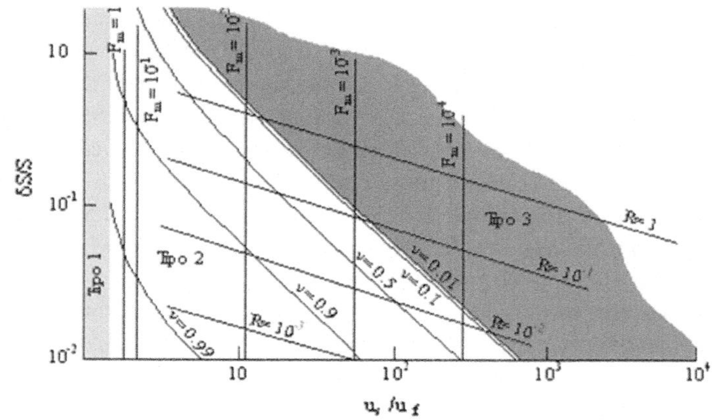

Fig. 3.13 Stratification-circulation diagram with the isolines of the parameter ν ($0.01 < \nu < 0.99$), the Froude number ($10^{-4} < F_m < 1$), and the Richardson estuarine number ($10^{-3} < Ri_e < 1$) (denoted in the figure as R), according to Fischer (1972). Estuarine Types 1, 2 and 3, according to Hansen and Rattray (1966), were included. Diagram axis in log-log scales

period of sampling. The results were in good agreement with the relationships between theoretical parameters and those proposed by Hansen and Rattray.

The Stratification-circulation diagram was re-plotted by Fischer (1972) choosing the Richardson number (Ri_e—Eq. 2.35, Chap. 2), and the Froude number, F_m, as parameter in place of P, as presented in Fig. 3.13. This figure shows that the stratification parameter ($\delta S/\bar{S}$) also depends primarily on the estuarine Richardson number Ri_e (denoted as R in the figure). According to this parameter definition, an increase in R_{ie} towards 1, correspond with increasing vertical estuarine stratification, and, as a first approximation, the transition between the well-mixed and partially mixed estuaries occurs when $0.08 < Ri_e < 0.8$.

Alternative studies were made by Prandle (1985) focusing on the circulation in narrow estuaries that are subject to a predominant tidal forcing. The results were also applied to improve the classical Stratification-circulation diagram with an analytical delimitation of the estuaries types 3 and 4, and 1 and 2. In relation to the formulations of Eqs. (3.1) and (3.2) it was suggested the following alternative expressions to the energy taxes that were introduced in the pioneer paper of Ippen and Harleman (1961): $G = (4/3\pi)k\rho(U_0)^3L$ and $J = (1/2)\Delta\rho g(H_0)^2u_f$, and its ratio (Eq. 3.3) by Eq. (3.8). In these formulations k, U_0 and L are the bottom shear stress coefficient, the tidal current amplitude and the estuary length, respectively. These quantities have dimensions of $[MLT^{-3}]$ and the stratification number $G/J = S_t$ is now calculated by a simple equation, in comparison to the previous one;

$$\frac{G}{J} = S_t = 0.85\frac{kU_0^3L}{g\frac{\Delta\rho}{\rho}H_0^2u_f} = \frac{kU_0^3L}{g'H_0^2u_f}. \tag{3.8}$$

This result indicates that the vertical stratification depends on the cubic power of U_0, the length L, and is inversely proportional to the velocity driven by the river discharge (u_f), the reduced gravity (g'), and the second power of the estuary depth (H_0). Computational results of this new formulation of the stratification number, using experimental data from natural estuaries and from physical models, were approximately twice the values of those obtained using the Eq. (3.3). This enabled the following stratification criteria to be established: $S_t < 100$ and $S_t > 400$ indicate estuaries with high and low stratification, respectively. Estuaries with a stratification number between 100 and 400 are classified as partially mixed.

With the purpose to introduce a classification diagram, experimental data from estuaries and laboratory experiments were used by Prandle (1985), to determine the correlation of the stratification parameter $\delta S/\bar{S}$ and the stratification number S_t. The following exponential expression was obtained, adjusted by mathematical regression:

$$\frac{\delta S}{\bar{S}} = 4(S_t)^{-0.56}. \tag{3.9}$$

This equation confirmed the conclusions of Ippen and Harleman (1961) and Hansen and Ratray (1966) related to the use of the stratification parameter $\delta S/\bar{S}$, and it was also used as a classification diagram from the correlation $\delta S/\bar{S} \times S_t$ shown in Fig. 3.14. This figure indicated that the stratification number variations in

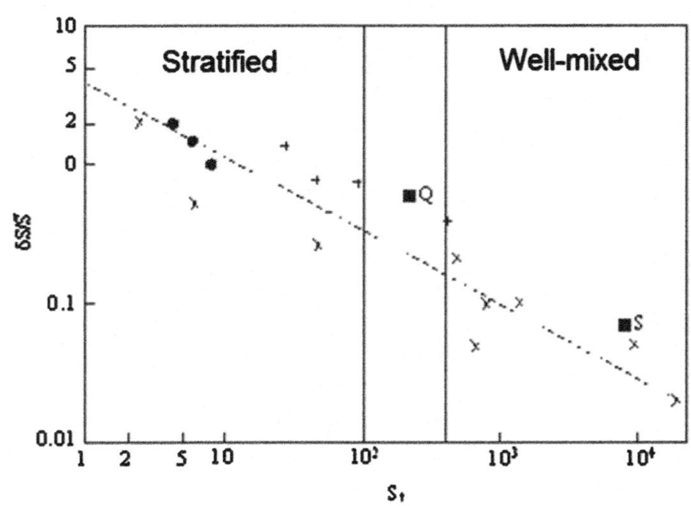

Fig. 3.14 Correlation of the stratification parameter ($\delta S/\bar{S}$) and the stratification number S_t (dissipated energy/potential energy gain). Estuarine values (x), physical models (•/+), according to Prandle (1985). In the diagram are also indicated results for the Bertioga river estuary of the Santos-São Vicente Estuary System (São Paulo, Brazil)) in spring (S) and neap (Q) tidal cycles. Diagram axis in log-log scales

the interval $20 \leq S_t \leq 400$ is used to separate the classification of the estuaries partially stratified from those well-mixed, and it is double than that obtained in the Ippen and Harleman's article ($20 \leq S_t \leq 200$). However, the original theoretical principle, related to the use of the ratio G/J, in the classification of estuaries was confirmed. This figure also shows the use of this diagram in the classification of natural estuaries and laboratory data to indicate the separation of the stratified and well-mixed estuaries. Exemplifying the use of this diagram, experimental results of the Bertioga estuarine channel (Santos-São Vicente Estuary System Estuarine System, Fig. 1.5, Chap. 1) which were calculated for neap Q, and spring S tidal conditions, indicate that there was a change in the estuary classification from partially mixed and highly stratified ($S_t < 100$) to well-mixed ($S_t > 400$).

The Prandle (1985) results were also applied to improve the original Stratification-circulation diagram of Hansen and Rattray (1966), with the analytic delimitation of estuaries highly stratified types 3 and 4, and 1 and 2. For the first delimitation, a well-defined layer thickness (d) may be assigned under the halocline, and it was theoretically demonstrated that it may be obtained with the following relationship of the stratification ($\delta S/\bar{S}$), circulation parameters ($u_s/<\bar{u}>$) and the normalized depth $D = d/H_0$:

$$\frac{\delta S}{\bar{S}} = \frac{\frac{u_s}{\langle \bar{u} \rangle}}{\frac{u_s}{\langle \bar{u} \rangle} - 1.26}, \text{ and } \frac{\delta S}{\bar{S}} = \frac{1}{D}. \quad (3.10)$$

These relationships, reproduces analytically the separation line of the estuarine types 3 and 4 as indicated by the upper thin line in Fig. 3.12.

The delimitation line of the estuarine types 1 and 2 (Fig. 3.12) have also been theoretically obtained with the following equations:

$$\frac{u_s}{\langle \bar{u} \rangle} = 1.14 + 0.036 \frac{S_\rho}{F_u}, \quad (3.11)$$

and

$$\frac{u_B}{\langle \bar{u} \rangle} = 0.70 - 0.029 \frac{S_\rho}{F_u}, \quad (3.12)$$

with u_S and u_B denoting the surface and the bottom velocity, respectively, and the non-dimensional quantities S_ρ and F_u are defined by,

$$S_\rho = \frac{H_0}{\rho} \frac{\partial \rho}{\partial x}; \quad F_u = \frac{kU_0 < \bar{u} >}{gH_0}. \quad (3.13)$$

In this equation k ($k = 2.5 \times 10^{-3}$) is the bottom friction coefficient taken as constant. In this theory, as shown by Eqs. (3.11) and (3.12) the transitions, between types 1 and 2, always occur when $S_\rho/F_u > 24$ or $(u_s/\langle\bar{u}\rangle) > 2$. This result is in agreement with the Hansen and Ratray theory, and the relationships (3.12 and 3.13) enable the introduction of an alternative axis to the circulation parameter: the ratio $S\rho/F_u$, (Fig. 3.12), which is an indicator deduced from the dynamic structure of the vertical velocity profile.

Another alternative axis was suggested by Scott (1993). In this paper, the classical mean estuarine circulation forced by the river discharge and the longitudinal mean salinity gradient $(\frac{\partial\bar{S}}{\partial x})$ was used to obtain an equation similar to Eqs. (3.8) and (3.11),

$$\frac{u_s}{\langle\bar{u}\rangle} = 1.15 + 0.036\frac{u_g}{\langle\bar{u}\rangle}, \tag{3.14}$$

where $u_g = gH_0^2\beta(\partial\bar{S}/\partial x)/kU_0$ is the gravitational circulation intensity, and the ratio $(u_g/\langle\bar{u}\rangle)$ may be used as an alternative axis to the circulation parameter. Details on the theoretical results may be found in Scott (1993) or in Dyer (1997).

Another estuary classification methodology was developed by Jay and Smith (1988), based on the perturbation analysis associated with finite amplitude wave theory by mean of a perturbation analysis of the time-dependent dynamic equations. This theoretical development takes into account the non-linear estuarine characteristics, barotropic and baroclinic mechanisms resulting in three distinct circulation types for shallow and estuarine systems (highly stratified, partially mixed and weakly stratified). The following five criteria where used in the system:

(i) The estuary classification should be related to the dominant vertical turbulent exchange and residual flow mechanisms.
(ii) The first order properties of estuaries should be used for classification, so that the major dynamical distinctions are clear.
(iii) Changes in river flow and tidal range should cause different and characteristic movements of the position of the estuary on the classification plot.
(iv) Baroclinic and barotropic nonlinearities should each be represented by a separate parameter.
(v) The entire along-channel extent of most systems should occupy a single position in the classification plot. The very real temporal heterogeneity of most estuaries and the dominating effects of topography in creating multiple basins in some other systems cannot and should not, however, be eliminated.

To satisfy these conditions, two new parameters were chosen to quantify the internal or interfacial processes associated with the baroclinic circulation, and the bottom-frictional and convective nonlinearities, related to the barotropic tidal wave in the estuarine basin. The first parameter is an *internal Froude number* (F_B) (Fig. 3.15), that is, a measure of baroclinic non-linearity (*baroclinic Froude number*), being a generalization of that used as the expansion parameter in the two-layer perturbation (Jay and Smith (1988):

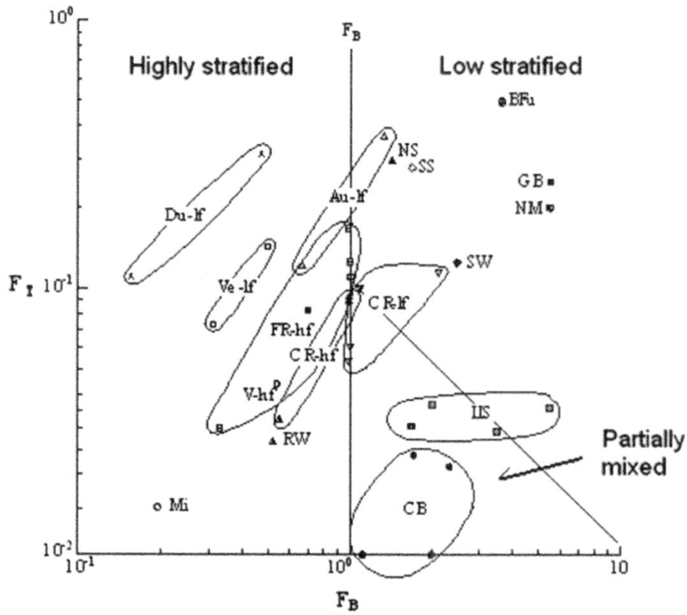

Fig. 3.15 Classification diagram for swallow estuary systems based on the barotropic and baroclinc Froude numbers F_T and F_B, respectively. Parameters were calculated and plotted with observational data from various natural estuaries, in high (hf) and low (lf) river discharge. Du = Duwamish; Ve = Vellar; Au = Aulne; Mi = Mississippi; Fr = Fraser; RW = Rotterdam channel; CR = Columbia; CB = Chesapeake Bay; LIS = Long Island Sound Channel; SW = Southampton Water; NS and SS = North and South Santee; BFu = Fundy Bay; GB = Great Bay and NM = Mersey Narrows (according to Jay and Smith 1988). Diagram axis in log-log scales

$$F_B = \frac{d_e}{D_e}\left(\frac{\Delta\rho_H}{\Delta\rho_V}\right)^{\frac{1}{2}}. \quad (3.15)$$

where d_e/D_e is the ratio of the excursion of the salt-wedge (d_e) by its mean depth during the tidal cycle (D_e), expressed in scaled variables, to indicate the baroclinic stability of the salt-wedge intrusion. $\Delta\rho_H$ and $\Delta\rho_V$ are the horizontal density difference between the estuary's mouth and head, and the vertical density difference mid-estuary, respectively. F_B is a measure of the stability or nonlinearity of the internal oscillation of the density field. It ranges from close to zero in arrested salt wedge, like the Mississippi (because the ratio d_e/D_e is small), to 0.5 to 1.0, in active salt wedge as the Fraser river estuary, and to much greater than 1.0 in weakly stratified estuaries.

A second parameter of the Jay and Smith (1988) classification diagram is the barotropic Froude number (F_T),

$$F_T = \frac{\langle \bar{\eta}_o \rangle}{H_0}, \tag{3.16}$$

where $\langle \bar{\eta}_o \rangle$ and H_0 are the time mean tidal amplitude and the mean depth over the basin. This number is a measure of barotropic tidal non-linearity; it is close to zero in deep estuaries with weak tides and approaches 1 in bays forced by hyper-tides.

The resulting correlation $F_B \times F_T$ plane is shown in Fig. 3.15, with the results plotted for various estuaries, based on data available in the literature. The semi-planes of the highly stratified ($F_B < 1$), and partially and weakly stratified ($F_B > 1$) estuaries are delimited by the $F_B = 1$. The boundary between the partially mixed and weakly stratified estuaries is a line of constant gradient of the gradient Richardson number $Ri_g \approx 1/4$, along which $F_T \approx (F_B)^{-1}$. The proportionality constant, which determines the intercept of the Ri_g line with $F_B = 1$, was set using the critical tidal amplitude of Columbia river estuary (Columbia, USA). Another system's geometry might lead to a different proportionality constant and therefore to a slightly different intercept, but the slope and general location of the line are well established (Jay and Smith, op. cit.).

The estuarine classification in terms of their longitudinal and secondary circulation structures and exchange flow was given by Valle-Levinson's (2008, 2010). The following characteristics of estuary circulation were considered: (i) the structure of the secondary flow is strongly influenced by bathymetry variations and may exhibit vertically and laterally sheared net exchange flows, with outflows over shallow parts of a cross-section and inflows in the channel (e.g. Dyer 1977; Wong 1994), and (ii) the lateral structure of exchange flows may ultimately depend on the competition of Coriolis deflection and frictional effects (Kasai et al. 2000). Two non-dimensional parameters were used to link these characteristics in the classification diagram (Fig. 3.16): the vertical Ekman (E_K)—the ratio of the friction coefficient to the Coriolis parameter, and Kelvin number (K_e)—the ratio of the estuary width to the internal radius of deformation (see, Eqs. 2.44 and 2.45, Chap. 2).

The Fig. 3.16 indicate that E_k low values imply that frictional effects are restricted to a thin bottom boundary layer, while high E_K values indicate that friction affects the entire water column. The lateral structure density-driven exchange flows may be described in terms of whether they are vertically sheared or unidirectional in the deepest part of the cross-section (Valle-Levinson 2008). Following his analysis, under low E_K (<0.001, i.e., <−3 in the abscissa of Fig. 3.16), the lateral structure of exchange flows depends on the dynamic width of the system. In wide systems ($K_e > 2$, i.e., 0.3 in the ordinate of Fig. 3.16), outflows and inflows are separated laterally according to the Earth's rotation, i.e., the exchange flow is laterally sheared. In narrow systems ($K_e < 1$, i.e., <0 in the

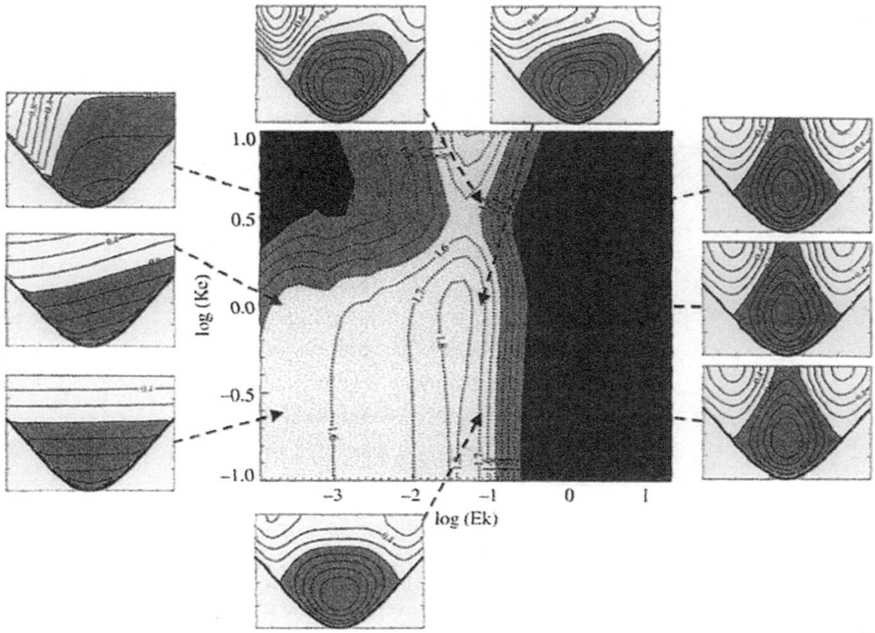

Fig. 3.16 Classification of estuarine exchange on the basis of Ekman (E_K) and Kelvin (K_e). The subpanels around the central figure denote cross-sections looking into the estuary, of exchange flows normalized by the maximum inflow, which are *negative* and *shaded*. The vertical and horizontal axes are non-dimensional depth and width from 0 to 1, respectively. The central figure illustrates contours of the difference between maximum inflow and outflow over the deepest part of the channel and for different values of E_K and K_e, both in logarithm scale. *Dark-shaded* contour regions denote net inflow throughout, i.e., laterally sheared exchange flow as portrayed by subpanels whose *arrows* point to the corresponding non-dimensional numbers in the *dark shaded* regions. *Light* contour regions illustrate vertically sheared exchange in the channel as portrayed by the subpanels whose *arrows* point to the corresponding E_K and K_e numbers in the *light-shaded* regions. Intermediated-shaded regions represent vertically and horizontally sheared exchange flow, similar to the second subpanel on the left, for $\log(K_e) = 0$ and $\log(E_k) \approx -3.7$ (after Valle-Levinson (2008, 2010)

ordinate Fig. 3.16) and low E_K (still <0.001, i.e., <−3 in the abscissa of Fig. 3.16), exchange flows are vertically sheared. In contrast, under high E_K (>0.3, i.e., >−0.5 in the abscissa of Fig. 3.16) and for all K_e, the density-driven exchange is laterally sheared independently of the width of the system. Finally, under intermediate E_K ($0.01 < E_K < 0.1$, i.e., between −2 and −1 in the abscissa of Fig. 3.16), the exchange flow is preferentially vertically sheared but exhibiting lateral variations.

The classification diagram developed by Geyer (2010) introduced a *prognostic* approach, in which the estuarine is classified based on forcing variables, and its purpose is to predict the estuarine regime based on these forcing conditions. It is recognized that due to the complexity and variability within and among estuaries, this approach could at best provide a rough estimate of the conditions of a particular

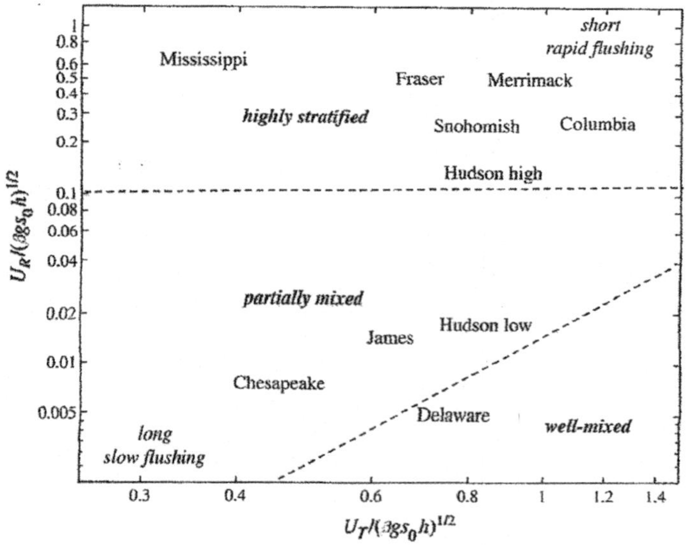

Fig. 3.17 A framework for prognostic estuarine classification. Tidal (U_T) and freshwater (U_R) velocities are non-dimensionalized by a densimetric velocity scale $(\beta g S_o h)^{1/2}$ (according to Geyer 2010). Diagram axis in log-log scales

estuary, due to the inherent difficult to predict estuarine processes. In the steady-state salt balance, the longitudinal salinity stratification ($\partial S/\partial x$) depends on the *tidal velocity* ($U_T = U_0$) and the *fresh water velocity* ($U_R = u_f$), providing the *master variables* for the prognostic approach for the estuary classification developed by Geyer (op. cit). This is illustrated in Fig. 3.17 which uses theoretical and experimental values of natural estuaries along the USA coastline. In the figure axes, the master variables (U_T and U_R) are non-dimensionalized by a densimetric scale $(\beta g S_o h)^{1/2}$ and β is the coefficient of saline contraction (7.0×10^{-5} is a typical value), S_o and $H_0(h)$ are salinity and depth mean values, respectively The prognostic diagram is subdivided to take into three regions: *highly stratified*, *partially mixed* and *well-mixed* estuaries. More strongly stratified estuaries appear on the upper part of the diagram, and weakly stratified on the lower part.

In the diagram three main estuarine types may be distinguished: (i) *highly stratified* characterized by the non-dimensional river discharge and tidal velocity varying in the intervals 0.1–1.0 and 0.2–1.4, respectively, and the *short rapid flushing*, with the non-dimensional tidal velocity higher than 1.0; (ii) *partially mixed* characterized by the non-dimensional river discharge lower than 0.1, and those classified as *long slow flushing*, with non-dimensional river discharge less than 0.005; and (iii) *well-mixed* estuaries in the lower right side delimited by the sloping dashed line, with the non-dimensional river discharge and tidal velocity less and higher than 0.04 and 0.47, respectively.

3.4 Estuarine Zone

Estuarine zone has been the label applied to an environmental system consisting of transitional areas which are consistently influenced or affected by water from an estuary such as, but not limited to, salt marshes, mangroves, coastal and intertidal areas, bays, lagoons, inshore areas and channels. Bights and inlets are coastal environments which may be generically classified as estuarine zones (Geophysics Study Committee 1977). These zones may have a large free surface (>100 km^2) and depths may vary between 20 and 100 m. The bights and inlets are usually sheltered from the influence of coastal processes; however, there are other coastal systems that receive these influences more intensively, as the coastal lagoons.

Large bays are common in the coast lines of all continents, particularly in regions that were tectonically active. If a bay is forced by moderate tidal height and the fresh water volume transported by the river is relatively small compared to the tidal prism, its thermohaline characteristics may be similar to those of the water masses of the continental shelf. Usually, the thermohaline effect is weak and the influence of the tidal and wind forcing may predominate. When the river discharge into the bay is relatively high it may present similar features to coastal plain estuaries, as was demonstrated by Kjerfve et al. (1997) and Bérgamo et al. (2006), applying the Stratification–circulation diagram to a set of observational data of the Guanabara Bay (Rio de Janeiro, Brazil). However, these water bodies are too large and complex to be defined as an estuary, and, due to the tectonic component of their formation, they should be classified in another category of an estuarine environment category as stated in the Pritchard (1952b) article. The largest bays along the Brazilian coast are: São Marcos Bay (2025 km^2) in the state of Maranhão, Todos os Santos Bay (1233 km^2) in the state of Bahia; Paranaguá Bay (677 km^2) in the state of Paraná, Guanabara Bay (384 km^2) in the state of Rio de Janeiro; the Santos Bay (54 km^2) in the state of São Paulo state is the smallest. The bays receive river discharge from complex catchment basins, and the ones located along the southern coast receive weak rivers discharges, compared to the bays in the northern and northeastern coast. In the almost rectilinear coastline of southern Brazil, which is a former non-tectonically active region, there are no bays.

Comprehensive study on the oceanographic characteristics of the Todos os Santos Bay was presented by Cirano and Lessa (2007) and Lessa et al. (2009), and a descriptive analysis of the seasonal variation of physical characteristics in the northern region of the bay have been presented by Miranda et al. (2011). Land-ocean fluxes in the Paranaguá Bay, which comprises an important subtropical estuarine system, have been investigated by Marone et al. (2005) and ecological studies have been presented by Lana et al. (2001).

Channels and straits are natural or human-made marine environments that interconnect two water bodies; they may be shallow or deep and with a simple or complex geometry. In the estuaries, mainly those used as shelter for harbors, we find artificial channels used for navigation purposes. These channels are constructed by bottom dredging, causing serious impacts to the environment. The sediment removal, besides being detrimental to the benthonic flora and fauna, may change the tidal wave characteristics, the current intensities and the tidal prism. The biological communities from the intra and intertidal regions also may be deeply affected by dredging.

The majority of artificial channels have resulted from the management of fluvial water to the hydroelectric power development for industry and urbanization. Most of the artificial channels have drastically affected the natural estuarine characteristics. The Valo Grande (former named *Vala do Rocio*), for example, which interconnects the Ribeira de Iguape river to Mar Pequeno near the town of Iguape, with a length of ≈4.0 km (Fig. 1.5, Chap. 1), was constructed, in 1884, to enable docking of greater ships. This artificial channel resulted in a gradual decrease in the salinity in the Cananéia-Iguape Lagunar Estuarine System due to the increase in the input of fresh water.

The channel project specified a width of 4.40 m and a mean depth of 2–3 m; however, by the time that the channel opened in 1884, a gradual increase in its depth (up to ≈7.0 to 16.0 m) and width (with an estimated erosion tax of 3.3 m/year) reaching up to 235 m was observed. Due to the increase in the fresh water input to the estuarine system, and subsequent increase in the water velocity in the channel, the erosion and sediment transport towards the Iguape harbor drastically decreased its depth.

In contrast to the situation observed in the Valo Grande, we should mention the channel construction in the Eastmain River, flowing into the James Bay, on the west Canada coast. In a great engineering project, this subarctic shallow river was diverted into the La Grande River for hydroelectric development, reducing the river water input into the eastern portion of the bay by 80%. The project was supported with monitoring of experimental oceanographic conditions during a 3-year program, and the response of the estuary to this major reduction of freshwater input was simulated with a one-dimensional finite difference explicit scheme numerical model, validated with the observational data. According to the final results, the tidal river zone (TRZ) had a salinity increase up to 18‰, and at the diversification end, the estuary conditions were formed in just 30 days. The low frequency estuary response to the rapid change in the river discharge occurred much more rapidly for the velocity regime than for salinity, with its intrusion occurring over approximately one month. After this transitional period, the estuary presented the classical estuarine circulation and was classified as transitional from types 2a to 2b during late summer (Lepage and Ingram 1986).

3.5 Coastal Lagoons

Coastal lagoons are inland water bodies, found in all continents, usually oriented parallel to the coast. They have been formed as the result of rising sea level, mostly during the Holocene, and marine processes generating coastal barriers. They reached the present elevation level about 2000 years B.P., and evolved seaward both during the slow rise of the sea level and their present elevation. They border more than 13% or 32,000 km of the world's continental coasts and play a substantial role in the transport, modification, and accumulation of matter at the land-sea interface (Kjerfve and Magill 1989; Knoppers 1994). They are often highly productive and ideal systems for aquaculture projects, and are often impacted by both natural and anthropogenic influences (Mee 1978; Sikora and Kjerfve 1985). Depending on local climatic conditions, lagoons exhibit salinities ranging from completely fresh to hypersaline (Moore and Slinn 1984; Kjerfve 1986; Merino et al. 1990, Knoppers et al. 1991). A comprehensive discussion on the geological processes controlling the origin and evolution of these coastal environments, and a synthesis on the physical and biogeochemical characteristics of coastal lagoons in southeastern of Brazil are presented by Martin and Dominguez (1994) and Knoppers and Kjerfve (2010), respectively.

The coastal lagoon environments are not the main subject of this book. However, as they are of great social and economic importance, some of their main geomorphologic characteristics and definitions found in the literature will be presented. These coastal systems occupy about 13% of coastal areas worldwide, and to enable their use for future generations, the utilization of these systems has been planned using experimental observations and taking into account the physical properties of estuaries, the dynamics of which are reasonably well known. According to Pheleger (1969) a coastal lagoon has been defined as:

An inner marine water body, usually oriented in the direction parallel to the coastline, separated from the ocean by a barrier, and connected to the ocean by one or more restricted inlets.

To this definition, Kjerfve (1994) added that the ocean(s) entrance can at times be closed off by sediment deposition as a result of wave action and littoral drift, and suggested the following definition for a coastal lagoon:

A shallow coastal water body separated from the ocean by a barrier, connected at least intermittently to the ocean by one or more restricted inlets, and usually oriented shore-parallel.

Coastal lagoons experiences forcing from river input, wind stress, tides, precipitation to evaporation balance, and surface heat balance, and respond differently to these driving forces. Water and salt balance, lagoon water quality and eutrophication depend critically on lagoon circulation, salt and material dispersion, water exchange through the ocean channel(s), and turn over, residence and flushing times. The understanding of physical, chemical, geological, and ecological dynamics of lagoons is important for the planning and implementation of coastal management strategies in coastal lagoons Kjerfve (op. cit.).

It is very convenient to characterize coastal lagoons according to the volume transport exchanged with the adjacent coastal region, and also with the tidal activity in the inner region. These dynamic characteristics are related to the geometric

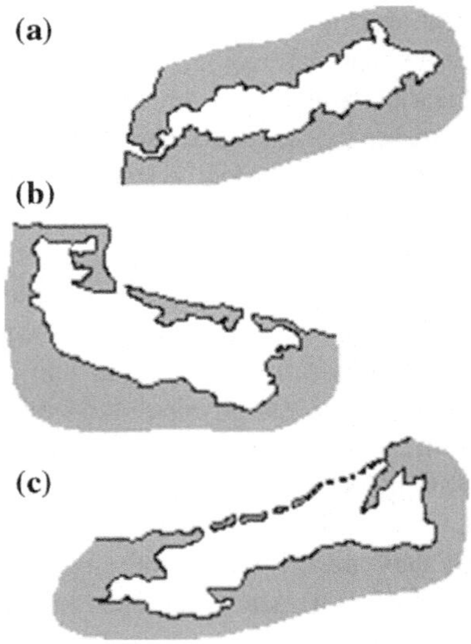

Fig. 3.18 Choked (**a**), Restricted (**b**) and Leaky (**c**) coastal lagoons from the wide geomorphic spectrum found in coastal regions (after Kjerfve 1986)

characteristics of the channel openings to the ocean. The rate and magnitude of oceanic exchange reflects both the dominant forcing(s) and the time-scale of the hydrologic variability. Kjerfve (1986a) sub-divided coastal lagoons into three geomorphic types according to their exchange with the coastal ocean (Fig. 3.18).

Choked Lagoons usually consist of a series of connected elliptical cells, connected by a single long narrow entrance channel, along coasts with high wave energy and significant littoral drift.

Although this lagoon type (Fig. 3.18a) experiences tides that co-oscillate with the coastal ocean, the entrance channel serves as a dynamics filter which largely eliminates tidal currents and water-level fluctuations inside the lagoon. Tidal oscillations in choked lagoons are often reduced to 5% or less compared to the adjacent coastal tide. They are characterized by long flushing times, dominant wind forcing, and intermittent stratification events due to intense solar radiation or runoff events.

Examples of choked lagoons in Brazil include: Mundaú and Manguaba (Alagoas), Maricá and Guarapina and the Saquarema system (Urussanga, Fora and Araruama (Rio de Janeiro), Lagoa dos Patos (Rio Grande do Sul). (Kjerfve et al. 1990; Kjerfve and Knoppers 1991).

Restricted Lagoons consist of a large and wide body of water, usually oriented parallel to shore, and exhibit two or more entrance channels or inlets. As a result, restricted coastal lagoons have a well-defined tidal circulation, are influenced by winds, are mostly vertically well-mixed, and exhibit salinities from brackish water to oceanic levels.

In this lagoon type (Fig. 3.18b) flushing times are usually considerably shorter than for choked coastal lagoons. Examples of restricted lagoons include Laguna de Terminos (Mexico), and Lake Pontchartrain (USA).

Leaky Lagoons are elongated shore-parallel water bodies with many ocean entrance channels, along coasts where tidal currents are sufficiently strong to overcome the tendencies by wave action and littoral drift to close the channel entrances.

Leaky lagoons (Fig. 3.18c) are at the opposite end of the spectrum to choked lagoons. They are characterized by numerous wide tidal passes, unimpaired water exchange with the ocean on wave, tidal, and longer time scales, strong tidal currents, and salinities close to that of the coastal ocean. Examples are Mississippi Sound (Mississippi, USA) and Wadden Zee (The Netherlands-Denmark).

The Lagoa dos Patos, located in the southern Brazilian coast (30°–32°S), is one of the largest lagoons in the world. It has a length of 250 km, a mean width of 40 km, a mean depth of 5 m, and is connected to the coastal ocean by a narrow channel with a length of 20 km (1 to 2 km wide). This coastal lagoon is classified as a *choked lagoon* (Kjerfve 1986; Möller et al. 1996), may exhibit estuarine characteristics, and the MZ may be localized up to 70 km from its mouth; however, research has shown that during dry seasons or periods high river discharge, the MZ boundary may be displaced northward or southward, and may be found very close to the channel entrance in occasions of great floods. Therefore, the basic mechanisms causing the formation and displacement of the MZ in the lagoon are: the baroclinic component of the gradient pressure force, water balance inside the lagoon forced by river discharge, and the southwestward wind stress forcing (Möller and Castaing 1999).

Coastal lagoons are environments very vulnerable to anthropogenic influence. Their existence is dependent on sea level variations and the inner and coastal sedimentation processes, which are usually intensified due to urban development

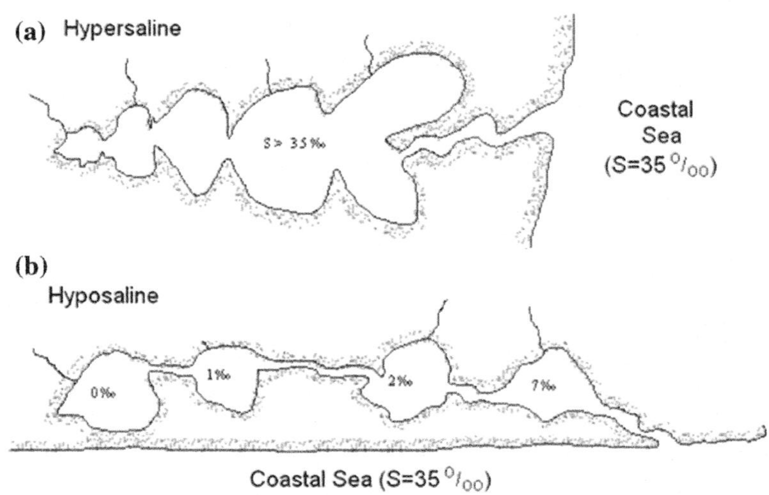

Fig. 3.19 Sketch of hyper (**a**) and sub-saline (**b**) coastal lagoons

and farming near the lagoons. As they are shallow and typical of coastal plain regions, their water mass responds quickly to synoptic and seasonal meteorological conditions.

The choked coastal lagoons may evolve, according to the climatic conditions, into concentrated or diluted marine environments. A coastal lagoon that has evolved to become concentrated is termed as hypersaline; the negative fresh water balance (precipitation + river discharge-evapotranspiration < 0) generates a water body with higher salinity than the coastal ocean (Fig. 3.19a). For example, the Araruama coastal lagoon (Rio de Janeiro, Brazil) is a hypersaline coastal lagoon as the result of the semi-arid climate conditions, a small drainage basin and choked entrance channel. This lagoon has been continuously hypersaline for at least 4–5 centuries, but the salinity has varied substantially; it decreased from 57 to 52‰ between 1965 and 1990, as indicated by salinity (density) measurements. The flushing half-life measures 83.5 days, which is considerably longer for most others coastal lagoons (Kjerfve et al. 1996).

When climatic conditions generate a fresh water balance positive (precipitation + river discharge-evaporation > 0), the lagoon becomes sub-saline and, in this condition, the salinity in the lagoon is less than in the coastal region (Fig. 3.19b).

In coastal regions with predominant arid or semi-arid climate, severe wind wave's regime, plenty of marine sediments, accentuated coastal drift, and dominated by wind generated dunes, it is possible to observe the formation of a system of shallow coastal lagoons known as *tidal sandbanks or shoal*. These environments have a negative fresh water balance evolving to a hyper-saline condition due to the mouth closure, which occurs under extreme coastal sedimentation. The closure time interval may be seasonal or remain for long periods. During some heavy rain periods the water exchange of the lagoon with the coastal ocean is re-established by the fluvial erosion forced by the hydraulic gradient. This type of coastal lagoon, which is usually of great beauty as a coastal environment, is found in the west coast of Australia and Mexico, and in the northeastern coast of Brazil such as the Macau lagoon (Rio Grande do Norte), whose planktonic communities have been studied by Chellappa (1991).

The case study of the Mundau-Manguaba system (Oliveira and Kjerfve 1993) is an example of a choked sub-saline coastal lagoon. Localized in Maceió (Alagoas, Brazil), and strongly affected by the input of wastes from the sugar cane industry, the system was studied as an environmental problem related to ecological processes to emphasize the inter-relationships of planning and industrial management.

During the rainy season (June to July, 1984), the Mundau lagoon was dominated by fresh water discharge, and the salinity variation was from zero to 9‰, with weak vertical variation. The Manguaba lagoon retained a salinity of less than 1‰. Due to the predominance of water mass with low salinity during this time of the year, events of great mortality of the abundantly occurring benthic estuarine mussel (Mytella falcata) were observed, popularly named in the region as *sururu*.

In the dry season (January to February, 1985) salinities in the Mundau-Manguaba lagoons system increased and varied in the interval (6–24‰) and (0.5–8‰), respectively, and the vertical stratification with the larger gradient occurred in the channel mouth. The abnormal growth of bacteria (cyanobacteria blooms) caused by eutrophication due to the contamination of the river discharge with high concentrations of industrial wastes, occurred frequently and resulted in mass mortality of fish and shellfish. In these lagoons, the circulation and mixing processes were dominated by the tidal forcing, but may also be modified by the wind stress and the river discharge. Maximum current velocities were observed in a transversal section across the external channel leading into Manguaba lagoon: maximum flood and ebb currents reaching speeds of 0.6 and 0.4 m s^{-1}, respectively, were measured during the rainy season (July, 1991).

The external lagoon of the Maricá system, directly interconnected with the coastal sea, is the Guarapina lagoon. It has been studied with a time series of current and hydrographic properties by Kjerfve and Knoppers (1991). During this experiment, the circulation was predominantly seaward and the direction of the tidal modulated current was reversed only in a few occasions. The occurrence of this reversing circulation was during the spring tidal cycle, and caused a quickly input of salt water into the lagoon system. During these events the salinity increased rapidly from $\approx$7.0 to 34‰. However, during the longer ebbing time intervals the salinity remained almost constant ($\approx$7.0‰) because the volume transport of the river discharge is small, in comparison with the total volume of the lagoon system.

The hydrographic characteristics and the trophic state of the Maricá coastal lagoon (the most interior of the Maricá system), has also been studied by Esteves (1992), using seasonal measurements of physical-chemical properties. In the sampled time period (December, 1989 to April, 1991) the salinity structure indicated low vertical stratification with bottom minus surface values of less than $\approx$1.0‰ and up to $\approx$3.0‰ in the sampling stations located in the western and eastern lagoon borders, respectively. Mean estimated salinity values varied from $\approx$4.1‰ to 7.4‰, indicating small seasonal variation. However, in February, 1990, the salinity value in the eastern sampling station reached up to $\approx$14.5‰, almost the same time as the occurrence of an anomaly occasioning a great fish mortality. This lagoon system presented trophic state during almost the sampling period; however, a hypertrophic environmental condition was observed in the summer.

The trophic state and water turn-over time in six choked coastal lagoons, located along the coast of Rio de Janeiro (Brazil), have been compared using the total phosphorus and chlorophyll-a concentrations and nutrient loading. The results indicated a clear relationship between the flushing half-life and the trophic state, based on a scatter diagram analysis of the total phosphorus and chlorophyll-a. Scatter in the relationships is most evident for the interior lagoons where the composition of primary producers differs. In this lagoon system the flushing half-life and nutrient loading was also observed to be related to the total phosphorus and chlorophyll-a concentrations (Knoppers et al. 1991).

References

Andutta, F. P. 2011. O Sistema Estuarino dos Rios Caravelas e Peruipe (BA): Observações, Simulações, Tempo de Residência e Processos de Difusão e Advecção. Tese de Doutorado. São Paulo, Instituto Oceanográfico, Universidade de São Paulo. 121 p.

Andutta, F. P., Miranda, L.B.; Schettini, C.A.F.; Siegle, E.; Silva, M.P.; Izumi, V.M. & Chagas, F.M. 2013. Temporal variations of temperature, salinity and circulation in the Peruípe river estuary (NovaViçosa, BA). Cont. Shelf Res., 70:36–45.

Bérgamo, A.L.; Miranda, L.B. & Fontes, R.F.C. 2006. Current Measurements and Volume Transport in the Baía de Guanabara Entrance (Rio de Janeiro, RJ). (Org. Elisabete de Santis Braga). In: SIMPÓSIO BRASILEIRO DE OCEANOGRAFIA, 3, Oceanography and Global Changes, São Paulo, p. 357–369.

Brandini, F. P.; Thamm, C. A. & Ventura, I. 1988. Ecological Studies in the Bay of Paranaguá III. Seasonal and Spatial Variations of Nutrients and Chlorophyll-a. Nerítica, Pontal do Sul, 3(1):1–30.

Cameron, W. M. & Pritchard, D. W. 1963. "Estuaries". In: Hill, M. N. (ed.). The Sea. Ideas and Observations on Progress in the Study of the Seas. New York, Interscience, pp. 306–324.

Chellappa, M. T. 1991. Studies on Microalgae of Rio Grande do Norte 3. Phytoplankton Communities from Hypersaline Lagoons of Macau-RN. Bolm. Inst. Oceanogr. Limin., Centro de Biociências, UFRGN, 8: 41–51.

Cirano, M. & Lessa, G. C. 2007. Oceanographic characteristics of Baía de Todos os Santos, Brazil. Rev. Bras. Geofis.. v.25, n.4:363-387.

Conomos, T.J. 1979. Properties and circulation of San Francisco Bay waters, in: Conomos, T.J. ed., San Francisco Bay-The urbanized estuary: San Francisco, California, Pacific Division, American Association for the Advancement of Science, p. 47–84.

Dawson, W. A. & Tilley, L. J. 1972. Measurement of Salt Wedge Excursion Distance in the Duwamish River Estuary, Seattle, Washington, by Means of the Dissolved-Oxygen Gradient. Geological Survey Water-Supply. Washington, D. C., U. S. Department of Interior, Paper 1873-D, pp. D1–D27.

Dyer, K. R. 1973. Estuaries: A Physical Introduction. London, Wiley. 140 p.

Dyer, K. R. 1977. Lateral Circulation Effects in Estuaries. Estuaries, Geophysics and the Environment. Washington, D. C., National Academy of Sciences, pp. 22–29.

Dyer, K. R. 1986. Coastal and Estuarine Sediment Dynamics. New York, Wiley. 342 p.

Dyer, K. R. 1997. Estuaries: A Physical Introduction. 2. ed., Chichester, Wiley. 195 p.

Dyer, K & Ramamoorthy, K. 1969. Salinity and Water Circulation in the Vellar Estuary. Limnol. Oceanogr., 14(1):4–15.

Esteves, P. C. D. 1992. Variação Sazonal da Qualidade da Água da Laguna de Maricá, Rio de Janeiro. Dissertação de Mestrado. Rio de Janeiro, Instituto de Geoquímica, Universidade Federal Fluminense. 163 p.

Fairbridge, R. W. 1980. The Estuary: Its Definition and Geodynamic Cycle. In: Olausson, E. & Cato, I. (eds.). Chemistry and Biogeochemistry of Estuaries. New York, John Wiley & Sons, pp. 1–35.

Fischer, H. B. 1972. Mass Transport Mechanisms in Partially Stratified Estuaries. J. Fluid Mech., 53:672–687.

Geophysics Study Committee. 1977. Overview and Recommendations. Estuaries, Geophysics and the Environment. Washington, D. C., National Academy of Sciences, pp. 1–10.

Geyer, W. R. 1986. The Advance of a Salt Wedge Front: Observations and a Dynamical Model. In: Dronkers, J & Van Leussen W. (eds.). Physical Processes in Estuaries. Berlin, Springer-Verlag, pp. 181–195.

Geyer, W. R. & Farmer, D. M. 1989. Tide-Induced Variations of the Dynamics of a Salt Wedge Estuary. J. Phys. Oceanogr., v.19, pp. 1060–1072.

Geyer, W.R. 2010. Estuarine salinity structure and circulation. In: ed. Valle-Levinson A. Contemporary Issues in Estuarine Physics. Cambridge University Press, pp. 12–26.

Hamilton, P. & Rattray Jr., M. 1978. Theoretical Aspects of Estuarine Circulation. In: Kjerfve B. (ed.). Estuarine Transport Processes. Columbia, University of South Carolina, pp. 37–73. (Belle W. Baruch Library in Marine Science, 7).

Hansen, D. V. & Rattray Jr., M. 1965. Gravitational Circulation in Sraits and Estuaries. J. Mar. Res., 23(1):102–122.

Hansen, D. V. & Rattray Jr., M. 1966. New Dimensions in Estuary Classification. Limnol. Oceanogr., 11(3):319–325.

Harleman, D. R. F & Abraham, G. 1966. One-Dimensional Analysis of Salinity Intrusions in the Rotterdam Waterway. Delft, Holanda, Delft Hydraulics Laboratory. 35 p. (Publication, 44).

Ippen, A. T. & Harleman, D. R. F. 1961. One-Dimensional Analysis of Salinity Intrusion in Estuaries. Committee on Tidal Hydraulics. Tech. Bull. Corps of Engineers U. S. Army, n. 5. 120 p.

Jay & Smith, J. D. 1988. Residual Circulation and Classification of Shallow, Stratified Estuaries. In: Dronkers, J. & Van Leussen, W. (eds.). Physical Processes in Estuaries. Berlin, Springer-Verlag, pp. 22–41.

Kasai, A.; Hill, A,E.; Fujikawa, T. & Simpson, J.H. 2000. Effect of the Earth's rotation on the circulation of regions of freshwater influence. J. Geophys. Res. 105(C7), 16,961:16,969.

Keulegan, G. H. 1949. Interfacial Instability and Mixing in Stratified Flows. J. Res. U. S. Geol. Surv., 43:487–500.

Kjerfve, B. 1986. Circulation and Salt Flux in a Well Mixed Estuary. In: Van de Kreeke, J. (ed.). Physics of Shallow Estuaries and Bays. Berlin, Spring-Verlag, pp. 22–29.

Kjerfve. B. 1986a. Comparative Oceanography of Coastal Lagoons. In Estuarine Variability (Wolf, D.A., ed.). Academic Press, New York, U.S.A., pp. 63–81.

Kjerfve, B. 1989. Estuarine Geomorfology and Physical Oceanography. In: Day Jr., J. W., C. H. A. S. Hall, Kemp W. M. & Yáñez-Arancibа, A. (eds.). Estuarine Ecology. New York, Wiley, pp. 47–78.

Kjerfve, B. 1994. Coastal Lagoons. In: Kjerfve, B. (ed.). Coastal Lagoon Processes. Elsevier Oceanographic Series, 60, pp. 1–8.

Kjerfve B. & Magill, K.E. 1989. Geographic and hydrographic characteristics of shallow coastal lagoons. Marine Geology 88, 187–199.

Kjerfve, B., Knoppers, B. 1991. Tidal Choking in a Coastal Lagoon. In: Parker, B. (ed.). Tidal Hydrodynamics. John Wiley & Sons, New York, pp. 169–179.

Kjerfve, B., Ribeiro, C. H. A.; Dias, G. T. M.; Fillippo, A. M. & Quaresma, V. S. 1997. Oceanographic Characteristics of an Impacted Coastal Bay: Baía de Guanabara, Rio de Janeiro, Brazil. Continent. Shelf Res., 17(13):1609–1643.

Kjerfve, B.; Knoppers, B.; Moreira B. & Turcq B. 1990. Hydrological regimes in Lagoa de Guarapina, a shallow Brazilian coastal lagoon. Acta Linologica Brasiliensis III, 931–949.

Kjerfve, B.; Schettini, C. A. F.; Knoppers, B.; Lessa, G. & Ferreira, H. O. 1996. Hydrology and Salt Balance in a Large, Hypersaline Coastal Lagoon: Lagoa de Araruama, Brazil. Estuar. Coast. Shelf Sci., 42:701–725.

Knoppers, B. 1994. Aquatic primary production in coastal lagoons. In: Kjerfve, B.(ed.) Coastal lagoon processes. Elsevier, Amsterdan, pp. 243–286.

Knoppers, B.A.; Brandini, F. P. & Thamm, C. A. 1987. Ecological Studies in the Bay of Paranaguá II. Some Physical and Chemical Characteristics. Pontal do Sul, Nerítica, 2(1):1–36.

Knoppers, B.A. & Kjerfve, B. 2010. Coastal lagoons of Southeastern Brazil: Physical and Biogeochemical Characteristics. In: eds. Perillo, G.M.E.; Piccolo, M.C. & Pino-Quivira, M.. Estuaries of South America: Their Geomorphology and Dynamics, Springer-Verlag. Cap. 3, p. 35–66.

Knoppers, B.A., Kjerfve, B. & Carmouze, J-P. 1991. Trophic State and Water Turn-Over Time in Six Choked Coastal Lagoons in Brazil. Biogeochemistry, 14:149–166.

Lana, P.C.; Marone, E.; Lopes, R.M & Machado, E.C. 2001. The Subtropical Estuarine Complex of Paranaguá Bay, Brazil. In: U. Seeliger and B. Kjerfve (eds.) of Coastal Marine Ecosystems of Latin America. Springer-Verlag, Berlin, Heidelberg, pp. 131–143.

Lepage, S. & Ingram, R. 1986. Salinity Intrusion in the Eastmain River Estuary Following a Major Reduction of Freshwater Input. J. Geophys. Res., 91(C1):909–915.

Lessa G.C.; Cirano, M.; Tanajura, C.A.S.; Silva, R.R. 2009. Oceanografia Física. In: Hatje, V.; Andrade, J.B. de (org). Baía de Todos os Santos: aspectos oceanográficos. v.1, pp. 68–119.

Mantovanelli, A. 1999. Caracterização da Dinâmica Hídrica e do Material Particulado em Suspensão na Baía de Paranaguá e em sua Bacia de Drenagem. Dissertação de Mestrado. Curitiba, Universidade Federal do Paraná. 151 p.

Mantovanelli, A., Marone, E.; Silva, E.T.; Lautert, L.F. Klingenfuss, V.P.; Prata V.P., Noernberg, B.A.; Knoppers, B.A. & Angulo, R.J. 2004. Combined Tidal Velocity and Duration Asymmetries as a Determinant of Water Transport and Residual Flow in Paranaguá Bay Estuary. Estuarine, Coastal and Shelf Science 59, 523–537.

Marone, E.; Machado, E.C.; Lopes, R.M. & Silva, E.T. 2005. Land-ocean fluxes in the Paranaguá Bay estuarine system, southern Brazil. Braz. j. oceanogr. v.53 no.3–4, São Paulo, Brazil. doi:10.1590/S1679-87592005000200007.

Martin, L. & Dominguez J.M.L. 1994. Geological History of Coastal Lagoons. In: Kjerfve, B. (ed.). Coastal Lagoon Processes. Elsevier Oceanographic Series, 60. pp. 41-??.

Mee, L. D. 1978. Coastal lagoons. In: Chemical Oceanography, Vol. 7 (Riley, J. & Skirrow, O. eds.). Academic Press, New York, U.S.A., pp. 441–490.

Merino, M.; Czitrom, S.; Jordán E.; Martin, E.; Thomé P. & Moreno, O.; 1990. Hydrology and rain flushing of the Nichupté lagoon system, Cancún, Mexico. Estuarine, Costal and Shelf Science 30, 223–237.

Miranda, L. B.; Castro, B. M. de. & Kjerfve, B. 1998. Circulation and Mixing in the Bertioga Channel (SP, Brazil) Due to Tidal Forcing. Estuaries, 21(2):204–214.

Miranda, L. B.; Dalle Olle, E.; Bérgamo, A.L.; Silva, L.S. & Andutta, F.P.; Silveira, I.C.A,; Fernandes, F.P.A.; Ponsoni, L. & Costa, T.P. 2011. A Descriptive Analysis of the Seasonal Variation of Physical Oceanographic Characteristics in the Northern Region of the Todos os Santos Bay. Braz. J. Oceanogr. 59(1): 9–26.

Miranda, L. B., Dalle Olle, E.; Bérgamo, A.L.; Silva, L.S. & Andutta, F.P. 2012. Circulation and salt intrusion in the Piaçaguera Channel, Santos (SP). Braz. J. Oceanography, 60(1):11–23.

Möller, O. O. Jr., 1996. Hydrodynamique de la Lagune dos Patos (30°S, Brésil). Measures et Modélisation. Tese de Doutorado. Bordeaux, L'Université de Bordeaux I. École Doctorale des Sciences de la Terre et de la Mer. 204 p.

Möller, O. O. Jr. & Castaing, P. 1999. Hydrographical Characteristics of the Estuarine Area of Patos Lagoon (30°S, Brazil). In: Perillo G. M. E; Picollo, M. C. & Pino-Quivira M. (eds.). Estuaries of South America: Their Geomorphology and Dynamics. Berlin, Springer-Verlag, pp. 83–100 (Environmental Science).

Möller, O. O. Jr.; Lorenzzetti, J. A.; Stech, J. L. & Mata, M. M. 1996. The Patos Lagoon Summertime Circulation and Dynamics. Cont. Shelf Res., 16(3):335–351.

Moore, N.H. & Slinn, D.J. 1984. The Physical Hydrology of a Lagoon System on the Pacific Coast of Mexico. Estuarine, Coastal and Shelf Science, v. 19:413–426.

Morris, A. W. 1985. Estuarine Chemistry and General Survey Strategy. In: Head, P. C. (ed.). Practical Estuarine Chemistry: A Handbook. Cambridge, Cambridge University Press, pp. 1–60.

Officer, C. B. 1976. Physical Oceanography of Estuaries (and Associated Coastal Waters). New York, Wiley. 465 p.

Officer, C. B. 1977. Longitudinal Circulation and Mixing Relations in Estuaries. Estuaries, Geophysics, and the Environment. Washington, D. C., National Academy of Sciences, pp. 13–21.

Officer, C. B. & Kester, D. R. 1991. On Estimating the Non-advective Tidal Exchanges and Advective Gravitational Circulation Exchanges in an Estuary. Estuar. Coast. Shelf Sci., 32: 99–103.

Oliveira, A. & Kjerfve, B. 1993. Environmental Responses of a Tropical Coastal Lagoon System to Hydrological Variability: Mundaú-Manguaba, Brazil. Estuar. Coast. Shelf Sci., 37: 575–591.

Phleger, F. B. 1969. Some General Features of Coastal Lagoons. In: Ayala-Castañares A. (ed.). Lagunas Costeras, un Simposio. Mexico, Universidade Nacional Autónoma de Mexico, pp. 5–26.

Pickard, G. L. 1961. Oceanographic Features of Inlets in the British Columbia Mainland Coast. J. Fish. Res. Bd. Can., 18:907–999.

Pickard, G. L. 1971. Some Physical Oceanographic Features of Inlets of Chile. J. Fish. Res. Bd. Can., 28: 1077–1106.

Prandle, D 1985. On Salinity Regimes and the Vertical Structure of Residual Flows in Narrow Tidal Estuaries. Estuar. Coast. Shelf Sci., 20:615–635.

Pritchard., D.W. 1952a. Salinity Distribution and Circulation in the Chesapeake Bay Estuarine System. J. Mar. Res. 11(1):106–123.

Pritchard, D. W. 1952b. Estuarine Hydrography. Advances in Geophysics. New York, Academic Press, vol. 1, pp. 243–280.

Pritchard, D. W. 1955. Estuarine Circulation Patterns. Proc. Am. Soc. Civ. Eng., 81:717:1–11.

Pritchard, D. W. 1967. What is an Estuary: Physical View Point. In: Lauff G. H. (ed.). Estuaries. Washington, D. C., American Association for the Advance of Science, pp. 3–5.

Pritchard, D. W. 1989. Estuarine Classification – A Help or a Hindrance. In: Neilson, B. J.; Kuo, A. & Brubaka, J. (eds.). Estuarine Circulation. Clifton, Humana, pp. 1–38.

Rattray Jr., M. & Uncles, R. J. 1983. On the Predictability of the 137Cs Distribution in the Severn Estuary. Estuar. coast. Shelf Sci., 16:475–487.

Saelen, O.D. 1967. Some features of the hydrography of Norwegian fjords. In: G.H. Lauff (ed.). Estuaries. American Association for the Advancement of Science, Washington, DC, pp. 63–70.

Sankarankutty, C.; Medeiros, G. F. & Lins Oliveira, J. E. 1995. Diurnal and Seasonal Fluctuations of Estuarine Zooplankton of Potengi Estuary, Natal, Northeast of Brazil. J. Mar. Biol. Ass. India, 37(1–2):201–211.

Scott, C. F. 1993. Canonical Parameters for Estuarine Classification. Estuar. Coast. Shelf Sci., 36:529–540.

Sikora, W.B. & Kjerfve, B. 1985. Factors Influencing the Salinity of Lake Pontchartrain, Louisiana, a Shallow Coastal Lagoon: Analysis of a Long–term data set. Estuaries, v.8: 170–180.

Simmons, H. B. 1955. Some Effects of Upland Discharge on Estuarine Hydraulics. Proc. Am. Soc. Civ. Engrs., 81(792).

Stommel, H. 1951. Recent Development in the Study of Tidal Estuaries. Tech. Rept., Massachusetts, Woods Hole Oceanographic Institution, n. 51–33. 18 p.

Stommel, H. 1953. The Role of Density Currents in Estuaries. Proc. Minnesota International Hydraulics Convention, Minneapolis, University of Minnesota, pp. 305–312.

Stommel, H. & Farmer, H. G. 1952. Abrupt Change in Width in Two-layer Open Channel Flow. J. Mar. Res., 11:203–214.

Tommasi, L. R. 1979. Considerações Ecológicas sobre o Sistema Estuarino de Santos (SP). Tese de Livre-docência. São Paulo, Instituto Oceanográfico, Universidade de São Paulo, 2 vols. 489 p.

Tundisi, J. G. & Matsumura-Tundisi, T. 2001. The Lagoon Region and Estuary Ecosystem of Cananeia, Brazil. In: Seeliger, U. & Kjerfve, B. (eds.). Coastal Marine Ecosystems of Latin America. Berlin, Springer-Verlag, p. 119–130 (Ecological Studies, 144).

Valle-Levinson, A. 2008. Density-driven exchange flow in terms of the Kelvin and Ekman numbers. J. Geophys. Res. 113, C04001, doi:10.1029/2007JC004144.

Valle-Levinson, A. 2010. Definition and classification of estuaries. In: ed. Valle-Levinson A. Contemporary Issues in Estuarine Physics. Cambridge University Press, pp. 1–11.

Wong, K.-C. 1994. On the nature of transverse variability in a coastal plain estuary. J. Geophys. Res. 99(C7), 14,209–14,222.

Wright, L. D. 1970. Circulation, Effluent Diffusion and Sediment Transport, Mouth of South Pass, Mississippi River Delta. Baton Rouge, Louisiana State University Press. 56 p.

Yáñez-Arancibia, A. 1987. Lagunas Costeras y Estuarios: Cronologia, Criterios y Conceptos para una Classificación Ecologica de Sistemas Costeros. In: Simpósio de Ecossistemas da Costa Sul e Sudeste Brasileira: Subsídios a um Gerenciamento Ambiental. São Paulo, Serra Negra, Aciesp, (54-III):1–38.

Chapter 4
Physical Properties and Experiments in Estuaries

The investigation of processes and how estuarine systems function, presents distinct aspects, being conducted with various purposes and objectives. The experimental procedure may include sampling to perform the analysis of spatial and temporal variations of physical measurements of hydrographical properties (salinity, temperature and pressure), currents, sea level oscillations (waves and mainly tides) are of fundamental importance in Physical Oceanography. The experimental procedure may include sampling to perform the analysis of spatial and temporal variations of chemical substances (natural and anthropogenic), sediments in suspension, dissolved gases and marine organisms in the water column and at the bottom.

The physical processes common to all estuaries are their cyclical motions and the mixing of the water masses with contrasting origins: the fresh water discharged by rivers and the seawater from the ocean. As result of this process, associated with generating forces of the motions, the estuaries are non-homogeneous water bodies and their properties may vary in wide time and spatial scales presented in Chap. 2.

Taking into account the comprehensive and systematic article of Kjerfve (1979) the basic principles of measurements, reduction and data edition of estuarine properties will be presented. These data are the result of measurements of vector and scalar physical properties, and the determination of dependent properties necessary to the estuarine knowledge to study its environmental and dynamical properties. According to the technical report of Unesco (1985), the physical properties must be expressed, whenever possible, in the Standard of International Unity System, usually abbreviate as SI unity.

4.1 Research Planning

Investigations into the behavior and estuaries characteristics take many forms and are initiated with varying purposes. Field studies could include water samples to determine concentrations of nutrients, ATP, plankton, suspended sediments, pH,

L. Bruner de Miranda et al., *Fundamentals of Estuarine Physical Oceanography*, Ocean Engineering & Oceanography 8, DOI 10.1007/978-981-10-3041-3_4

dissolved gases, and many others properties. However, independent of the discipline, all field-oriented estuarine studies include current velocity (speed and direction), water temperature and salinity measurements, associated with river discharge, tide measurements and meteorological variables, as wind speed and direction, and echo sounder measurements. Remote sensing may also be included in the investigation.

Estuaries typically experience great spatial variations, seasonal, fortnightly, semi-diurnal and/or diurnal oscillations in water level, velocity and material concentrations. This is primarily due to a combination of tidal influence, fresh-water river discharge, meteorological forcing, and the constraints imposed by the configuration and morphology of the estuary. Thus, the representativeness of any set of estuarine measurements is highly dependent on the sampling design (choice of sampling locations, sampling rate, and study duration), as well as a rational procedure of analysis and synthesis of the data. These results are used not only for the spatial and temporal variations, but also to the estuary classification and determination of flux and transports, as well as to validate theoretical results generated by analytical and numerical models.

Before any experimental investigation, a detailed analysis of the objectives and the theoretical aspects must be accomplished; this makes it possible to decide what properties must be measured, and to establish the procedure analysis of each variable for its convenient reduction.

In the classical book of Dyer (1973), examples are presented of typical estuaries illustrating the propertie's variability under the influence of different tidal amplitude, river discharge and geomorphology characteristics. In these examples, the property descriptions (temperature, salinity and current velocity) were made based in averaged values in transversal and longitudinal sections, taking into account the local variability. Other articles (Stommel 1953a; Dyer 1977; Kjerfve and Proehl 1979 and Uncles and Kjerfve 1986) were able to show that transversal motions induced by bottom topography and channel irregularities may also be as important as the longitudinal variations. In these works it was also shown that the experiment time length, to calculate the mean circulation during tidal cycles, takes into account subtidal frequency variations due to the synoptic and the seasonal meteorological forcing, as well as the time variability of the river discharge.

The tidal currents and the mixing process generate problems in data reducing and analysis which can only be solved by calculation of time mean values. The selection of a suitable sampling duration to determine the time-averaged estuarine conditions is as critical as the spatial sampling. Elliott's (1976) study points out extreme variability in time-averaged estuarine currents from one tidal cycle to another, and shows how the time-averaged current direction frequently reverses. It is likely that in most estuaries the greatest portion of the variability, on time scales from two to 20 days, occurs in response to meteorological forcing such as wind stress and atmospheric pressure fluctuations (Kjerfve 1979).

In calculating the mean flow in the Providence river estuary (Rhode Island, USA), frequently forced by the wind, Weisberg (1976) observed that, in general, it is not enough to calculate time mean values from measurements of only a few tidal cycles. Considering wind-induce current fluctuations Weisberg derived, using spectral analysis techniques, the following equation to arrive a meaningful time interval (Δt_m), which is necessary to the calculation non-tidal circulation or flux estimates,

$$\Delta t_m = \frac{\Psi^2}{2B_e \varepsilon^2 u_m^2}. \quad (4.1)$$

In this equation both $\Psi(\Psi = 2 \times 10^{-2}\,U$ is the variance of the axial current), and B_e (effective spectral bandwidth) are primarily dependent on the local wind intensity $U(\mathrm{cm\,s^{-1}})$. The symbol ε indicates the normalized error and u_m, the mean current velocity ($\mathrm{cm\,s^{-1}}$) may be anticipated from previous estuarine measurements published data, river discharge or conservations considerations for an idealized two-layered mean flow.

Let us assume, for example, that the wind speed is $500\,\mathrm{cm\,s^{-1}}$ ($\Psi^2 = 10^2\,\mathrm{cm\,s^{-2}}$), $\varepsilon = 0.2$, $B_e = 0.03$ cph, and $u_m = 10\,\mathrm{cm\,s^{-1}}$. Then, in this case, $\Delta t_m = 417$ h or 33 semi-diurnal tidal cycles. The main conclusion to be drawn from Weisberg's (1976) analysis is that if a too short sampling duration is selected, the resulting time-averages may not be representative of the typical conditions for a particular estuary (Kjerfve 1979); these results must be carefully interpreted, especially if during the data sampling the estuary has been forced by strong winds or abnormal meteorological events. These difficulties may be overcome with the use of continuous record instrumentation which may be operated during great time intervals. In the absence of these abnormal meteorological events and in estuaries with relatively small surface area, experiments conducted during a few tidal cycles may give reliable results.

Another example, on the correlation of the low-frequency response of estuarine sea level to non-local forcing variability, is found in the article of Kjerfve (1978). The analysis of one-year time series records of sea level, atmospheric pressure, and wind (speed and direction), representative to the well-mixed North Inlet estuary (South Caroline, USA), indicated two important sea level variability due to the forcing from the coastal ocean: a 3.2 cm high sea level wave at 6.0 days period is highly correlated with changes in atmospheric pressure, and 6.4 cm high sea level wave at 9.2 days periods is attributed to continental shelf waves driven by the along-shore wind stress.

The formulation of a project must be preceded by a scientific hypothesis, and it is not recommended to start the field sampling without one or more hypotheses, because they are the basis of the scientific method. As an environmental project, their stages must be carefully planned taking into account the following activities:

1. Planning: before the measurements and hypothesis extracted from previous studies and from the theoretical knowledge of the problem, a decision must be taken about the measurements and what should be done after data quality control.

2. Field work: as this stage involve data collection spatially distributed (with an oceanographic boat), and/or time series measurements of properties (fixed stations or moorings) with calibrated instruments, it is necessary to take into account the sampling details: geographic location and geometry (nautical charts), stations number and/or moorings, sample distribution in the water column, and logistic aspects related to the time interval between the measurements. It is advisable to start and finish the measurements at high or low water, mainly when estuary circulation is very low. The field work may also be associated with remote sensing (airplane or satellite) to observe the estuary in a time spot or sequentially in time.
3. Control and editing: at this stage, which may be partially accomplished on board, the experimental data must be carefully examined, to prevent observational errors and the one due to the malfunction of the equipment sensors. After that data control, the observational data are reduced and filed in a convenient format to enable its analysis.
4. Numerical treatment: analytical and numerical modeling are very useful to theoretically simulate the spatial and temporal properties, currents and transport forced by the river discharge, tidal oscillation, wind stress and density gradients. The comparison of theoretical simulations with the observational data is necessary to validate the theoretical results. This stage may also include statistical treatment of the experimental data with mathematical regressions, time series analysis (time domain) and spectral analysis (frequency domain).
5. Analysis and synthesis: this stage includes the synthesis of the experimental and theoretical results in tables and graphics, and its interpretation with known theories. This stage isn't a trivial one in a big project.
6. Reports and articles: this is the final step activity and its products are technical reports and articles submitted to specialized magazines.

Percentage estimation of these activities, in comparison to the work efforts to the financial cost and their contributions to the final product of the project are presented in Table 4.1.

Table 4.1 Project components and percentage estimation relative to the work effort, financial cost and the final product

Component	Effort (%)	Cost (%)	Product (%)
i-Planning	05	05	00
ii-Field work	20	60	05
iii-Control and editing	10	05	05
iv-Numerical treatment	40	10	05
v-Analysis and synthesis	10	05	10
vi-Report and articles	10	15	75

When there are no previous studies it is necessary to calculate theoretically the magnitudes of such variables as: current velocity, tidal height and excursion. With this information it is possible to specify correctly the equipment to be used in the measurements, in the planning and logistics to reach the objectives.

Field works, demanding lateral sampling to the knowledge of the mean vertical structure in the cross-section of a laterally non-homogeneous estuary require at least in three stations across. It also required the sampling time interval of one or half-hour ($\Delta t = 1$ h, or 0.5 h), and take care if the shallow depths near the margins are adequately sampled. To well defined vertical profiles in the water column a minimum of selected depths must be sampled from the surface down to the bottom, with intervals of $\Delta z = 1.0$ or 0.5 m. If a continuous property profiling is used in the measurements, the interpolation may be made at the same selected time intervals. When the instruments have no pressure sensors, the sampling depth (z) must be made taking into account the wire angle (ϕ, measured by a clinometer) and the wire length (L) of the hanging instruments, generally calculated by $z = L\cos(\phi)$. In longitudinal or transversal sections, the distance between the sampling stations needs to satisfy the following: (i) the difference of the property mean value between stations must be higher than the measurement error; and (ii) the property gradient must vary linearly with the distance. An analysis of the decrease in the committed error in the computation of transport or flux properties, in vertical sections with number of sampled stations, is presented in detail in the article of Boon III (1978).

The sampling time interval (Δt) during measurements of one or more tidal cycles is of crucial importance to prevent *aliasing*; if the sampling frequency is not adequately made in relation to the temporal scale of the phenomenon, it will not be sampled with the necessary detail.

The description of the motions and physical characteristics of an estuary may be obtained in two different basic methods which are related to the theoretical aspects of Fluid Dynamics, namely the Euler and Lagrange formulations. In the Eulerian description, the properties are measured in the time domain, in a fixed point of estuary, and in the Lagrangian the measurements are made in a drifting volume and in the time domain. The first description is the most used in estuarine research. For instance, in the estuary classification, using the stratification-circulation diagram, the set of measurements must be made in a fixed station and during the time, during one of more tidal cycles. Ideally, Eulerian measurements of hydrographic properties and currents should be made in a set of oceanographic stations distributed in the estuary space for its adequate sampling.

When, in the experiment, only instrumentation which needs an observer is used, it is very difficult to perform measurements during several tidal cycles. In these experiments a series of oceanographic stations may be occupied sequentially in space and time with only one equipped boat, or several stations may be simultaneously sampled with several boats aligned in the estuary cross section, as in the experiment described in the article of Kjerfve and Proehl (1979), as part of a multi-disciplinary investigation of the material transport between the cross section and the coastal ocean. In this experiment, in a the well-mixed tidally driven North Inlet estuary (South Caroline, USA), a 320 m wide cross section, with typical channel depths of 5 m, 11 stations were occupied with simultaneously sampling taken every 30 min. However, with only one instrumented boat and this sampling rate, the experimental work may cover partially the estuary and this limitation may be crucial in large estuaries. The detailed results of this investigation, related to the total material flux estimates was part of a multi-disciplinary investigation.

4.2 Current Measurements, Tide and Hydrographic Properties

4.2.1 Current Velocity

As in others branches of Physical Oceanography, among the physical properties of interest, the most difficult one to be measured is current velocity. It is a vector property which presents great spatial and time variability both in intensity (speed) and direction.

The observational procedure of current measurements in estuaries is not trivial, typically, the speed increases from zero, at slack low water, to maxima values which can be in excess of 2–3 ms^{-1} (or higher in some estuaries) at mid tide before decelerating, reversing at slack high water, and then accelerating to achieve similar or higher values in the opposite direction (Hardisty 2007).

Velocity measurements may be made using a moored boat or moored equipment in a fixed position. In the case of measurements with the sailing boat, depending on the equipment used, it is necessary to know the boat velocity and its direction, to extract the real current velocity.

In velocity measurements manual, mechanical, electronic, electromagnetic and automated techniques may be used to measure the speed and direction of the current in vertical profiles or in time series recording. Usually, current metering is made on board of a moored boat, performing velocity profiles from the surface down to the bottom at programmed depths and time intervals. Time series velocity measurements also may be made with more sophisticated electronic equipment as: autonomous moorings hanging on surface or bottom buoys, and with Acoustic Doppler Current Profiles fixed at the bottom. Mechanical devices embodying some form of rotating element which are used for water velocity measurements are called current meters, and The Proceedings of the Royal Society of London records descriptions of such devices from Newton's time (Hardisty op cit.).

The operation autonomy of this equipment in the field is determined by the time sampling rate (Δt), storage capacity (memory) and battery life duration. Although the last generation instruments are very expensive, there are less expensive versions which work at limited depths (<200 m) and can adequately be used in most estuaries.

Current meters record the current velocity (intensity and direction) by two sensors: one for intensity and other to the direction. However, others sensors may also be installed in this equipment to simultaneously measure properties as temperature, electrical conductivity, turbidity, oxygen concentration, for example. The current intensity sensors may be of three types: rotor, acoustic and electromagnetic, each one with its vantages concerning the resolution. The current direction is measured in relation to magnetic field of the Earth, with a compass (magnetic or electronic) and the equipment orientation through a steering device, or an orthogonal set of sensors which indicate its orientation with the flow. Consequently, in the data edition, it is necessary to take into account in the direction angle the magnetic declination, which depends on the geographic position and time of the measurement.

In current-meters with propellers or rotors the current intensity is measured by the time rate of the rotation, calibrated in a experimental channel with the controlled flow velocity. A type of Savonius rotor has curved plates mounted around a vertical axis. However, instruments with these types of intensity measurement respond excessively to the up and down fluctuations of the hanging cable of the instrument. This undesirable noise is known as *wave bombing*, resulting in a major intensity in all frequency energy bands. Equipments with this type of rotor must not be used hanging on board of boats due to its oscillations, mainly in the presence of moderate and high wave conditions. When hanging in moored buoys in regions which may be reached by high wind intensities, or in the proximity of the estuary mouth, which may be more strongly affected by waves, the time series record must be filtered to eliminate the undesirable fluctuations.

The rotor with six straight blades mounted around a vertical axis, is being turned to be a standard one in oceanographic instruments. Although less sensitive to the *wave bombing* the conventional equipment which uses this rotor type are not adequate to be used on the surface layer, due to the motions induced by the gravity waves, mainly due to the delay into respond to the direction changes. Under this layer, they may be moored because are less sensitive to the cable oscillations.

There are current meters using data sampling with a system of vector means in the *burst* connection. This methodology has a high temporal resolution and alternates between short and relatively long time, with and without measurements, respectively. The time series of intensity and direction are internally processed to calculate the mean values of intensity and direction during the sampling time intervals. This equipment, even when equipped with blade rotors, may be moored near the water surface mainly because the direction sensor, which consists of a small blade fixed in the compass has a small inertia.

The acoustic current meter measure the propagation time interval of a high frequency pulse between a source and the receptor separated by a fixed distance. As much as is the water velocity in the sound direction propagation, less will be the measured time interval. These instruments may use two or three pairs of this transmitter-receptor system disposed in perpendicular axes, two measure the horizontal component or the horizontal and vertical velocity components, respectively. A compass is used to measure the earth's magnetic field, and a tilt sensor measures the instrument's angle, and with these data the direction of the instrument is determined and consequently the current direction. The current direction and speed is thus measured in relation to the equipment, and an internal compass is use to determine the true direction. These types of current sensor may be moored even near the surface layer subjected to the wave motion.

Electromagnetic current meter has its functioning principle base in the Faraday's electromagnetic law, and the seawater functions as the electric conductor in motion. Once crossing the electromagnetic field created by the instrument, the sea water induces the generation of an electromagnetic force proportional to the current intensity and perpendicular to the current direction. This equipment has a system with two axes and an internal compass which measures the velocity horizontal components. They also may be moored nearby the surface.

The Doppler Current Meter system of acoustic profiling, generally referred as ADP or ADCP, uses the principle that the sound wave propagating in the seawater is modified when reflected by an object in motion. The equipment sends pulse sounds with different intensities which propagate into the water column, and these pulses are reflected back to the equipment by the reflecting particles in the seawater. These reflectors may be materials or organisms in suspension in the water, as sediments in suspension or by planktonic organisms transported by the currents and by the Doppler effect the frequency of the reflected sound wave length is different from that originally emitted. The transducer receptor is projected to pick up these anomalous frequencies and the relative velocity between the reflector and the instrument is calculated. This equipment has three sets of transducers, each one to do the measurement of the vector velocity component (two horizontal and one vertical). As the pulse intensities are calibrated to travel different distances in the water column, Doppler current meter measures vertical velocity profiles in discrete cells in the water column. The sensitivity of these instruments is very high to oscillations, and they have sensors to measure and compensate these oscillations. They may be used fixed in the boat or in its hull, sampling velocity current during the ship's track. There are also ADP versions to be moored for an autonomous operation, being the substitute of a conventional mooring line of a set of instruments typically used.

The main difficulty for measurements of current in estuarine regions is associated with the high frequency oscillations of the free surface due to the gravity waves, intense currents and the biological fouling, accidents and vandalism. Due to the low depth of coastal plain estuaries, the wave energy usually may reach the bottom, and it is very important to choose an adequate equipment to measure the current velocity operated in boats or in equipment installed in moorings. Further, using sub-superficial moorings, with an underwater buoy in the main branch, as the classical U type shown in Fig. 4.1 must be preferred, to minimize the noise due to the waving motion in the surface in the buoy and equipment. This scheme permits also the use of warning surface flashing signals, preventing boats and fishing boats with arrested nets to catch and destroy the mooring.

The relatively high speeds which may occur during the sampling produce a drag in the boat, cables, current meters and sampling bottles. Then, when using a boat as

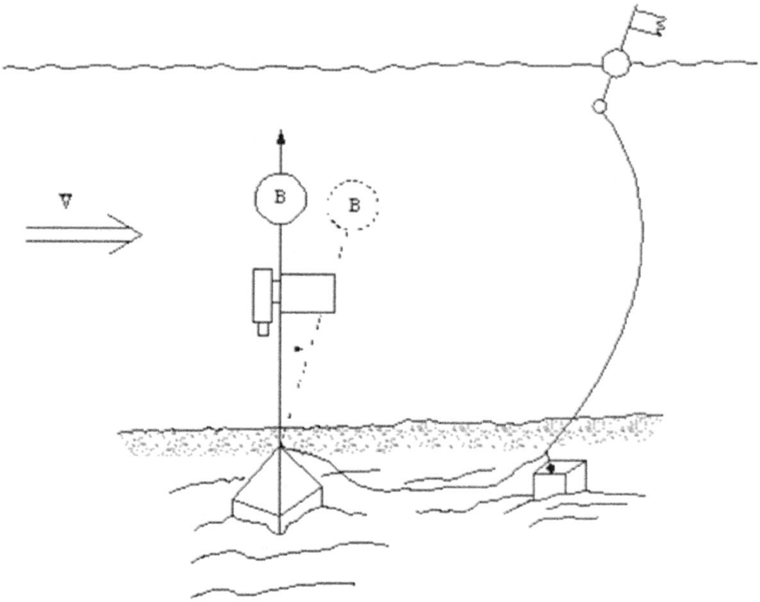

Fig. 4.1 Schematic displacement of the classical U mooring with a current meter and others sensors to measure the Eulerian variability of estuarine water mass

working platform, it must be anchored in two positions (*aft* and *fore*) to minimize its motions during the tide inversion from flood to ebb, and vice versa. Attention also must be paid in the dimension of the buoy lift that will sustain the main mooring cable. This lift drag must be strong enough for the equilibrium angle, between the vertical and the main branch inclination due to the current drag, must be less than the one specified by the current meter manufacturer. It is also desirable that the current meter has pressure sensors, so, the observed depth variations of the initial project may be established.

An instrumented mooring must receive frequent maintenance, mainly to cleaning of bio-fouling which may alter the sensors sensibility and the current meter calibration (speed and direction). The fouling may be so critical that may stop the rotor function, causing not only a gap in the time series measurement, but also money expenditure; the maintenance time interval depends on the investigated region and the season of the year. As a general rule, when measurement of current are made in a shallow water estuary and in warm season, the maintenance to clean the equipment must be more frequent, sometimes at every one or two weeks' time intervals. Recent advances in high sensors performance and anti-fouling technology is being applied to a new generation of equipment, improving the estuarine long-term data acquisition.

The equipment lost through accident with fishing nets, fault in the correct use of cable and launching procedures, inadequate chains and weights, and vandalism is relatively high in estuaries. However, using better technologies and maintenance it is possible to reaches indexes of recoveries of good measurements higher than 80–90%.

There is now underway in several estuaries around the world an up to date online measurements of velocity, meteorological, tidal height measurements and others properties. One source of these estuarine experiments is the estuarine flow data long-term acquisition PORTS (Physical Oceanographic Real Time System), operated in several USA harbors as: New York/New Jersey, Anchorage, Tampa Bay, Chesapeake Bay, Narragansett Bay, and others. These data will be very helpful for a better understanding of the Physical Oceanography of estuaries and the related components as Biological, Chemical and Geological.

4.2.2 Tide

The estuarine water surface oscillates horizontally and vertically forced by several distinct processes. In temporal scales from seconds to year the main are gravity waves, tidal co-oscillation, wind shear, seasonal river discharge and atmospheric pressure, as well as the circulation due to the wind shear stress on the continental shelf.

The measurement of the sea level oscillation in the intratidal domain are made by tidal gauges, installed in coastal stations, or rigidly moored on the estuary bottom. There are two main types of tide gauges: the mechanical driven by a surface buoy, and the acoustic.

In the mechanical instruments the sensor is a fluctuating buoy installed in the interior of a tube vertically displaced. The connection between the buoy and the water is made through a small diameter hole in the lateral, or situated at its bottom. The correct relationship between the tube and the diameter acts as a filter to the oscillating non-desirable noise of the gravity waves. The buoy, through a steel cable, senses the up and down tidal motion which is transferred to a mechanical devise and a pen that records analogically the tidal motion in function of time. Once installed, it must be calibrated by leveling in relation to a known *datum level*, and a periodic maintenance by cleaning the tube and buoy due to the bio-fouling. In remote places, it is necessary to take into account the possibility of vandalism.

Electronic pressure tide gauges are usually settled at the bottom in rigid platforms to prevent undesirable motions. The majority pressure sensors use the crystal piezoelectric effect whose precision is adequate to estuarine investigations. The equipment configuration must be set previously (time, date, sampling time rate, maximum tidal height) and, if necessary, signalization surface buoys may be launched to prevent the mooring to be drag by fishing nets.

The tidal height sampled by tide gauges are used to perform tidal analysis (components determination) and prediction. The first devices for tidal prediction were the tide prediction machine invented by Lord Kelvin, who devised the method of harmonic analysis in the second half of the XIX century. Between the several

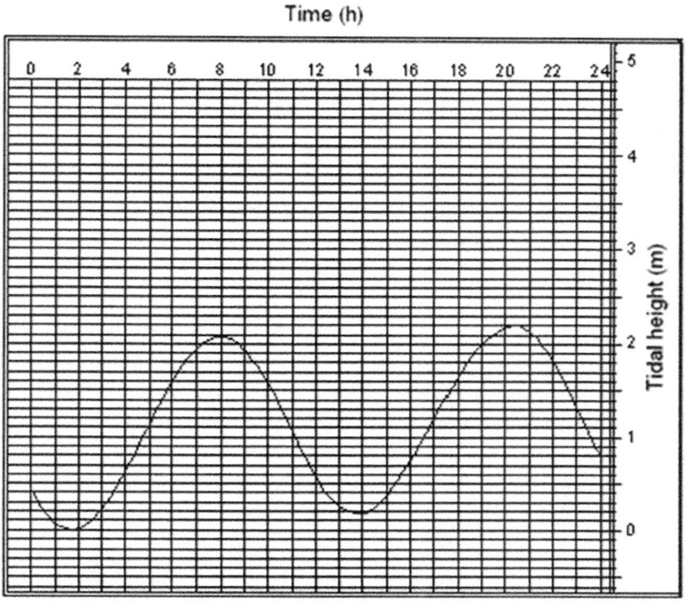

Fig. 4.2 Tide forecast for 24 h (two tidal semi-diurnal cycles) for the Natal harbor using the Pacmaré developed by Franco (2000)

methods we have already mentioned (Chap. 2) the software developed by Franco (2000), which uses the spectral analysis to tidal and current analysis and prediction (Pacmaré), using discrete tidal heights or current sampled at different time intervals (usually $\Delta t = 1.0$ or 0.5 h). The *Pacmaré* has a set of programs which compute the following correlated tasks: (i) Harmonic analysis for current, and tide and prediction, processing time series up to 3–4 years; (ii) Long time series analysis and prediction, processing up to 18.6 years. The computed results of set of harmonic tidal (current) components are used to predict the tidal heights or current speeds for any desired time and may be extracted in tables or graphic format. To exemplify one of the results which may be obtained with this software, the tide predicted to the Natal harbor (Rio Grande do Norte, Brazil) was calculated from an hourly time series of tidal heights with 1.8 years (Fig. 4.2).

4.2.3 Hydrographic Properties

The temperature and especially the salinity are hydrographic properties to be sampled in an estuary investigation. As shown in Chaps. 1 and 2, the salinity is a fundamental property, but also because its longitudinal gradient (or density gradient) is capable to generate the up-estuary longitudinal circulation due to the baroclinic pressure gradient and its vertical gradient indicates the vertical stratification (stability). Knowing the distribution of this property enables its classification according to the salinity stratification and the calculation of the Stratification-circulation Diagram, however, for that the longitudinal steady-state motion also must be known.

The density of the seawater is dependent on the temperature (T), salinity (S) and pressure (p), usually known as independent variables, and the density as function of these properties, $\rho = \rho(S, T, p)$, is calculated by the Equation of State of Sea Water. The pressure influence on the density is important to be considered only in type fjord estuaries, because of the great depth, which may reach 1000 m. As the most common are coastal plain and shallow estuaries, the pressure influence on the density is of minor importance, and the density may be considered only as function of salinity and temperature, $\rho = \rho(S, T)$, it may be determined by the equation state of seawater at atmospheric pressure ($p = 0$), or with simplified equations. As studied in the Chap. 2, the density is the physical property necessary to the determination the baroclinic component of the gradient pressure force (Eq. 2.10a); this component, associated with the river discharge and the vertical mixing processes, generates the *gravitational circulation*.

Temperature is the thermodynamic property to indicate if two physical bodies are or are not in thermodynamic equilibrium, or it may be taken as a measure of the heat content of a volume element. This property has variation in space and time, $T = T(x, y, z, t)$, due to the advective and diffusive processes and the exchange of sensible and latent heat with the atmosphere. The temperature may be measured at different depths in the water column with the classical protected *reversing thermometer*, having as sensor the differential coefficient of cubic expansion mercury in glass, was first manufactured in Italy by Negretti and Zambra, in 1874. This instrument was improved in Germany and, in about ten years, reached the high precision (±0.02 °C and ±0.005 °C). The detailed description of this thermometer may be found in classical books of Oceanography (Sverdrup et al. 1942; Defant 1961; Neumann and Pierson 1966, among and others). Nowadays, these classical thermometers are being replaced by the high precision *electronic reversing thermometer* with platinum temperature sensors. These thermometers were specially projected to be installed in bottles of Nansen, Ninskin or Van-Dorn bottles, to enable water sampling simultaneously to determination of salinity, and others chemical components and micro-biological micro-specimens of estuarine water mass.

The temperature (T') registered in a thermometer of the mercury in glass type is read in the boat laboratory or on the deck and, at a different temperature measured in situ (T). Then, it is necessary to apply a correction (ΔT), due to the volumetric expansion, because the temperature difference of the water in situ and the one in the boat (t_a), at the reading time of the reversing thermometer. The temperature (t_a) is measured by a thermometer named auxiliary, located at the same protecting glass of the reversing thermometer. Besides this influence, it is necessary to algebraically add to the reading T' (made with a magnifying glass) an experimental correction (I), named *index error* obtained in laboratory during the thermometer calibration against a standard thermometer, furnished by the manufacturer. Then, the in situ temperature is determined by:

$$T = T' + \Delta T + I. \quad (4.2)$$

A detailed revision on the published equations to calculate the volumetric expansion error (ΔT) was published in the article of Keyte (1965). To calculate this error it is necessary the following physical quantities furnished by the manufacturer: (i) V_0—volume of mercury in the capillary tube of the reversing thermometer, at the temperature 0 °C; (ii) K_T the coefficient of volumetric expansion of the thermometric system. For the water mass of the coastal plain estuaries it is enough to use the simplest correction formula, as the one deduced by G. Ferruglio in 1912:

$$\Delta T = (T' + V_0)(T' - t_a)/K_T. \quad (4.3)$$

This ΔT value combined with the Eq. (4.2), is the accurate value of the in situ temperature (T).

The great advantage of the electronic thermometers, in relation to the mercury in glass, is that the reading is made in a display digital and there is no volumetric expansion of the system. These thermometers have the same characteristics as the classic ones (they may be fitted to any sampling bottle) and have the following advantages:

- They are programmed in three operation modes (wait, continuous and sampling), which are selected by a magnetic key;
- There is no correction in the reading and its precision is ±0.015 °C;
- The display electronic can't be erased by mistake.

The salinity (S) is a physical-chemical property calculated as the ratio of the salt concentration mass (m, in grams) dissolved in a given mass of seawater (M, in kilograms): then, it is a non-dimensional property, $[S] = [MM^{-1}]$. Besides its importance to the ecologic characterization of the estuary this property it is used to calculate the density of the seawater. In the estuary the salinity presents great variability in time and space, $S = S(x, y, z, t)$, mainly due to the process of mixing (advection and diffusion) and the of river discharge; with some exceptions, the direct exchanges of the estuarine water mass with the atmosphere (through the processes of precipitation-evaporation) generally do not have an important contribution to the salinity variations, with exception to the hypersaline estuaries which may be formed in regions of arid climate.

The history of the salinity definitions and the methodology to its determinations dates back to 1693, thanks the early work of Robert Boyle, on measuring the saltiness of the seawater, evaporating the water, and weighing the solid residue (Hardisty 2007). Without going into the details, several definitions of this property evolved, and are from the dominion of the physical-chemistry, it follows that the main results are internationally known.

(a) *Classical definition of Martin Knudsen*

The traditional parameter used for estimate salinity is the chlorinity (Cl) concentration chemically determined in g/Kg (symbolically ‰), which measures chloride and bromide concentrations in the seawater usually by a volumetric procedure using a standard sea water as reference (*normal seawater*). The conversion of Cl concentration in salinity is made with the equation $S = 0.03 + 1.805 \times Cl$, named Knudsen equation, and may be used to in the range from 2.0 to 42.0‰, with an accuracy of ±0.03‰. This procedure was published in the Hydrographic Table of Knudsen et al. (1902), where the salinity has been defined as: "The total amount of solid material in grams contained in 1 kg of sea water when all carbonate has been converted to oxide, the bromine and iodine replaced by chlorine, and all organic matter completely oxidized". In the above equation the constant 0.03 represent approximately the solid content of river water flowing into the Baltic Sea, being the dominant influence in determining the ratio of ions in the solution of low salinity water.

(b) *Inductive scale*

With the advent of the electronic instruments to measure accurately the *conductivity ratio* (R_t), defined as the conductivity in situ in relation to that of a standard, in the 1960 decade, the salinity was redefined and the oceanographic community has started to use this accurate method to obtain this property of seawater more quickly and with higher precision.

To maintain the continuity of Knudsen scale and redefine the salinity as an addictive property, the conversion of chlorinity in salinity was made with the following equation: $S = 1.80655 \times Cl$. For $Cl = 19.374$ it follows, from these equations and the one of Knudsen, the same salinity $S = 35.0$‰. Using a set of 135 samples of seawater collected in the oceans which were carefully and precisely analyzed of its chlorinity content (Cl) and conductivity ratio (R_t) values, which were correlated with multiple correlation techniques, and a 5th° polynomial equation was fitted to determine salinity as variables R_t and the T, as independent variables, $S = S(R_t, T)$, with $T \geq 10$ °C. This equation was used to calculate the salinity in the inductive scale, in the same salinity interval of the Knudsen scale, and the algorithms for its determination were published in the Unesco Technical Papers in Marine Sciences (UNESC0 1966), enabling the salinity precision of ±0.003‰. The conductive ratio (R_t) and the simultaneous temperature of the sample (T) to be converted in salinity are measured with an equipment named Salinometer.

Until 1979, the salinity was reported in the same unity as in the classical unity (grams per kilogram or ‰). In the General Assembly of the International Association for the Physical Sciences of the Sea (IAPSO), held in December, it was made the recommendation that the symbol ‰, should be replaced by 10^{-3}; thus, for example, a salinity value of 35.120‰ should be expressed as 35.120×10^{-3} and, in non-dimensional formulation (kg/kg). However, due to the inconvenience of the unity change this recommendation has not been used by the oceanographic community.

(c) *Practical Salinity Scale*

As pointed out by Lewis (1980) the *conductivity ratio* (R_t) defines the salinity scale better than *chlorinity* scale for density determinations, and the new Practical Salinity Scale (1978) has been defined to eliminate the following difficulties of the former definitions: (i) The *standard seawater* may the reproduced in laboratory, independent on the ionic composition of seawater; (ii) The same algorithm approved by the Joint Panel on Oceanographic Tables and Standard (JPOTS) may be used for the calculation of the practical salinity from conductivity at all temperature and pressure over the ranges of oceanographic interest, in laboratory equipment or *in* with Conductivity-Temperature-Depth instruments; (iii) It turns to be a conservative property (Unesco 1981a).

Independent of the measured property (chlorinity or electrical conductivity) the Practical Salinity Scale and the former scales reproduces the same value corresponding to the value S = 35‰. And this value of S = 35‰ has by definition the conductivity ratio of unity ($R_t = 1.0$) at 15 °C, with a potassium chloride solution with concentration of 32.4356 g KCl/Kg. For conductivity ratio measurements in laboratory (at atmospheric pressure), and the salinometer has been standardize with a sub-standard of the KCl solution, the determination of the salinity is also made with a 5th° polynomial equation, but having as independent variable the square root of the conductivity ratio and the temperature, $S = S(R_t^{1/2}, T)$ with an accuracy of ±0.003‰. This polynomial expression may also be used to the determination salinity values at the atmospheric pressure, which may be accomplished using the algorithms published in the Technical Reports of Unesco (Unesco 1981a).

As in those former scales, accurate values may be calculated in the range of 2 and 42‰. Further details on the polynomials fitting may found in Perkin and Lewis (1980), and the algorithms for the salinity determination as function of R_t and T have been programmed in the MatLab® computational environment by Morgan (1994). It should be pointed out that it is possible to apply this new salinity definition to hyper-saline seas, estuaries and coastal lagoons, because the upper limit of the Practical Salinity Scale (42‰) was increased up to (50‰) by Poisson et al. (1991).

The described salinity definitions have as fundamental hypothesis the constant composition of the seawater. However, due to the river and runoff discharges into estuarine waters, other ions may be found discharged, and this hypothesis may not be true, and the accuracy of the methods (±0.03‰ Knudsen) and (± 0.003‰ Inductive and Practical scales) are only for oceanic waters. For coastal and estuarine waters there is no yet detailed information on its correct ionic composition and errors of ±0.04‰ in salinity and $\pm 5.0 \times 10^{-5}$ g cm^{-3} = $\pm 5.0 \times 10^{-2}$ kg m^{-3} in density may be tolerated. Thus, the Practical Salinity Scale and the International Equation of State of Seawater can be used for estuarine systems even without the detailed knowledge of their ionic composition (Millero 1984).

With the advance of the Electronic Engineering applied to Physical Oceanography, small equipment from the type of Conductivity-Temperature-Depth (CTD) were developed to operate in estuarine waters. With these instruments it is possible to sample continuous vertical salinity profiles, and concentration of others properties. The electrical conductivity or the conductivity ratio measured with these instruments, are automatically converted in salinity with the algorithms of the Practical Salinity (PSS-1978).

Salinity determination in hyper-saline estuaries and coastal lagoons presents difficulties when values to be sampled are higher than 50‰. These difficulties may be overcome under the hypothesis that it has the same ionic compositions as seawater. If so, to reduce the salinity to a determinable value, the sample may be dissolved with a certain amount of distilled non-ionic water, and the analysis of the new sample may be made with the methodologies already known, and the salinity of the original sample may be calculated by applying a correction factor.

Although less precise, the hyper-saline waters (S > 50‰) may have its salinity determined indirectly using measurements with refractive instruments and an hydrometer, both graduated with known high salinity samples. The hydrometer, in the Baumé scale, has the measurements in Be degrees; a measurement value may be converted in density with the following equation (CRC 1979):

$$\rho = \left[\frac{145}{(145 - \mathrm{Be}}\right] \mathrm{x}\, 10^3, \tag{4.4}$$

with the density in SI unities. For a distilled sample water at 4 °C and Be = 0°, it follow from this equation $\rho = 10^3$ kg m^{-3}. This method has been used to determine the salinity in a study related to the hydrology and salt balance in the hyper-saline Araruama lagoon (Rio de Janeiro, Brazil). In this study, the following linear correlation between salinity and the Be degrees was determined: $S = -2.9 + 11.0 \times \mathrm{Be}$ (Kjerfve et al. 1996).

Time series of the longitudinal velocity component (velocity decomposition will be presented in the next chapter), salinity and tidal measurements were sampled in the Cananéia-Iguape Estuarine System (Fig. 1.5). The Eulerian temperature and salinity time series were measured in a self recorder current-meter, moored at 6 m depth from the bottom and equipped with temperature and salinity sensors. The hydrographic properties were recorded, and are presented comparatively to the tidal record (Fig. 4.3). The visual analysis of the *intratidal* time variability of the current component (v) indicates the influence of the longitudinal circulation forced by the barotropic component of the gradient pressure force (tidal forcing). Visually it may be observed that there is a phase difference of approximately two hours between current and tidal oscillations, with the tide in advance of the velocity oscillation, indicating a non-progressive tidal wave. It is also possible to visualize that the ebb current ($v < 0$) is more intense than the flood current ($v > 0$), which indicate the superposition of a seaward residual motion generated by the system of rivers discharges empting into the estuary. The salinity variability also presents a local variability that, although out of phase with the tidal oscillation, oscillates closely to the tidal current. Mainly due to the advective salt flux this property increases and decreases during the flood and ebb tidal oscillations, respectively. Low frequency

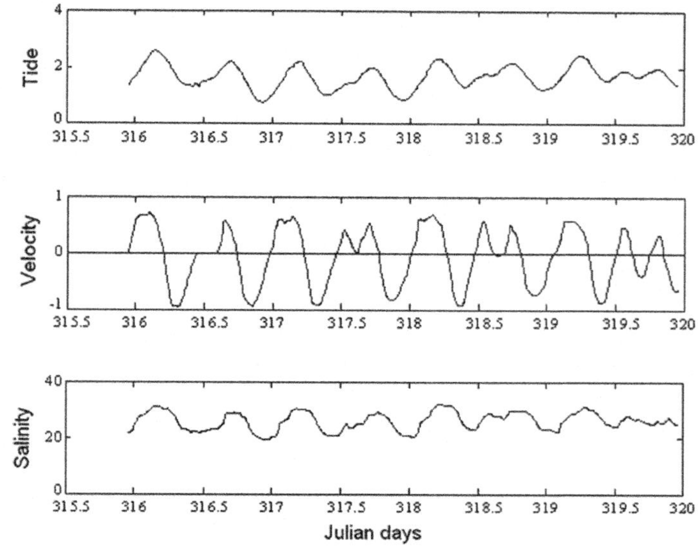

Fig. 4.3 Four day time series of tidal height, longitudinal velocity component and salinity recorded in the Cananéia-Iguape Estuarine System (São Paulo, Brazil) (November, 1996). The abscissa axis is graduated in Julian days (according to Miranda and Castro 1998)

variations, superimposed to the tidal, current and salinity oscillations may also be seen in this four day time series.

The accuracy of temperature and salinity measurements in estuarine waters are not so critic as in the open ocean, mainly due to the strong advective and diffusive influence of the tidal oscillation, and the occurrence of strong horizontal gradients (mainly salinity) in comparison to the open ocean. However, it should be pointed out that the longitudinal density gradient, necessary to calculate the baroclinic component of the gradient pressure force has a great time variability, turning it to be a very difficult physical quantity to be determined in the estuary. Then, although every measurement must be made as accurate as possible, and the high accuracy as in oceanic sea water may not be reached, it is very important in the estuarine research that measurements should be made as fast as possible to minimize the error induced by its relative high variability.

(d) *Pressure*

Around the year 287 (BC), Archimedes formulated the laws of hydrostatics. He also combined numbers and experiments and gave the principle of the surface level. These historical facts, which were of fundamental importance to the development of the Fluid Dynamics and Oceanography, were cited in the von Arx (1962) book relying on secondary sources of the Mc-Grow Hill series of history books.

Under ideal hydrostatic conditions, the pressure at every level of a water column is assumed to be equal to the weight of the fluid per unit area $[ML^{-1}T^{-2}]$.

Departures from hydrodynamic balance along the estuary and under normal conditions, it is mainly a consequence of the combined effects of the non-uniform density distribution and tidal oscillations. The analytical expressions of these components were presented in Chap. 2 (Eq. 2.10a, b).

The units of pressure usually employed in Oceanography are generally derived from the c.g.s. (centimeter, gram, second) system of units. In this system the force is expressed in dyne and, because the average atmospheric pressure (p_a) is $\approx 1.01 \times 10^6$ dynes cm^{-2}, it become a commonplace to consider pressure in terms of the *bar* (1.0 bar = 1.0×10^6 dynes cm^{-2}). The bar unity and its decimal parts as the decibar (1.0 decibar = 10^{-1}bar = 1.0×10^5 dynes cm^{-2}) is a very convenient unity of pressure, because the depth in meters in a seawater column may be numerically approximate to the pressure in decibars. This may be easily understood by the hydrostatic law written as $\Delta p = \rho g \Delta z = 1.03 \times 9.8 \times 10^2 \times 1.0 \times 10^2$ which is $\approx 1.0 \times 10^5$ dines cm^2 = 1.0 decibar; then, for an increase in the depth of 1 m (1.0×10^2 cm) the increase in pressure is 1.0 decibar, with an error less than 2%. This numerical equality is very useful and pressure sensors in CTD's usually have a piezoelectric as pressure sensor calibrated in decibar's.

4.3 Density and Equations of State

The seawater density (ρ) and the volume specific ($\alpha = 1/\rho$), representing physically the ratio of the mass per volume unity and its inverse, with dimension $[ML^{-3}]$ and $[L^3M^{-1}]$, respectively, are dependent on the salinity (S), temperature (T) and pressure (p). To the majority of estuaries (coastal plain), with the exception to the fjord estuaries, the density (volume specific) may be considered as dependent only on the salinity and temperature. The salinity variation interval usually is great as compared to the temperature, and the pressure effects on low pressure variations ($p < 100$ decibars$\approx$100 m) on the density may be disregarded.

Although CTD's (equipment with conductivity, temperature and pressure sensors) have high sensitivity and their measurements have good precision, as indicated above, this precision is not adequate to the determination of the horizontal component of the pressure gradient (Chap. 2, Eq. 2.9c) which is used to the determination of the gradient pressure force per mass unit, and it is impossible to separate with this measurement, the barotropic and baroclinic components (Eq. 2.10a, b Chap. 2). Then, this problem may only be resolved if the pressure is calculated in function of the density and under the assumption of the hydrostatic equilibrium, $dp = -\rho g dz$, remembering that the signal minus indicate in this equation that the Oz axis is oriented against the gravity acceleration).

There are in some tropical regions that, according to the season of the year, the river discharge into the estuary is too low, and there are two possibilities: (i) high evaporation and the estuary turns to be hyper-saline, and (ii) the surface heating is too intense and capable to cause density gradients, and thus influencing the estuarine circulation. In this last condition, due to the diurnal cycle of the temperature variation, these effects usually are transitory. In the fjords type estuaries occurring in high latitudes, the intense cooling on the surface layer during the winter time may generate deep convection, and thus the increase in the oxygen concentration in these layers. Hence, according to the estuary characteristics the temperature influences may not be disregarded (Dyer 1973).

The equation of state of seawater at the atmospheric pressure in function of the independent variables S and T enabling the determination of the density, $\rho = \rho(S, T)$ was based in laboratory experiments. In between these equations we may detach the classical Knudsen equation and the International Equation of Seawater (IESS-1980) (Knudsen 1902; Unesco 1979, 1981b), which have non-linear dependence on the variables S and T. The classical Knudsen equation is composed of a set of relations obtained in picnometer measurements under controlled conditions of S and T, using also the well-established state equation of pure water of E. H. Amagats (1893, quoted in Mamayev (1975)), and the seawater state equation at 0 °C. Originally it was resolved in relationship to one seawater parameter (Sigma–t or σ_t) associate to the density, $\rho = \rho(S, T)$, and defined by $\sigma_t = [\rho(S, T) - 1] \times 10^3]$, with the density expressed in the *c.g.s* system of unity.

The International Equation of State for Seawater, 1980 (IESS-1980) was determined with highly precision experimental data which were algebraically manipulated to result a set of equations in the polynomial format (algebraic power series in the S and T variables). The first equation of this set is the equation of pure water. The final result of this equation is the density value in units of the International System of Units (SI), and density is expressed in kg m^{-3}; due to the use of the SI system the parameter σ_t (also named density anomaly at the atmospheric pressure) is defined as $\sigma_t = \rho(S, T) - 10^3$.

Due to the great variability of the ionic composition of the estuarine water mass and the expected accuracy of ±0.04‰ in the salinity determination, with the PSS (Millero 1984), the precision in the determination of the seawater density with the IES-1980 is ≈0.05 kg m^{-3}. Due to the great variability of the density in estuaries this accuracy is satisfactory to the solutions of problems related to their hydrodynamics. These equations are widely found in technical papers and oceanographic tables (Fofonoff and Millard 1983; Unesco 1987), and all algorithms have been programmed in the computational MatLab® environment by Morgan (1994).

In the theoretical and numerical treatment of the circulation and mixing processes in estuaries, it is necessary to solve a closed hydrodynamic system of equations, and the state equation, $\rho = \rho(S, T, p)$ or $\rho = \rho(S, T)$ must take part of this system. In analytical solutions, simplified linear and non-linear equations that is reasonably efficient numerically, and has a wide range of application is one developed by Mellor (1991); this is an equation to calculate density whose independent variables are salinity, potential temperature, and pressure, and cover the small range of pressure and the large range of temperature and salinity found in estuaries, as well as, the large pressure range for deep basins application.

Due to the difficulties in analytical solutions it is convenient the use of equations of state in simplified linear and non-linear formulations, so disregarding the pressure effect. For deduction of linear equations, according to Mamayv (1975), the total differential ($d\rho$) of the functional relation $\rho = \rho(S, T)$, must be obtained,

$$d\rho = \left(\frac{\partial\rho}{\partial S}\right)_{T,p} dS + \left(\frac{\partial\rho}{\partial T}\right)_{S,p} dT. \tag{4.5}$$

As the saline ($\partial\rho/\partial S$) and thermal ($\partial\rho/\partial T$) gradients, according to their definitions, are related to the coefficients contraction of salinity (β) and thermal expansion (α_e) by,

$$\beta(S,T) = \left(\frac{1}{\rho_o}\right)\left(\frac{\partial\rho}{\partial S}\right), \tag{4.6}$$

and

$$\alpha_e(S,T) = -\left(\frac{1}{\rho_o}\right)\left(\frac{\partial\rho}{\partial T}\right), \tag{4.7}$$

where ρ_o is a density reference. Then, the combinations of Eqs. (4.6) and (4.7) with Eq. (4.5), and the differential $d\rho/\rho_o$ is expressed by:

$$\frac{d\rho}{\rho_o} = \beta dS - \alpha_e dT. \tag{4.8}$$

As known from the Thermodynamic of Seawater the coefficients β and α_e are dependent on the S and T. In the assumption of a mean value for these coefficients for the estuarine water mass, the differential expression (4.8) turns to be an equation with constant coefficients, and may be easily integrated,

$$\rho(S,T) = \rho_o(\beta S - \alpha_e T) + C. \tag{4.9}$$

the integration constant (C) may be, for simplicity, taken as $\rho_o = \rho(0,0)$, equal to the density of pure water at 0 °C ($\approx 1.0 \times 10^3$ kg m^{-3}), and the linear equation of state of the estuarine water is given by:

$$\rho(S,T) = \rho_o(1 + \beta S - \alpha_e T). \tag{4.10}$$

When the temperature effects may be disregarded, it follows the linear expression of the equation of state,

$$\rho(S, T) = \rho_o(1 + \beta S). \tag{4.11}$$

or, in terms of the specific volume,

$$\alpha(S, T) = \alpha_o(1 + \beta S)^{-1}. \qquad (4.12)$$

The constant values of the saline contraction and thermal expansion in the preceding equations must be mean values of the thermohaline characteristics of the estuary under investigation. These coefficients, determined with the analytical expressions have the following mean values: $\beta = 7.5 \times 10^{-4}$ and $\alpha_e = 2.0$ 10^{-4} °C^{-1}; these order of magnitudes are valid for the following variation intervals of salinity and temperature: 15 < T < 30 °C, and S > 10‰. However, the variation of β with S and T is much less than the dependency of α_e with these variables, and their values must be altered according to the problem to be studied. These values may be calculated with the Morgan's (1994) MATLAB® routines for calculating properties of seawater.

These linear Eqs. (4.11 and 4.12) have been used in solutions the of analytical steady-state models on gravitational circulation as, for example, in the classical article of Hansen and Rattray (1965), which resulted in theoretical profiles of the longitudinal velocity and salinity in estuaries partially mixed. In some solutions, the longitudinal density gradient $\left(\frac{\partial \rho}{\partial x}\right)$ may e substituted by the longitudinal salinity gradient using the relationship $\left(\frac{\partial \rho}{\partial x} = \rho_0 \beta \frac{\partial S}{\partial x}\right)$.

The non-linear approximations of the equation of state of seawater have been introduced in order to maintain the main non-linear dependence of the density with the properties S and T. The first simplified non-liner equation was presented by N. E. Dorsey in 1968, and later modified in the following more convenient formulation for practical applications (Mamayev, 1975):

$$\rho(S,T) = \rho_o + b(T - T_o) + c(T - T_o)^2 + [f + g(T - T_o)](S - S_o). \qquad (4.13)$$

In this equation the coefficients b, c, f and g are constants to be determined. S_o and T_o are also constant values of salinity and temperature, which may be chosen according to the variation of these properties. For convenience, the first term of the equation is $\rho_o = \rho(S_o, T_o)$; with this approximation, the thermal $(\partial\rho/\partial T)$ and saline $(\partial\rho/\partial S)$ gradients, with the first calculated with $S = S_o$, turns to be linear functions of the temperature.

Starting from the generic non-linear formulation of Eq. (4.13), and with S_o = 35‰ and T_o = 0.0 °C, Mamayev (1964, quoted in Mamayev (1975)), calculated the following equation to the density determination at atmospheric pressure, using values of the classical Knudsen equation, expressed in terms of the Sigma-t (σ_t) parameter:

$$\sigma_t(S,T) = 28.152 - 7.35 \times 10^{-2}T - 4.69 \times 10^{-3}T^2 + \left(0.802 - 2.0 \times 10^{-3}T\right)(S - 35), \qquad (4.14a)$$

and the density, in g cm^{-3}, is determined by;

$$\rho(S, T) = 1 + 10^{-3}\sigma_t(S, T). \tag{4.14b}$$

This equation may be applied for S and T varying in the following intervals: 0 < S < 40‰ and 0 < T < 30 °C, to calculate the density of the estuarine water mass. The comparison the results of this Eqs. (4.14a, 4.14b), with the Knudsen equation used in the determinations of its coefficients, indicate mean deviations varying from $\pm 5 \times 10^{-5}$ g cm^{-3} to $\pm 1.0 \times 10^{-4}$ g cm^{-3}.

Applying the same procedure, but using the thermal and saline gradients calculated by the IESS-1980, the following simplified non-linear state equation was obtained:

$$\begin{aligned}\rho(S,T) &= 1028,106 - 7.18575 \times 10^{-2}T - 4.54944\text{x}10^{-3}T^2 \\ &\quad + \left(7.99667 \times 10^{-1} - 1.84981 \times 10^{-3}T\right)(S - 35.0).\end{aligned} \tag{4.15}$$

This equation, with the density expressed in the SI units system (kg m^{-3}), may be applied to the following intervals of S and T: 0 < S < 40.0 ‰ and 0 < T < 40.0°C, which may be used to calculate the density when the salinity is measured in the practical scale (PSS-1978). The deviation in comparison with the IESS-1980, are near the deviations calculated by Millero (1984) which may be expected due to the different ionic composition of the seawater and coastal water masses (± 0.05 kg m^{-3}). This precision is adequate to the purposes of the Physical Oceanography of coastal plain estuaries.

The analytical expressions of dependent properties of seawater presented in this chapter, as the equation of state at atmospheric pressure $\sigma_t = \sigma_t(S, T)$ or $\rho = \rho(S, T)$, the coefficients of saline contraction $\beta = \beta(S, T)$, the thermal expansion $\alpha_e = \alpha_e(S, T)$ and the algorithmic of the PSS-1978, among others fundamental of sea water properties, may be easily determined with the Morgan's (1994) sub-routines.

An up to date item of information on salinity and the state equation of seawater is that, in 2010, the Intergovernmental Oceanographic Commission (IOC) and others associations, jointly adopted the new standard for the calculations of the absolute salinity (S_A), and a new standard for the calculation of the thermodynamics properties of sea water. This new standard, called Thermodynamic Equation of Ocean Seawater (TEOS-10), has been adopted in substitution the former equation of state of seawater (IESS-1980). The absolute salinity is defined in function of the currently used methodology of salinity measurements (PSS-1980), based on conductivity ratio measurements, and depends on the ionic composition of seawater at a geographical position of latitude (φ) longitude (λ) and pressure (p), expressed by Pawlowicz (2010);

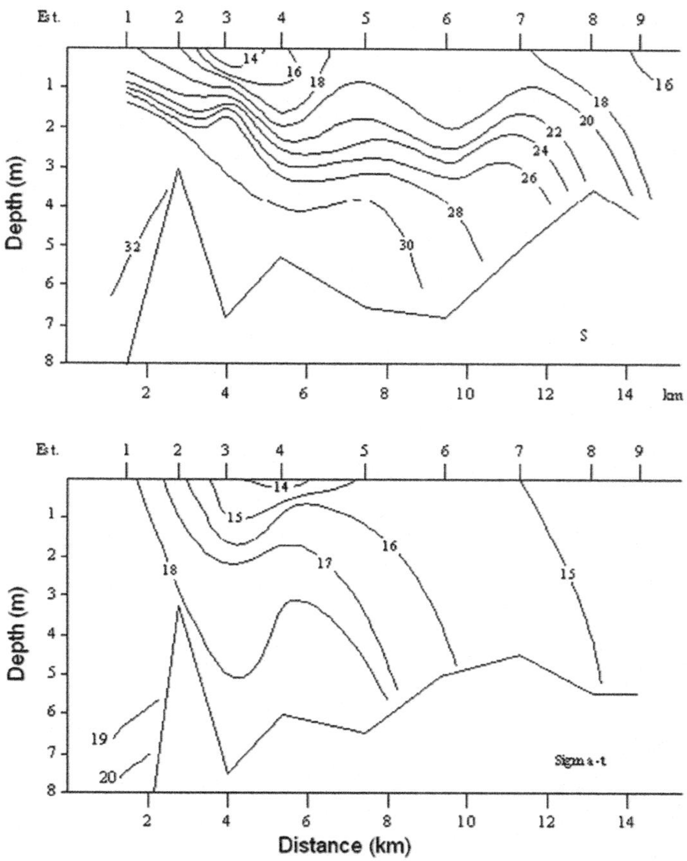

Fig. 4.4 Longitudinal salinity and density (Sigma-t) distributions in the Bertioga channel in the Santos-São Vicente Estuarine System (Fig. 1.5)

$$S_A = \left(\frac{35.16504}{35.0}\right) S_P + \delta S_A(\varphi, \lambda, p). \tag{4.16}$$

However, due to the difficulties for the accurate determination of the ionic composition of the estuarine water mass, which is necessary to know according to the S_A expression (4.16), its determination with the classical PSS-1980 salinity algorithmic will continue to be used in the estuarine research.

The longitudinal salinity and density (σ_t) distributions, in the estuarine channel of Bertioga (Fig. 1.5, Chap. 1) are presented in Fig. 4.4. The longitudinal salinity and density (σ_t) distributions in the estuarine channel are characterized by the isohalines and isopicnals with configurations with some similarities, showing that between the salinity and temperature properties, the first is the main one responsible to influence the density of the estuarine water mass in this environment.

References

Boon III, J. D. 1978. Suspended Solids Transport in a Salt Marsh Creek—An Analysis of Errors. In: 014, B. (ed.). Estuarine Transport Processes. University of South Carolina Press, Columbia, pp. 147–159. (The Belle W. Baruch Library in Marine Science, 7). 11 e final. indd 413 30/9/2011 10:13:53.

CRC (Chemical Rubber Company). 1979. Handbook of Chemistry and Physics. 60. ed., CRC Press, Boca Raton, p. F-3.

Defant, A. 1961. Physical Oceanography. Oxford, Pergamon Press, vol. 1. 729 p.

Dyer, K. R. 1973. Estuaries: A Physical Introduction. London, Wiley. 140 p.

Dyer, K. R. 1977. Lateral Circulation Effects in Estuaries. Estuaries, Geophysics and the Environment. Washington, D. C., National Academy of Sciences, pp. 22–29.

Elliott, A. J. 1976. A Study of the Effect of Meteorological Forcing on the Circulation in the Potomac Estuary. Spec. Rept. n. 56, reference 76–9, Chesapeake Bay Institute, The Johns Hopkins University, 67 p.

Fofonoff, N. P. & Millard Jr., R.C. 1983. Algorithms for Computation of Fundamental Properties of Seawater. Unesco Tech. Pap. Mar. Sci. n. 44. 53 p.

Franco, A. S. 2000. MARÉS: Programa para Previsão e Análise. In: Manual, BSP, São Paulo. 36 p.

Hansen, D. V. & Rattray Jr., M. 1965. Gravitational Circulation in Sraits and Estuaries. J. Mar. Res., 23(1):102–122.

Hardisty, J. 2007. Estuaries: Monitoring and Modeling the Physical System. Blackwell Publishing. 157 p.

Keyte, F. K. 1965. On the Formulas for Correcting Reversing Thermometers. Deep-Sea Res. 12(2):163–172.

Kjerfve, B.; Schettini, C. A. F.; Knoppers, B.; Lessa, G. & Ferreira, H. O. 1996. Hydrology and Salt Balance in a Large, Hypersaline Coastal Lagoon: Lagoa de Araruama, Brazil. Estuar. Coast. Shelf Sci., 42:701–725.

Kjerfve, B.; Greer, J. & Crout, R.L. 1978. Low-frequency response of estuarine sea level to non-local forcing. In M. Wiley ed., Estuarine Interactions. Academic, pp. 497–515

Kjerfve, 1986. Comparative Oceanography of Coastal Lagoons. In Estuarine Variability (Wolf, D. A., ed.). Academic Press, New York, U.S.A., pp. 63–81.

Kjerfve, 1979. Measurement and Analysis of Water Current, Temperature, Salinity, and Density. In: Dyer, K. R. (ed.). Estuarine Hydrography and Sedimentation. Cambridge, Cambridge University Press, pp. 186–226.

Kjerfve & Proehl, J. A. 1979. Velocity Variability in a Cross-Section of a Well-Mixed Estuary. J. Mar. Res., 37(3):409–418.

Knudsen, M. H. C. (ed.). 1902. Hydrographical Tables. Copenhagen, G. E. C. Gad., (press in 1962 by G. M. Manufacturing). 63 p.

Lewis, E. L. 1980. The practical salinity scale 1978 and its antecedents. In: Background papers and supporting data on the practical salinity scale. Unesco Tech. Pap. Mar. Sci., n. 37.

Mamayev, O. I. 1975. Temperature, Salinity Analysis of World Ocean Waters. Amsterdam, Elsevier. 374 p.

Mamayev, O.I. 1964. A simplified relationship between the density, temperature and salinity of seawater. Izv. Akad. Nauk S.S.S.R., ser. Geofiz., 2 (in Russian) (quoted in Mamayev (1975), p. 42).

Mellor, G.L. 1991. An Equation of State for Numerical Models of Ocean and Estuaries. J. Atmospheric and Ocean Technology. V.18, 00. 609–611.

Millero, J. M. 1984. The Conductivity-Density-Salinity-Clorinity Relationships for Estuarine Waters. Limnol. Oceanogr., 29(6):1317–1322.

Miranda, L. B.; Castro, B. M. & Kjerfve, B. 1998. Circulation and Mixing in the Bertioga Channel (SP, Brazil) Due to Tidal Forcing. Estuaries, 21(2):204–214.

Morgan, P. P. 1994. SEAWATER—A Library of MATLAB Computational Routines for the Properties of Sea Water. Rept. CSIRO Marine Laboratories, Australia. 222, 28 p.
Neumann, G. & Pierson, W. J. Jr. 1966. Principles of Physical Oceanography. London, Prentice-Hall. 545 p.
Pawlowicz, R. 2010. A model for predicting changes in the electrical conductivity, practical salinity and absolute salinity of seawater due to variations in relative chemical composition. Ocean Sci., 6, 361–378.
Perkin, R.G. & Lewis, E.L. 1980. The Practical Salinity Scale 1978; Fitting the Data. In IEEE Journal of Oceanic Engineering, v. OE-5, No. 1, pp.
Poisson, A.; Gadhoumi, M. H. & Morcos, S. 1991. Salinity and Density of Seawater: Tables for high salinities (42 to 50). Unesco Tech. Pap. Mar. Sci., n. 62. 85 p.
Stommel, 1953a. The Role of Density Currents in Estuaries. Proc. Minnesota International Hydraulics Convention, Minneapolis, University of Minnesota, pp. 305–312.
Svedrup, H. U.; Johnson, M. W. & Fleming, R. H. 1942. The Oceans, their Physics, Chemistry and General Biology. New Jersey, Prentice-Hall. 1042 p.
Unesco. 1966. Second Report of the Joint Panel on Oceanographic Tables and Standards. Unesco Tech. Pap. Mar. Sci., n. 4. 9 p.
Unesco. 1981a. The Practical Salinity Scale 1978 and the International Equation of State of Seawater 1980. Unesco Tech. Pap. Mar. Sci., n. 36. 25 p.
Unesco. 1979. Ninth Report of the Joint Panel on Oceanographic Tables and Standards. Unesco Tech. Pap. Mar. Sci., n. 30. 32 p.
Unesco. 1981b. Background papers and supporting data on the International Equation of State of Seawater. Unesco tech. Pap. Mar., Sci., v. 38, 192 p.
Unesco. 1987. Internacional Oceanographic Tables. Unesco Tech. Pap. Mar. Sci., n. 40. 195 p.
Unesco. 1985. The International System of Units (SI) in Oceanography. Unesco Tech. Pap. Mar. Sci., n. 45. 124 p.
Weisberg, R. H. 1976. A Note on Estuarine Mean Flow Estimation. J. Mar. Res., 34(3):387–394.
von Arx, W. S. 1962. An Introduction to Physical Oceanography. Addison-Wesley, Massachusetts. 422 p.

Chapter 5
Reduction and Analysis of Observational Data: Flux and Transport of Properties

Aspects related to the numerical treatment and analysis of observational data, which were included as items of the project component list (Table 4.1, Chap. 4) necessary to the project development, will be presented in this chapter. This comprises data reduction and analysis of scalar (hydrographic properties and tide) and vector (current velocity) data sampled in the water column (vertical profiles) and/or as temporal time series.

At this stage is important not to characterize the estuarine environment only in terms of its spatial and temporal variation of hydrographic properties and circulation. It is also important to value the theoretical interpretation of results. The hydrographic properties are very important in determining the flux and the transport of volume, concentration of salt, nutrients, pollutants and suspended sediments, and to establish the main characteristics of the importation or exportation of these concentrations. However, they do not provide the full dynamical understanding about the estuary, which is a much large picture.

5.1 Decomposition of Velocity

The magnetic (or electronic) compass inside the velocity measuring device is oriented to the North magnetic field of Earth, and the measured angle indicates the direction and its orientation relative to this magnetic field. As with the estuaries, special attention must be given to the longitudinal and transversal (secondary) velocity components as well as to the decomposition of the velocity vector.

Before considering the decomposition of velocity measured in an estuary, some elementary considerations will be given in relation to a vector denoted by $\vec{v}$, relative to a plane orthogonal reference system (Oxy). According to the reference system in Fig. 2.8 (Chap. 2), the u and v-velocity components in relationship to the Ox and Oy axes are calculated, respectively, as

L. Bruner de Miranda et al., *Fundamentals of Estuarine Physical Oceanography*, Ocean Engineering & Oceanography 8, DOI 10.1007/978-981-10-3041-3_5

$$u = V\cos(\theta), \tag{5.1}$$

and

$$v = V\sin(\theta). \tag{5.2}$$

In these equations V is the intensity of the velocity vector ($\vec{v}$), and θ is the trigonometric angle formed between the abscissa (Ox), measured in an anti-clock wise rotation. Then, this vector decomposition will result in u and v-velocity components that are positive, negative or null, according to the angle.

If the vector ($\vec{v}$) is the velocity at a given position in the estuary, measured by a current-meter, it has an intensity (V) and a direction denoted by the angle ($\underline{dd}$). As the direction of the current velocity is measured in the clockwise rotation, with its origin in the North magnetic field (Fig. 5.1b), and the origins of the angles θ and $\underline{dd}$ aren't coincidental, it is necessary to answer the following question: how to achieve the decomposition of the vector velocity in the components u and v, with the system of Eqs. (5.1) and (5.2)?

To answer this question, the first thing required is to make the origins of these angles ($\underline{dd}$ and θ) the same, because the trigonometric Eqs. (5.1) and (5.2), may only be applied for angles with that origin. Secondly, it must be taken into account whether the rotation angle has opposite directions (clockwise and anti-clockwise). In Fig. 5.1a it is possible to verify that these origins will be the same if the trigonometric angle θ is calculated by:

$$\theta = 90^{\circ} - dd \tag{5.3}$$

Now, it is necessary to adjust the magnetic North (NM) to the true North. This adjustment is of great practical importance when we need to plot the vector velocity in a nautical chart, because they are displayed in relation to the true north. This may be done without difficulty if the local magnetic declination angle (D), which is a deviation of the true North to east or west, is known. Magnetic declination varies

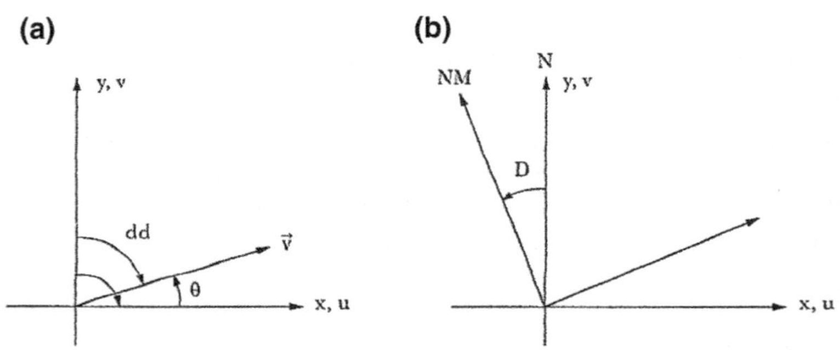

Fig. 5.1 **a** Decomposition of a velocity vector ($\vec{v}$) in an orthogonal reference system; **b** the same velocity vector in relation to the magnetic North (NM) and the true North (N), respectively

both with time and with geographical location due to astronomic and geophysical phenomena. Nautical charts present this angle and its annual variation along with the chart printing date, to enable its extrapolation and correction for the date when the experiment was performed. Thus, to adjust the magnetic direction, $\underline{dd}$, to the true North, in the case of a declination to the west (anti-clockwise), as shown schematically in Fig. 5.1b, it is necessary to change the direction angle $\underline{dd}$ by $(\underline{dd} - D)$, in Eq. 5.3. If the magnetic declination is to the east (clockwise), the substitution should be made by the angle $\underline{dd} + D$. Then, it follows that

$$\theta = 90^\circ - (\underline{dd} \pm D) \quad (5.4)$$

and the signals + and − (between parenthesis) are applied when the magnetic declination is to the east or west, respectively.

Finally, let us consider an estuary with its longitudinal axis oriented according to an angle, γ, in relation to the true North, such as in Fig. 5.2. This angle (γ) corresponds to an anti-clockwise rotation for the Ox axis to be in the longitudinal direction and oriented positively seaward. Thus, for Eqs. 5.1 and 5.2 which are used in this decomposition, the value θ of Eq. 5.4 must be subtracted from the rotation angle (γ); in the case of a clockwise rotation this angle (γ) must be added to the angle (θ). Then, the final angle θ angle for the decomposition of the velocity vector in an estuary is:

$$\theta = 90^\circ - (dd \pm D) \pm \gamma \quad (5.5)$$

Substitution of the angle (θ), into Eqs. (5.1) and (5.2), will enable calculate the velocity components (u and v) which are necessary for estuarine circulation

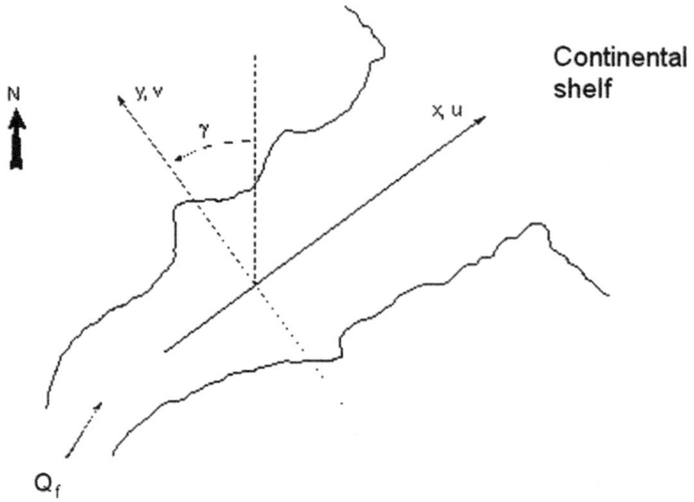

Fig. 5.2 Decomposition of a current velocity vector ($\vec{v}$) into u and v components (longitudinal and transversal, respectively), in relation to the local coordinate system (Oxy)

analysis, and calculation of advective and diffusive transports. According to the longitudinal axis orientation (Fig. 5.2), positive and negative values of the u-velocity component indicates seaward (ebb) and landward (flood) motions, respectively, or, in relation to the tide motions ebb and flood, respectively.

Wind stress is one of the main forces generating the circulation in the oceanic and coastal seas and may influence the circulation and vertical stratification in estuaries (Weisberg 1976; Elliott 1976; Geyer 1997; Valle-Levinson et al. 1998; and others). It is also necessary to calculate the wind-velocity components which are required to determine the correlation of wind stress with current velocity measured on the open sea and in estuaries, to investigate the strength of its influence on the circulation. However, wind velocity measured with an anemograph often refers to the direction in which the wind is coming from. This is the opposite convention to current velocity. Therefore, in such cases it is necessary to rotate the wind velocity by 180° to match the current velocity convention, so that correlation between the wind velocity components (U_v, V_v) and the current velocity components (u, v) can be measured. Then, the wind velocity decomposition can be made with the following set of equations:

$$U_v = U\cos(\vartheta), \tag{5.6}$$

and

$$V_v = U\sin(\vartheta), \tag{5.7}$$

where U indicate the wind intensity and the angle ϑ is given by:

$$\vartheta = 270° - dd_V \pm \gamma, \tag{5.8}$$

with dd_V indicating the wind direction. The 270° angle is 180° out of phase in relation to Eq. (5.4), and the magnetic declinations have not been taken into account because, generally, they usually are compensated during the anemograph installation in coastal meteorological stations. If the wind vector direction is referenced as the current motions at sea, which has also been used in wind measurements, the ϑ angle of Eq. (5.8) must be adjusted to $\vartheta = 90° - dd_V \pm \gamma$.

As an example of decomposition of a velocity vector measured in an estuary, consider the intensity and direction of velocity presented in Table 2.2 (Chap. 2), in the assumption that the estuary is oriented in the north-south direction and that the estuary mouth is located to the south. Then, by rotating the Ox axis 90° clockwise, it will be oriented southward and towards the estuary mouth. At the position of the original measurements, the magnetic declination was 20° towards west (D = 20°), and the final rotation angle is $\theta = 90° - (dd - 20°) + 90° = 180° - (dd - 20°)$. The results obtained by applying Eqs. (5.1) and (5.2) to calculate the vertical profiles of the u and v-velocity components are presented in Table 5.1. This table shows that $|u| \gg |v|$ at times t_1 and t_2; however, in the experiment, at time t_2 (current with low speed near slack water), the signal of the u-component changes

Table 5.1 Longitudinal (u) and transversal (v) velocity components calculated with vector velocities and direction from Table 2.2 (Chap. 2)

Depth (m)	u (m s^{-1}) (t_1)	v (m s^{-1}) (t_1)	u (m s^{-1}) (t_2)	v (m s^{-1}) (t_2)
0.0	1.09	−0.02	0.22	0.01
1.0	1.07	0.23	0.20	0.05
2.0	0.92	0.21	0.10	0.00
3.0	0.84	0.13	−0.02	0.01
4.0	0.74	0.04	−0.19	−0.03
5.0	0.59	0.02	−0.30	−0.02
6.0	0.40	0.19	−0.32	−0.06
7.0	0.26	0.14	−0.47	−0.02

Time measurements on the ebb tide are indicated by t_1 and t_2

direction at 3 m depth from seaward to landward, indicating that this direction change was forced by the intensity increase of the baroclinic component of the gradient pressure force (Eq. 2.10, Chap. 2).

5.2 Vertical Velocity Profiles

Continuous or discrete profile measurements of hydrographic properties and current velocity must be interpolated at discrete depths, usually at equidistant depths between the surface and the bottom. These discrete values may be obtained by graphical or numeric methods of interpolation, the latter being the best for minimizing errors and when the number of profiles is large. Among the numerical methods, we may cite the Lagrange and the *cubic spline*.

In the cubic splines, cubic polynomials are found to approximate the curve between each pair of data points, and the data adjustment is made by a third degree polynomial enabling the interpolations at pre-selected points. In the language of splines, these data points are called the breakpoints, and, since a straight line is uniquely defined by two points, an infinite number of cubic polynomials can be used to approximate a curve between two points. If the discrete profile has N measured quantities from the surface down to the bottom, this method assumes that the extreme polynomial points have no curvature, that is, the second derivative in relation to these points is null, or the curvature is constant. Thus, if the angular coefficient is known, extrapolations may be performed along the water column, from the surface and down to the bottom (Pennington 1970; Hanselman and Littlefield 1998). To obtain best results with this method, two conditions must be satisfied: (i) the experimental measurements must be made as close as possible to the surface and bottom; and (ii) the number of experimental data points must be higher than the number of depths to be interpolated in the water column. Further details on cubic splines processing in MatLab® computational environment may be found in Hanselman and Littlefield (op. cit).

According to the basic principles of hydrodynamics, the stress is proportional to the velocity component perpendicular to the motion, and the maximum friction is related to the water molecules at the bottom solid surface, with no horizontal significant movement. The friction at the bottom is estimated as a function of the velocity, called the *friction velocity* (to be defined in this chapter), or as a function of the amplitude of the tidal velocity and the water column depth. Due to this bottom characteristic, the velocity shear has a distinct structure called *boundary layer*, where the fluid velocity goes to zero or has a small value indicating a no-slip and a slippery bottom conditions, respectively.

Under simplified conditions, the velocity intensity increases from the bottom towards the surface until the motion occurs as if the bottom was a smooth surface (Chriss and Caldwell 1984), as schematically shown in Fig. 5.3a. Assuming this figure illustrates the unidirectional motion in an estuary during the ebb tide, in a later time the flow may be in the opposite direction (tidal flood), or may even be a bidirectional turbulent motion (flood and ebb), because in these coastal environments the circulation can be very complex, and turbulent fluctuations of velocity across the main flow may also occur causing vertical instabilities.

The estuarine water mass circulation has a free surface Newtonian fluid behavior and its intensity decreases with depth due to internal friction and frictional shear stress at the boundaries. The turbulence transmission due to velocity shears in the water column is caused by momentum exchange between layers, which may be parameterized by an eddy viscosity coefficient. As may be observed, near the bottom (which is plane and smooth by hypothesis), the fluctuating velocity profile is gradually damped by the fluid viscosity (Fig. 5.3b). Near the bottom, the shear stress imposed by the water motion to the solid bottom is transmitted almost entirely by the molecular viscosity.

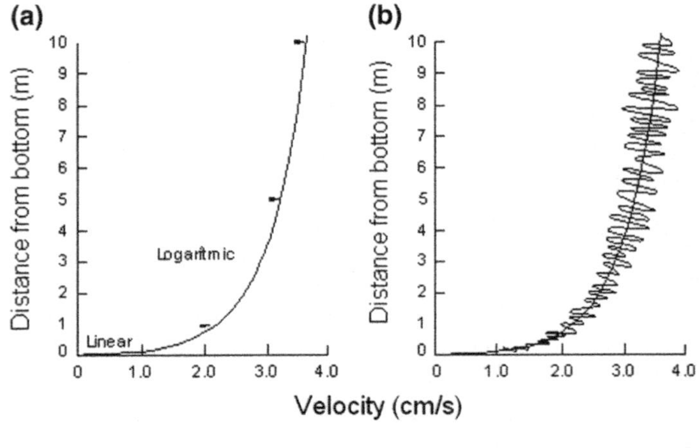

Fig. 5.3 **a** Velocity profile over a smooth bottom surface characterized by the linear viscous sub-layer near the bottom and the upper logarithmic layer. **b** Vertical profile showing the fluctuations and the turbulent motions decrease and turbulent motions in the sub-layer near the bottom (adapted from Chriss and Caldwell 1984)

Figure 5.3a shows a theoretical velocity profile composed of two layers: (i) the logarithmic layer, where the process of vertical diffusion is controlled primarily by small scale motions (Reynolds stress), characteristic of stable stratified estuarine water; and (ii) the viscous sub-layer close to the bottom. To describe these layers, two equations are necessary Chriss and Caldwell (1984); the shear stress (τ) between adjacent fluid layers moving with different velocities is given by:

$$\tau = \rho \nu_c \frac{du}{dz}, \tag{5.9}$$

where ν_c, $[\nu_c] = [L^2T^{-1}]$, is the kinematic coefficient of molecular viscosity, related to the dynamic coefficient calculates by the product $\rho\nu_c$, $[\rho\nu_c] = [ML^{-1}T^{-1}]$, is the dynamic viscosity coefficient.

As the density and the shear stress may be approximated by constant values in this boundary layer, it follows that the ratio $(\tau/\rho)^{1/2}$ is also constant. As this quantity has dimension of velocity $[LT^{-1}]$, it is conventionally defined as the *friction velocity* (u_*), and $(\tau/\rho)^{1/2} = u_*$, or $\tau = \rho(u_*)^2$, which was introduced to represent the shear strength. With the origin of the Oz axis on the bottom and positively oriented upwards, integrating Eq. (5.9), with the adherence principle as the bottom boundary condition, $u|_{z=0} = u(0) = 0$, up to a generic vertical position (z), the vertical velocity profile as a function of the friction velocity (u^*) is:

$$u(z) = \frac{(u_*)^2}{\nu_c} z. \tag{5.10}$$

This equation shows that the horizontal velocity component varies linearly with distance from the bottom (Fig. 5.3a). The viscous sub-layer was always present in the field experiments of Chriss and Caldwell (1984), off the Oregon coast (Oregon, USA) using a fine resolution velocity profiler, and it may be scaled roughly with the ratio (ν_c/u_*) which has dimension of length $[\nu_c/u_*] = [L]$.

Above the viscous sub-layer, the vertical mixing brought about by the turbulence, associated with the bottom friction or shear flows at mid-depth, is of interest itself in the dispersion of a pollutant discharge at a given depth (Bowden 1978). The deflecting influence of the Coriolis acceleration in this layer is not dominant, and may be disregarded. Experiments indicate that in this layer, the horizontal velocity increases in proportion to the logarithm of the distance over the bottom (z). This logarithmic layer, above the viscous sub-layer near the bottom (Fig. 5.3b) is simulated in steady-state conditions, neutral stability and moderate bottom roughness by the following equation (Sverdrup et al. 1942):

$$u(z) = \frac{1}{\kappa}\sqrt{\frac{\tau}{\rho}}\ln\left(\frac{z}{z_o}\right) = \frac{u_*}{\kappa}\ln\left(\frac{z}{z_o}\right), \tag{5.11}$$

assuming $u = 0$ at $z = z_o$. In this equation, $\kappa = 0.40$ (or $\kappa = 0.41$) is a non-dimensional constant of *von Kármán*, z_o is the depth above the bottom or sea floor where the velocity is zero (named *roughness dynamic length* or *roughness length*), and is related to the average height of the roughness elements on the bottom (Sverdrup et al. 1942), and u_* (friction velocity) scale the turbulence of velocity. This profile is named the logarithm profile *Kármán-Prandt* and is used to simulate velocity profiles of one-dimensional motions in the continental shelf and estuaries. From this equation we can verify that the velocity profile u(z) and the vertical velocity shear (du/dz) vary linearly with the natural (or neperian) logarithm of the distance z ($\ln(z/z_o)$ down to the bottom, and with the inverse of this distance (1/z), respectively. In both cases, the angular coefficients of the correlations are equal to (u_*/κ); however, the intersection of the straight line with the ordinate axis is equal to the roughness length (z_o) in the first case (Fig. 5.4a), and in the second case it is independent of this length (Fig. 5.4b).

Equations (5.10) and (5.11) assume that the bottom is a plane surface with little roughness. However, in estuaries the bottom is not perfectly plane, and irregularities or roughness elements (tunnels, holes, ripples, sand and gravel) due to erosion, sedimentation, and transport, and benthonic communities generate turbulent shear motions in the bottom viscous sub-layer. If this turbulence is not dissipated by viscosity, the motion regime becomes turbulent and the sub-layer disappears. Despite the erosion of this viscous sub-layer the vertical velocity profile may be approximated by a logarithmic profile which is only related to the vertical exchange of the turbulent momentum. The influence of turbulent motion on the estuary bottom is very important to the sediment dynamics and in solving practical problems related to harbor navigation.

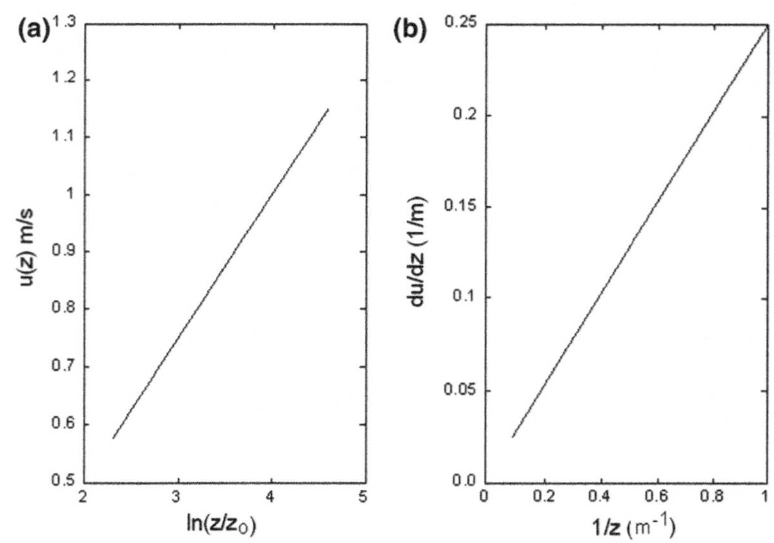

Fig. 5.4 **a** Linear correlations of the logarithmic profile of von Kármán-Prandt, with the ordinate axis representing u = u(z) and **b** du/dz = (1/z), respectively, calculated with $u_* = 0.1\ \text{m s}^{-1}$ and $z_o = 0.1$ m

The rough dynamic length (z_o) is higher in a turbulent flow regime than in the laminar flow regime, and its value has been used to estimate dimensions of irregularities at the bottom. There are several empirical formulations to estimate the z_o value as a function of geometric dimensions, for example, the obstacles mean height and the bottom sand wave slope.

The characteristics of the friction at the bottom (τ_o) in channels forced by the tide were examined by several investigators aiming to relate this shear to a non-dimensional drag coefficient (C_{100}), usually calculated with velocity measurements one meter (100 cm) above the bottom (u_{100}). By analogy with the theoretical result $\tau = \rho(u_*)^2$, obtained for the viscous sub-layer, experimentation indicates that a good quadratic approximation to the coefficient for τ_o is: $\tau_o = \rho C_{100} u_2^{100}$; then, with the assumption that $\tau_o \approx \tau$, it follows that $u_* = (C_{100})^{0.5} u_{100}$. Combining this result with Eq. (5.11), and taking into account that z = 100 cm, it follows that $C_{100} = [\kappa/\ln(100/z_o)]^2$. This expression for the coefficient C_{100} depends on z_o, which may be obtained with knowledge of the longitudinal component of the vertical velocity profile plotted as a function of ln (z/z_o) (Fig. 5.4a). There are also published tables where this quantity is related to the type of the bottom characteristics (Soulsby 1983; Dyer 1986).

Typical z_o values and drag coefficients C_{100} for different bottom types (from mud to gravel), using data drawn from several sources (Lesht 1979 and Heathershaw 1981; quoted in Soulsby 1983), are listed in the Soulsby table according to their observations numbers. The results indicate that the mean z_o and C_{100} values varied from 5.0×10^{-3} to 0.6 cm, and from 1.6×10^{-3} to 6.1×10^{-3}, respectively.

The relationship of C_{100} with z_o was also used by Sternberg (1968) in studies related to friction factors in six tidal channels in the north-west of the USA (Puget Sound and the Strait of Juan de Fuca) divided in two regions (transitional and rough). Among his conclusions, it was identified that the C_{100} value was not very sensitive to the bed characteristics, although the roughness elements of bed types varied from rocks and gravel with maximum heights from 2 to 10 cm, and the mean C_{100} values varied by less than a factor of 2 (2.3×10^{-3}–4.0×10^{-3}). The mean C_{100} value for fully turbulent flow, $C_{100} = 3.1 \times 10^{-3}$, is in agreement with the ranges found in the Soulsby (op. cit) table.

Another interesting conclusion of Sternberg's article is that the transition between fully rough and transitional flow appears to be related to the bottom configuration, and for simple topographically beds the flow becomes fully rough at lower Reynolds numbers (less than 1.5×10^5) than those with complex seabed topographical seabed, for which rough flow conditions occurred at Reynolds number (R_e) greater than 3.6×10^5.

The *profile method* uses Eq. (5.11) with experimental data of vertical velocity profiles to calculate the friction velocity (u_*) and the rough length (z_o) and uses the relation $\tau = \rho u_*^2$ to estimate the shear stress on the bottom. This method is the most commonly used to estimate the shear stress of geophysical fluids in shallow waters (channels, estuaries and continental shelf).

This method has been applied to several oceanographic conditions and coastal environments. In estuaries, the pioneering experiments were performed after the

1920 decade, thanks to the works of Merz (1921) and Mossby (1947). The experimental data used by these investigators (vertical velocity profiles) were sampled in the Dardanelles fjord (Denmark) and in the Avaerströmmem fjord (Bergen, Norway), respectively. The historical data of A. Merz published in the Defant (1961) book and reproduced in Fig. (5.5a) is an example of the method used to determine the analytical expression of the logarithmic velocity profile. The first member of the left-hand side of Eq. (5.11) is known at discrete points in the water column; however, in the second term, there are two unknowns, u_* and z_o. An alternative way to eliminate one of these unknowns is to calculate the derivative of this equation in relation to z, resulting in the following expression u_* being the only unknown:

$$\frac{du}{dz} = \frac{u_*}{\kappa}\frac{1}{z}. \tag{5.12}$$

This equation shows that the vertical velocity shear (du/dz) is inversely proportional to the distance from the bottom (z). As the first member of this equation may be determined by finite differences ($du/dz \approx \Delta u/\Delta z$) with experimental data (Fig. 5.5a), the angular coefficient of the linear correlation of this quantity with 1/z is equal to the ratio u_*/κ (Fig. 5.5b) This procedure results in the following value for this ratio: 2.5×10^{-2} m s^{-1}. As the constant k = 0.40, it follows that the value for $u_* = 1.10 \times 10^{-2}$ m s^{-1}.

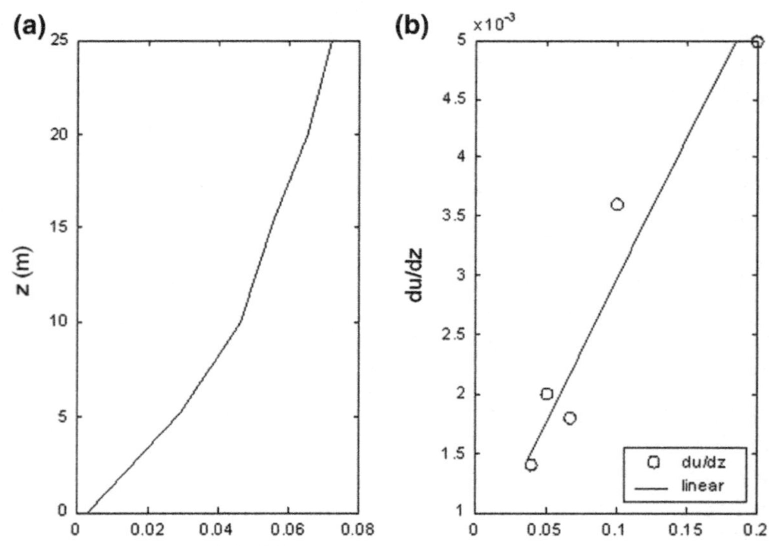

Fig. 5.5 **a** Vertical velocity profile in the south entrance of the Dardanelles estuary (Denmark), with data sampling by Merz (1921). **b** Correlation of the vertical velocity shear ($\Delta u/\Delta z$) as a function of (1/z) (*open dots*) and the linear adjustment

With the calculated value of u_*, the *rough dynamic length*, z_o, of the logarithmic profile (Eq. 5.11) may be obtained with successive adjustments, up to the best linear fit in comparison with the experimental data. Several values were tested, with the best fit found to be $z_o = 1.5$ m as shown in Fig. 5.6. Then, with this method, the experimental data is analytically formulated by the following logarithmic profile:

$$u(z) = 2.5 \times 10^{-2} \ln\left(\frac{z}{1.5}\right), \tag{5.13}$$

with the ordinate z and u(z) in SI units (m and m s^{-1}), respectively. As the shear stress ($\tau = \rho u_*^2$) is proportional to the square of the friction velocity, the value of this physical quantity was estimated to be ≈0 0.11 N m^{-2}, with the assumption $\rho = 1.02 \times 10^3$ kg m^{-3}.

Theoretical simulation of the logarithmic profile is not always possible from the surface down to the bottom, and the water layer from the bottom up to the best fit is named *height of the logarithmic layer*, $h_{L,}$ (Lueck and Lu 1997). In the exemplified adjustment of the logarithmic profile (Fig. 5.6) this height reached 15 m.

The mean vertical velocity profile used in this example is unidirectional, which is characteristic of a well-mixed estuary. However, due to the gravitational circulation forcing, the time variability of the velocity during a tidal cycle may be unsteady, and so the vertical velocity profiles are as shown in Fig. 2.9 (Chap. 2). In partially mixed estuaries, the vertical velocity profiles indicate the occurrence of seaward and landward motion, and it will be impossible to simulate their logarithmic velocity profiles.

Vertical velocity profiles in estuaries are often difficult to be sample over several tidal periods, particularly in harbors which may experience intense maritime traffic, or due to the weather conditions. However, investigations in estuaries increased substantially in the decades following on the historical experiment in fjords, as described above, improving the knowledge of these transitional water bodies. Further results based on logarithmic profile adjustments are presented in Dyer (1986).

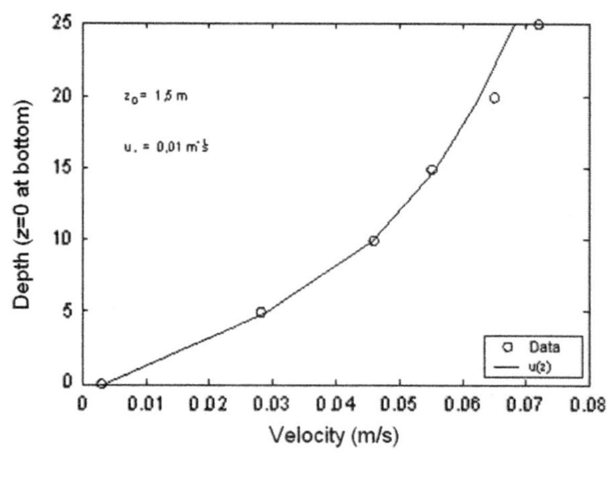

Fig. 5.6 Adjustment of the logarithmic profile to experimental data (o) measured in the Dardanelles estuary (Denmark), presented in the classical article of Merz, in 1921, in which $h_L = 15$ m

To describe results of the application of the profile method, based on detailed time series of bottom-mounted acoustic Doppler profiles (ADP), this topic is complemented with results from the experiments of Lueck and Lu (1997). In this experiment 20-min time averaged velocity profiles at 30 m depth were made in the Cordova channel near Vancouver Island (Canada). The time series of current measurement where made during 4.5 days, with the objective to study the local variability in the bottom-boundary layer at 3.6 m and the current profile variability above this depth. In this investigation, the friction velocity (u_*) based on the logarithmic profiles, the time variability of the logarithmic height (h_L) and the rough dynamic length (z_o) were calculated.

The vertical velocity profiles measured during a time interval of 1.5 days are one-directional during the ebb (>0) and flood (<0) tides, and their logarithmic adjustments are presented in Fig. 5.7. A well-defined logarithmic layer is observed during the events of intense tidal currents, where height (h_L) is in the top half of the water column; however, logarithmic adjustments were not possible during the time intervals of low current intensity.

An alternative method to calculate the shear stress is possible with the introduction of a non-dimensional drag coefficient, C_D. This coefficient is often used in analytical and numerical models to parameterize the bottom frictional shear as a function of a velocity of reference (U_r) according to the following expression:

$$\frac{\tau}{\rho} = C_D U_r^2, \tag{5.14}$$

or, taking into account that $\tau/\rho = u_*^2$,

$$C_D = \left(\frac{u_*}{U_r}\right)^2. \tag{5.15}$$

The friction velocity (u_*) may be determined from the logarithmic profile and can be used to calculate C_D (Eq. 5.15). As the u^* velocity is known, it is adequate to select the reference velocity in the viscous sub-layer from the velocity profile. The

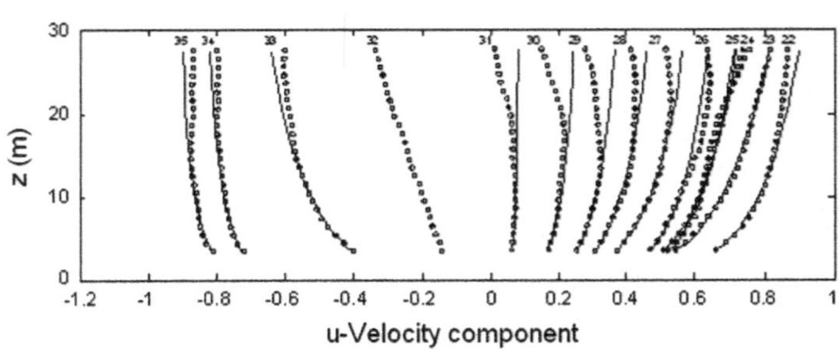

Fig. 5.7 Vertical velocity profiles of the longitudinal component (o) in the Cordoba channel (Vancouver, Canada). The analytical simulations of the logarithmic layer are indicated by the *continuous lines* (according to Lueck and Lu 1997)

ebb and flood velocity profiles (Fig. 5.7) were used to calculate this coefficient during a tidal cycle, using as reference velocity (U_r) the following velocity values: the mean velocity in the water column (u_m), the velocities at the heights of 1.0 m (u_{100}) and 3.6 m (u_{360}) above the bottom. According to the results, the lower values of this coefficient were observed during the ebb tide currents ($u > 0$): $3.5 \times 10^{-3} < C_D < 8.8 \times 10^{-3}$, with the extreme values obtained for the u_m and u_{100} as reference velocities. During the flood tide currents ($u < 0$), the variation interval of this coefficient was higher: 4.0×10^{-3}—2.5×10^{-3}.

As indicated above, simulation of logarithmic velocity profiles in tidal estuaries is very important to provide clues on the friction shear at the bottom. However, it may only be applied for estuaries with predominating longitudinal velocity profiles (seaward and landward), which are characteristic of well-mixed estuaries. Analytical simulations of well-mixed, partially mixed and salt wedge estuaries, in steady-state conditions will be studied later using the hydrodynamic equations of motion. The vertical profiles of scalar properties, such as temperature, salinity and density usually have no simple analytical and numerical simulations, and for these properties numeric values at non-sampled depths may be determined at selected depths by numeric (*cubic splines*) or graphical methods.

5.3 Temporal and Spatial Averages

The superposition of motions generated by tide, river discharge, baroclinic gradient pressure force and wind, create difficulties in the experimental data treatment and processing, and demonstrates the convenience of using mean values in time and space to calculate an estuary's nearly steady-state condition. To better understand the estuarine processes, we must be able to separate those forcing influences (Dyer 1997).

Measurements of hydrographic properties and current velocity are sampled at selected positions (stations) and discrete time intervals, or continuous in time and space. The selection of a suitable sampling duration to determine the time-averaged estuarine condition is as critical as the spatial sampling. Then, to obtain instantaneous or mean values of these properties and the steady-state circulation, it is necessary to adequately reduce these measurements taking into account the following criteria and results to be accomplished:

(i) Make the data analysis easier;
(ii) Obtain average properties for estuary classification;
(iii) Validate theoretical results of analytical and numerical models;
(iv) Determine the advective and diffusive components of salt flux and transport, and the net transport of natural substances and pollutants.

In the determination of mean velocity profiles and other properties (hydrographic, chemical, biological, and suspended sediment), it is necessary to take into account that the local depth is dependent on the position and time, which may vary greatly due to the tidal oscillation, η(x, y, t), during the measuring period. This is because the local depth at any given time is h(x, y, t) = H_0(x, y) + η(x, y, t), with H_0 and η representing mean water level and the tidal oscillation, respectively. The non-dimensional number defined by the ratio of the tidal height (H_o) to the local depth h(x, y, t) varies in the interval $0 < |H_o|/H_0| \leq 1$. If the tidal height (H_o) is greater than 30% of the $local_{depth}$ (H_0),

$$\frac{|H_o|}{|H_0|} > 0.3, \tag{5.16}$$

this ratio is relatively great in comparison to the water depth, and the time variation of the sampling depths must be minimized in the determination of the mean vertical profiles of the properties, as prescribed in the articles of McAlister et al. (1959) and Kjerfve (1975).

Thus, the periodic fluctuations of the water column layer due to the tidal oscillation cause variations in the sampling depths along the water column. If these depth variations are not taken into account when determining the mean values in space and time, undesirable errors may occur when the inequality expressed in (5.16) is reached. This correction may be accomplished by determining data values at equally spaced distances between the surface and bottom, with the non-dimensional depth Z(t), which depends on the origin (z = 0) and the orientation of vertical ordinate Oz:

(i) If this origin is at the bottom (positively oriented towards the surface level), h(t) is the water layer depth and z is the ordinate of the sampling depths, Z(t) is defined as,

$$Z(x, y, t) = \frac{z - h(t)}{h(x, y, t)}, \tag{5.17a}$$

and varies from zero (Z = 0) for one observation at the surface, z = h(t), to (Z = −1), for the observation at the bottom, z = 0.

(ii) If the origin of the vertical ordinate is at the surface and positively oriented against the gravity acceleration, the dimensionless depth is defined by,

$$Z(x, y, t) = \frac{z}{|h(x, y, t)|}. \tag{5.17b}$$

Thus, the non-dimensional depth, Z(x, y, t), and varies from 0 (surface) up to Z = ±1 (bottom), and the signal depends on the Oz axis orientation. In numerical modeling Z(x, y, t) is called the *Sigma* coordinate.

As one of the objectives of estuarine research is to describe the spatial and local distribution of properties, a common survey goal is to obtain the circulation of a particular estuary and the net movement, flux and transport of dissolved or suspended constituent. In both cases it is necessary to compute time-averages of tidal cycles over at least one complete tidal cycle. Due to the expansion and contraction of the water column height in the flood and ebb tide, this is better accomplished with data processing in terms of the non-dimensional depth, Z(x, y, t). This procedure is illustrated in the temperature and salinity profiles measured in an estuarine channel that is partially mixed estuary (Fig. 5.8). The comparative analysis of these properties as a function of the dimensional (z) and non-dimensional (Z = z/7.5) depths, shows the conservation of the profile configurations from the surface down to the bottom.

With the changes from dimensional (z) to the non-dimensional depth (Z), mean depths values of properties may be calculated even in the most unfavorable conditions (Eq. 5.16). After, considering the non-dimensional depth form, each measured property may be interpolated at each non-dimensional depths (Z = 0; Z = 0.1; … Z = 0.9 and Z = 1.0), as illustrated in Fig. 5.9.

Discrete measurements should be made at a constant time interval (Δt), over at least one complete tidal cycle. However, if at all possible, sampling should be continued for the sampling duration indicated by Eq. (4.1, Chap. 4) as net values may vary drastically from one tidal cycle to the other. It is suggested that the initial time measurement (t_0) should begin at slack water (close to the low or high tide)

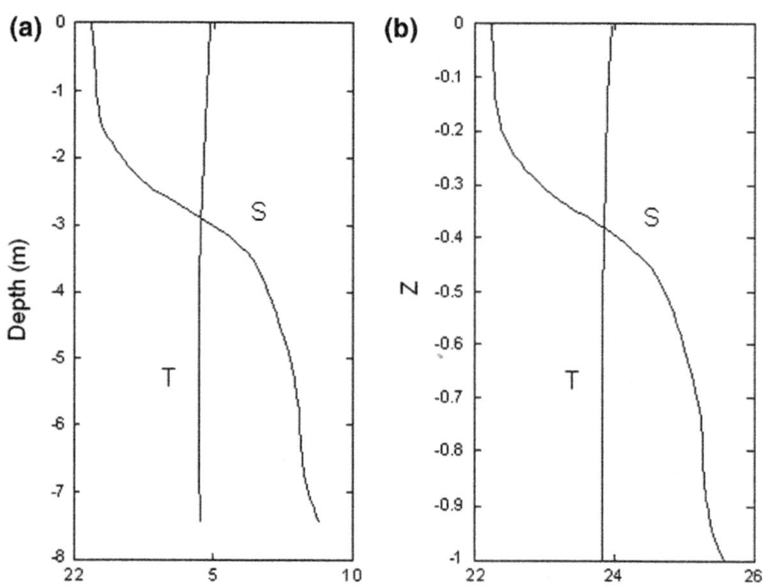

Fig. 5.8 Vertical salinity, S = S(z) and S = S(Z), and temperature T = T(z) and T = T(Z) profiles as functions of dimensional (**a**) and non-dimensional depths (**b**). Oz is positively oriented upward

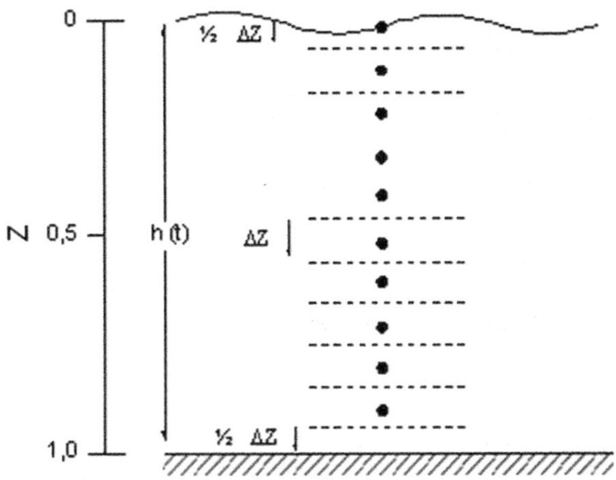

Fig. 5.9 Water column with discrete equidistant non-dimensional depth intervals ΔZ and (1/2)ΔZ of the non-dimensional depth (Z)

and at time intervals, Δt, of one or half hour (1.0 or 0.5 h) to minimize possible averaging errors. In the case of continuous sampling, the time series must be interpolated for analytical analysis at the Δt interval just specified. Usually, for a particular tidal cycle period (T_P), we have $T_P = n\Delta t$ (and $n = T_P/\Delta t$), where (n + 1) is the number of required profiles to compute the net profile for a given tidal cycle. Typically, in semi-diurnal or mixed tide regimes, if *n* is selected to be 12.4 h, the sampling rate equals 1.035 h or one *lunar hour* (Bowden 1963; Kjerfve 1975). In such cases, it is necessary to plot curves of each variable at all Z-depths as functions of time and then divide each time series into *n* equal increments. The interpolated values for each Z, at n + 1 times would then be used in computing time averages.

If the net value of one measured property, P, at a given non-dimensional depth, Z (Z = 0.0; 0.1; 0.2;…, 1.0) is expressed as a function of time P(Z, t), its time mean value during a tidal period (T_P) and at each depth is indicated generically by Z_j (j = 0, 1, 2, …, 10), and are calculated by:

$$\langle P(Z_j \rangle = \frac{1}{T_P}\int_0^{T_P} P(Z_j, t)dt, \tag{5.18}$$

where the symbol ⟨⟩ indicates the time mean value of the property. Taking into account that the values $P(Z_j, t)$ were obtained in *n* discrete interpolations at constant time intervals, Δt ($T_P = n\Delta t$), the integral of Eq. (5.18) may be easily calculated for each depth (Z_j) by the following sum:

$$\langle P(Z_j)\rangle = \frac{1}{n}\left[\frac{P(Z_j, t_0)}{2} + \sum_k P(Z_j, t_k) + \frac{P(Z_j, t_n)}{2}\right], \tag{5.19}$$

where k = 1, 2, …, n − 1, and t_1 and t_n are the first and last time measurements and $(t_n - t_0) = T_P$ is the tidal period. In this equation, the property values at the initial and final time (t_0 and t_n) are multiplied by the factor ½, because it is assumed that each of these values are representative for only (½ Δt), as shown in Fig. 5.10. It should be noted that the time mean value at a generic depth, $\langle P(Z_j \rangle$, may be almost independent of time (nearly steady-state).

Equation (5.19) applied for j = 0, 1, …, 10, may be used to calculate the mean vertical profiles of any scalar property: velocity component, temperature, salinity, and concentrations of nutrients and suspended sediments. When applied to the salinity, the time mean values simulate nearly steady-state conditions and the surface (S_s) and bottom (S_f) values may be used to determine the stratification parameter, $\delta S / \langle \overline{S} \rangle = (S_f - S_s)/\langle \overline{S} \rangle$ of the Stratification-circulation Diagram. Measurements of properties along the transverse section of the estuarine channel may also be processed in the same way to calculate mean property profiles.

As the velocity is a vector, the method used to calculate its mean value during tidal cycles is applied to the longitudinal (u) and transversal (v) components, which may be obtained with the procedure described in this paragraph. For instance, to find the longitudinal component, the temporal mean value is determined for each non-dimensional depth Z_j with an equation similar to (5.19):

$$\langle u(Z_j) \rangle = \frac{1}{n}\left[\frac{u(Z_j, t_0)}{2} + \sum_k u(Z_j, t_k) + \frac{u(Z_j, t_n)}{2}\right], \tag{5.20}$$

where k = 1, 2, …., n − 1.The mean u-velocity component on the surface corresponds to the u_s value, which simulates a nearly steady-state value and is used to calculate the circulation parameter (u_s/u_f) of the Stratification-circulation Diagram.

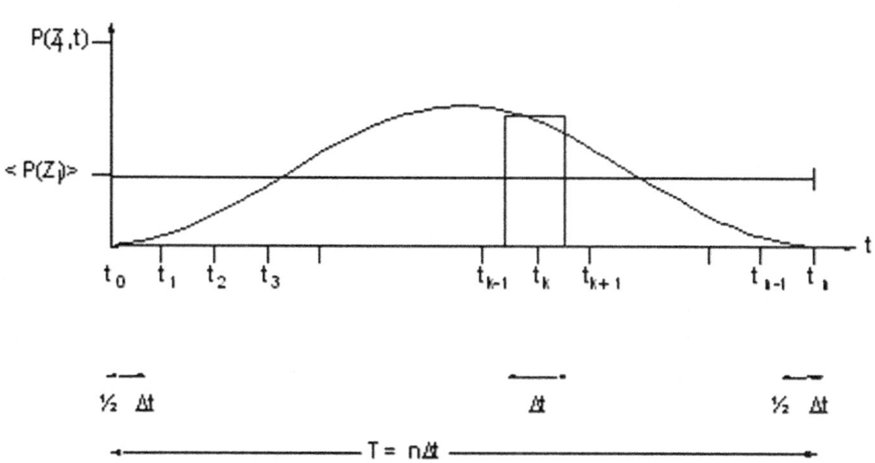

Fig. 5.10 Schematic sequence of time measurements. At the initial (t_0) and final (t_n) times the measured property is assumed to be representative for ½ (Δt) time interval

To calculate the time-mean value of the transversal velocity component (v) this value must be used in Eq. (5.20) as a substitute for the u component.

Once calculated, the time-mean velocity u and v components for the velocity vector at each depth along the water column may be calculated by the vectorial composition:

$$\langle \vec{v}(Z_j) \rangle = \langle u(Z_j) \rangle \vec{i} + \langle v(Z_j) \rangle \vec{j}, \tag{5.21}$$

where $\vec{i}$ and $\vec{j}$ are the unity vectors of the coordinate system used. This computation of time mean velocity components during tidal cycles is very important because: (i) it indicates the net flow which is a characteristic of import and export of property concentrations, and; (ii) it may be used to validate analytical and numerical models (for a bi-dimensional numerical model this procedure has been used by Blumberg 1975).

Let's now consider a property profile P = P(z, t) or P = P(Z, t) in a determined time. The mean value ($\overline{p}(t)$) in the water column ($0 \leq z \leq h$ or $0 \leq Z \leq 1$) is given by:

$$\overline{P}(t) = \frac{1}{h}\int_0^h P(z, t)dz = \int_0^1 P(Z, t)dZ, \tag{5.22}$$

where the over bar indicates a spatial mean, in this case the depth z and Z, and taking into account that, by definition, dz = |h|dZ.

As this integral (Eq. 5.22) must be calculated from known values at discrete points along the water column, for example, spaced 0.1ΔZ, its mean value is calculated by an equation similar to (5.19),

$$\overline{P}(t) = \frac{1}{10}\left[\frac{P(0, t)}{2} + \sum_j (P(Z_j, t) + \frac{P(-1, t)}{2}\right], \tag{5.23}$$

where j = 1, 2, …, 9. When the property P(Z, t) is substitute by the velocity components (u, v) and the no-split condition of the adherence principle at the bottom adopted for the bottom friction (null velocity), the extrapolation down to the bottom of the last parcel in this equation is null, because u(−1, t) = v(−1, t) = 0.

5.4 Reduction and Analysis of Temporal Data Series

Time series of properties' measurements of short or long duration may have their analysis performed in the time domain and in the frequency domain with spectral techniques. For example, in the first case the analysis of the current may be made at selected depths in the form of vector plotted sequentially in time in the graphic known as a *current rose* and as a *progressive vectorial diagram*, usually applied for experiments of only a few tidal cycles. The theory evolved in the spectral analysis is not presented in this text, and may be found in the following books: Bendat and Piersol 1966; Rayner 1971; Jenkins and Watts 1968; Moretin 1999. Its importance will be exemplified with hourly time series of tidal height, salinity and temperature.

To give an example, the time series analysis was made for a half-hour current measurements (intensity and direction) at an anchor station during one spring tidal cycle in the Caravelas river estuary (Bahia, Brazil). The station is in the position where the channel orientation is approximately in the E-W direction (Chap. 12, Fig. 12.7). The current roses (Fig. 5.11) show the velocity vectors plotted during the semi-diurnal tidal period at two selected depths $Z = 0$ and $Z = -0.9$, at surface (left) and near the bottom (right). The analysis of these results indicate the gradual decrease of the intensity with depth, and changes in the current directions during the tidal cycle; from the surface down to mid-depths the current is mainly towards east (90°), indicating an ebb motion with maximum speed of $\approx 1.3\ \mathrm{m\ s^{-1}}$ (Fig. 5.11-left). At the middle of the tidal cycle the tidal current direction changed towards the west (270°), indicating the tidal flood with maximum intensity of $\approx 0.9\ \mathrm{m\ s^{-1}}$. Close to the bottom ($Z = -0.9$) the maximum current intensity is $\approx 0.6\ \mathrm{m\ s^{-1}}$ seaward, and due to the baroclinic gradient pressure force, the direction of the current changes towards west (270°), indicating the motion forced by the tide flood with slightly higher intensity ($\approx 0.7\ \mathrm{m\ s^{-1}}$) than in the surface (Fig. 5.11-right).

The current roses also indicate that the transverse circulation is very weak, which also is shown in the comparison of the time variation of the u- and v- velocity components (Fig. 5.12), which clearly indicates that the main advective influence is in the estuary longitudinal direction.

As should be expected, the progressive vector diagram shown in Fig. 5.13, plotted with the u-and v-velocity components of the previously data used in figure

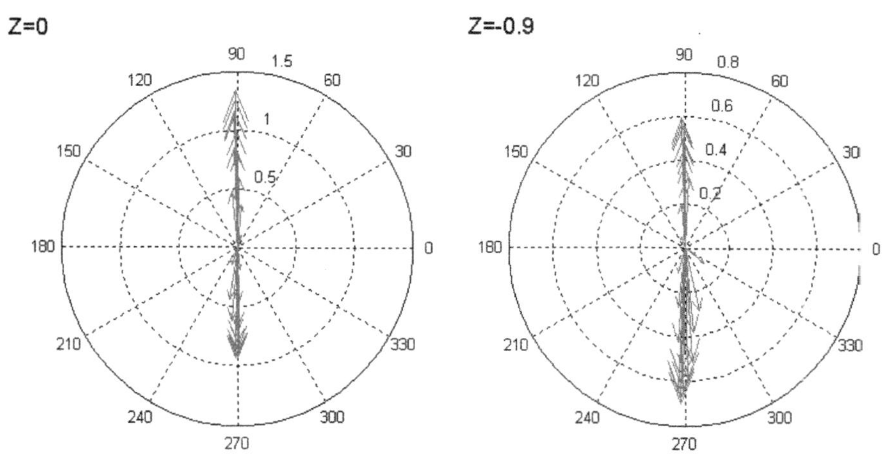

Fig. 5.11 Current roses from half-hourly measurements at the surface (*left*) and near the bottom (*right*), in the Caravelas river estuary (Bahia, Brazil) during spring tide

(Fig. 5.12), indicates the predominance of the longitudinal motion and almost negligible transverse motion. At the surface the particle excursion during the flood and ebb are approximately 22 and 17 km, respectively, and near the bottom the particle excursion is shorter (≈9 km) in the ebb than in the flood (≈13 km).

The time variation of the scalar properties temperature, salinity and density (Sigma-t) at the surface (Z = 0) during the spring tidal cycle, associated with the velocity (Fig. 5.11), are presented in Fig. 5.14. The small temperature variation (ΔT ≈ 0.8 °C) indicated that the salinity is the main property responsible for the density variation, as also shown in the Sigma-t time variation. Near the bottom the hydrographic properties indicates small variations in comparison to that observed in the surface.

To give an example of time series spectral analysis in practice, three simultaneous records measured during 20.8 days in the estuarine channel of the Cananéia sea (Fig. 1.5, Chap. 2) located in the southern São Paulo State (Brazil) have been analysed. The tidal height time series was registered in a recording buoy tidal gauge and the tidal height values (cm) were digitalized with half hour time intervals (Δt = ½ h). Simultaneous temperature (°C) and salinity (‰) measurements were recorded at the same time interval in digital format by the equipment positioned 6.0 above the bottom. These Eulerian measurements were sampled in 10 m mean water depth, and the time variability of these properties (In Julian days) is presented in Fig. 5.15.

The tidal height oscillations show semidiurnal variations superimposed to fortnightly tidal modulations, with amplitudes higher in the spring tide than in the neap tide. The visual time series analysis of salinity and temperature records follow the general trend of the tidal oscillations, showing a quick response to the advection of salt and heat transport generated by the tidal currents. During the spring tidal oscillations, the amplitudes of these properties vary more than during the neap tide, and it is possible to visualize low frequency variations within periods of several days. However, these temporal variations do not show details of the correlations between tidal, salinity and temperature oscillations related to its time, periodicity and phase variations.

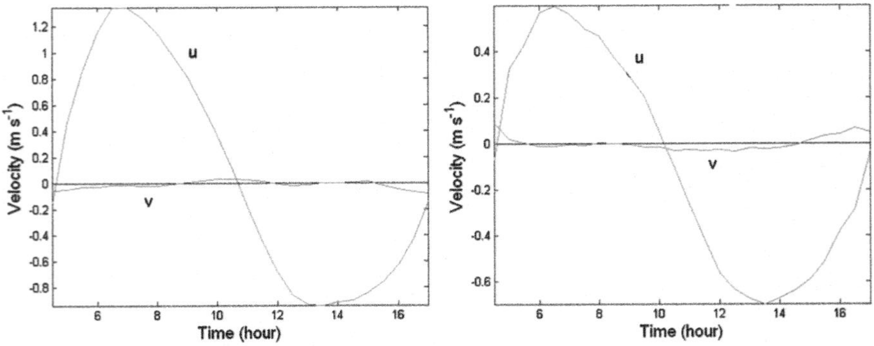

Fig. 5.12 Time variation of the u- and v-velocity components at the surface (*left*) and near the bottom (*right*) measured in the Caravelas river estuary (Bahia, Brazil) during spring tide

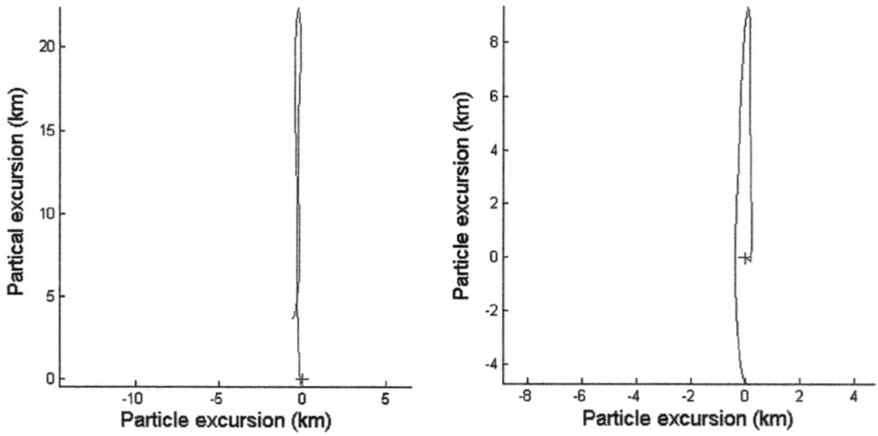

Fig. 5.13 Progressive vector diagram at the surface (*left*) and near the bottom (*right*), based in current measurements measured in the Caravelas river estuary (Bahia, Brazil) during spring tide. The initial position of the diagram is indicated by the plus (+) symbol

Fig. 5.14 Time variation of the hydrographic properties at the surface (Z = 0) measured in the Caravelas river estuary during the spring tidal cycle (Bahia, Brazil)

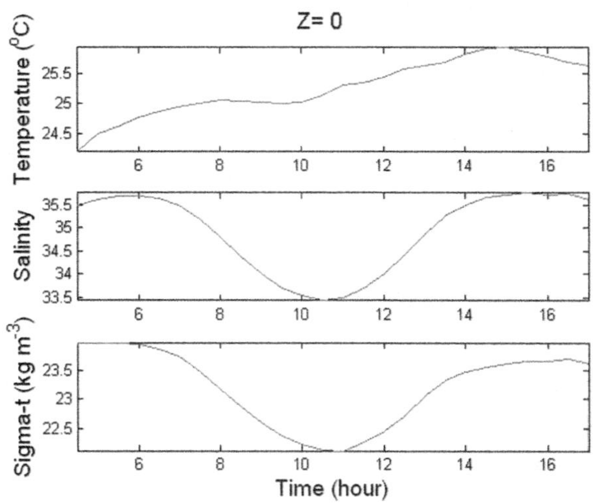

The spectral analysis determines the variance in frequency bands. The variance is a statistical quantity which describes the dynamic component of the time series, numerically calculated by the quadratic mean value around its mean value. In the *variance spectra*, it is possible to do the analysis of the dynamical component intensity of these properties as function of frequency. The frequency band for which the variance may be estimated has two limiting factors: the time series length, T = nΔt, and the sampling interval, Δt. For the records of Fig. 5.15, these values are equal to 20.8 and 0.5 h, respectively.

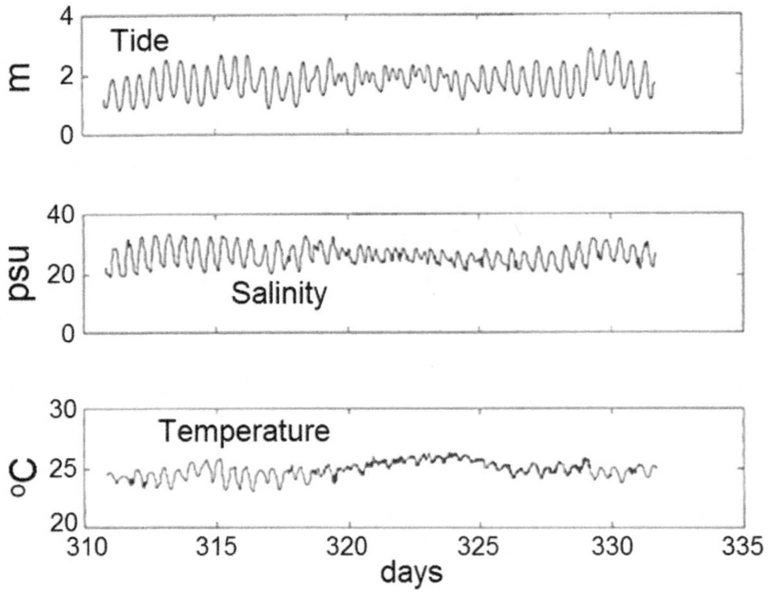

Fig. 5.15 Time series of tidal height, salinity (psu) and temperature (°C) in the microtidal Cananéia estuarine channel (time scale in Julian days)

The time series length determines the lowest frequency for which the variance may be estimated, which is equal to the inverse of the series time length (1/T), and, in our example, corresponds to the frequency of 0.048 cph. The highest frequency for determining the variance is called the *Nyquist* frequency, which is half of the sampling frequency (f_s). As the sampling interval is 0.5 h, it follows that $f_a = 2$ cph, and the *Nyquist* frequency is equal to ($\frac{1}{2}$ $f_a = 1.0$ cph); the variance calculated for this lower frequency may not be representative because, for this quantity to have statistical meaning, it must be determined taking into account three or four complete tidal cycles. Due to this limiting factor, for the 20.8 day time series under analysis, the longest period that may be adequately resolved is approximately five days (period of 120 h and frequency of 0.008 cph).

The sampling time interval chosen is very critical. If the property presents variability with frequencies higher than the *Nyquist*, the spectra may present a doubtful increase in low frequencies. This change of the variance from high to low is called *aliasing*. Then, the correct use of the sampling time interval (Δt) is of fundamental importance to apply the techniques of the spectral analysis to observational oceanographic data. In estuaries the Δt equal to one or half an hour is usually satisfactory, because the more energetic variance signals are associated with tidal oscillations, with diurnal od semi-diurnal cyclic oscillations (frequencies of 0.04 and 0.08 cph, respectively).

The variance determinations are calculated at regular frequency intervals between the extreme points of the domain. The number of variance estimates in this interval is fixed by the total number of points in the time series divided by the degrees of freedom. The choice of variance requires a compromise between the following conflicts of interest: (i) an increase in the degrees of freedom determines narrows confidence intervals, (ii) a decrease in degrees of freedom increases the frequency resolution. The latter compromise is favorable when analyzing long time series. In choosing degrees of freedom, it is necessary to take into account the frequency interval which has the greatest interest to the processes being investigated. For better resolution at lower or higher of frequencies, it is necessary to choose the data processing with low or high degrees of freedom, respectively. Details for obtaining confidence intervals as functions of the degrees of freedom, which are of fundamental importance to the verify the statistical meaning of the variance results, may be found in the book of Bendat and Piersol (1966).

In the tidal variance spectra (Fig. 5.16a), it is possible to understand what processes influence this dynamical component of the sea level oscillations. In this figure, the extreme points (peaks) in the variance, A, B, C, D and E, between the domains of low and high frequency, respectively, were inserted to facilitate the understanding of the analysis of this spectra calculated with 5 degrees of freedom. In these spectra peaks, the variances are statistically confident within the 95% interval. The estimated variances for the frequency band A (0.012 cph—period of 83 h) are too low to be adequately solved. This peak (A), although with little significance, is usually associated with meteorological forcing causing storm surges against the coastline (Csanady 1982). The remaining peaks are oscillations with the following frequencies and periods, respectively: diurnal (B) with 0.04 cph and period of 25.0 h, semidiurnal (C) with 0.08 cph and period of 12.5 h, and the frequencies in the sub-tidal domain (D) and (E), with frequencies of 0.12 and 0.16 cph, and periods of 8.2 and 6.2 h.

The salinity variance spectra (Fig. 5.16b) is very similar to the tidal variance spectra, and conclusively indicates the advection process forced by the barotropic gradient pressure force (tidal forcing) in the salinity redistribution in the Cananéia main channel.

The generation of internal overtides at multiples of the dominant tidal frequency is termed *barotropic tidal asymmetry*, because it distorts the free surface and causes flood or ebb dominant currents, depending on the relative phases of the tides and its overtides (Fisher et al. 1972; Ianniello 1977; Simpson et al. 1990; Jay and Musiak 1996). Factors such as friction and channel morphology generate shallow water over tides such as M_4 and M_6. When these tidal components are added to M_2 tidal current, maximum ebb and flood may be shifted close to high and low water, resulting in a strong tidal current that is distorted from that generated by the semi-diurnal M_2 tidal component.

Another important result is the cross-correlation between the tidal and salinity time series. From this correlation, two spectra results: (a) *covariance* or *normalized coherence*, calculated by the ratio of the product of covariance to the individual square roots, and; (b) the *phase spectra* (Fig. 5.17a, b). The covariance is a non-dimensional quantity, which varies between zero and one, and measures any linear relationship between the individual series, and values equal to zero and one indicate no correlation or a strong linear correlation between the series and

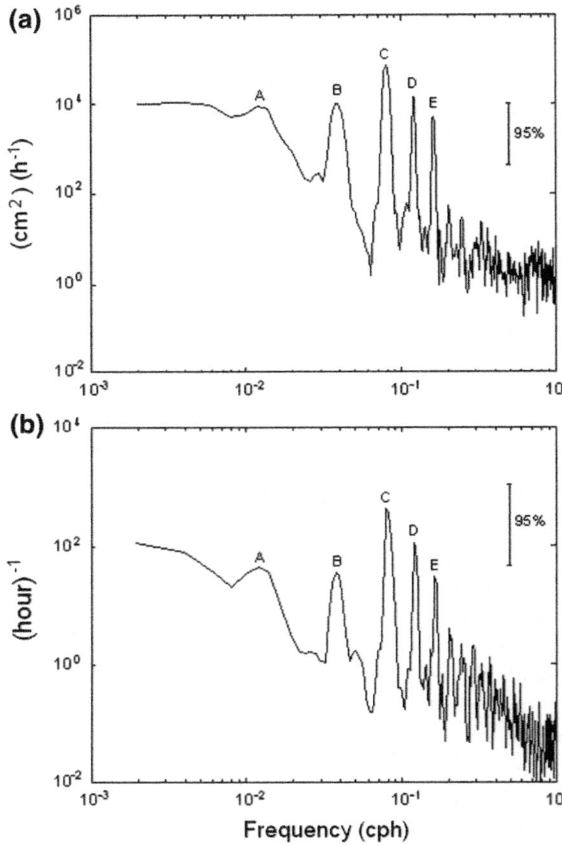

Fig. 5.16 Variance spectra of tidal height oscillations (**a**) and salinity (**b**) as functions of the frequency in cph in the main channel of the Cananéia-Iguape estuarine lagoon (Fig. 1.5, Chap. 1) showing variability of the tidal oscillation and salinity

processes, respectively. This quantity may be identified as a correlation coefficient in the frequency domain.

In the covariance spectra, exemplified by the cross correlation of tidal height versus salinity (Fig. 5.17a), the 95% statistically significant variance with 5 degrees of freedom has a coherence of over 0.9 in the low frequency domain (0.012 cph and period of 83 h) and for diurnal frequencies (0.04 cph and period of 25 h). In the semi and three-diurnal frequencies (periods of 0.08 and 0.12 cph), the coherence is close to 1.0, decreasing just a little near the frequency of 0.16 cph (fourth-diurnal) as shown in Fig. 5.17b. These results indicate that in these frequency bands there are a very strong linear correlation between the tidal height and the salinity.

The final result of the spectral analysis is the phase spectra (Fig. 5.17b), which gives the phase differences between the series. The phase angle (ϕ) may be converted into time intervals when divided by the product of the angular frequency (ω) by $360°/2\pi$, $\Delta t = \phi/(\omega \times 360°/2\pi)$. Then, for example, the diurnal component of the salinity oscillation is in phase with the tide ($\phi \approx 0°$) and the semi-diurnal component is 20° out of phase, or $\Delta t = 0.7$ h in relation to the tide.

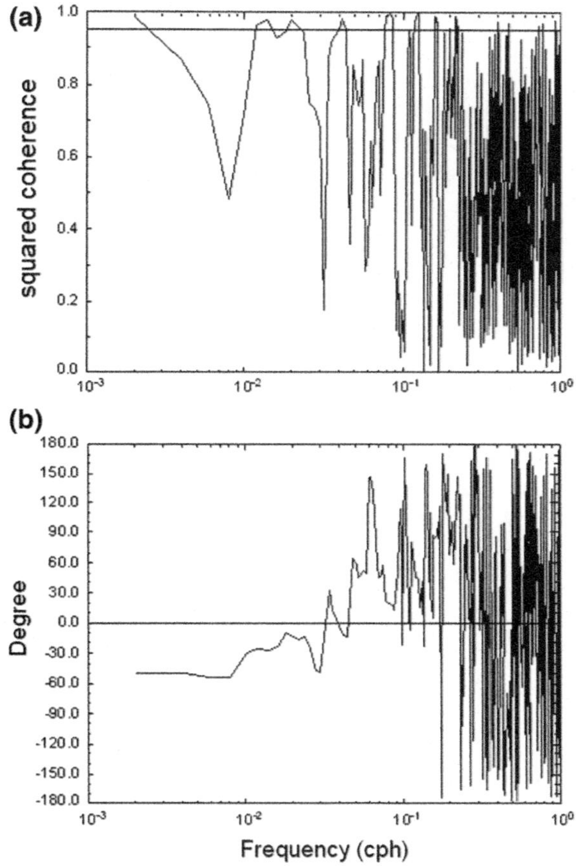

Fig. 5.17 Coherence normalized spectra (**a**) and phase spectra (**b**) of the cross-correlation of tidal heights and the salinity in the main channel of the Cananéia-Iguape estuarine lagoon (Fig. 1.5, Chap. 1)

The spectral analysis is a powerful mathematic tool for quantifying the linearity (or non-linearity) between time series with complex variability. However, the interpretation of its results must be made with caution, because statistically significant coherence values may not necessary imply cause and effect in the occurrence of linear relationships. Small coherence values suggest non-linearity, but there is the possibility of the existence of relationships between the physical forcing processes.

5.5 Isopleths Method and Mean Vertical Profiles

After the reduction and final data processing of hydrographic variables and current velocity components (u, v), graphical representation is necessary to enable interpretation and analysis of the experimental results. When the measurements are made at a fixed station along the water column for a duration of at least one complete tidal cycle (Eulerian sampling), the isopleths method is convenient for studying the variability of properties under investigation.

In Fig. 5.18, results are presented for the local variability of the u-velocity component and salinity in the Bertioga estuarine channel (Fig. 1.5, Chap. 1), during two semi-diurnal spring tidal cycles (≈25 h), measured at a fixed station at hourly time intervals. The tidal oscillation and the u-velocity time variations are asymmetric (Fig. 5.18-upper), and the highest velocities (≈0.8 m s^{-1}) during the ebb have a phase difference of ≈2.5 h in relation to the HW; however, at the flood the intensities are very low (≈0.2 m s^{-1}).

The temporal salinity variation (Fig. 5.18—lower) indicates that the estuarine channel is highly stratified, and at HW and LW the salinity values are 36.0 and 22.0‰, respectively. The salinity differences between the bottom and the surface are up to 14 and 6‰ at HW and LW, respectively, and nucleus with maximum values (36‰) indicate the Tropical Water mass (TW) intrusion into the estuary. The phase difference between the u-velocity component and the tidal oscillations usually observed in partially-mixed estuaries (Hunt 1964), is mainly due to frictional energy dissipation at the bottom. It is also possible to observe at low tide the

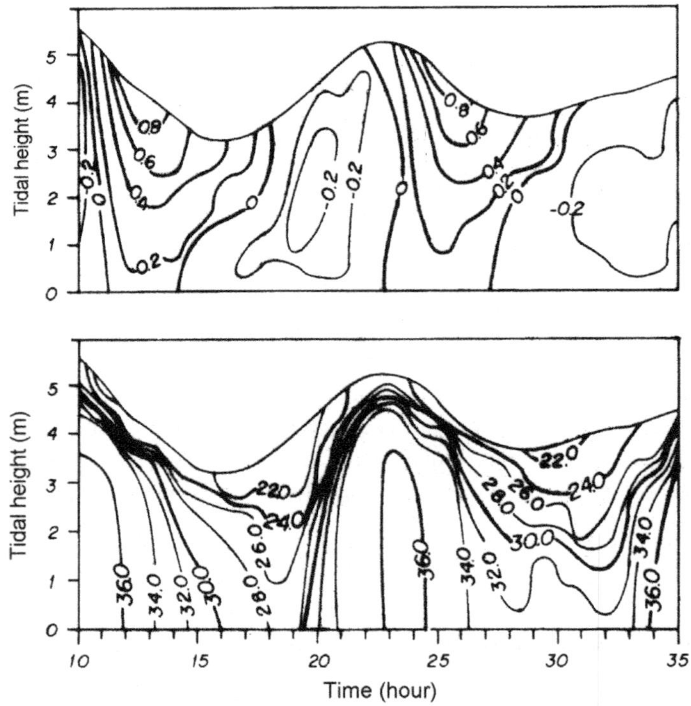

Fig. 5.18 Isopleths of the u-velocity component (m s^{-1}) (*upper*), and salinity (‰) (*lower*) in the Bertioga estuarine channel (São Paulo, SP, Brazil) during spring tide, which are in phase with the slack water (u = 0). Positive and negative velocity values indicates flood and ebb currents, respectively (after Miranda et al. 1998)

occurrence of bidirectional motions due by the baroclinic pressure gradient force at the neighborhood of the u = 0 velocity.

The sub-superficial velocity nucleus of low velocity (0.2 m s^{-1}—Fig. 5.18 upper) during the flood, which is associated with an increase in the halocline stratification (Fig. 5.18 lower), may be due to the low vorticity shear generated by the bottom friction stress. An opposite occurrence (halocline erosion during the tidal ebb) due to the intensification of the vertical mixing is generated by an increase in the bottom stress vorticity as demonstrated by Zhou (1998).

The time mean salinity and the u-velocity profiles for two semidiurnal tidal cycles, calculated by Eqs. (5.19) and (5.20), are presented in Fig. 5.19. These results simulate nearly steady-state conditions and may be used to classify estuaries with the Stratification-circulation Diagram. The salinity profile has an accentuated halocline with values varying from the surface (S_s = 15.84‰) to (S_f = 29.95‰) at the bottom (Fig. 5.19, left), indicating a highly stratified estuary. Its mean depth average is equal to $\overline{S} = 25.8$‰, and the stratification parameter $S_P = \delta S/\overline{S} \approx 0.55$. The u-velocity component, without the influence of the barotropic gradient pressure force indicates a bidirectional motion due to the gravitational circulation (seaward and landward in the upper and lower layer, respectively), which is another characteristic of the partially mixed estuary, with the no-motion depth at $Z \approx -0.5$ ($z \approx 3.0$ m). The velocity values to calculate the circulation parameter u_s = 0.15 m s^{-1} and the residual or net velocity $u_f \approx u_a \approx 0.04$ m s^{-1}. Thus, the circulation parameter, $C_P = u_s/u_f \approx u_s/u_a = 3.7$; finally, with these parameters the investigated estuarine channel can be classified as type 2b. Finally, it should be pointed out that although the low net value of the velocity it is responsible for the seaward advective flux and transport of the concentrations of any property natural or pathogenic.

5.6 Flux and Transport of Properties

The mixing in the estuarine water mass (river + seawater), is physically determined by the local variation of a property concentration ($\partial C/\partial t$) due to the simultaneous action of turbulent diffusion (dispersion) and the advective processes. As the diffusive and the advection are inherent to the motion of non-homogeneous fluids, it is opportune to include in this chapter the fundamental concepts related to the terminologies and determinations of flux and transport of volume and mass (salt), which may also be applied for any conservative property.

Taking into account the physical principles of Hydrodynamics, it is well known that the *volume and mass transports* of a property is the volume and mass of the fluid flow through a transversal section per unit of time. Then, according to this concept, the *instantaneous* volume transport, $T_V = T_V(t)$, $[T_V] = [L^3T^{-1}]$, and mass $T_M = T_M(t)$, $[T_M] = [MT^{-1}]$, are expressed in volume and mass per time unity. In mathematical terms these quantities are calculated, respectively, by the following surface integrals extended to an area $A = A(x, t)$:

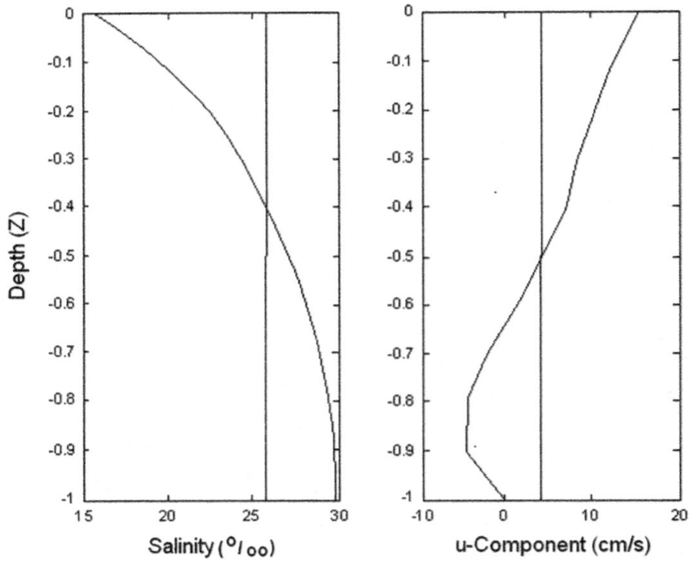

Fig. 5.19 Vertical time mean profiles of salinity and the u-velocity component, determined with hourly values measured during two tidal cycles, sampled during neap tide in the Bertioga estuarine channel (São Paulo, Brazil). The vertical mean depth values of salinity and velocity are shown by *vertical lines* (adapted from Miranda et al. 1998)

$$T_V = \iint_A \vec{v} \cdot \vec{n} \, dA = \iint_A u dA = \bar{u}A, \tag{5.24}$$

and

$$T_M = \iint_A \rho\vec{v} \cdot \vec{n} \, dA = \iint_A \rho u dA = \overline{\rho u}A, \tag{5.25}$$

where the dot, •, indicates the scalar product, and in the last term on the right-hand-side of these equations the *mean value theorem* was used, in Eqs. (5.24) and (5.25), the quantities $\bar{u}$ and $\bar{\rho}$ indicate the mean values of the u-velocity component and density in the area, A, respectively. The transports, T_V and T_M, in the SI system of units are given in $m^3 s^{-1}$ and kg s^{-1}, respectively. The same equations are applied for a transverse cross-sectional area, A, in an estuary; however, the experimental fields of velocity ($\vec{v}$) and density, $\rho = \rho(S, T, p)$, may only be measured at discrete points or oceanographic stations, distributed in a cross-section, A, and the area integrals of these equations must be calculated numerically, because the functional relationship $u = u(x, y, z, t)$ are not analytically known. Further details on the transport determination methods will be given later.

The *instantaneous volume transport*, divided by the cross section area (A), is named *volume flux* (ϕ_V), and it is numerically equal to the mean transport at the transversal section. Then, this physical quantity is calculated by ϕ_V/A, or by combining this definition with Eq. (5.24),

$$\Phi_V = \frac{1}{A}\iint_A \vec{v}\cdot\vec{n}\,dA = \frac{1}{A}\iint_A u dA = \bar{u}. \quad (5.26)$$

Then, the *volume flux* is numerically equal to the mean velocity value in the transversal section A.

By analogy, it is possible to define the *mass flux* ($\phi_M = T_M/A$) from Eq. (5.25),

$$\Phi_M = \frac{1}{A}\iint_A \rho\vec{v}\cdot\vec{n}\,dA = \frac{1}{A}\iint_A \rho u dA = \overline{\rho u}. \quad (5.27)$$

The *salt flux* (Φ_S) may be calculated by inserting the salinity ($S \times 10^{-3}$, converted in kg/kg units) into the integrand of Eq. (5.27). Hence, the salt flux is $\Phi_S = \rho v S \times 10^{-3}$, $[\Phi_S] = [ML^{-2}T]$ which, integrated in the area, A, of the transversal section, results in the salt transport T_S, $[T_S] = [MT^{-1}]$, and in SI unities, this quantity is expressed in kg s^{-1}.

Determinations of volume and mass transport in estuarine studies is always important, particularly when the objectives of the research are to investigate the import or exportation of natural concentrations of biological, chemical substances, suspended sediments, and pathogenic substances. In practice, this determination merits special attention due to the temporal and spatial variability of the evolved properties and the cross-section area (A).

Let us use an example of the determination of the volume transport across a transversal section, with the following data known: the area (A) of transverse section area, and the steady-state velocity field based on measurements at three oceanographic stations A, B, C shown in Fig. 5.20. After the vector velocity decomposition, corresponding to the scalar product $\vec{v}\cdot\vec{n} = u$ (the function being integrated in the first term of Eq. 5.24) with known u-velocity component profiles at stations A, B and C, it is possible to drawn the vertical velocity field $u = u(y, z)$, as illustrated in Fig. 5.20.

Figure shows the occurrence of a bi-directional motion, and, as may be observed in the signal changes of the velocity field, the motion has a layer of null velocity ($u = 0$). In the assumption that the motion is occurring in an estuarine channel, it is characteristic of a partially-mixed estuary (type 2 or B, according to the classification criteria) in nearly steady-state, with motions down and up estuary in the upper and lower layers, respectively. With the velocity isolines (u = const.) drawn in this figure, it is possible to numerically calculate the volume transport (Eq. 5.24) with the following steps:

(a) With a planimetry technique, the area $[L^2]$ between the velocity isolines may be determined;
(b) The mean value of the area between the isolines is multiplied by the mean velocity value between them $[L^2LT^{-1}]$

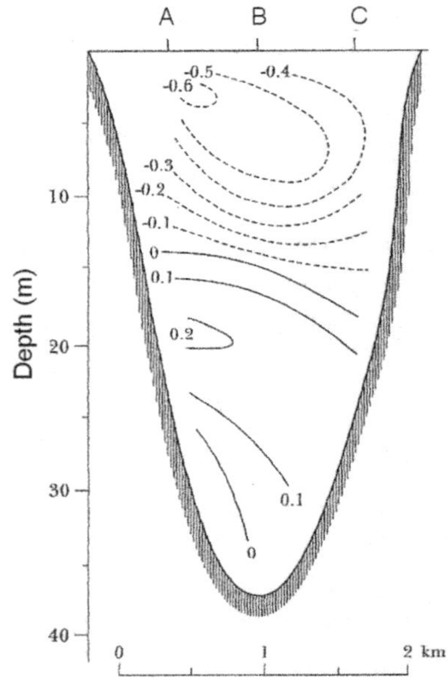

Fig. 5.20 Steady-state vertical structure of the u-velocity component, u = (y, z), in m s^{-1}, orthogonal to the vertical section. *A*, *B*, and *C* are positions of oceanographic stations. The Ox axis is oriented in the landward direction (u > 0 and u < 0) indicate flood and ebb, respectively

(c) The sum of all these numerical value of the integral and the volume transport through the vertical section.

As the velocity is given in m s^{-1} and the area calculated in m^2, the net volume transport is approximately—$1.2 \times 10^3\ m^3\ s^{-1}$ ($-3.2 \times 10^3\ m^3\ s^{-1}$ and $2.0 \times 10^3\ m^3\ s^{-1}$, seaward and landward, respectively). The same procedure may be used to calculate the mass and salt transport; however, in these cases, in the transversal section, the isolines of the ρu and ρuS quantities must be drawn, and the mean velocity multiplied by the corresponding isolines values.

Acoustic Doppler Current Profilers (ADCP) may perform velocity profile measurements accurately and at short time intervals along estuarine cross sections, enabling accurate determination of volume transport across transversal sections. The software of the ADCP equipment may also automatically compute the volume transport; however, the displayed results must be checked against other methodologies.

Let's now consider a non-stationary velocity profile, u = u(x, t), in a known fixed position in an estuarine transversal cross section. With this profile, it is possible to calculate its mean velocity value $\bar{u} = \bar{u}(t)$ in the water column, with depth h = h(t), which oscillate periodically with time. With these data it is possible to calculate the volume transport, $(T_V)_L$, per unit of the section width, which is representative of the neighboring geographic position. Then, the net volume transport is given by:

$$\langle (T_V)_L \rangle = \frac{1}{T} \int_0^{T_P} \bar{u}(t)h(t)dt = \langle \bar{u}h \rangle. \quad (5.28)$$

In this equation T_P is the tidal period, and the volume transport dimension is $[(T_V)_L] = [L^2T^{-1}]$. Under normal meteorological conditions and for a laterally homogenous estuary with a width, B, the product B $\langle (T_V)_L \rangle$ is the numerical approximation of the fresh water discharge (Q_f) at the estuary head.

Because, in general, the integrand of Eq. (5.28) is not known analytically, this volume transport may be numerically calculated with an equation similar to Eq. (5.20);

$$\langle (T_V)_L \rangle = \frac{1}{n} \left[\frac{\bar{u}(t_0)h(t_0)}{2} + \sum_k \bar{u}(t_k)h(t_k) + \frac{\bar{u}(t_n)h(t_n)}{2} \right], \quad (5.29)$$

where n is the number of lunar hours of the tidal cycle and k = 1, 2, …, n − 1.

When the functions $\bar{u} = \bar{u}(t)$ and $h = h(t)$ of Eq. (5.28) are known, this equation may be integrated by analytical methods. Under the assumption velocity and level oscillations may be simulated by sinoidal oscillations, similar to the solutions obtained for the propagation of a tidal wave in a channel with infinity length (Eqs. 2.21 and 2.22, Chap. 2),

$$h(t) = H_0 + \eta_0 \sin(\omega t - \Phi), \quad (5.30)$$

and

$$u(t) = u_0 + U_0 \sin(\omega t), \quad (5.31)$$

where the angle, Φ, is the phase difference between the velocity and the water depth variations. In these equations, H_0 and u_0 are the mean sea-level depth and the mean velocity, which are superimposed to the values η_0 and U_0, respectively.

Replacing Eqs. (5.30) and (5.31) in Eq. (5.28), simplifying the resulting expression with trigonometric identities and completing the resulting analytical integration, it follows that the expression for the mean volume transport is,

$$\langle (T_V)_L \rangle = u_0 H_0 + \frac{U_0 \eta_0}{2} \cos(\Phi), \quad (5.32)$$

This final result indicates the volume transport, per unit of the transversal section, which may be determined only by the product of the mean values (u_0 H_0) when the phase difference is equal to $\pi/2$, and the tidal oscillation in the estuarine channel is a stationary wave. This phenomenon has been observed in some estuaries where time differences of ($T_P/4$) or phase difference of $\pi/2$, between the tidal height oscillations and the longitudinal velocity differences were detected (Dyer and Ramamoorthy,

1969; Kjerfve 1970, 1973). However, when the tidal wave propagates as a progressive wave, $\Phi = 0$ (Eq. 2.24, Chap. 2), the volume transport is determined taking into account the tide and velocity amplitudes (u_oH_o). In partially stratified estuaries (type 2 or B), the phase differences vary in the interval $0 < \Phi < \pi/2$, and the volume transport also depends on the product $U_0\eta_0$. If the estuary width may be approximated by a constant value (B), the volume transport across the transversal section may be calculated by $B\langle (T_V)_L \rangle$.

5.7 Advective Salt Transport Components

The landward salt transport is driven by the current velocity generated during the tidal flood and gravitational circulation, while the seaward salt transport is driven by the reversal of the tidal oscillation, the ebb tidal current, the fresh water discharge and gravitational circulation. To gain a better understanding of these processes, Pritchard (1954) studied the salinity and current velocity measured in the James river estuary (Virginia, USA), averaged over several tidal cycles. This study confirmed the hypothesis that the mixing processes are related primarily to tidal forcing, and suggested the possibility of predicting the eddy diffusion terms from the tidal velocities. Pritchard (op. cit.) also showed that the horizontal advective flux and the vertical non-advective (diffusive) flux of salt are the most important factors in maintaining the salt balance; however, although the vertical advective flux is of secondary importance but still significant, the longitudinal non-advective (diffusive) salt flux is small. Confirming these results, in studies on the salt dispersion in the Hudson river estuary (New York, USA), Hunkins (1981) stated that at the simplest level, an estuary may be considered as a *black box* which pumps salt upstream against the mean river flow, and the overall landward mixing is better termed dispersion, rather than diffusion, and that dispersion is produced primarily by the effects of winds, tides and gravitational circulation. Thus, the process termed dispersion has advection and vertical turbulent diffusion as main components.

For studies on the advective and non-advective salt transport components through an estuary transverse cross-section, measurements of current velocity and salinity must be taken at Δt intervals from 30 min to 1 h for a duration at least one complete tidal cycle, and the profiles of these properties must be interpolated at the non-dimensional depth (Z). To simplify the mathematical treatment, consider a lateral homogeneous estuary, which is a simplified version of the non-laterally homogeneous estuary studied by Hunkins (1981). Under this simplification, it is assumed that the experimental data are from a single fixed station in the middle of the channel, and the instantaneous salt transport (M_S), per width unit, is determined by:

$$M_S = \int_0^h \rho S\vec{v} \cdot \vec{n}.dz = \int_0^h \rho uSdz = \langle \overline{\rho uSh} \rangle, \quad (5.33)$$

In this equation, the Oz axis is oriented in the $\vec{g}$ direction, and in the last term of this equation $(\overline{\rho uS})$ is the mean value of the flux salt, and h is water thickness. The dimensional analysis of this equation shows that M_S, $[M_S] = [ML^{-1}T^{-1}]$ is calculated in the SI system of units, in kg m^{-1} s^{-1}.

The mean advective salt transport (T_S) during one (T_P) or more tidal cycles (nT_P) is calculated by:

$$T_S = \langle M_S \rangle = \frac{1}{T}\int_0^T M_S dt = \langle \overline{\rho uSh} \rangle. \quad (5.34)$$

The mean density in the water column, appearing in Eqs. (5.33 and 5.34), is calculated by a State Equation of Sea Water, at atmospheric pressure.

The term $\langle \overline{\rho uSh} \rangle$ in Eq. (5.34) is the mean values in space (depth) and time. Considering u-velocity component and salinity as examples, the time mean depth value, are calculated by:

$$\langle \overline{u} \rangle = u_a = \frac{1}{T}[\frac{1}{h}\int_0^h u(x, z, t)]dt, \quad (5.35a)$$

and

$$\langle \overline{S} \rangle = S_a = \frac{1}{T}[\frac{1}{h}\int_0^h S(x, z, t)]dt. \quad (5.35b)$$

These time-mean values are function of the longitudinal distance and its value varies according to the cross-sectional area, and u-velocity component may be considered in first approximation to the velocity component generated by the river discharge,

$$\langle \overline{u} \rangle = u_a \approx \frac{Q_f}{A}. \quad (5.36)$$

As demonstrated in the pioneer article of Bowden (1963), the advective salt transport also has contributions of a diffusive nature that don't explicitly appear in Eq. (5.34). This phenomenon may be investigated with the separation of the periodic tidal forcing (barotropic), the gravitational circulation (baroclinic) and other effects, such as the turbulence generated by the wind. The main objective of the following theoretical treatment is to separate the longitudinal salt transport in dominant parcels. For this purpose, the longitudinal velocity component (u) the salinity (S) and the water depth (h) must be decomposed in determinate parcels to make identification of the various correlations possible, which indicate the advective and dispersive physical processes responsible for the landward and seaward salt transport. This decomposition process may be used in the determination of the transport components of any conservative substance dissolved in the seawater.

Using the articles of Bowden (1963), Fischer (1976), Hunkins (1981), Dyer (1974) and Kjerfve (1986), as references for a laterally homogenous estuarine channel, the instantaneous velocity and salinity profiles are decomposed into mean, tidal, steady (subscripts a, t, s) and deviation terms ('):

$$u(x, z, t) = u_a(x) + u_t(x,t) + u_s(x,z) + u'(x, z, t), \tag{5.37}$$

$$S(x, z, t) = S_a(x) + S_t(x, t) + S_s(x, z) + S'(x, z, t). \tag{5.38}$$

In the decompositions, the first term on the right-hand-side are the mean values $u_a(x) = \langle \overline{u} \rangle$ and $S_a(x) = \langle \overline{S} \rangle$ due to the dominant influence of the river discharge (advective process). The second and third terms of these equations, $u_t(x, t)$, $u_s(x, z)$, $S_t(x, t)$, $S_s(x, z)$, mathematically simulate the cyclic tidal influence and the stationary influence of the gravitational circulation, respectively, which also are dominant. These components are defined as,

$$u_t = \overline{u} - u_a, \tag{5.39}$$

$$S_t = \overline{S} - S_a, \tag{5.40}$$

$$u_s = \langle u \rangle - u_a \tag{5.41}$$

$$S_s = \langle S \rangle - S_a, \tag{5.42}$$

where the expressions (5.39 and 5.40) and (5.41 and 5.42) are the tidal and the steady-state components, respectively. The last terms, $u'(x, z, t)$ and $S'(x, z, t)$, of Eqs. (5.37) and (5.38) are the deviation components due to the dispersive small scale physical processes, and are calculated by,

$$u' = u(x, z, t) - u_t(x, t) - u_s(x, z) - u_a(x), \tag{5.43}$$

and

$$S' = S(x, z, t) - S_t(x, t) - S_s(x, z) - S_a(x). \tag{5.44}$$

Figure 5.21 shows the velocity decomposition in the parcels u_t, u_s and u' from the vertical profiles; profile (a) the nearly steady-state profile (averaged during one or more tidal cycles) and, (b) the instantaneous profile. The same schematic profiles may be used for the salinity components (S_t, S_s and S').

As the water column thickness varies with the tidal oscillation it must be separated into the following components,

$$h(x,t) = h_a + h_t(x,t), \quad (5.45)$$

where $h_a = \langle h \rangle = H_0$ is the time mean value of the local depth, and $h_t(x, t) = \eta(x, t)$ is the tidal height.

Replacing the decompositions of Eqs. (5.37), (5.38) and (5.45) in Eq. (5.34), the salt transport T_S, $[T_S] = [ML^{-1}T^{-1}]$ will be decomposed into 32 terms. Disregarding the small terms and others without well-defined physical meaning, results in only seven terms to calculate the time mean salt transport (per unity width) during one or more tidal cycles, described by the equation:

$$\begin{aligned} T_S = \bar{\rho}.(u_a.h_a.S_a + \langle u_t.h_t \rangle S_a + \langle u_t.S_t \rangle h_a + h_a.\overline{u_s.S_s} \\ + h_a \left\langle \overline{u'.S'} \right\rangle + \langle u_t.S_t.h_t \rangle + u_a.\langle S_t.h_t \rangle), \end{aligned} \quad (5.46)$$

or

$$T_S = A + B + C + D + E + F + G. \quad (5.47)$$

The terms A–G are related to the processes responsible for the time-mean salt transport or net salt transport.

The first term (A) represents the seaward advection of salt by the mean u-velocity component. The term B is the salt transport generated by the tidal wave propagation in the estuary, due to the inclined topography of the estuarine channel. According to the pioneer article by G.G. Stokes, published in 1847, the orientation of this wave transport is opposite to its propagation, which is known as the Stokes drift phenomenon (Longuett-Higgins 1969). Thus, its contribution to the salt transport is generally seaward, and, like the term A, it constitutes an advective contribution to the salt transport which may be important in macro-tidal estuaries.

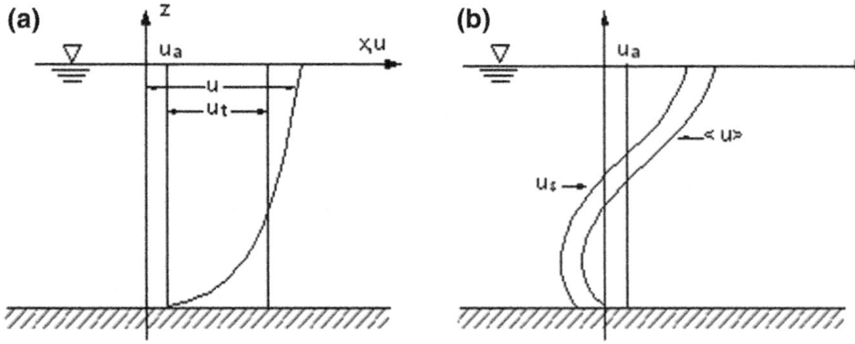

Fig. 5.21 Vertical velocity profiles: the instantaneous (**a**) and time-mean values (**b**), showing the decompositions in the parcels $u_t = \bar{u} - u_a$ and $u_s = \langle u \rangle - u_a$ s(according to Fischer et al. 1979)

The advective processes, A and B, tend to sweep the estuary clear of salt and sharpen the frontal gradient between the river and ocean (Hunkins 1981).

The remaining terms (C–G) are generally considered to represent up-estuary dispersion of salt through mixing of various processes. For the physical interpretation and the orientation (up or down estuary) of these terms, it will be necessary to anticipate the laterally integrated salt conservation equation from Chap. 7 (Eq. 7.80). This equation may be further simplified: with constant width (B = const.), steady-state conditions and vertical velocity component equal to zero (w = 0), because only the longitudinal salt transport is being calculated. In these conditions, the salt balance (advective salt flux = diffusive salt flux) is expressed by,

$$\rho u S = \rho K_E \frac{\partial S}{\partial x}, \tag{5.48}$$

where $u = u_a + U_s$; U_s is the Stokes velocity and K_E is the kinematic dispersive longitudinal coefficient of salt. If all members of this equation are multiplied by the water layer thickness (h), the first member of this equation is the advective salt transport per width unit [$ML^{-1}T^{-1}$], generated by the river discharge and the Stokes drift (terms A + B). Then, this advective salt transport is in balance with the up-estuary salt transport generated by others mechanisms, because under steady-state conditions there is no net transport of salt. Hence, from Eqs. (5.47) and (5.48) it follows (Fischer et al. 1976):

$$-K_E \frac{\partial S}{\partial x} = C + D + E + F + G. \tag{5.49}$$

The terms in the second member of this equation are forced by the combined influences of the tidal stirring and vertical turbulent fluctuations, which have been defined as dispersive mechanisms.

Some of the suggested physical mechanisms connected with the terms of Eq. (5.46) were obtained by Hunkins (1981), through analysis of current and salinity observations in the lower Hudson estuary (New York, USA), providing a basis for assessing the relative importance of the following physical processes:

(i) (A) and (B) total discharge and Stokes wave transport, which have the fresh water discharge as a physical process;
(ii) (C) tidal correlation, with topographic trapping as a physical process;
(iii) (D + E) steady shear dispersion due to the gravitational circulation, bathymetric tidal pumping and steady wind effect;
(iv) (F + G) oscillatory dispersion due to tidal shear and unsteady wind.

The sum A + B is positive and represents the advective salt transport by river discharge. In the case of a standing wave, tidal height and tidal velocity would be 90° out of phase, and the contribution of the term B, which is averaged over the tidal cycle, would vanish. In a long progressive wave, high water occurs at maximum flood and low water at maximum ebb, therefore water is carried up the estuary along with and its up-estuary salt content, and this term has a negative contribution. However, term A incorporates a compensation current which offsets the landward transport by term B.

The term defined as *tidal correlation* (C), is determined by the mean correlation of the product of velocity (u_t) and salt (S_t) multiplied by h_a, which may be positive, indicating a process that acts to carry salt out of the estuary, such as in the Hudson estuary. In an idealized well-mixed estuary, maximum salinity would be reached at the end of the flood tide, so there will be a phase difference of 90° between u_t and S_t with no net contribution from this term since the integral of these terms in quadrature would be zero. It has been suggested by Fischer et al. (1979) that the trapping of water by topographic irregularities (topographic trapping) along the edge of an estuary with its later release during a different tidal stage, could lead to a phase difference of less than 90°. This would be a dispersive process, leading to a negative value for term C. It is generally observed in partially mixed estuaries that the longitudinal salinity gradients are less in the lower layer than they are in the upper layer; consequently, the tidal salinity oscillation is substantially reduced in the lower layer. Also, the tidal current near the bottom affects the tidal current in the upper layer due to the frictional effects. Due to these two well documented characteristics of estuarine circulation, the tidal salinity must lag behind the tidal current by a phase angle greater than 90°, which makes the term C positive.

The component D, defined as *steady shear dispersion*, is due to the vertical gravitational circulation minus the circulation generated by the river discharge. As it acts up-estuary it is not only the largest contribution to the salt dispersion but is also subjected to the largest fortnightly and seasonal changes in partially stratified estuaries. It has a small contribution in well-mixed estuaries. The component E results from the oscillatory turbulent shear with a time scale less than the tidal oscillation generated by the wind; it usually has a small contribution to the salt balance.

The oscillatory dispersion components (F and G) were investigated by Hunkins (1981) and Kjerfve (1986) and have tidal shear and unsteady wind effect as physical processes. They were introduced in the decomposition of the advective salt transport because they may be important in well-mixed estuaries forced by mesotides. The component F, calculated by the triple correlation of the temporal variations of the u-velocity component (u_t), salinity (S_t) and tide (h_t), is by definition the *oscillatory dispersion*. In the component G, the mean value of salinity (S_t) and tidal height (h_t) correlations are multiplied by the river discharge velocity (u_a) and is, in general, dispersive and named oscillatory dispersion. The lack of consistency of its determinations in the Hudson estuary suggests that they were too small to be adequately sampled.

Equations (5.34) and (5.46) are different mathematic expressions of the same physical quantity, the salt transport. Consequently, the comparison of their results may be used to verify the computational results, and also to indicate that the neglected terms in Eq. (5.46) are negligible.

Although there are difficulties in identifying reliable measurements for the application of this method for the decomposition of the advective salt transport, it has been used in studies related to the contribution of various mechanisms in estuaries with different salinity stratifications (Bowden 1963; Dyer 1974; Hunkins 1981; Lewis and Lewis 1983; Kjerfve 1986, and others). As an example of this method applied to a partially mixed estuary (type 2b or B), Table 5.2 presents the results obtained from measurements made in the Bertioga estuarine channel (Fig. 1.5, Chap. 1) during neap tide.

In this experiment, the salt transport was dominated by the advective component (A) and the shear dispersion component (D), generated by the gravitational circulation. In a steady-state condition, the salt transport (T_S) should converge to zero. In this example, $T_S \neq 0$, due to seaward net salt transport equal to 3.54 kg m^{-1} s^{-1}, calculated by the sum of all individual terms; almost the same value (3.61 kg m^{-1} s^{-1}) was obtained applying Eq. (5.34), which confirms the results of the individual terms.

Among the advective salt transport components in the entrance channel of the Patos lagoon, which were calculated from the same data set that was used to classify the estuary as well-mixed (type 1b), the following components were predominant: the advective term A due to the fresh water discharge, and the oscillatory dispersion terms (F + G) generated by the up-lagoon salt transport forced by a meteorological frontal zone (Möller and Castaing 1999).

In laterally non-homogeneous estuaries it is necessary to include the cross-section variations, A = a(x, t), due to the tidal height oscillation, which is decomposed into two components, $A(x, t) = \langle A \rangle + A_t(x, t)$, a time mean area and its time variation, respectively, to take into account its transverse variation. This procedure will increase the complexity of the mathematical treatment and, for further details on this subject, we recommend the Hunkins (1981) article, where advective and dispersive components of salt transport were calculated in transversal sections of the Hudson estuary (New York, USA).

Table 5.2 Advective salt transport components (in kg m^{-1} s^{-1}) for an experiment with hourly measurements during two neap tidal cycles in the Bertioga estuarine channel (SP) (according to Miranda et al. 1998)

Definition	Physical process	Formulation	Transport
A-Total discharge	Fresh water discharge	$\overline{\rho}.u_a.h_a.S_a$	5.97
B-Stokes wave transport	Fresh water discharge	$\overline{\rho}.\langle u_t.h_t\rangle S_a$	−0.16
C-Tidal correlation	Topographic trapping	$\overline{\rho}.h_a\langle u_t.S_t\rangle$	−0.49
D-Steady shear dispersion	Gravitational circulation	$\overline{\rho}.h_a.\overline{u_s S_s}$	−1.32
E-Steady shear dispersion	Bathymetry, tidal pumping, steady wind effects	$\overline{\rho}h_a\langle\overline{u'S'}\rangle$	−0.45
F-Oscillatory dispersion	Tidal shear	$\overline{\rho}\langle u_t.S_t.h_t\rangle$	−0.04
G-Oscillatory dispersion	Unsteady wind effect	$\overline{\rho}.u_a\langle S_t.h_t\rangle$	0.03
T_S = A + B + C + D + F + G			3.54
T_S (Eq. 5.34)			3.61

5.8 Advective Concentration Transport

One of the objectives for measuring velocity, salinity and temperature in estuaries is to use these properties to calculate the flux and transport concentrations of these properties. In this case, let us consider a conservative substance concentration (C), which has been simultaneously measured with the hydrographic properties and current velocity. As this property may be determined by the ratio of its mass or volume per unit of mass, its dimensions are $[C] = [MM^{-1}]$ or $[L^3M^{-1}]$, these concentrations must be taken into account in the determinations of the corresponding flux or transport.

For instance, if C is the oxygen concentration dissolved in the estuarine water mass, which being a non-dimensional property, is usually determined volume of oxygen per volume $[L^3L^{-3}]$. Then, the mathematical expression to calculate the time mean flux (Φ_{O_2}) across an area, A, during the tidal cycle is determined by:

$$\Phi_{O_2} = \frac{1}{T}\int_0^T [\frac{1}{A}\iint_A (C\vec{v} \cdot \vec{n}dA)]dt = \langle \overline{cu} \rangle. \quad (5.50)$$

From this equation, it follows that Φ_{O_2} has dimension of $[L^3L^{-2}T^{-1}] = [LT^{-1}]$. This result represents the physical dissolved oxygen volume contents crossing the unity of the transversal section per time unity (oxygen flux). If in Eq. (5.50) the factor (1/A) is eliminated, the result is the oxygen transport, T_{O_2} (volume per time unity, and $[T_{O_2}] = [L^3T^{-1}]$).

Under the assumption that the concentration, C, of a given property is now expressed as a generic property (Pr) per mass unity $[PrM^{-1}]$, then its flux Φ_C is calculated by,

$$\Phi_C = \frac{\overline{\rho}}{T}\int_0^T [\frac{1}{A}\iint_A (C\vec{v} \cdot \vec{n}dA)]dt = \overline{\rho}\langle \overline{cu} \rangle, \quad (5.51)$$

where the flux Φ_C and its dimension is expressed by $[\Phi_C] = [PrL^{-2}T^{-1}]$. If the factor (1/A) is eliminate from this equation, the result is property transport (property per time unity, in dimension $[PrT^{-1}]$.

Equations (5.50 and 5.51) may also be used to calculate the flux or transport of any property if the measured concentration is expressed as the salinity unity, which is non-dimensional. Equations (5.34) and (5.51) are similar, and the only difference is that the first equation gives salt transport per unit of the transversal section. Then, it is always possible, with the same treatment as described in the previous topic (5.7), to perform decomposition of the property transport into advective and dispersive components and investigate the main mechanisms and physical processes responsible for the exportation or importation of substances in estuaries.

River discharge and Stokes drift are exportation mechanisms which decrease the concentrations of harmful substances introduced into rivers of directly into estuaries. Due to the potential danger to the estuarine environment, the monitoring of harmful substances is of fundamental importance to estuary management, because estuaries' natural conditions may be drastically altered by construction of sand bars, navigational channels, river diversion and fresh water used for urban, agriculture and industrial purposes.

The main methods of reduction, and processing hydrographic and velocity data presented in this chapter have their algorithms programmed in the computational environment MatLab®, described in Bergamo et al. (2002).

5.9 Tidal Prism Determination

When the u-velocity component, u = u(y, z, t), as presented in Fig. 5.20, is measured at regular time intervals in a transversal section at the estuary mouth from the low tide (t = 0) to the high tide ($t = T_P/2$), it is possible to determine the tidal prism T_{PR}, defined in Chap. 2, by the following mathematical expression:

$$T_{PR} = \int_0^{T_P/2} [\iint_A u(y, z, t)dA]dt. \tag{5.52}$$

In this equation the surface integral over the area A (whose numerical integration has been described), is the volume transport $[L^3T^{-1}]$. The integration in the time domain between the time interval [0—T_P/2] may also be determined numerically using an algorithm similar to those presented in Eq. (5.20). This integration (Eq. 5.52) may also be to calculate automatically with an Acoustic Doppler Current Profiler (ADCP), performing velocity profile measurements accurately and at short time intervals, but these results must be checked against others methodologies.

References

Bendat, J. S. & Piersol, A. G. 1966. Measurement and Analysis of Random Data. New York, Wiley. 390 p.

Bérgamo, A. L.; Miranda, L. B. de & Corrêa, M. A. 2002. Estuário: Programas para Processamento e Análise de Dados Hidrográficos e Correntográficos. Rel. Téc. Instituto Oceanográfico, São Paulo, (49):1–16.

Blumberg, A. F. 1975. A Numerical Investigation into the Dynamics of Estuarine Circulation. Tech. Rept. Chesapeake Bay Institute, The Jonhs Hopkins University. n. 91. 110 p. + Apêndices.

Bowden, K. F. 1963. The Mixing Processes in a Tidal Estuary. J. Air Wat. Pollut., 7:343–356.

Bowden, K. F. 1978. "Mixing Processes in Estuaries". In: Kjerfve, B. (ed.). Estuarine Transport Processes. University of South Carolina Press, Columbia, pp. 11–36. (Belle W Baruch Library in Marine. Science, 7).

Chriss, T. M. & Caldwell, D. R. 1984. Universal Similarity and Thickness of the Viscous Sublayer at the Ocean Floor. J. Geophys. Res., 89(C4):6403–6414.

Csanady, G. T. 1982. Circulation in the Coastal Ocean. D. Reidel Publishing, Dordrecht, 279 p. (Environmental Fluid Mechanics).

Defant, A. 1961. Physical Oceanography. Oxford, Pergamon Press, vol. 1. 729 p.

Dyer, K. R. 1974. The Salt Balance in Stratified Estuaries. Estuar. Coast. Mar. Sci., 2:273–281.

Dyer, K. R. 1977. Lateral Circulation Effects in Estuaries. Estuaries, Geophysics and the Environment. Washington, D. C., National Academy of Sciences, pp. 22–29.

Dyer, K. R. 1986. Coastal and Estuarine Sediment Dynamics. New York, Wiley. 342 p.

Dyer, K. R. & Ramamoorthy, K. 1969. Salinity and Water Circulation in the Vellar Estuary. Limnol. Oceanogr., 14(1):4–15.

Elliott, A. J. 1976. A Study of the Effect of Meteorological Forcing on the Circulation in the Potomac Estuary. Spec. Rept. n. 56, reference 76-9, Chesapeake Bay Institute, The Johns Hopkins University, 67 p.

Fischer, H. B. 1976. Mixing and Dispersion in Estuaries. Ann. Rev. Fluid Mech., 8:107–133.

Fisher, J. S.; Ditmars, J. D. & Ippen, A. T. 1972. Mathematical Simulation of Tidal Time Averages of Salinity and Velocity Profiles in Estuaries. Massachusetts Institute of Technology, Mass., Rept. MITSG 72–11, 157 p.

Fischer, H. B.; List, E. J.; Koh, R. C. Y.; Imberger, J. & Brooks, N. H. 1979. Mixing in Inland and Coastal Waters. New York, Academic Press. 483 p.

Geyer, W. R. 1997. Influence of Wind on Dynamics and Flushing of Shallow Estuaries. Estuar. Coast. Shelf Sci., 44:713–722.

Hanselman, D. & Littlefield, B. 1998. Mastering MatLab 5. A comprehensive Tutorial and Reference. Prentice Hall, The MatLab® Curriculum Series, 638 p.

Heathershaw, A. D. 1981. Comparisons of measured and predict sediment transport rates in tidal currents. Mar. Geol., 42:75–104.

Hunt, J. N. 1964. Tidal Oscillations in Estuaries. Geophys. J. R. Astr. Soc., 8:440–455.

Hunkins, K. 1981. Salt Dispersion in the Hudson Estuary. J. Phys. Oceanogr., 11:729–738

Ianniello, J. P. 1977. Tidally induced residual currents in estuaries of constant breadth and depth. Journal of Marine Research, 35(4), 755–786.

Jay, D. A. & Musiak, J. D. 1996. Internal Tidal Asymmetry in Channel Flows: Origins and Consequences. In: Pattiaratchi C. ed. Coastal and Estuarine: Mixing in Estuaries and Coastal Seas. American Geophysical Union, pp. 211–249

Jenkins, G. M. & Watts, D. G. 1968. Spectral Analysis and its Application. San Francisco, Holden-Day. 525 p.

Kjerfve, B. 1970. Description of a Georgia Estuary. M. S. Essay. Seattle, Dept. of Oceanography, University of Washington. 43 p.

Kjerfve, B. 1973. Volume Transport, Salinity Distribution and Net Circulation in Duplin Estuary. Tech. Rept. ERC 0273. Marine Institute, Georgia, University of Georgia, 30 p.

Kjerfve, B. 1975. Velocity Averaging in Estuaries Characterized by a Large Tidal Range to Depth Ratio. Estuar. Coast. Mar. Sci., 3:311–323.

Kjerfve, B. 1986. Circulation and Salt Flux in a Well Mixed Estuary. In: Van de Kreeke, J. (ed.). Physics of Shallow Estuaries and Bays. Berlin, Spring-Verlag, pp. 22–29.

Lesht, B. M. 1979. Relationships between sediment resuspension and the statistical frequency distribution at bottom shear stress. Mar. Geol.:M19–M27.

Lewis, R. E. & Lewis, J. O. 1983. The Principal Factors Contributing to the Flux of Salt in a Narrow, Partially Stratified Estuary. Estuar. Coast. Mar. Sci., 16:599–626.

Longuet-Higgins, M. S. 1969. On the Transport of Mass by Time-Varying Ocean Currents. Deep-Sea Res., 16:431–447.

Lueck, R. G. & Lu, Y. 1997. The Logarithmic Layer in a Tidal Channel. Continent. Shelf Res., 17 (14):1785–1801.

McAlister, W. B.; Rattray Jr, M. & Barnes, C. A. 1959. The Dynamics of a Fjord Estuary: Silver Bay, Alaska. Department of Oceanography. Tech. Rept. n. 62, University of Seattle, WA.

Merz, A. 1921. Die Strömungen von Bosphorus un Dardanellen. Verh. d. 20. Dtsch. Geogr.-Tag Juni, 1921.

Miranda, L. B.; Castro, B. M. de. & Kjerfve, B. 1998. Circulation and Mixing in the Bertioga Channel (SP, Brazil) Due to Tidal Forcing. Estuaries, 21(2):204–214.

Möller, O. O. Jr., 1996. & Castaing, P. 1999. Hydrographical Characteristics of the Estuarine Area of Patos Lagoon (30°S, Brazil). In: Perillo G. M. E; Picollo, M. C. & Pino-Quivira M. (eds.). Estuaries of South America: Their Geomorphology and Dynamics. Berlin, Springer-Verlag, pp. 83–100 (Environmental Science).

Morettin, P. A. 1999. Ondas e Ondaletas: Da Análise de Fourrier à Análise de Ondaletas. Edusp, São Paulo. 272 p. (Acadêmica 23).

Mossby, H. 1947. Experiments of turbulence and friction near the bottom of the sea. Bergens Mus. Aarb. 1946/1947, Naturv. Rek. No. 3, Bergen, 1947.

Pennington, R. H. 1970. Computer Methods and Numerical Analysis. 2. ed., Macmillan. 452 p.

Pritchard, D. W. 1954. A Study of Salt Balance in a Coastal Plain Estuary. J. Mar. Res., 13 (1):133–144.

Rayner, J. N. 1971. An Introduction to Spectral Analysis. London, Pion. 174 p. (Monographs in Spatial and Environmental Systems Analysis, 2).

Simpson, J. H.; Brown, J.; Mattews, J. & Allen, G. 1990. Tidal Straining, Density Currents, and Stirring in the Control of Estuarine Stratification. Estuaries, 13(2), p. 125–132.

Soulsby, R. L. 1983. The Bottom Boundary Layer of Shelf Seas. In: Johns B. (ed.). Physical Oceanography of Coastal and Shelf Seas. Amsterdam, Elsevier, pp. 189–266. (Oceanography Series, 35).

Sternberg, R. W. 1968. Friction Factors in Tidal Channels with Differing Bed Roughness. J. Mar. Geol., 6:243–260.

Sverdrup, H. U.; Johnson, M. W. & Fleming, R. H. 1942. The Oceans, their Physics, Chemistry and General Biology. New Jersey, Prentice-Hall. 1042 p.

Valle-Levinson, A.; Miller, J. L. & Wheless, G. H. 1998. Enhanced Stratification in the Lower Chesapeake Bay Following Northeasterly Winds. Continent. Shelf Res., 18:1631–1647.

Weisberg, R. H. 1976. A Note on Estuarine Mean Flow Estimation. J. Mar. Res., 34(3):387–394.

Zhou, M. 1998. Influence of Bottom Stress on the Two-layer Flow Induced by Gravity Currents in Estuaries. Estuar. Coast. Shelf Sci., 46:811–825.

Chapter 6
Mixing Processes in Estuaries: Simplifyed Methods

The classical definition of an estuary establishes that it is a partially closed water body with openings to the adjacent ocean, where the seawater is diluted by fresh water of the fluvial drainage basin. The input of fresh water decreases the potential energy of the water column, which is supplied by tidal energy through the mixing process produced on the bottom and internal shear instabilities.

The presence of denser water at the estuary mouth generates a system which constantly pushes seawater into the estuary. The water mass in the mixing zone (MZ) is composed of fresh and salt water, which varies along the estuary. Due to the seaward salinity increase, the horizontal salinity gradient generates the baroclinic component of the gradient pressure force. The barotropic and the baroclinic components of the gradient pressure force, the fresh water input, and the wind shear cause agitation of the estuarine water mass and generate the mixing processes (advection and turbulent diffusion).

In this chapter, we take a semi-empirical approach to estuarine processes. The estuary will be considered as a *black box* and one-dimensional system, and the salinity and the fresh water will be used as tracers under steady-state conditions. The estuary geometry, river discharge (Q_f), tidal height (H_o), and the non-diluted salinity of the adjacent coastal ocean (S_0) will be considered as known. The conservation of salt and volume, with the assumption of a well-mixed estuary, will be used to calculate the longitudinal salinity distribution as well as the *flushing time,* which indicate the estuary capacity to flush out the salinity and concentrations of any conservative property.

The non-diluted salinity of the adjacent coastal ocean (S_0) is measured outside of the estuarine plume and, due to mixing processes on the continental shelf, may be subjected to slow seasonal variation. Then, it is advisable to take monthly time mean representative values for the seasonal period, if possible. For estuaries on the Southeastern Brazilian continental shelf (23°S–28°S), S_0 values may be found in Castro and Miranda (1998).

L. Bruner de Miranda et al., *Fundamentals of Estuarine Physical Oceanography*,
Ocean Engineering & Oceanography 8, DOI 10.1007/978-981-10-3041-3_6

6.1 Fundamental Concepts

The definition of an estuary may follow from the experimental evidence. In fact, the longitudinal salinity gradient in the mixing zone, shown in Fig. 4.4 (Chap. 4), indicates conclusively that, due to the measurable dilution of the seawater by the river fresh water, a parcel of this fresh water remains in the estuarine water body. According to Ketchum (1950), the *flushing time* (t_q) is the ratio of the fresh water volume retained in the MZ and the river discharge (Q_f), thus,

$$t_q = \frac{V_f}{Q_f}. \tag{6.1}$$

This property, with dimension $[t_q] = [T]$ has also been called *mean detention time* by Fischer et al. (1979), and is dependent on two main quantities intimately related: the fresh water volume retained in the MZ and the river discharge that dilutes the seawater entering the estuary. It should be noted that the tidal height, which determines the mixing intensity, and the direct and remote wind stress, which may dam or remove the estuarine water mass, may be important to determine this time interval.

In normal conditions, the river discharge (Q_f) may be considered constant during the tidal cycle; however, the fresh water volume (V_f) usually increase and decrease during the ebb and flood tide, respectively, but in the intratidal time scale it is very difficult to calculate this quantity. Therefore, it is more common to calculate a global *flushing time*, representative for a tidal period; it is an important quantity because it measures the time interval necessary for the fresh water volume retained in the mixing zone to be removed from the estuary, along with concentrations of other substances in the estuarine water mass.

The input of fresh water volume (R) into the estuary, during the time interval of a complete diurnal or semi-diurnal tidal cycle (T_P) is calculated by $R = T_P Q_f$. If the MZ of the estuary had already accumulated a fresh water volume, V_f, the flushing time for the tidal period is the ratio,

$$t_q = \frac{V_f + R}{R} T. \tag{6.2}$$

This result indicates that t_q is higher than or equal to the tidal period, and $t_q = T_P$ only when $V_f \to 0$. Small values of t_q indicate that the removal of all fresh water is due to macro or hyper tides, and the mean estuarine depth is similar to the tidal height. In these conditions, the estuarine water mass is completely renewed (the MZ is flushed out of the estuary) at each tidal cycle, because almost the whole water in the estuary is flushed out. Therefore, this environment is less susceptible to water pollution by pathogenic substances.

Let's now consider influences on the flushing time, which may occur due to the fortnightly tidal modulation, under the assumption that the river discharge and the salinity in the continental shelf don't vary and the wind forcing is negligible.

During neap tides, the estuarine circulation is less intense and the vertical stratification is high; however, during spring tides, the conditions are opposite (intense circulation and less vertical stratification). Thus, during the spring tidal cycle, less fresh water is retained than during in the neap tide, $(V_f)_S < (V_f)_N$. Applying the flushing time definition (Eq. 6.1) the corresponding values at the spring $(t_q)_S$ and neap $(t_q)_N$ tides are given by:

$$(t_q)_S = \frac{(V_f)_S}{Q_f}, \quad \text{and} \quad (t_q)_Q = \frac{(V_f)_Q}{Q_f}. \tag{6.3}$$

By combining these results it follows that:

$$(t_q)_S = \frac{(V_f)_S}{(V_f)_Q}(t_q)_Q. \tag{6.4}$$

Consequently, taking into account the inequalities of the fresh water volumes retained in the spring and neap tidal conditions, it follows from Eq. (6.4) that $(t_q)_S < (t_q)_N$. Then, according to the simplified conditions, the fresh water volume retained in the MZ is removed quickly during the spring tide than the neap tide.

The fresh water volume (V_f) necessary to calculate the flushing time may be obtained from the knowledge of the non-diluted salinity value at the continental shelf (S_0). But, to do so, it is necessary to define the mean *fresh water fraction* or *concentration* indicated by $\overline{f}$. This quantity is defined as the ratio of the fresh water volume in the MZ (V_f) and the corresponding steady-state geometric volume (V) of the estuarine water mass,

$$\overline{f} = \frac{V_f}{V}. \tag{6.5}$$

The fresh water fraction (concentration) is a non-dimensional quantity which varies between the following extreme values: $\overline{f} = 1$, when the estuary is completely filled with fresh water, and $\overline{f} = 0$, when there is no fresh water in the MZ (meaning $Q_f \approx 0$). As this quantity is function of the fresh water volume (V_f), which is the unknown required to calculated the flushing time (Eq. 6.1), then it is necessary to find an alternative way to calculate $\overline{f}$ (Eq. 6.5), using another quantity. In this case we can use the salinity.

Considering a small control volume dV, it is possible to define fresh water fraction as $f = dV_f/dV$. Following its displacement from the estuary head (where $S = 0$ and $dV = dV_f$) to the estuary mouth, where the fresh water influence is small $(dV_f \approx 0)$, the fresh water fraction of this volume varies from 1 to almost zero $(0 < f \leq 1)$. This variation interval is the same as that presented above. Then, using the salinity definition and the conservation principle of the mass of salt in seawater, the fresh water fraction may be calculated as function of this physical-chemical property. Consider a volume of salt water V_i with the total mass, M, a salinity S_i, and density ρ_i. Then, from the salinity definition,

$$S_i = \frac{m}{M} = \frac{m}{\rho_i V_i}, \tag{6.6}$$

where m indicates the mass of the dissociated salts in the volume V_i. Adding to a the fresh water volume ΔV_f, to the initial water volume, the resulting salinity value S ($S < S_i$) due to this dilution is calculated by,

$$S = \frac{m}{\rho(V_i + \Delta V_f)}, \tag{6.7}$$

where ρ indicates the new density value of the solution. Solving these Eqs. (6.6) and (6.7) for the mass m and equating the results gives,

$$\rho_i S_i V_i = \rho S(V_i + \Delta V_f). \tag{6.8}$$

Disregarding the density variation ($\rho \approx \rho_i$) gives the ratio:

$$\frac{\Delta V_f}{V_i} = \frac{S_i - S}{S}. \tag{6.9}$$

The first member of this equation is the fresh water fraction in relation to the initial volume. Then, this final result indicates that is possible to calculate the fresh water fraction if the initial and final salinity values are known. By analogy, and considering that in estuaries this fraction varies from 1 to 0, when the control volume of water is displaced along the estuary from its head down to the mouth, the mean value of this quantity may be calculated by (Ketchum 1950):

$$f = \frac{V_f}{V} = \frac{(S_0 - S)}{S_0}, \quad \text{and} \quad V_f = (1 - \frac{S}{S_0})V. \tag{6.10}$$

As previously mentioned, the undiluted salinity value (S_0) or the salinity at the salt source, is known in this equation, and is a characteristic value of the water mass of the adjacent continental shelf without influence of the estuarine plume.

If the salinity distribution in the estuarine MZ is in steady-state condition, its spatial distribution depends only its spatial coordinates, $S = S(x, y, z)$. Then under this condition, if the salinity field is known, it follows from Eq. (6.10) that:

$$f(x,y,z) = \frac{S_0 - S(x,y,z)}{S_0} = 1 - \frac{S(x,y,z)}{S_0}, \tag{6.11}$$

and the fresh water concentration is also a function of the spatial coordinates. This equation may be solved for $S = S(x, y, z)$, if $f = f(x, y, z)$ is known, and we have $S(x, y, z) = S_0[1 - f(x, y, z)]$.

Applying Eq. (6.5) to a small differential volume dV, the corresponding fresh water volume is calculated by: $dV_f = f(x, y, z)dV$. Then, integrating in the finite geometric volume (V) of the estuarine MZ,

$$V_f = \int_V f dV = \iiint_V f(x,y,z)dxdydz. \tag{6.12}$$

In this equation the fresh water fraction is calculated by the Eq. (6.11). Applying the *Mean Value Theorem* of calculus Eq. (6.12) may be rewritten as,

$$V_f = \bar{f}V = (1 - \frac{\bar{S}}{S_0})V, \tag{6.13}$$

and the fresh water volume is obtained as a function of the mean salinity and the geometric volume of the MZ.

Combining the flushing time definition (Eq. 6.1) and Eq. (6.13), it follows that:

$$t_q = \frac{V_f}{Q_f} = \frac{\bar{f}V}{Q_f} = \frac{(S_0 - \bar{S})}{S_0} \frac{V}{Q_f}. \tag{6.14}$$

This equation indicates that the flushing time is directly proportional to the difference $(S_0 - \bar{S})$ and the fresh water volume V, and is inversely proportional to the fresh water discharge Q_f.

As an example of the flushing time determination, let us consider a laterally homogeneous estuary, with known values for its stationary salinity field $S = S(x, z)$ and non-diluted salinity (S_0) at the coastal region. Then, it is possible with Eq. (6.11) to convert the isohalines into the corresponding isolines of fresh water fraction $f(x, z) = \text{const.}$, presented in Fig. 6.1. With the assumption that the MZ width (B) may be considered constant, the fresh water volume (V_f) is calculated with an equation similar to (6.12):

$$V_f = B \iint_A f(x,z)dxdz, \tag{6.15a}$$

or if the estuary has a length, L, and $\bar{A}$ is the mean cross section area, the fresh water volume is determined by,

$$V_f = \bar{A} \int_0^L f(x)dx. \tag{6.15b}$$

With the assumption that the estuary width is 500 m, and numerically calculating the area integral (Eq. 6.15a) with the data of Fig. 6.1, the computed fresh water volume (V_f) is approximately 61.0×10^5 m^3. In the case of a river discharge

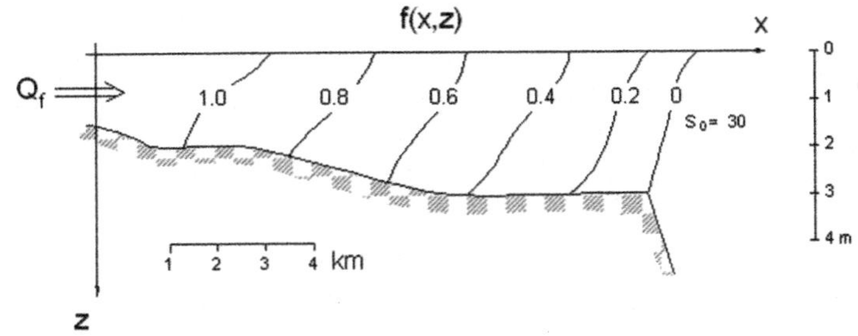

Fig. 6.1 Longitudinal steady-state distribution of the fresh water fraction f = f(x, z) in a partially mixed and laterally homogeneous estuary

of 50 $m^3\ s^{-1}$, the flushing time of this estuary is approximately 34.0 h, which is almost three semi-diurnal tidal cycles.

As another example of flushing time and flushing rate calculation is the following from Fischer et al. (1979). "A well-mixed estuary has a constant cross-sectional area $A = 10^4\ m^2$, a length of $L = 30 \times 10^3$ m, and a constant kinematic longitudinal diffusion coefficient $K_H = K = 10^2\ m^2\ s^{-1}$. The fresh water inflow is 30 $m^3\ s^{-1}$. Find the flushing time and the flushing rate according to Eqs. (6.1) and (6.15b), respectively".

To simulate the longitudinal salinity distribution from its mouth, $S = S_0$ at $x = 0$, towards its head, $S = 0$ for $x \rightarrow -\infty$, a possible solution of the one-dimensional salinity variation may be expressed by,

$$S(x) = S_0 \exp[u_f(\frac{x}{K})] = S_0 \exp(3.0 \times 10^{-5}),$$

and the corresponding longitudinal distribution of the fresh water fraction is

$$f(x) = 1 - \frac{S(x)}{S_0} = 1 - \exp(3 \times 10^{-5}).$$

The fresh water volume in the estuary volume (Eq. 6.15b) is calculated by the following integral:

$$V_f = A \int_0^L [1 - \exp(-3 \times 10^{-3})]dx = 1.02 \times 10^8 m^3,$$

and the flushing time t_q, and the *flux rate* F (to be defined in Eq. (6.16), are calculated by

$$t_q = \frac{V_f}{Q_f} = \frac{1.02 \times 10^8}{30} = 3.41 \times 10^6\,\mathrm{s} \approx 39.4\,\mathrm{days}$$

and

$$F = \frac{V}{t_q} = \frac{V}{V_f} Q_f = 90\,\mathrm{m^3 s^{-1}}.$$

In the classical and recent literature there are several examples using the concepts and equations presented here to calculate the flushing time of estuarine systems: For example, Ketchum et al. (1951) calculated the flushing time near the Hudson river mouth in the New York Bight (USA) during high (488 $m^3\ s^{-1}$) and low (197 $m^3\ s^{-1}$) river discharge volumes, and the flushing time corresponding values where 6.0 and 10.6 days, respectively. Another classical result was published by Hughes (1958), who analyzed data collected during low river discharge (25.7 $m^3\ s^{-1}$) in the Mersey Narrows estuary (Liverpool, England) and the calculated flushing time to be 5.3 days. The annual variations analysis by Pilson (1985), using various monthly time intervals of river discharge and salinity obtained in the Narrangansett Bay (Rhode Island, USA) during 1951 and 1977, indicated large flushing time variations, from approximately 12 and 40 days, with the extreme values occurring during periods of high and low river discharges, respectively. Miranda and Castro (1993) investigated the flushing times associated with fortnightly tidal modulation; using data sampled during two spring and neap tide tidal cycles, in the Bertioga Channel (Chap. 1, Fig. 1.5), and obtained values of 2.5 and 3.2 days, respectively.

This methodology was applied by Geyer (1997) using moored measurements and along-estuary hydrographic stations in flushing time studies in two small and shallow (1–2 m depth) sub-estuaries (the Child and the Quashnet) in Waquoit Bay (Cape Cod, USA). These sub-estuaries were forced by different wind directions and intensities, had low average rivers discharges of 0.1 and 0.4 $m^3\ s^{-1}$, and had weak spring-neap tidal modulation a tidal height of approximately 0.5 m. This study demonstrates the strong influence of wind forcing on the salinity structure and flushing characteristics of these shallow estuaries. According to the article's conclusions, onshore winds inhibit estuarine circulation, increasing the along-estuary salinity gradient and reducing the flushing rate, due to the landward freshwater accumulation. Offshore winds enhanced the surface outflow, flushing out the freshwater and reducing the along-estuary salinity gradient. The flushing time of the Childs varied from less than one day, in offshore wind conditions, to 2.7 days during strong onshore winds (6.0 $m\ s^{-1}$), with a significant correlation at the 95% confidence level. Because onshore wind flushes water into the estuary increasing the MZ geometric volume, the flushing time may be explained by this volume increase, according to the Eq. (6.14), in which the mean salinity increase compensates for the decrease in the $S_0 - \overline{S}$. The flushing time of the Quashnet was shorter, typically 15 and 17 h, with only one observational in which it was more than one day, and showed little wind-induced variability.

The flushing time may only be applied with rigor to a conservative pollutant that is adequately discharged into the estuary head; however, if the substance is discharged at another longitudinal position, its flushing time will be different (Bowden 1967a, b).

Another physical quantity related to the mixing process—combined effects of salt dilution due to the advection and diffusion, is the time rate exchange of the MZ volume (V) during the flushing time interval (t_q). This quantity (F), named the *flux rate,* is calculated by the ratio $F = V/t_q$, and, according to the Eq. (6.14), it may be calculated as (Officer 1976) and Officer and Kester (1991):

$$F = \frac{V}{t_q} = \frac{S_0}{(S_0 - \overline{S})} Q_f = \frac{Q_f}{\overline{f}}. \quad (6.16)$$

This equation indicates that the flushing rate is directly proportional to the river discharge and inversely proportional to the mean fresh water fraction, which is dependent on the mixing intensity (non-advective tidal processes and advective gravitational exchanges). The determination of the flushing rate with the data of the exercise of this topic (i.e. 50 $m^3\ s^{-1}$), and its definition or in function of the Q_f and $\overline{f}$ (Eq. 6.16), then $F = 117\ m^3\ s^{-1}$, that is approximately to 2.4 times of the river discharge.

The mean values ($\overline{S}$) and ($\overline{f}$) are dependent on the diffusive up-estuary salt transport generated by the tidal forcing, the fresh water discharge and the gravitational circulation. When the estuary is dominated by the river discharge, and the mixing zone (MZ) is advected to the coastal region, the mean salinity and the fresh water fraction tend to zero and one ($\overline{S} \to 0$, $\overline{f} \to 1$), respectively. Then, from Eq. (6.16), the flushing rate (F) is equal to the river discharge, and, under this condition, the angular coefficient of this correlation, $F = f(Q_f)$ tends to one. Another limiting condition is: for $Q_f \to 0$ also $F \to 0$. When the turbulent diffusive process generated by the tide is predominant (as in a well-mixed estuary), the mean salinity, $\overline{S}$, may be considered independent on the river discharge. As F is inversely proportional to ($S_0 - \overline{S}$) it is possible, in first approximation, to also consider this rate as independent of the river discharge and F = const. The correlations under these limiting conditions ($F = Q_f$ and F = const.) are illustrated in Fig. 6.2.

To first-order effects, the tidal exchange flux should be independent of the freshwater input into the estuary, disregarding the dependence of the tidal exchange flux on the vertical shear stratification. If, for example, there were no the tidal diffusion exchanges, a plot F versus Q_f should be a curve with a zero intercept for F at $Q_f = 0$, and increasing values of F corresponding with increasing river discharge Q_f (Fig. 6.2, continuous line). Thus, for a more general situation, where both tidal and gravitational circulation processes are operative, the intercept value F_I, for F at $Q_f = 0$, will represent the tidal exchange flux, F values in excess of the intercept value will represent the various freshwater input conditions, and Q_f will represent

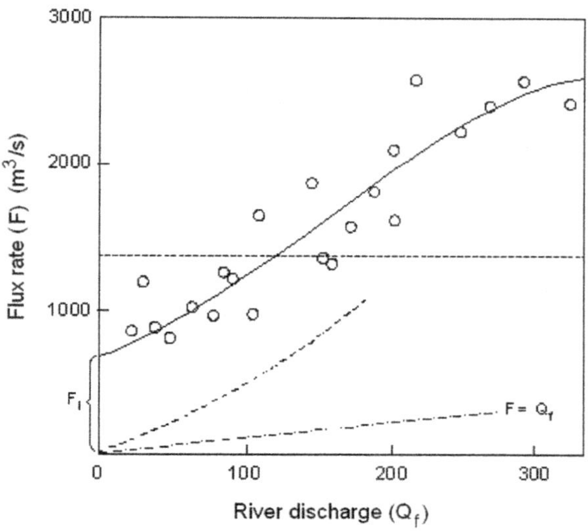

Fig. 6.2 Correlations of the *flushing rate* (F) and river discharge (Q_f) inferred under conditions $F = Q_f$ and F = const. and with experimental data (o), and the mean correlation, for the estuarine system of the Narragansett bay (Rhode Island, USA). The intersection with the ordinate axis (F_I, for $Q_f \rightarrow 0$) represents the diffusive parcel of tidal mixing. The *dashed* and *dashed-point line curves* indicate the flushing rate due to the diffusive and advective processes, respectively (adapted from Officer and Kester 1991)

the gravitational circulation influences on salt flux (Officer and Kester 1991). Then, according to this result, the physical quantities F_I (for $Q_f \rightarrow 0$) and F are related to the mixing parameter, ν, of the classical Stratification-circulation Diagram of Hansen and Rattray (1966) (Chap. 3), which indicates the ratio of the up-estuary salt transport due to diffusion (Φ_{dif}) and the longitudinal total salt flux due to diffusion and advection (Φ_{adv}) expressed by:

$$\nu = \frac{F_I}{F} = \frac{\Phi_{dif}}{\Phi_{dif} + \Phi_{adv}}. \tag{6.17}$$

From this ratio it follows that when $F = F_I$ or $F_I \rightarrow 0$ the parameter ν is equal to 1 or 0, respectively, and the up-estuary salt transport predominant to the mixing process is due to tidal diffusion or advection, respectively. This result is in close agreement with the physical interpretation of the ν parameter of the Hansen and Rattray (1966) (Eq. 11.96b, Chap. 11).

The Pilson (1985) data used to calculate flushing time variation, were revisited and the corresponding values of flushing rates (F) were calculated by Officer and Kester (1991). Figure 6.2 is a plot F versus Q_f, showing a well-defined dependence of these variables. An empirical curve has been drawn through the data points, with a zero intercept value for F, around 700 $m^3\ s^{-1}$; this value identifies F_I, which is the diffusive component of the tidal mixing. Values of the mixing parameter, ν, were determined from Eq. (6.17) with tabulated values of F and the zero intercept value, F_I, which were correlated with the river discharge. From this correlation, it was observed that for $Q_f \to 0$, the parameter $\nu \to 1$, showing that under this condition the estuarine bay, is dominated by tidal mixing. From the tabulated results of Officer and Kester (op. cit.), it is possible to observe that for $Q_f = 154\ m^3\ s^{-1}$, the parameter ν is equal to 0.52. Then, for this observational period, the dynamic exchange processes due to tidal diffusion and gravitational circulation forced by the river discharge had almost the same magnitude.

Although being a simple procedure, estimation of the relative contribution of the tidal and gravitational circulation to the salt flux, using an alternative methodology to calculate the estuarine parameter ν (Eq. 6.17), requires time series measurements of river discharge, and mean salinities of the estuary and in the coastal ocean.

The classical concepts we have introduced here are the basic concepts for the following topics related to the simplified mixing models, which, although semi-empirical, are important for providing the initial knowledge of the main estuarine characteristics influenced by the mixing processes (advection and turbulent diffusion). Their objectives are:

- To determine the salinity, fresh water fraction and the concentration of conservative substances in the estuary, in steady-state conditions.
- To calculate the time interval that the a small fresh water volume remains inside the estuary (flushing time).

6.2 Tidal Prism

The presentation of one-dimensional tidal prism mixing models must be initiated with the *tidal prism*, which is the simplest of the *box model*. The tidal prism will be applied to the salinity determination in an estuary with known tidal amplitude, estuarine surface area, fresh water discharge (Q_f) and coastal ocean salinity (S_0). It is applied to an ideal estuary (Fig. 6.3), with the assumption that the tidal prism ($T_{PR} = V_M$), defined in Chap. 2) with constant salinity (S_0) is completely mixed, with the fresh water from the river discharge introduced into the estuary during the flood tide. We also assume that the estuary is well-mixed (Type 1 or A). The fresh water volume at the disposal of the mixing at high tide is $(1/2)V_M Q_f = (1/2)R$ and, with the hypothesis that the low tidal water volume does not contribute to the mixture, it follow that the equality taking into account the salt mass conservation condition is:

$$\overline{\rho}\overline{S}(V_M + \frac{1}{2}R) = \rho_0 S_0 V_M, \tag{6.18}$$

where $V_M = T_{PR}$. In this equation $\overline{S}$ and S_0 are the mean salinity at high tide and that at the coastal sea, respectively.

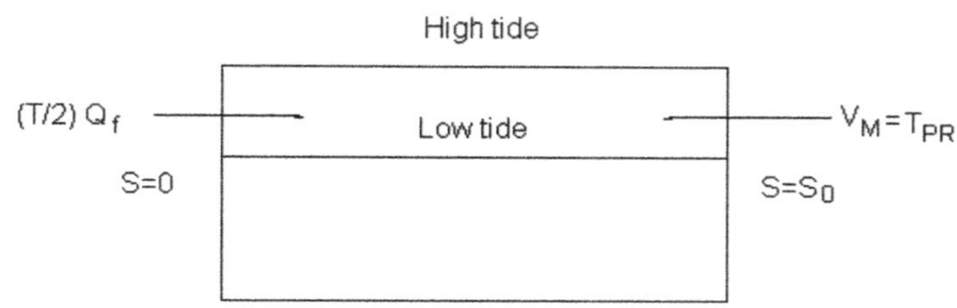

Fig. 6.3 Schematic diagram of the tidal prism model. Water exchanges and the salinities at the boundaries are indicated. For convenience the tidal prism (T_{PR}) will also be denoted by V_M

Solving Eq. (6.18) for the mean salinity value $\overline{S}$, it follows that the equation to calculate this property at high tide is:

$$\overline{S} = S_0 \frac{V_M}{V_M + \frac{1}{2}R}, \tag{6.19}$$

With this equation, it is possible to calculate the salinity mean value ($\overline{S}$) at high tide, knowing the tidal prism and the river water volume discharged into the estuary mixing zone (MZ) during half a tidal period, which are quantities that may be known for well-mixed estuaries. With this result, the mean fresh water fraction ($\overline{f}$) may be calculated by:

$$\overline{f} = \frac{S_0 - \overline{S}}{S_0} = 1 - \frac{V_M}{V_M + \frac{1}{2}R}, \tag{6.20}$$

or

$$\overline{f} = \frac{\frac{1}{2}R}{V_M + \frac{1}{2}R} = \frac{R}{2V_M + R}, \tag{6.21}$$

Taking into account Eqs. (6.1) and (6.5), the fresh water volume (V_f) and the flushing time (t_q) may be calculated by the following equations:

$$V_f = \overline{f}V = \frac{R}{2V_M + R}V, \tag{6.22}$$

and

$$t_q = \frac{V_f}{Q_f} = \frac{R}{(2V_M + R)Q_f}V = \frac{T_P}{2V_M + R}V. \tag{6.23}$$

This result indicates that the lowest flushing time interval T_P (tidal period) occurring when the geometric volume of the MZ (V) is equal to $2V_M$ + R, which corresponds to an ideal condition when this volume is entirely removed to the coastal zone during a complete tidal cycle. In this ideal condition, the mean salinity at the inner mixing zone tends to zero and the flushing rate is equal to the river discharge ($F = Q_f$). The flushing time may also be expressed in tidal period units (T_P). Then, it is suitable to divide Eq. (6.23) by T_P, and $t_q = \frac{V}{(2V_M + R)}$; when expressed in tidal period (T_P), the flushing time is directly proportional to the geometric volume of the MZ and inversely proportional to the tidal prism plus the fresh water discharged during the tidal cycle.

In Eqs. (6.19) and (6.20), the quantity (1/2)R may be substituted by R, resulting in approximate mean values for the salinity and fresh water fraction, ($\overline{S}$) and ($\overline{f}$), during a complete tidal cycle. In this condition, it is possible to demonstrate that the fresh water volume and the flushing time are calculated by: $V_f = [R/(V_M + R)]$ and $t_q = [V/(V_M + R)]T_P$, respectively.

Finally, with the Eq. (6.23) and the flux rate (F) definition (Eq. 6.16) it follow that: $F = (2V_M + R)/T_P$.

The solution of this simplest box model (tidal prism) to calculate $\overline{S}$ (Eq. 6.19), the flushing time t_q (Eq. 6.23) and the flux rate F may not give satisfactory results due to the following approximations:

- The fresh river water doesn't mix completely with the seawater during the flood tide or during a complete tidal cycle;
- The coastal region isn't a perfect sink and a water parcel flushed out to the near shore turbidity zone (NTZ) may return to the estuary in the next tidal cycle.

6.3 Segmented Tidal Prism Model

The tidal prism model hypothesis assumes a uniform steady-state salinity distribution at high tide, and can be applied to well-mixed estuaries. To eliminate this restriction, it has been re-worked by several researches aiming to enable its application to stratified estuaries. In the pioneering article by Ketchum (1951), a one-dimensional model was presented, where the mixing zone (MZ) was partitioned in segments or cells.

The main hypothesis of the segmented tidal prism model is complete mixing of the river and sea water in each segment or cell at high tide. The conservation equation for this model is based on the principle of volume continuity of fresh water volume in the estuary. According to Ketchum (op. cit.), complete mixing occurs at high tide in each segment, while in the Dyer and Taylor (1973) model, complete mixing may occur in the segments at high and low water levels. Then, the segmented tidal prism may be applied to one-dimensional well-mixed estuaries ($\nu = 1$, according to the Hansen and Rattray 1966 classification method), which are dominated by vertical turbulent tidal diffusion.

Semi-empirical models of Ketchum and Dyer and Taylor will be presented and applied to an ideal estuary to an ideal estuary, and inter-comparisons will also be made with observational data collected in the Winyah Bay estuarine system (South Carolina, USA).

The Ketchum paper presents a semi-empirical theory on the mixing of fresh river water with the seawater in selected segments distributed along the estuary (Fig. 6.4). The theory attempts to predict average conditions in successive volume segments for a constant river discharge and a mean tidal range. According to this theory, it is possible to calculate the one-dimensional mean salinity, the fresh water fraction distribution and the flushing time. This theory uses the fresh water as an indicator and may be easily adapted to include the one-dimensional variation of any conservative property concentration dissolved in the mixing zone. This theory assumes the following hypothesis:

- Steady-state river discharge and salinity field and a balance between inflow and outflow of sea water.
- Full mixing of fresh and salt water during flood tides.
- During a tidal cycle, a seaward volume of fresh water, equal to the input of fresh water discharged at the estuary head, must be moved.
- Salinity at the coastal sea has small temporal variation ($S_0 \approx$ const.).

The inner end of the estuary (segment 0) is defined as the section above which the volume required to raise the water level from low to high water is equal to the river discharge input during the tidal cycle (Fig. 6.4); during the ebb tide, there will be a loss through this section of one river flow volume per tide. Then, by this definition, there is no seawater interchange at the boundary between *segments* 1 and 0, segment 0 being completely fresh water. During the flood tide and above the boundary of segments 0 and 1, the salinity value and the fresh water fraction are 0 and 1, respectively. It should be noted that this is a dynamic boundary, not a geometric definition, since the boundary location will move corresponding to changes in river discharge (Ketchum 1951).

The segment volumes along the estuarine channel are calculated by their water volume at low tide (V_n), with their corresponding tidal prism (P_n) volumes added at high tide, as indicated in Fig. 6.4. Then, at high tide, the segment volume are equal

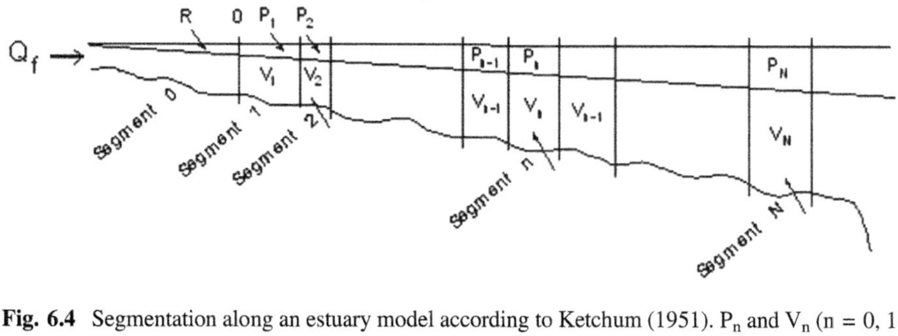

Fig. 6.4 Segmentation along an estuary model according to Ketchum (1951). P_n and V_n (n = 0, 1, 2, … N) indicate tidal prism and low tidal volumes in a generic segment n. In the inner most segment (n = 0) there will be only fresh water given by $R = T_P Q_f$ (adapted from Dyer and Taylor 1973). The original notation of the tidal prism as P_i (for i = 0 to N) has been maintained

to $V_n + P_n$. It should be noted that, according to this notation, the index n = 0, 1, 2, … N indicates the segments located along the estuary, from the segment 0 (with S = 0), up to the segment N, located at the estuary mouth; by hypothesis, this last segment has the same salinity as the coastal region, $S(N) = S_N = S_0$, and along the estuary the total number of segments is N + 1.

From the definition of the inner most tidal prism segment (*segment* 0), its tidal prism is equal to the fresh water volume accumulated during a complete tidal period, that is, $P_0 = T_P Q_f = R$. As the fresh water volume is filled during the flood tide, Dyer and Taylor (1973) suggested that this volume should be equal to (1/2)R, because in the Ketchum's original theory the corresponding time interval (1/2)T of the ebb tide has not been taken into account. However, the original theory will be presented, and later the changes which may be applied following the suggested correction. Then, the *segment 0* volumes are equal to V_0 and $V_0 + P_0$ at low and high tidal, respectively.

According to Ketchum (1951), consecutive volume segments are defined so that the distance between their inner and outer boundaries are equal to the average excursion of a water element on the flood tide. The average excursion is derived from the water volume entering each part of the estuary on the flood tide, as well as on the estuary topography. If the water volume entering with the flood tide was to act as a piston, displacing and pushing an equivalent volume of water upstream from the next landward segment, the distance moved would be the average excursion of a particle of water on the flood tide in that part of estuary. By definition, the high tide volume of any segment along the estuary length is equal to the low tide volume in the adjacent seaward segment. Consequently, along the estuary each segment is defined by the high tide volume in the landward segment, which is equal to the low tide volume in the adjacent seaward segment. Beginning with the *segment* 0, defined above, the entire estuary can thus be subdivided into a series of volume segments composed of low tide volume (V_n) for a generic segment, and the local intertidal or tidal prism volume (P_n), as shown in Fig. 6.4. Salinities and volumes in the segments are distributed as follows: the *segment 0* at high tide (P_0, with salinity zero), and the corresponding landward segments, *segment n*, for n = 1, 2, … N, with volumes equal to $V_n + P_n$ and salinities S(1), S(2), … S(N). By definition, the segment located at the estuary mouth has salinity equal to the adjacent coastal sea S_0, $S(N) = S_0$.

The tidal excursion, defined in Chap. 2 (Eq. 2.26) is directly proportional to the amplitude of the velocity generated by the tide, or to the tidal amplitude at the estuary mouth; high values of these quantities generate high salinity intrusion lengths, increasing the number of volume *segments* along the estuary.

With $R = T_P Q_f$ indicating the volume of the river water introduced during the tidal cycle, and from the above definitions of P_n and V_n as segment volumes, the following equations summarize these fundamental definitions (Ketchum op. cit.):

$$P_0 = R, \tag{6.24}$$

$$V_1 = V_0 + P_0, \tag{6.25}$$

$$V_2 = V_1 + P_1 = V_0 + P_0 + P_1, \tag{6.26}$$

$$\begin{aligned} V_3 &= V_2 + P_2 = V_0 + P_0 + P_1 + P_2, \\ &\dots \end{aligned} \tag{6.27}$$

$$V_N = V_{N-1} + P_{N-1} = V_0 + \sum\nolimits_i P_i, \quad (i = 1, 2, \dots, N-1), \tag{6.28}$$

or generically $V_n = V_{n-1} + P_{n-1}$, with n = 1, 2, ..., N.

The next step in performing the practical segmentation of the estuary is the geometric determination of the volume segments at low tide. A bathymetric nautical chart can be used to subdivide the estuarine channel into auxiliary cross-sections, and from this determine the segment areas and volumes. If there is no such nautical chart for the region being investigated, echo-sounding measurements of the estuarine channel must be made.

The volume of each segment is calculated as the product of the distance between the areas and the mean cross-section of its limiting areas. The chosen distance between the cross section areas along the estuarine channel and the depths variations must be as uniform as possible, to minimize errors in the calculations due to the non-uniformity of the limiting areas. With the data generated from this procedure, it will be possible to calculate the sum of the cumulative low tide volumes along the estuary, and these cumulative values may be plotted as function of the longitudinal distance from the head to the estuary mouth.

The tidal prism volumes from each auxiliary partition are calculated from the product of the surface area between limiting cross-sections and the tidal height; details of these volume calculations may be obtained in Anderson (1979). These volumes may also be cumulatively plotted as functions of the longitudinal distance between the head and estuary mouth. Finally, the sum of the low tide volumes (V_n) and their corresponding tidal prism volumes (P_n) are equal to the high water volumes ($V_n + P_n$), which may also be plotted as function of the estuary longitudinal distance.

The longitudinal variation of the cumulative sum of the high water volumes ($\sum_n (P_n + V_n)$), low ($\sum_n V_n$), and tidal prism volumes ($\sum_n P_n$), were calculated for the Winyah Bay estuarine system (South Caroline, USA), and indicated by curves A, B and C, respectively (Fig. 6.5).

If the fresh water volume discharged by the river during the tidal period (R) is known, it is possible to perform the estuary segmentation according to the equations system (6.24–6.28), and determinate its geometric limits along the estuary. In fact, by plotting the ordinate $P_0 = R$ in Fig. 6.5, the interception with curve C (tidal prism) determines the landward limit of the *segment 0*. With this abscissa value, it is possible to determine the ordinates corresponding to the volumes V_0 and $V_0 + P_0$

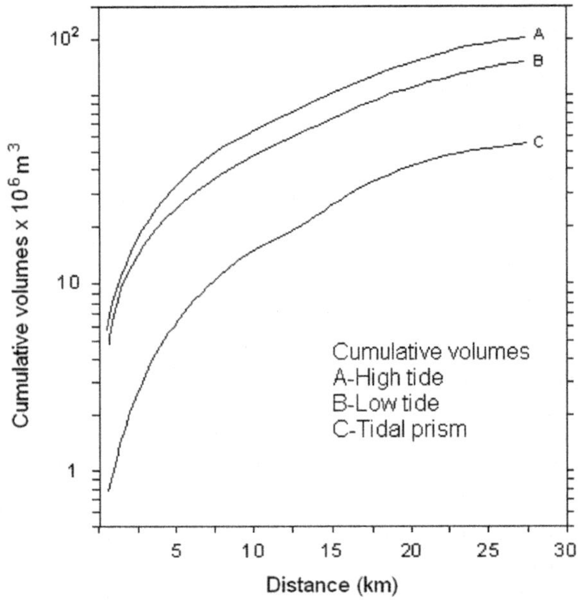

Fig. 6.5 Cumulative volumes at high (*A*) and low (*B*) tides and the tidal prism (*C*), as function of the longitudinal distance in the Winyah Bay (South Carolina, USA) estuarine channel

from curves B (low tide) and A (high tide), respectively. In the segmentation system of equations, the volume *segment 1* at low tide is equal to $V_0 + P_0$ (Eq. 6.25); then, on the cumulative curve B, this ordinate determines the landward limit of this segment in the abscissa axis. In turn, this abscissa value on curve A corresponds to the volume $V_1 + P_1$, which is equal to the volume V_2, according to the segmentation equation (Eq. 6.26). Using this V_2 value, the process may be repeated considering curve B.

This procedure will then be repeated until the last estuary segment is found in the abscissa axis of Fig. 6.5. Then, the volume of each segment at low tide is equal to the adjacent segment at high tide (Fig. 6.4). As the salt intrusion length limits the upper MZ position, the segmentation process is very important to this theory application, enabling the geometric limits of the segments their volumes at low (V_n) and high ($V_n + P_n$) tides, from the estuary head down to the mouth, to be obtained.

The estuary divided into volume segments as described above, indicates the limits of each segment and the average excursion of a water element with the flood tide. With the assumption that the water within such a volume segment is completely mixed at high tide, the proportion of water removed on the ebb tide will be given by the ratio of the local intertidal volume and the high tide volume of the segment. This proportion of river water will be removed by the ebb tide taking with it any particles dissolved or suspended in it. Thus, an exchange ratio (r) was defined by Ketchum (1951), for a generic segment n) as the ratio of the tidal prism (P_n) by the high tide element volume ($V_n + P_n$),

$$r_n = \frac{P_n}{V_n + P_n}. \tag{6.29}$$

This ratio quantifies the fraction of fresh water renewed of the total fresh water discharged into the estuary in a complete tidal cycle. Its extreme values are $r_n = 0$ and $r_n = 1$ due to the following conditions:

- $r_n = 1$, when the tidal height is equal to the estuary depth and all water is removed at the low tide ($V_n \rightarrow 0$ or $V_n \ll P_n$);
- $r_n < 1$ or $r_n \ll 1$, in estuaries forced by regular or micro tidal forcings ($P_n < V_n$ or $P_n \ll V_n$);
- $r_n = 0$, when tidal prism is to low ($P_n \rightarrow 0$).

In the first condition ($r_n = 1$), the estuary is an ideal system to renewing the salinity or concentration of any substance in its water body, because the water volume introduced during a complete tidal cycle is completely removed during the ebbing tide, acting as a perfect sink. In the third condition ($r_n = 0$) there is no water renewal in the estuary and it may accumulate salt (or other substances). The intermediate condition is the most common in partially or highly stratified estuaries.

The river water present in the estuary is a mixture of fresh and salt water, accumulated during many tidal cycles. In the condition of a constant input of fresh water, during each tidal cycle, each segment receives an influx of river water (R) equal to the total volume introduced into the estuary by the river during the tidal period. Taking R_1 as the volume of river water entering a segment during the current tidal cycle (*age of one tidal cycle*), then the amount removed on the ebb tide will be r_nR^1, and the amount behind will be $(1 - r_n)R$. Considering one step forward in time, the portion of river water that arrived in the previous time-step, R_2, was not fully removed during ebb currents. Therefore, the remaining portion of fresh water from previous time-step is required to be taken on the following time-step (e.g. *age of two tidal cycles*). For two time steps (or two tidal cycles), fresh water removed will be $r_n(1 - r_n)R_2$ and the remaining fresh water for two successive ebb tides, will be $(1 - r_n)^2R_2$. The proportion of water of various tidal ages which is removed (1, 2, 3, … m), or remaining behind within the segment, as a result of the exchanges on any given ebb tide, may be summarized as follow in Table 6.1 (Ketchum 1951).

Table 6.1 Water mass volumes removed and accumulated into the estuary as function of the tidal age (according to Ketchum 1951)

Tidal age	Removed fresh water (volumes)	Accumulated fresh water (volumes)
1	r_nR	$(1 - r_n)R$
2	$r_n(1 - r_n)R$	$(1 - r_n)^2R$
3	$r_n(1 - r_n)^2R$	$(1 - r_n)^3R$
…	…	…
m	$r_n(1 - r_n)^{m-1}R$	$(1 - r_n)^mR$
Σ	R	R/r_n

The parcels summation of the first line of Table 6.1 is the fresh water balance at the tidal age 1 and, as this result is equal to R, the principle of volume conservation in this first tidal cycle (*tidal age 1*) is satisfied. The second column corresponds to the total water volume removed during the successive tidal cycles. As the ratio of terms of this row is constant and equal to $(1 - r_n)$, its summation may be easily calculated by the formula of the geometric progression series. Considering a series with a great number of elements $(m \to \infty)$, it is convergent to R which confirms the fresh water volume conservation.

As, by hypothesis, the fresh water input is constant, all values R are equal and the steady-state condition can be assumed. The total volume (V_f) of river water accumulated within any volume segment (n) of the estuary at high tide, is calculated by the sum of the remaining volumes given in the final column of Table 6.1. Since the equation is written for the high tide condition, one volume of the river flow (R) which has not been depleted is also present, and the fresh water volume, $(V_f)_n$, accumulated in high tide is:

$$(V_f)_n = R\left[1 + (1 - r_n) + (1 - r_n)^2 + \cdots + (1 - r_n)^m\right]. \tag{6.30}$$

As the expression between the square bracket is the sum of a geometric progression with a ratio equal to $(1 - r_n)$, the fresh water volume accumulated in the segment, n, is calculated by:

$$(V_f)_n = \frac{R[1 - (1 - r_n)^m(1 - r_n)]}{[1 - (1 - r_n)]}. \tag{6.31}$$

As $(1 - r_n) \le 1$ and the number (m) of tidal cycles (m) is great, the final result for the fresh water volume is:

$$(V_f)_n = \frac{R}{r_n}, \tag{6.32}$$

and is determinate by the volume of the fresh water discharged by the river during the tidal cycle (R) divided by the exchanged ratio r_n (Eq. 6.29). This relationship states that the volume of fresh water discharge is flowing seaward during the tidal cycle, and is the product of the exchange ratio (r_n) and the accumulated volume of river water (rQ_f), satisfying the hypothesis of the steady-state condition.

The exchange ratio was defined on the assumption of complete mixing of the water mass in each segment at high tide. The average excursion of seawater during the flood tide is presumed to set the upper limit of the saline intrusion length, over which complete mixing was assumed.

Before using this method to calculate the fresh water fraction (f), the average longitudinal salinity, $S = S(x)$, and the flushing time (t_q), it is opportune to observe that Eq. (6.32) allows immediate determination of the flushing time in a generic segment, using its definition (Eq. 6.1):

$$(t_q)_n = \frac{(V_f)_n}{Q_f} = \frac{R}{r_n Q_f} = \frac{T_P}{r_n}. \qquad (6.33)$$

This equation agree with the result already obtained with the Eq. (6.2); the lower flushing time ($t_q = T_P$) occurs when the exchange ratio is equal to one, $r_n = 1$; also, $t_q \to \infty$ when $r_n \to 0$.

Combining the previously calculated values for the fresh water volume discharge ($R = T_P Q_f$), the tidal prism (C) and the low tidal volumes (B) obtained from Fig. 6.5, with Eqs. (6.29) and (6.32), it is possible to calculate the accumulated fresh water volumes in the generic segment (n) applying the fresh water fraction definition (6.5), and the result is:

$$f_n = \frac{V_{fn}}{V} = \frac{V_{fn}}{V_n + P_n} = \frac{R}{r_n(V_n + P_n)} = \frac{R}{P_n}. \qquad (6.34)$$

Using Eq. (6.10), which defines the fresh water fraction as a function of the salinity, it follows that:

$$f_n = \frac{S_0 - S_n}{S_0}, \qquad (6.35)$$

and combining this definition with Eq. (6.34), the mean salinity at the segment, n, may be calculated from the known undiluted salinity at the coastal ocean (S_0),

$$S_n = S_0(1 - f_n), \qquad (6.36)$$

and, combining with Eq. (6.34),

$$S_n = S_0\left[1 - \frac{V_{fn}}{(P_n + V)_n}\right] = S_0\left(1 - \frac{R}{P_n}\right). \qquad (6.37)$$

From this result, it is possible to calculate the mean salinity (S_n) at each segment, for n = 0, 1, 2, 3, … N. It then follows that for the segment 0 the salinity is zero, because at this segment $P_0 = R$ (Eq. 6.24).

The flushing time for the *segment n* may be calculated by Eq. (6.33) and its sum for each segment is equal to the estuary flushing time,

$$t_q = T_P \sum_n \frac{1}{r_n}, \qquad (6.38)$$

or in tidal period (T_P) units,

$$t_q = \sum_n \frac{1}{r_n}. \qquad (6.39)$$

To satisfy the assumption of the complete mixing of the fresh river water discharge and the seawater at high tide, this method must be applied to well-mixed estuaries or to low stratified partially mixed estuaries. Thus, the incomplete mixing in the segments implies difficulty in application of this. However, in such cases, the exchange ratio (r_n^*) will be dependent on the segment depth (h) and on the well-mixed thickness (D) (Ketchum 1951):

$$r_n^* = \frac{P_n}{V_n + P_n}\left(\frac{D}{H}\right). \tag{6.40}$$

In this equation, D is the height of the *segment n*, and H is the mixed layer thickness (or also its height). When the exchange ratio is larger ($r_n^* > r_n$), then the resulting accumulation of river water $(V_f)_n = R/r_n^*$ will be small. In cases, the segmentation of the estuary is also made using volumes computed to the mixed layer depth. The entire treatment is therefore developed, with the assumption that the water bellow the mixed layer takes no part in the tidal mixing.

The Ketchum's method has been applied for three different estuaries in almost all characteristics, the Raritan river and Bay (New Jersey, USA), the Alberni Inlet (Columbia, Canada) and Great Pond (Massachusetts, USA); however, the method was only described in detail for the Raritan river, and the theoretical results corresponded closely to the observed distributions of salinity and fresh water.

For simplicity, the method was applied for a model estuary with rectangular cross sections and constant depth, and with equal low tide and tidal prism volumes ($V_n = P_n$); then, the seaward variations of these cumulative volumes are equal, its longitudinal distributions are coincident (B = C), and the volumes at high tide ($V_n + P_n$) are indicated by (A), as shown in Fig. 6.6. For further simplification, for the tidal cycle, a fresh water discharge equal to one (R = 1) implying $P_0 = 1$, was adopted. This ordinate, plotted in the figure, starts the segmentation process, enabling the determination of the geometric limits of the estuary segments. As $V_n = P_n$ and R = 1, it is possible, using Eqs. (6.29) and (6.32), to calculate the exchange ratio (r_n) and the fresh water volume (V_f) retained in the segments, which are constants equal to ½ and 2, respectively.

The calculate values of the exchange rate, fresh water volume, relative salinity (S/S_0) and the flushing time (t_q) in tidal period units are presented in Table 6.2. The relative salinity in the *segment 0* is zero, and converges to one (1) at *segment 10*; this convergence is accentuated in the first segments and tends asymptotically to one (1) from *segment 4*. The flushing time of this model estuary, determined by the sum of the corresponding value of each segment (2), is 20 tidal periods.

The semi-empirical segmented tidal prism was applied to the estuarine system of Winyah Bay (South Caroline, USA) (Fig. 6.7). As this estuary is partially mixed, but with low vertical stratification, the exchange ratio was calculated with the assumption that it is well-mixed, and its segmentation was performed with the longitudinal variation of the cumulative tidal prism at low and high tide, as presented in Fig. 6.5. The results in Table 6.3 were calculated using the following hydrologic and hydrographic data: input of the average discharge of fresh water by the river

Fig. 6.6 Schematic diagram of the longitudinal variation of the cumulative volumes of high (*A*) and low (*B*) tide, and the tidal prism (*C*) of an estuary model with $V_n = P_n$., and R = 1, according to Miranda (1984)

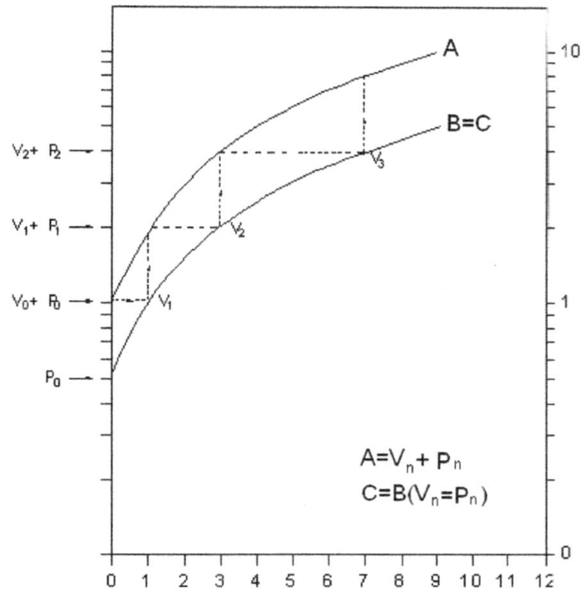

Table 6.2 Results of the estuary model ($V_n = P_n$ and R = 1): exchange ratio (r_n), fresh water volume $(V_f)_n$, relative salinity (S/S_O) and flushing time $(t_q)_n/T$

n	V_n	$V_n + P_n$	r_n	$(V_f)_n$	f_n	S/S_O	$(t_q)_n/T$
0	1	2	½	2	1	0	–
1	2	4	½	2	0.5	0.5	2
2	4	8	½	2	0.25	0.75	2
3	8	16	½	2	0.125	0.875	2
4	16	32	½	2	0.062	0.938	2
5	32	64	½	2	0.032	0.968	2
6	64	128	½	2	0.016	0.984	2
7	128	256	½	2	0.008	0.992	2
8	256	512	½	2	0.004	0.996	2
9	512	1024	½	2	0.002	0.998	2
10	1024	2048	½	2	0.000	1.000	2

during a tidal cycle, $R = 8.6 \times 10^6$ m^3, and non-diluted salinity at the coastal sea, $S_O = 34.0‰$.

Table indicates that, using this method, only a few segments were determined along the estuary, and the flushing time was calculated to be 14.4 semi-diurnal tidal cycles (7.2 days). Of course, the steady-state longitudinal variation of salinity, forced by river discharge and tide, must be validated with observational data.

To calculate the results presented in Tables 6.2 and 6.3, the estuary segmentation process from the longitudinal variation of segment volumes A, B and C (Fig. 6.5), associated with Eqs. (6.24–6.28), were used, along the following equations:

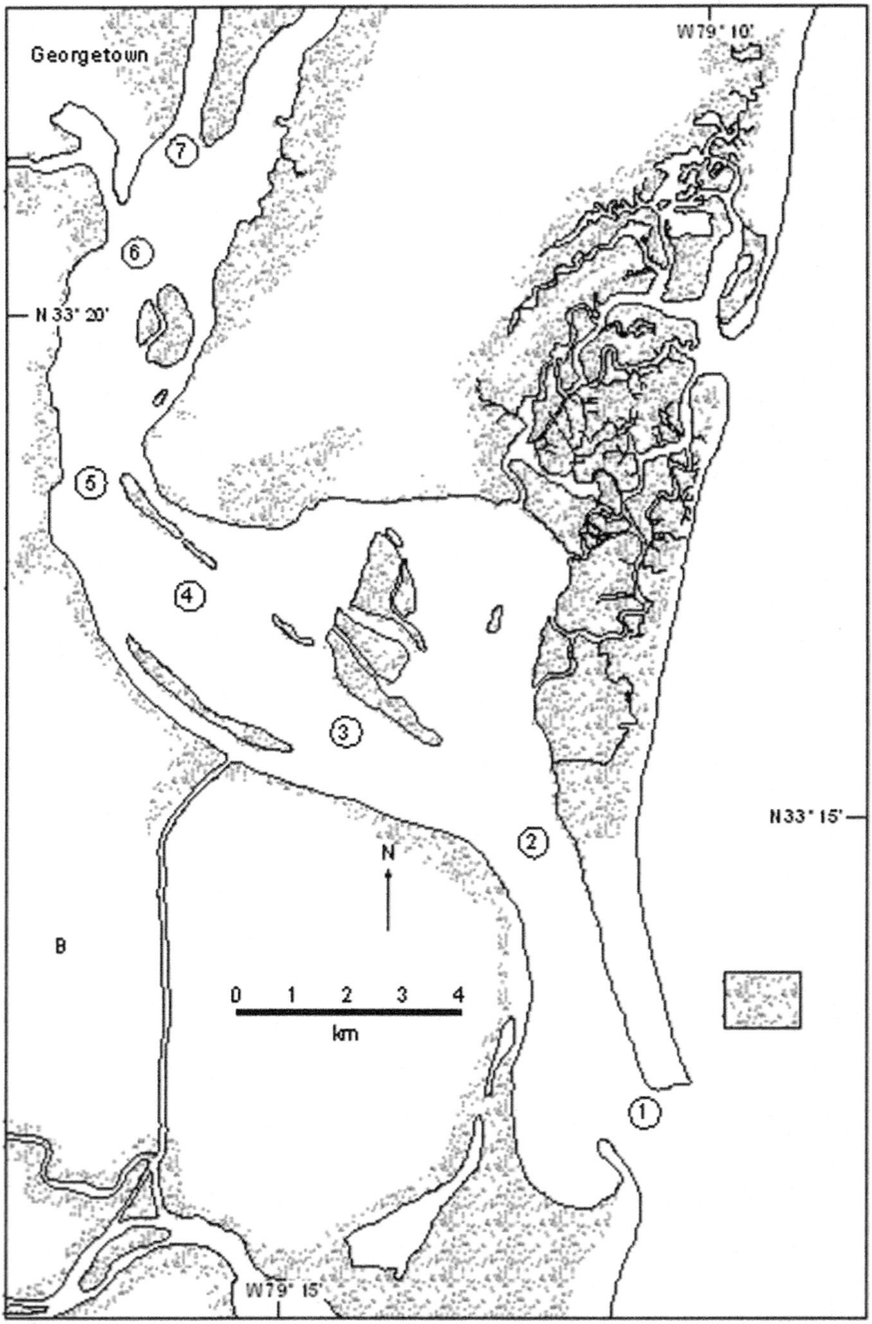

Fig. 6.7 Winyah Bay estuarine system located SE of South Carolina (USA). The along channel numbers (*1–5*) indicate the segments boundaries

Table 6.3 Results of the Ketchum (1951) method applied to the Winyah Bay estuary

n	$V_n + P_n$	r_n	$(V_f)_n$	f_n	S		$(t_q)_n/T$
0	25.3	0.17	25.3	1.0	0.0	–	–
1	60.5	0.28	30.7	0.87	4.4		3.6
2	111.7	0.31	27.7	0.54	15.6		3.2
3	182.7	0.28	30.8	0.43	19.4		3.6
4	–	0.25	34.4	0.36	21.8		4.0
5	–	–	–	0.00	34.0		–

Salinity (‰), flushing time $[(t_q)_n/T]$. With: $R = 8.6 \times 10^6\ m^3$, $S_0 = 34.0$‰ and V_n and P_n (in unities of $10^6\ m^3$)

$$r_n = \frac{P_n}{V_n + P_n}; \quad (V_f)_n = \frac{R}{r_n}; \quad f_n = \frac{(V_f)_n}{V_n + P_n}. \tag{6.41}$$

and

$$S_n = S_0(1 - f_n), \quad \text{and} \quad (t_q)_n = \frac{T_P}{r_n}. \tag{6.42}$$

In the Ketchum's theory, the tidal prism volume of the *segment 0* was taken as R ($P_0 = T_P Q_f = R$). However, as previously mentioned, Dyer and Taylor (1973) suggested a correction for this volume as half of the value in the original paper, $P_0 = (1/2)R$. Although this correction is applied at the very beginning of the segmentation procedure, and therefore alters the volume of segments along the MZ, all semi-empirical equations from Ketchum's original paper remain the same.

This method has been applied in several investigations, and, in some cases, the longitudinal mean salinity distribution values were acceptable, however, in others they were not. These inconsistent results indicated that further investigations should be sought, and a modified version of the original segmented tidal prism model was developed by Dyer and Taylor (1973). The model presented by Dyer and Taylor was based partly on a combination of Ketchum's method and Maximon and Morgan's (1955) concepts, allowing for additional inflows into the estuary from tributaries and outfalls, while keeping the method simple, with more consistent physical interpretations of the mixing processes and fresh water continuity.

In order to make this second method more comprehensible, the fundamental differences between the methods of Ketchum (1951) and the Dyer and Taylor (1973), will be described, including the terminology and notation of variables. According to Maximon & Morgan's (op. cit.) theory, the seaward mean salinity is calculated at high and low tide, allowing for time dependence of various quantities involved and the introduction of solutes (or salinity) into the estuary. Secondly, in Dyer and Taylor's segmentation equations, a non-dimensional *mixing parameter* (a) was included, enabling adjustments and validation based on experimental data. Concerning terminology and notation, the term *fresh water concentration* (C) in Dyer and Taylor's, was used instead *fresh water fraction* (f); thus C = f, and, according to Dyer and Taylor's original papeer, the conditions at high and low tides will be identified by upper letters H and L, respectively, and the following equalities will exist: $C_n^H = f_n^H = f_n$. In this case, the *fresh water concentration* at high tide in the segment, n, is numerically computed following similar approach to the fresh water fraction of Ketchum's paper.

The *segment 0* contains only fresh water, and its fresh water concentration will be denoted as C_0, which by definition is equal to one ($C_0 = 1$). For the segment located at the estuary mouth (n = N), the fresh water content is practically equal to zero, following the equality $C_N^H = C_{N+1}^L = 0$; then, the N + 1 index for the low tide concentration indicates the segment adjacent to the estuary mouth, located coastal region.

The segmentation of the Ketchum's model prescribe that the low tide volume of a generic segment (n + 1) is equal to the high tide volume of the adjacent *segment n*, located landward ($V_{n+1} = V_n + P_n$). This process implies that during the flood a volume equal to V_{n+1} crosses this segment boundary. Also, as R is the fresh water volume accumulated during the tidal cycle, the water volume transported through the segment boundary in the ebb tide is equal to V_{n+1}+R. Then, taking into account the fresh water concentration (and hence fresh water fraction) for a complete tidal cycle, the following identity to satisfies the principle of fresh water conservation (Dyer and Taylor 1973):

$$(V_{n+1} + R)C_n^H - V_{n+1}.C_{n+1}^L = RC_0, \quad n \geq 0. \tag{6.43}$$

This identity is satisfied only when $C_n^H = C_{n+1}^L = C_0 = 1$. As previously indicated, the fresh water concentration and fresh water fraction are equal numeric quantities at high tide, that is:

$$C_n^H = f_n^H = \frac{(V_f)_n}{(V_n + P_n)}. \tag{6.44}$$

However, according to Eqs. (6.29) and (6.32),

$$(V_f)_n = \frac{R}{r_n} = R\frac{(V_n + P_n)}{P_n}. \tag{6.45}$$

Combining Eqs. (6.44) and (6.45), it follows that:

$$C_n^H = f_n^H = \frac{R}{P_n} = C_0\frac{R}{P_n}. \tag{6.46}$$

With the fresh water balance expressed by Eq. (6.43), we have already concluded that $C_n^H = C_{n+1}^L = C_0 = 1$; this result is incompatible with Eq. (6.46), because it is true when R = P_n. Then, it was shown that the Ketchum's model doesn't agree completely with the principle of volume conservation, because the fresh water concentration is not taken into account during the transition from low to high tide.

Dyer and Taylor's model retains the simplicity of the Ketchum's method and assures a more consistent fresh water balance, applying the same simplifying hypothesis: stationary conditions of the mean salinity field and complete mixing at low and high tide.

The geometric limits of the estuary segments are also determined using prior knowledge of the cumulative volumes from the head and estuary mouth at low and high tides, exemplified for the estuary system of Winyah Bay (Fig. 6.5, curves B and C, respectively). Using the same notation for the identification of the volume segments at low tide (V_n) and the tidal prism (P_n), the estuary segmentation is schematically shown in Fig. 6.8.

Volumes $(1 - a)V_n$ at low tide (n = 2, 3, … N) between the segment (sections A and B in Fig. 6.8), limited by the dashed line in this figure are accounted in the mixing at high tide. Then, the total volume of this segment at high tide is equal to $(V_n + P_n)$ because:

$$(1 - a)V_n + aV_n + P_n = V_n + P_n, \qquad (6.47)$$

with n = 1, 2, 3 … N. The parameter associated with the mixing process (a) may vary from zero to one. It could, in principle, be determined from observational tidal excursion data, and potentially to vary from one segment to another.

Similar to Ketchum's method, the upstream end of the model is defined by the section across which there is no flow during the flood tide. If R is the river flow per tidal cycle, the tidal prism volume above the segment 0 will be R/2 (not R as stated by Ketchum). This definition is unaffected if the tidal limit is determined by a weir. The segmentation equations of this model are (Dyer and Taylor 1973):

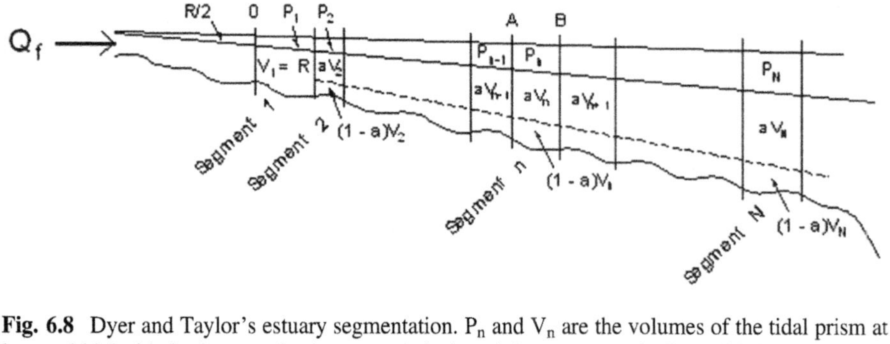

Fig. 6.8 Dyer and Taylor's estuary segmentation. P_n and V_n are the volumes of the tidal prism at low and high tide in the generic element n (a is the mixing parameter), $(1 - a)V_n$ is the low tide volume to be used for mixing at high tide. *A* and *B* are control boundaries [after Dyer and Taylor (1973)]

$$V_1 = R, \quad (6.48)$$

$$aV_2 = P_1, \quad (6.49)$$

$$aV_3 = aV_2 + P_2 = P_1 + P_2, \quad (6.50)$$

$$aV_N = aV_{N-1} + P_{N-1} = \Sigma_i P_i (i = 1, 2, 3, \ldots N-1), \quad (6.51)$$

or generally, $aV_n = aV_{n-1} + P_{n-1}$, for n = 3, 4, …,N. If the mixing parameter is equal to one (a = 1), these equations are equal to the Ketchum's segmentation (Eqs. 6.26–6.28). In Dyer and Taylor's analysis, this parameter was assumed to be constant (a = const.), giving reasonable agreement between computed and observed high and low tide mean salinity distributions.

Equations (6.48–6.51) indicate that the segments are defined as follows: on the flood tide, the water volume occupying aV_{n+1} of the segment n + 1 is moved up-estuary to occupy the volume $aV_n + P_n$ at high tide, or just the volume P_1 in the segment 1. During this process, it is assumed that the volume at high tide is mixed with the portion of water remaining in the segment, n, from the low, i.e., with $(1 - a)V_n$, or with the volume V_1, when n = 1, because at low tide this volume is entirely supplied by the river discharge. Then, the high tide volume at any segment is equal to $V_n + P_n$, for n = 1, 2, … N (Fig. 6.8).

This model may be applied for different river discharge volumes, and also taking into account additional fresh water contributions in the MZ boundaries. The volume of segments increases seaward, and if a volume aV_{n+1} crosses the segment boundary (B in Fig. 6.8) during the flood, due to volume continuity a water volume equal to $aV_{n+1} + R$ will cross this boundary during the ebb tide.

After defining the estuary segmentation, the following step is to find the equations to calculate the concentrations C_n^H and C_n^L for each segment. This may be established by applying the volume continuity, to assure that during each tidal cycle the fresh water volume transport out of the estuary is equal to R.

6.3.1 High Tide Fresh Water Balance

Consider a generic control segment nth, which occupies a volume aV_{n+1} ($aV_{n+1}=aV_n + P_n$, according to the segmentation Eqs. 6.48–6.51) at high tide, and is completely mixed with a water volume $(1 - a)V_n$ disposable at low tide (Fig. 6.8). Then, according to the volume conservation principle, during the flood tide the following fresh water balance will occur through boundary B of this segment:

$$(V_n + P_n)C_n^H - (1 - a)V_n C_n^L = aV_{n+1} C_{n+1}^L, \quad n \geq 2, \quad (6.52)$$

or

$$(V_n + P_n)C_n^H = aV_{n+1}C_{n+1}^L + (1 - a)V_nC_n^L, \quad n \geq 2. \tag{6.53}$$

For the segment n = 1, the fresh water balance, equivalent to the Eq. (6.52) is given by the following expression:

$$(V_1 + P_1)C_1^H - V_1C_1^L = aV_2C_2^L. \tag{6.54}$$

6.3.2 Low Tide Fresh Water Balance

Again considering the nth control segment, in the ebb tide the flow travels from the adjacent segment (n − 1) to the segment, n, through the control boundary A (Fig. 6.8). Then, a water volume (aV_n + R), with a concentration C_{n-1}^H, will enter the segment, n, and mix with the water volume $(1 - a)V_n - R$ that remained in this segment at low tide, with a concentration C_n^H, in the segment n. With this procedure, we are making the assumption that an additional volume, equal to R, is coming from the water volume $(1 - a)V_{n-1}$. Then, to establish the volume conservation it is necessary that:

$$V_nC_n^L = (aV_n + R)C_{n-1}^H + [(1 - a)V_n - R]C_n^H, \quad n \geq 2. \tag{6.55}$$

For *segment 1* holds the following conservation equation:

$$V_1C_1^L = RC_0. \tag{6.56}$$

As according to the segmentation process, V_1 = R (Eq. 6.48), if follows from this equation that $C_1^L = C_0 = 1$.

6.3.3 Fresh Water Balance During the Tidal Cycle

An additional relationship may be obtained with the assumption, according to the volume conservation principle, that after a complete tidal cycle the net water volume flow across any cross section boundary is equal to R = RC_0, then:

$$(R + aV_{n+1})C_n^H - aV_{n+1}C_{n+1}^L = RC_0, \quad n \geq 1. \tag{6.57}$$

Equations (6.53), (6.55) and (6.57) are not independent, and the unique relationship with the two unknowns C_n^L and C_n^H may be obtained by combining Eqs. (6.53) and (6.57). In effect, Eq. (6.57) may have its parcels rearranged as:

$$R(C_0 - C_n^H) = aV_{n+1}C_n^H - aV_{n+1}C_{n+1}^L, n \geq 1. \tag{6.58}$$

Now, the terms of Eq. (6.53) will be rearranged in order to isolate the last parcel of Eq. (6.58), and the result is:

$$aV_{n+1}C_{n+1}^L = (V_n + P_n)C_n^H - (1 - a)V_nC_n^L, \tag{6.59}$$

and Eq. (6.58) is rewritten as:

$$(1 - a)V_nC_n^L + [aV_{n+1} - (V_n + P_n)]C_n^H = R(C_0 - C_n^H). \tag{6.60}$$

The expression in brackets of the first member of this equation may be rewritten taking into account the following identity for the high tide volume of the nth segment (Eq. 6.47):

$$(V_n + P_n) = (1 - a)V_n + aV_n + P_n, \tag{6.61}$$

hence, from the segmentation equations system,

$$(V_n + P_n) = (1 - a)V_n + aV_{n+1}. \tag{6.62}$$

Then,

$$aV_{n+1} - (V_n + P_n) = -(1 - a)V_n, \tag{6.63}$$

and substituting Eq. (6.63) in Eq. (6.60), it follows that,

$$(1 - a)V_n(C_n^L - C_n^H) = R(C_0 - C_n^H), \tag{6.64}$$

or, rearranging its terms,

$$C_n^L = C_n^H + \frac{R(C_0 - C_n^H)}{(1 - a)V_n}, \quad \text{with} \quad n \geq 2. \tag{6.65}$$

As the quantities R, $\underline{a}$ and the volumes V_n are known, Eq. (6.65) has two unknowns C_n^L and C_n^H. However, the fresh water concentration, C_n^H, may be calculated by Eq. (6.58),

$$C_n^H = \frac{RC_0 + aV_{n+1}C_{n+1}^L}{aV_{n+1} + R}. \tag{6.66}$$

Now, calculating the above equation for n = N and assuming that pure water enters the estuary mouth on the flood tide ($C_{N+1}^L = 0$), it is possible to calculate the fresh water concentration at high water in the last segment (C_N^H),

$$C_N^H = \frac{RC_0}{(aV_{N+1} + R)}, \tag{6.67}$$

and from the segmentation equations $aV_{N+1} = V_N + P_N$, the final value of C_N^H is:

$$C_N^H = \frac{R}{(aV_N + P_N + R)}. \tag{6.68}$$

As all variables are known to calculate C_N^H with this equation, the value of C_N^L may be determined by Eq. (6.65) for n = N. Repeating this procedure, sequentially for n = N − 1, n = N − 2, …, n = 2, n = 1, Eqs. (6.66) and (6.65) correspond to a system with two equations and two unknowns (C_N^H and C_N^L). This equation system may be solved to obtain high and low volumes of fresh water concentration (or salinity) starting from the segment n = N, located at the estuary mouth. With these results, the equations to calculate the salinity values are:

$$S_n^H = S_0(1 - C_n^H), \tag{6.69}$$

and

$$S_n^L = S_0(1 - C_n^L). \tag{6.70}$$

The volumes of fresh water retained in the estuary at high and low tide during the flood (V_{fn}^H) and ebb (V_{fn}^L), respectively, may be calculated with known corresponding geometric volumes,

$$V_{fn}^H = C_n^H(V_n + P_n), \tag{6.71}$$

and

$$V_{fn}^L = C_n^L(V_n). \tag{6.72}$$

The flushing time (t_q) at high and low tide are functions of the fresh water volumes and may be calculated by:

$$t_{qn}^H = \frac{V_{fn}^H}{Q_f}, \text{or in tidal cycles } t_{qn}^H = \frac{1}{T_P}\frac{V_{fn}^H}{Q_f}, \tag{6.73}$$

and

$$t_{qn}^L = \frac{V_{fn}^L}{Q_f}, \text{or in tidal cycles } t_{qn}^L = \frac{1}{T_P}\frac{V_{fn}^L}{Q_f}. \tag{6.74}$$

In practical applications of the model, negative values of the fresh water volume concentration in the low tide segment (C_n^L) located near the seaward end of the estuary may be found when $(1 - a)V_n < R$. If this occurs, the appropriated interpretation is that any salt water passing upstream into the segment (n), on the flood tide is entirely removed on the ebb tide so that $C_n^L = C_0$, and in consequence $C_{n-1}^H = C_0$, $(C_{n-1}^L) = C_0 = 1$, which should only occur near the head of the estuary (Dyer and Taylor 1973).

In order to exemplify the application of this method, it was applied to the same ideal estuary previously used for Ketchum's tidal prism segment model with $V_n = P_n$, the mixing parameter (a) equal to $a = 0.8$ and $R = 1$. The calculated volumes V_n, P_n, aV_n and $(1 - a)V_n$, using Eqs. (6.48–6.51) are presented in Table 6.4.

With the results of Table 6.4, Eqs. (6.65) and (6.66) may be calculated successively for n = 10, 9, …, 2, 1, and fresh water concentrations at low (C_n^L) and high tide (C_n^H) for all segments will be obtained. With Eqs. (6.69) and (6.70), the relative salinity values for high, $(S_n/S_0)^H$, and low tide, $(S_n/S_0)^L$, can be easily calculated and are presented in Table 6.5. As fresh water volumes are calculated by the product of the fresh water concentration to the segments at low and high tides, respectively, the flushing times may also be calculated by Eqs. (6.73) and (6.74). The results of this table also indicate the convergence of the low (C_n^L) and high (C_n^H) fresh water concentrations to the value 1 (one), indicating the absence of salt water in segments 1 and 2. Therefore, these segments correspond to segment 0 in the Ketchum's model. Also, as may be observed, $(C_n^L) > (C_n^H)$ and the relative salinity values are higher at high tide than at low tide. In comparing the flushing times there is a great difference between results from the two methods. In tidal periods, these values are 20 and 52 for the Ketchum's and Dyer and Taylor's model, respectively (Tables 6.2 and 6.5).

Results of the longitudinal mean relative salinity variation for the model estuary (Tables 6.2 and 6.5) calculated with the Ketchum's (K) and Dyer and Taylor's

Table 6.4 Partial volumes of a simple estuary model ($V_n = P_n$, $R = 1$ and $\underline{a} = 0.8$), according to Dyer and Taylor (1973)

n	V_n	P_n	aV_n	$(1 - a)V_n$
1	1.0	1.0	0.8	0.2
2	1.25	1.25	1.0	0.25[a]
3	2.81	2.81	2.25	0.56[a]
4	6.33	6.33	5.06	1.26
5	14.24	14.24	11.39	2.85
6	32.04	32.04	25.63	6.41
7	72.08	72.08	56.67	14.42
8	162.20	162.20	129.70	32.44
9	364.90	364.90	292.00	73.00
10	821.12	821.12	656.90	164.22

[a]Note that $(1 - a)V_a < R$ and so $C_3^L = 1.0$

Table 6.5 Results of a simple estuary model ($V_n = P_n$, R = 1 and a = 0.8) of fresh water concentrations (C_n^L), (C_n^H), relative salinities $(S_n/S_0)^L$, $(S_n/S_0)^H$ and flushing times $(t_q)_n^H/T$, according to Dyer and Taylor (1973)

n	(C_n^L)	(C_n^H)	$(S_n/S_0)^L$	$(S_n/S_0)^H$	$(t_q)_n^H/T$
1	–	–	–	–	–
2	–	–	–	–	–
3	1.00	0.92	0.00	0.08	5.17
4	0.91	0.58	0.09	0.42	7.34
5	0.54	0.30	0.46	0.70	8.50
6	0.27	0.13	0.73	0.87	8.33
7	0,12	0.06	0.88	0.94	8.65
8	0.05	0.02	0.95	0.98	6.49
9	0.02	0.01	0.98	0.99	7.43
10	0.01	0.00	0.99	1.00	0.00

(D&T) methods are comparatively presented in Fig. 6.9a, b. Salinity variations indicates some differences, as should be expected. However, the longitudinal salinity distributions are very close (Fig. 6.9b).

Dyer and Taylor's method was also applied to the Raritan river estuary and bay, using the volumetric data given by Ketchum (1951), with different values of the mixing parameter ($\underline{a}$). A mixing parameter $\underline{a}$ = 0.5 gave reasonable comparison with the salinity distribution in high tide observed by Ketchum in the Raritan river. For further details on these comparisons, as well as for the Thames river estuary, may be found in the Dyer and Taylor's original paper.

Dyer and Taylor's method has also been applied to Winyah Bay (Fig. 6.7), using the previous volumetric data (Fig. 6.5). The results are in Table 6.6, calculated with hydrologic measurements and salinity at the coastal sea conditions, as previously indicated ($R = 8.6 \times 10^6$ m^3, and S_0 = 34.0‰), and used in the application of the first method, and the mixing parameter used $\underline{a}$ = 0.8. The comparative analysis of the mean theoretical salinity distribution along the bay, obtained with these

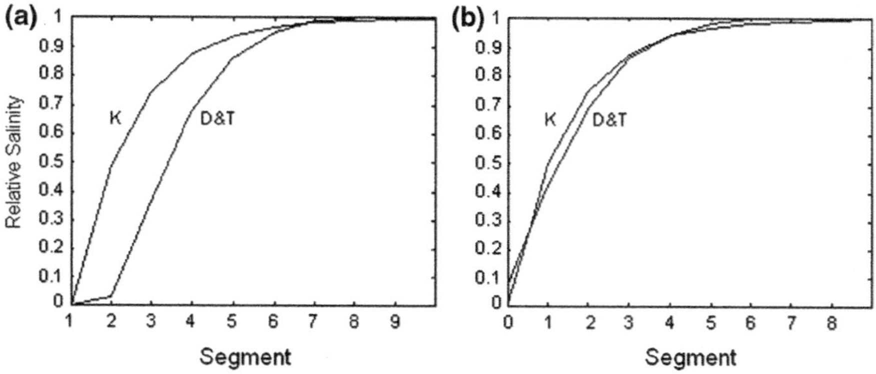

Fig. 6.9 Mean longitudinal salinity variation in the estuary model with $V_n = P_n$. **a** Values calculated with Ketchum's (K) and Dyer and Taylor's (D&T) methods, with the mixing parameter a = 0.8. **b** The best agreement was obtained displacing the second method to the left

methods, is shown in Fig. 6.10. Similar to the model estuary (Fig. 6.9), differences were observed in the salinity values, which were also minimized by displacing the first segment of the Dyer and Taylor's to the left. The results of this figure indicate variations which have some dependence on the used method, and, due to the higher number of segments in it, the salinity varies smoothly from the head down to the estuary mouth.

The results in Fig. 6.10 indicate the dependence of the longitudinal salinity variation on the chosen method. The confidence in the Ketchum's and Dyer and Taylor's methods may only be validated by comparing both results with experimental data. However, Dyer and Taylor's method satisfies the volume conservation of the fresh water input, and should presents longitudinal salinity distributions and flushing time more consistently. It may also be observe that the longitudinal salinity variation is nearly linear in the central MZ (Fig. 6.10), and this result has also been confirmed with observational data. Another observation of the results of these methods is that the flushing time from Dyer and Taylor's (28.9) is twice the duration calculated by Ketchum's method (14.4), in tidal periods.

Dyer and Taylor's model was adapted by Brown and Arellano (1980) for a branching estuary in order to study the mixing of salt within the Great Bay estuary (New Hampshire, USA). This estuarine system has two main branches with their own river discharge, and it was necessary to take into account this particular morphology. This estuary is classified as vertically well-mixed (type 1) most of the year, with a few exceptions of highest river discharge periods, when this estuary has been classified as partially mixed with low stratification (type 2a). In the application of this model, the mixing parameter ($\underline{a}$) was allowed to vary and was chosen on the basis of a calibration procedure using observational data. The predicted mean salinity distribution over a range of river discharges volumes were in agreement with observational data when the flux ratio was higher than 1 (one) (tidal prism much less than the river discharge per tidal cycle). As another result of the Brown &

Table 6.6 Results of the Dyer and Taylor (1973) method applied to the Winyah Bay estuarine system

n	V_n	P_n	aV_n	C_n^L	C_n^H	S_n^L	S_n^H	t_{qn}^H/T
1	8.6	1.5	6.9	–	–	–	–	–
2	21.2	4.2	17.0	–	–	–	–	–
3	26.5	6.5	21.2	1.0	1.0	0.00	0.00	3.8
4	34.6	9.7	27.7	1.0	0.99	0.00	0.34	4.6
5	46.7	14.3	37.4	0.99	0.91	0.34	3.10	6.4
6	64.6	21.4	51.7	0.90	0.71	0.34	9.86	7.1
7	91.4	34.8	73.1	0.68	0.40	10.88	20.40	5.9
8	134.8	50.0	107.8	0.35	0.05	22.10	32.30	1.1
9	197.4	–	157.9	0.00	0.00	34.00	34.00	–

Salinities S_n^L and S_n^H in ‰. With $R = 8.6 \times 10^6\ m^3$, $S_0 = 34.0$‰ and mixing parameter $\underline{a} = 0.8$. Volumes in units of $10^6\ m^3$

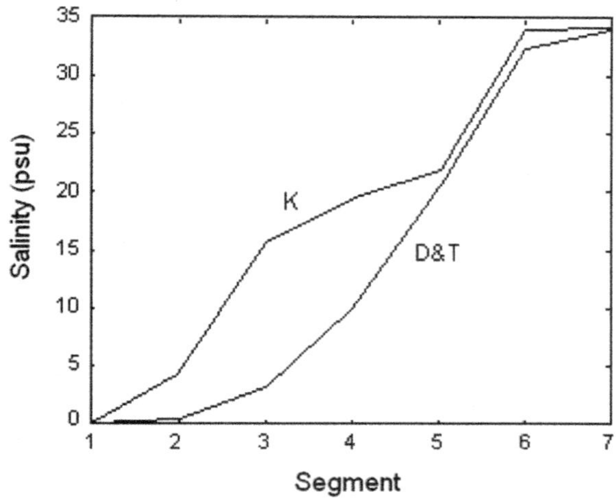

Fig. 6.10 Theoretical mean salinity variation in the MZ in the Winyah Bay estuary. Ketchum's (K) and Dyer and Taylor's (D&T) methods

Arellano (op. cit.) investigation was that related to the flushing times calculations; for water parcels entering at the estuary head during periods of low and high river flow the flushing times were 54.5 and 45.9 tidal cycles, respectively.

The Dyer and Taylor (1973) one-dimensional tidal prism model was also been used by Bradley et al. (1990) to simulate the changes in the longitudinal mean salinity distribution, which occurred in the Cooper River (South Carolina, USA), because a diversion in 1985 caused a reduction in the mean river discharge from 442 to 130 $m^3 s^{-1}$. The model simulation indicated that a salinity increased of 10–14‰, has occurred in the region of the river where the marsh plant community shifts from a virtual monoculture of *Spartina alterniflora* to a more diverse brackish community. The flow reduction, due to the river diversion, and the associated salinity increase are expected to result in the dominance of the halophyte, *S. alterniflora*, and a progressive exclusion of the less halotolerant species that currently inhabit the region.

A segmented tidal prism model has also been developed by Wood (1979) and presented comparatively with the previously described methods by Miranda (1984), and we encourage the reader to follow the analysis of the Wood's model.

The one-dimensional segmented tidal prism models gives better results to estimate the fresh water, salinity, flushing times in well-mixed estuaries, and could be also applied to other conservative properties, as long as their input rates are known. These models are convenient because it is only necessary to know the basic estuarine data, such as tidal height, river discharge, geometric characteristics of the estuary and the salinity in the coastal sea. Of course, to achieve validation, observational data for the steady-state salinity distribution must also be known.

6.4 Concentration Estimates of a Conservative Pollutant

The concepts and semi-empirical models related to the steady-state mean salinity and fresh water estimates in an estuary, may be applied to other conservative chemical constituents or pollutants introduced into estuaries, provided their flux or transport inputs are known. Consider a one-dimensional estuary partially mixed (type 2 or B), forced by fresh water discharge, with tidal mixing due to horizontal flow associated with the flood and ebb tidal currents. Its mixing zone (MZ) may also be schematically segmented according to Fig. 6.11, and $R = T_P Q_f$ is the fresh water volume disposable to mixing during a complete tidal cycle with period (T_P). In this type of estuary, salinity increases with depth, as well as progressively increasing seaward due to the mixing process related to advection and turbulent diffusion. To maintain the volume (mass) conservation, this seaward transport must be compensated by an equal up-estuary fresh water volume (Q) in the sub-surface layer; for steady-state volume conservation Q = R.

Due to the tidal forcing attenuation towards the estuary head, in the uppermost segment the advective influence of the river discharge predominates, and the entrainment is the main process transporting water into the surface layer. The landward mass transport of salt in the bottom layer is equivalent to $6Q^{(1)}$, and $1Q^{(2)}$ is the compensating upward transport due to the entrainment between the bottom and the surface layers (Fig. 6.11). Due to dilution of the upward subsurface water by the fresh water volume, R, the salinity of this layer increase from 0 at the estuary head to 3‰ in the upper layer of the adjacent segment; in fact, due to this salt balance transport it follows that: 6Q = S(2R) and S = 3‰. We must also observe that in this segment the longitudinal volume and salt balance are equal to 2R − Q = R and 3(2R) − 6Q = 0, respectively.

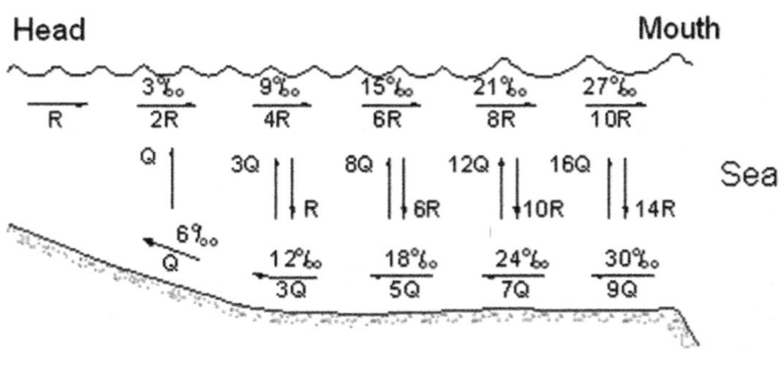

Fig. 6.11 Schematic changes in the mean salinity and in the volume of fresh water transported by advection in order to maintain the steady-state balance in an estuary during a complete tidal cycle. Salinity in ‰. River (nR, n = 1, 2, 4, … 10) and fresh water (mQ, m = 1, 3, 5, … 9) indicate its contents in seawater volumes, respectively. Landward and seaward, *up* and *downward arrows* indicate interchanges of water volumes (adapted from Ketchum 1953)

(1) In the salt balance the mass transport is calculated by ρSQ. Adopting $\rho = 1.0 \times 10^3$ kg m^{-3}, S in ‰ (g kg^{-1}) and Q in m^3, and the mass of salt is calculated by $1.0 \times 10^3 \times S \times 10^{-3} \times Q = S \times Q$.
(2) The compensating upward transport (upwelling) of the seaward surface flow has been observed for the first time by F. Ekman, in 1876, during his studies on the circulation and salt observations at the mouth of the Gotaelf river flowing into the Elfsborgsfjord (Sweden).

The tidal mixing increase in the adjacent seaward segment due to turbulent diffusion surpasses the entrainment, and the net volume transport between the bottom and surface layers becomes 2Q + (3Q − R); consequently, the seaward volume transport on the surface layer increase to 4R (Fig. 6.11). By the mass of salt conservation principle, the seaward salinity in the surface layer increases to 9‰ due to the salt balance: $12 \times 3Q = S \times 4R$ and S = 9‰ on the upper layer. By analogy, as in the previous segment, the longitudinal volume and salt balance are equal to: 4R − 3Q = R and 9(4R) − 12(3Q) = 0, respectively, and the seaward transport of water increases in proportion to its salt content.

This process is repeated in all segments located seaward and, according to the continuity principle, the net volume and salt mass across any cross section are equal to R and zero, respectively. Also, as illustrated in Fig. 6.11, there is an increase in seaward transport of mixed water and the compensating landward transport of salt. A direct consequence of this simple relationship is that the total circulation in the estuary increases enormously in volume as the water moves from the river towards the sea. This volume increase associated with the mixing process is called the *equivalent down-estuary transport*, which is a fictitious quantity and would only be measured under unusual conditions (Officer 1978).

The process just described is related to the volume and salt mass conservation principle under steady-state conditions or near steady-state conditions, within the time frame of the tidal period. In the cross section located at the estuary head, where $\overline{S} = 0$ and f = 1, the equivalent down-estuary transport (Q_d) is equal to the river discharge, and $Q_d = Q_f$, according to Eq. (6.75), and shown in Fig. 6.12. In any other section located seaward, the net volume transport is equal to Q_f. However, if f = 0.5 in this section, the equivalent transport is equal to $2Q_f$, to compensate for the water parcel retained in the system due to the mixing process. Thus, the ratio $Q_d/R = 1/f$, is a measure of the total process of removing a pollutant from an estuary compared with the advection effect due to the river discharge R (Officer 1978).

$$Q_d = \frac{R}{\overline{f}} = \frac{S_0}{S_0 - \overline{S}} Q_f = \frac{Q_f}{\overline{f}}. \tag{6.75}$$

As Eq. (6.75) is equal to the Eq. (6.16), which defines the *flushing rate* F, it has been proved that this *flushing rate* and the *equivalent down-estuary transport* are the same physical quantity. Hence, the mixing zone (MZ) volume is exchanged in the time interval equal to the flushing time (t_q). Another interpretation is that the

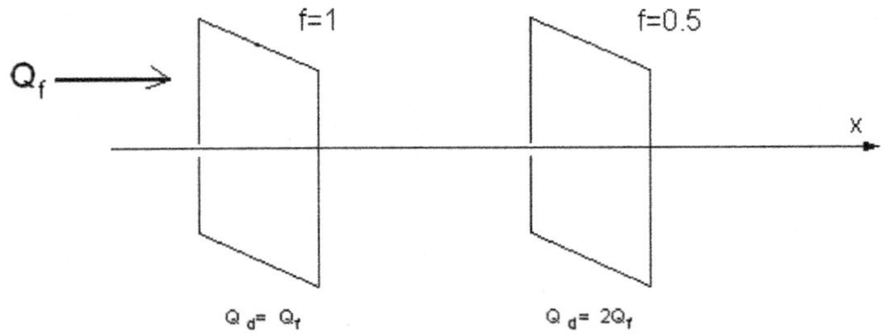

Fig. 6.12 Diagram showing the relationship of the volume transport, Q_f, and its *equivalent* (Q_d), and the fresh water fraction (f) (according to Officer 1978)

ratio, Q_d/R, is a measure of the total process for removing a conservative pollutant from an estuary compared to the simple advection effect of the river discharge, Q_f.

Let's consider now a mass transport, W, $[W] = [MT^{-1}]$, of a conservative effluent that is discharged into a river cross-section (Fig. 6.13). By hypothesis, this discharge is made through a multiport diffuser system to increase the effectiveness of the dilution of the less dense ascending plume located at the bottom (not indicated in the figure), extended across the estuarine channel. Then, the initial cross sectional average concentration per unit volume of sea water (c_o^*) is calculated by:

$$c_0^* = \frac{W}{Q_f}, \tag{6.76}$$

where Q_f is the steady-state river discharge, and $[c_o^*] = [ML^{-3}]$, and kg m^{-3} in units of the SI.

For an estuary, the river advection must be replaced by the equivalent downstream transport (Q_d) at the outfall, and the effluent concentration (c_0) is determined by,

$$c_0 = \frac{W}{Q_d} = \frac{W}{F}, \tag{6.77}$$

or, taking into account Eq. (6.75):

Fig. 6.13 Schematic diagram of the input of a discharge, W, of a conservative effluent into a cross-section of a non-tidal river, or estuary (according to Officer 1978)

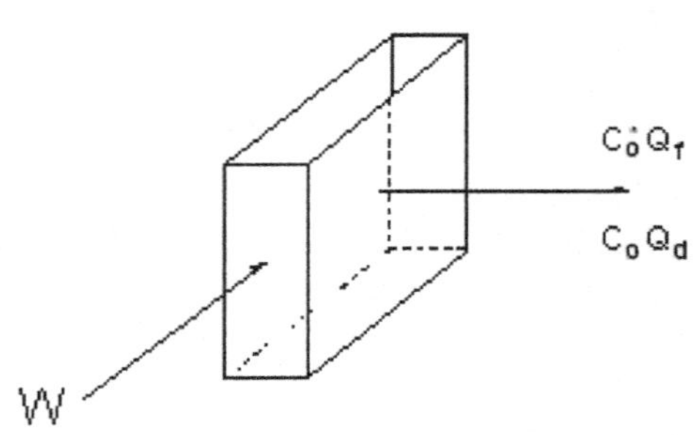

$$c_0 = \frac{W}{Q_f} f_W, \tag{6.78}$$

where c_0 has dimension of $[ML^{-3}]$, and f_w is the average fresh water fraction at the outfall. This result of Ketchum (1955), obtained with the implementation of the segmented tidal prism model, indicates that under steady-state conditions the initial concentration is directly proportional to the fresh water fraction, and inversely proportional to the river discharge which is assumed to be constant.

From Eqs. (6.76) and (6.78) it follows that the relationship between river (c_c^*) and estuary (c_0) concentrations is,

$$c_0 = (c_o^*) f_W. \tag{6.79}$$

As $0 \leq f_W \leq 1$, this implies that $c_0 < c_o^*$. Also, it should be noted that $f_W \to 0$, and also $c_0 \to 0$ at the estuary mouth.

Downstream of the outfall, when steady-state conditions are achieved, the pollutant must pass a cross section at the same rate it is discharged from the source, and its concentration is (Officer 1978):

$$(c_x)_d = \frac{W}{Q_f} f_x, \tag{6.80}$$

where $(c_x)_d$ and f_x are the average concentration of pollutant and the fresh water fraction at the cross-section located at the longitudinal position x, respectively.

Combining Eqs. (6.80) and (6.78) gives the following expression to calculate the pollutant concentration at the position (x) downstream of the pollutant introduction:

$$(c_x)_d = c_0 \frac{f_x}{f_W}. \tag{6.81}$$

This result indicates that the average concentration in the transversal section is directly proportional to the initial concentration (c_0) and the fresh water fraction at position x, and inversely proportional to the fresh water concentration at the position of the pollutant discharge. Using the expressions to calculate the fresh water fraction as a function of salinity (Eq. 6.10), Eq. (6.81) may be written as:

$$(c_x)_d = c_0 \frac{(S_0 - S_x)}{(S_0 - S_W)}. \tag{6.82}$$

In this equation, S_0 and S_x are the salinities at the adjacent coastal sea and in the cross section downward of position x, respectively.

The pollutant will also be carried upstream from the outfall by the diffusion and advection of tidal currents during tidal flood and ebb, and above the outfall there will be no net exchange across any boundary when the steady-state condition is reached. The pollutant quantity carried up-estuary will be exactly balanced by the quantity carried down-estuary. This is the same criterion that applies to the salt distribution up-estuary from the outfall. Thus, the up-estuary distribution of a conservative pollutant will be directly proportional to the salinity distribution, as given by:

$$(c_x)_u = c_0 \frac{S_x}{S_0}. \tag{6.83}$$

It is clear from these relationships that the knowledge of the distribution of salinity is essential in order to predict the expected steady-state distribution of conservative pollutants. These derivations were originally given by Ketchum (1955) and are a simple and direct method for estimating the distribution of a conservative pollutant or other index quantity in an estuary, with the knowledge of the salinity distribution alone. The pollutant distribution is calculated directly in terms of the salinity distribution without recourse. However, as stated in Officer (1978), it is important to emphasize that only the longitudinal effects have been considered and the definition contains the implicit assumption that the ocean at the mouth of the estuary is a perfect sink.

An observed fresh water concentration and the expected distribution of a conservative pollutant has been derived from the salinity distribution and fresh water fraction in the Raritan river and bay (Fig. 6.14). Four locations (A, B, C and D) have been arbitrarily selected for its position at an outfall. The horizontal distribution of pollutant concentration in percentage is obtained, assuming that the

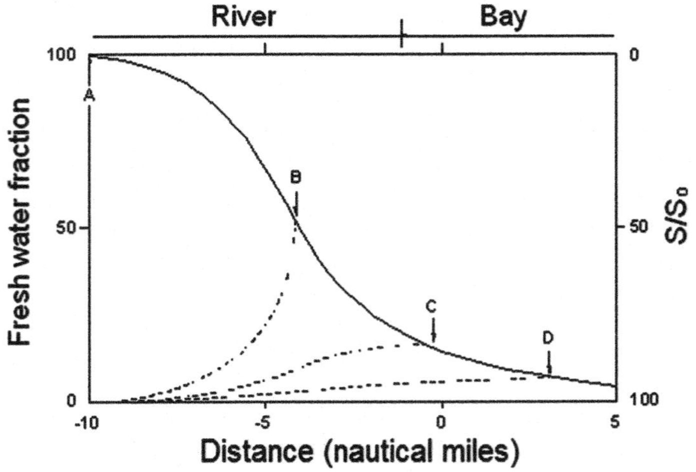

Fig. 6.14 The observed average distribution of fresh water fraction (in %) and the calculated distribution of a conservative pollutant (*continuous* and *dashed lines*) in Raritan river and bay for four possible outfall locations (*A*, *B*, *C* and *D*) (according to Ketchum 1955)

pollutant is uniformly mixed in the estuarine water through a multiport diffuser system.

From Fig. (6.14) and Eq. (6.81), at the pollutant releasing location $f_x = f_w$, the pollutant concentration decreases towards the estuary mouth from the maximum concentration (point A), as shown by the solid line of the fresh water concentration. The concentration distribution is directly dependent on its initial value. Therefore, different release location of pollutant may be less effective to the environment (e.g. points B,C and D). Up-estuary, from each outfall location, the pollutant distribution will follow the salinity distribution curve, or its inverse (the fresh water fraction curve), indicated by the dashed lines in Fig. 6.14). Two important consequences of moving the outfall downstream or upstream, respectively are (Ketchum 1955; Officer 1978):

1. As the outfall is progressively moved further down the estuary, the pollutant concentration up-estuary from the outfall is decreased while the concentration down-estuary remains the same.
2. The concentration of pollutants upstream from the outfall is decreased.

This method estimates the longitudinal variation of the concentration of a conservative pollutant discharged into an estuarine channel, under steady state conditions, and uses the salinity and/or the fresh water concentration as indicators. Thus, the method sensitivity is dependent on the river discharge and the salinity distribution, with the assumption that the coastal sea is a perfect sink, which is a very simplified boundary condition.

Pollutants that decay or decrease with time will be less concentrated throughout the estuary than the predicted concentrations of the conservative pollutants. This degradation of pollutants over time, which is superimposed on the circulation and diffusion has also been presented in the Bowden (1955) classical article.

To exemplify some concepts of this topic a practical example will be given based on Fischer et al. (1979). "During one complete tidal cycle, the multiport outfall diffuser of an industry discharges 200 $m^3\ s^{-1}$ of effluent, containing fifteen parts per thousand of a toxic material, into an estuary that is less dense than seawater. The mean river fresh water discharge is 500 $m^3\ s^{-1}$, the salinity at the coastal sea and the mean salinity value, at the outfall point are 34.0 and 20.0‰, respectively". Estimate the initial concentration of this substance at the transverse section, as well as the flushing time and rate, with the assumption that the estuarine MZ has a volume of 10^8 m^3.

With the Eq. (6.10) the fresh water concentration at the outfall (f_W) is determined by:

$$f_W = 1 - \frac{20}{34} = 0.41. \tag{6.84}$$

Knowing the exiting transport, and the pollutant concentration within this transport, then $W = 0.015 \times 200.0 = 3.0\ m^3\ s^{-1}$, and the initial pollutant concentration (c_0) is calculated by:

$$c_0 = \frac{W}{Q_f} f_W = 2.46\,g/kg = 2.46 \times 10^{-3}\,kg/kg, \tag{6.85}$$

The concentration $c_0 = 2.46 \times 10^{-3}$ is an average for the transverse section and representative if the outfall diffuser was adequately projected. The flushing time (t_q) is calculated by,

$$t_q = \frac{V_f}{Q_f} = \frac{(S_0 - \overline{S})}{S_0}\frac{V}{Q_f} = 23\,h\ (\approx \text{two semi-diurnal tidal cycles}).$$

As an exercise, the reader may solve the following expression for t_q,

$$t_q = \frac{V_f}{Q_f} = \frac{V_f\,c_0}{W\,f_W} = \frac{Vc_0}{W}. \tag{6.86}$$

The *flushing rate* (F) may also be calculate, combining its definition (Eq. 6.16) with Eqs. (6.1) and (6.85):

$$F = \frac{V}{t_q} = \frac{VQ_f}{V_f}, \quad \text{or} \quad F = \frac{VQ_f}{V_f} = \frac{Vf_WW}{V_fc_0}. \tag{6.87}$$

Introducing the known numerical values $F \approx 1214\ m^3\ s^{-1}$.

6.5 Water Mass Exchange at the Estuary Mouth

For application in the analysis of water exchange through the estuary mouth, this method uses steady-state mean salinities in transversal sections to calculate a non-dimensional parameter named *tidal exchange ratio,* defined by Fischer et al. (1979). This method was implemented with the introduction of a second parameter (*volumetric exchange ratio*), enabling its use for others coastal transition environments, such as bays (MacDonald 2006).

A portion of the water volume that enters an estuary forced by tidal flood currents is composed of water that left the estuary the previous ebb but retained in the estuarine plume. The remainder is water that we may think of as “new” ocean water volume (V_O); this water volume is what contributes to the dilution of pollutants inside the estuary, and knowledge of this new ocean water volume is important in the one-dimensional analysis of dilution of the concentration of substances or pollutants introduced into estuaries (Fischer et al. op. cit). The *tidal exchanged ratio* (TER) has been defined as the ratio of new ocean water (V_O) to the total volume of water that enters the estuary during a flood tide (V_f), which has already been defined as the tidal prism ($V_f = T_{PR}$, Chap. 2),

$$\mathrm{TER} = \frac{V_O}{T_{PR}}. \quad (6.88)$$

This ratio varies between the limits TER = 0 (when $V_O = 0$, in the absence of new water) and TER = 1 when $V_O = T_{PR}$ (i.e., the new water volume is equal to the total volume of water entering the estuary during the flood); the condition TER = 1 or $V_O = T_{PR}$ is the most favorable to dilution of pathogenic substances discharged in natural coastal environments.

The new water volume (V_O), is usually is not possible to be predicted theoretically; it is dependent on coastline circulation, which transports the estuarine plume, controls the ebb flow along the coast and delivers the supply of new ocean water for the flood. Without a favorable water mass renewal, the water mass exchange with the continental shelf offshore of the estuary mouth will be ineffective ($V_O \ll V_{PR}$), and eventually estuarine polluted water may return to the estuary. As previously discussed, using ADCP equipment, the tidal prism T_{PR} may be determined with the u-velocity component normal to the cross section of the current velocity at the estuary mouth, u = u(y, z, t). With vertical velocity profiles measured at time intervals during the food ($0 \leq t \leq T/2$), this volume may be calculate by numerical integration (Chap. 5, Eq. 5.52).

The tidal exchange ratio, defined in Eq. (6.88), may be determined with known mean volumes, salinities and densities at the transverse cross section at the estuary mouth, and tidal exchange ratio is solved taking into account the volume and mass conservation during the ebb and flood through the estuary mouth. To achieve this, let us follow the Fischer et al. (1979) empirical determination, using the same symbols and definitions:

T_{PR}	Total volume of seawater entering the estuary on the flood tide (tidal prism);
V_O	Volume of the new ocean entering the estuary during the flood tide;
V_e	Volume of the estuarine water leaving the system on the ebb tide;
$V_Q = T_P Q_f$	Volume of river discharge (fresh water) entering the estuary during the tidal cycle;
V_{fe}	Part of the volume (V_e) which flowed out of the estuary on the previous ebb;
$S_f(\rho_f)$	average salinity (density) of water entering the estuary on the flood tide;
$S_e(\rho_e)$	Average salinity (density) of water leaving the estuary on the ebb tide;
$S_O(\rho_O)$	Salinity (density) of ocean water

In steady-state conditions, the salinity and the water volume in the estuary remain constant. Then, the salt mass balance of water entering and leaving the estuary will be:

$$\rho_e S_e V_e = \rho_f S_f T_{PR}. \quad (6.89)$$

The dimension of each side of the equation is units of mass [M], because the product (ρS) numerically represents the salt concentration per volume unit. The salt balance of the water mass that enters the estuary during the flood, which has an increased salinity from the new ocean water (V_0), and includes the volume of water that is re-entering the estuary having previously flowed out of the estuary on the ebb tide, is given by:

$$\rho_e S_e V_e = \rho_f S_f T_{PR} + \rho_0 S_0 V_0. \quad (6.90)$$

The second member of this equation is the mass of salt that flows into the estuary mixed with the new water mass but without the estuarine plume influence. In Eqs. (6.89) and (6.90), the quantities ρ_e, ρ_v and ρ_0 are densities of the water masses which were introduced to maintain the equation with its dimension correctly [M]. However, the following simplification will be made $\rho_e \approx \rho_f \approx \rho_0$.

The water volume entering the estuary on the flood (T_{PR}), added to the volume of fresh water discharged into the estuary during the tidal cycle (V_Q) must be equal to the volume of water leaving the estuary during the ebb tide, then

$$T_{PR} + V_Q = V_e. \quad (6.91)$$

As the flood water (V_{PR}) is composed with some water volume that flowed out of the estuary on the previous ebb (V_{fe}) plus the new water volume (V_O) entering into the estuary, another relationship may be written for V_f,

$$V_{PR} = V_{fe} + V_O. \quad (6.92)$$

Combining Eqs. (6.89), (6.90) and (6.92), the new water volume is given by:

$$V_O = S_e \frac{(V_e - V_f)}{(S_0 - S_e)}, \quad (6.93)$$

and inserting this result in the definition of tidal exchange ratio (6.88),

$$TER = \frac{V_O}{T_{PR}} = \frac{S_e}{(S_0 - S_e)}\left(\frac{V_e}{T_{PR}} - 1\right), \quad (6.94)$$

and using the Eq. (6.89) transformed as $(\frac{V_e}{T_{PR}} = \frac{S_f}{S_e})$, the result for the tidal exchange ratio is given by:

$$TER = \frac{(S_f - S_e)}{(S_0 - S_e)}. \quad (6.95)$$

This result indicates that the TER is directly dependent on the differences in the mean salinity values at the estuary mouth in the flood (S_f) and ebb (S_e) tides, and is inversely proportional to the difference between the non-diluted salinity at the adjacent coastal ocean (S_0) and the average salinity leaving the estuary mouth during the ebb (S_e). $S_f = S_0$ or $S_f = S_e$ implies that TER = 1 and TER = 0, which corresponds to the best and the worst tidal exchange, respectively.

Solving Eq. (6.89) for the flood salinity, S_f, and combining with the TER of Eq. (6.95) and the equality $V_Q = V_e - T_{PR}$ (Eq. 6.91), it follows that a useful expression of the tidal exchange ratio when in function of $V_Q = TQ_f = R$ (Fischer et al. 1979) is

$$\text{TER} = \frac{S_e}{(S_0 - S_e)}\frac{V_Q}{V_f}, \qquad (6.96)$$

where V_Q is the volume of the river discharge entering the estuary during the tidal cycle.

This result indicates that besides the river discharge (Q_f), to determine TER and simulate nearly steady-state conditions during the flood and ebb tides, accurate observational data are required (hydrographic and current velocity), measured at the cross section at the estuary mouth during one of or more tidal cycles. Exemplifying, from salinity values measured at a cross section with an area, A, the averaged value (S_e) leaving the estuary during the ebb tide interval ($0 \leq t \leq T/2$) is calculated as follows: firstly, its cross-section mean value $S_e = S_e(t)$ is calculated by:

$$S_e(t) = \frac{1}{A(t)}\iint_A S_e(y,z,t)dydz, \qquad (6.97a)$$

Then, it follows that,

$$S_e = \frac{2}{T}\int_0^{\frac{T}{2}} S(t)dt = \frac{2}{T}\int_0^{\frac{T}{2}} [\frac{1}{A(t)}\iint_A S(y,z,t)dA]dt, \qquad (6.97b)$$

where dA indicates the area element at the mouth cross-section. Similar procedure may be used to calculate mean velocities and the corresponding values of the tidal prism ($V_f = V_{PR}$).

Pioneering studies Nelson and Lerseth (1972), quoted Fischer et al. (1979) describe measurements of the tidal exchange ratio at the entrance of San Francisco Bay (California, USA). Salinity and current velocity were measured throughout the tidal cycle at a number of positions along transect at the Golden Gate Bridge. Measurements were made on two occasions with different tide conditions, and TER values were calculated by Eq. (6.95). In this article, the authors found that increasing the flood tide in the range from 0.30 to 2.3 m increased the tidal exchange rate (TER) from 0.1 to 0.5, respectively, and thus reducing the effectiveness of pollutant discharges.

To illustrate TER results estimates in the Curimataú river estuary (Rio Grande do Norte, Brazil) were analysed by Miranda et al. (2005, 2006), using observational data measured during two neap and spring semi-diurnal tidal cycles in the vicinity of the estuary mouth. Due the strong river discharge, during the neap tide, the estuary was classified as partially mixed and highly stratified (type 2b), evolving in the spring tide to a partially mixed and low stratification (type 2a). In the first attempt to classify the estuary, $S_f < S_e$ and the TER < 0 (Eq. 6.95) has no physical meaning due to the abnormal river discharge. In the spring tide, the estimated mean salinity values were $S_f = 34.97‰$, $S_e = 34.03‰$ and $S_O = 36.8‰$, and the calculated TER = 0.3. This result indicates a small new oceanic water volume intrusion (V_O) into the estuary and therefore less effective conditions for the dilution of pollutants input into the estuary.

6.6 Mixing Diagrams

M*ixing diagram*s are very useful for investigating the presence of sources and sinks of natural components and/or pollutant concentrations in the estuarine water, tidal river or in the adjacent coastal sea. This diagram is a Cartesian orthogonal coordinate system used to correlate a given concentration versus salinity, which has the coastal sea as its main source and has a well known longitudinal variation.

The classical T-S diagram was introduced in the oceanographic literature by Björn Helland-Hansen in papers published in 1916 and 1918, as a pioneering study on classification, distribution and mixing of oceanic water mass. It is a diagram with temperature and salinity (heat and salt concentrations) in the ordinate and abscissa axis, respectively. It has also been used as a basic mixing diagram for estuarine water mass classification.

If the water body is homogeneous in salinity and temperature the image of these properties on the T-S diagram is a single point, representing the final stage of the irreversible mixing generate by the advection and diffusion processes. If there is no homogeneity in the water mass, due to the variations of these properties, the S and T pairs of points will appear as a set of aligned points on the diagram; the point distributions indicate the occurrence of changes in the heat and salt concentrations during mixing.

As the density anomaly at atmospheric pressure (Sigma-t or σ_t) is dependent only on the salinity and temperature, it is possible to drawn in the T-S diagram a set of parametric curves, which represents the state equation of seawater at atmospheric pressure; this diagram is named *state diagram* of *seawater*.

With a few exceptions, salinity and temperature in estuarine water respond more quickly to mixing processes (advection and turbulent diffusion) than to air-sea interaction processes. Although these properties have small temporal variability, a

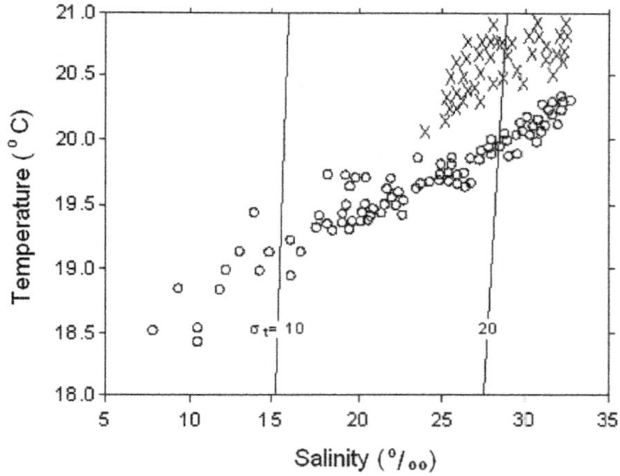

Fig. 6.15 Scatter T-S diagram with salinity (S) and temperature (T) values sampled during neap (o) and spring (x) tidal conditions in the estuarine channel of Bertioga, in July, 1991, showing estuarine water mass with high and low stratification, respectively. *Continuous lines* indicate the density in Sigma-t units

well defined correlation may exist in the T-S diagram, if the set of (S, T) pairs are sampled during complete tidal cycles. The set of sampled data is named *scatter* T-S diagram, which may be used to identify whether or not the tidal river zone (TRZ), the mixing zone (MZ) or the adjacent coastal sea have heat and salt sources or sinks.

The scatter T-S diagram of the Bertioga estuarine channel (Fig. 5, Chap. 1) with hydrographic data sampled during two complete tidal cycles, in neap and spring tidal conditions, 5 km landward from its mouth is shown in Fig. 6.15.

In the neap tide experiment, the temperature interval change was 3 °C (from 18.5 to 20.8 °C); however salinities varied over a large interval from ≈8 to 33‰. In

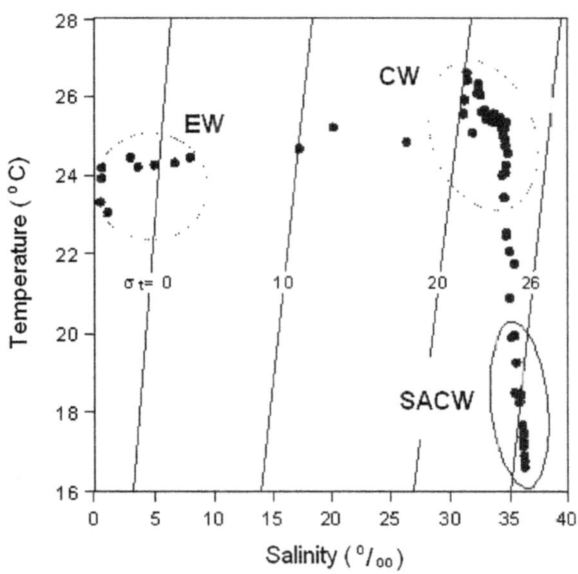

Fig. 6.16 Scatter T-S diagram in the mixing (MZ) and near-shore turbidity (NTZ) zones of the Itajaí-açu estuarine system, in a longitudinal section showing thermohaline characteristics of the Estuarine Water Mass (EW), the Coastal Water Mass (CW) and the South Atlantic Central Water Mass (SAW) (adapted from Schettini et al. 1998)

the spring tide, the temperature and salinities varied from 20 to 21.0 °C, and ≈24 to 33‰, respectively (Fig. 6.15). In the spring, tide the tidal currents were more intense than in the neap tide, and more energy was available to raise the potential energy of the water column and non-isopicnal vertical turbulent diffusion.

This diagram is also an indicator of the vertical salinity stratification, and we may observe that the estuarine water mass changed from highly to moderately stratified, from the neap and spring tides, respectively, with the (S,T) points presenting a relatively large and small scatter, respectively. The positive correlation between temperature and salinity indicates that the main source and sink of the heat and salt concentrations in the estuary were the adjacent coastal waters and the river water, respectively.

Thermohaline characteristics of the mixing zone (MZ) and the near-shore turbidity zone (NTZ) of the Itajaí-açu river (Santa Catarina State, Brazil) estuarine system were almost synoptically sampled, and the analysis using the scatter T-S diagram (Fig. 6.16) was presented in Schettini et al. (1998.

In this diagram (Fig. 6.16), the following water masses were identified: the Estuarine Water (EW), Coastal Water (CW) and the South Atlantic Central Water (SACW). It can also be observed that the less dense water formation of the estuarine plume is due to the non-isopicnal mixing of the EW and CW water masses, and the upper part of the oceanic water (SACW) with 20 °C < T < 16 °C and 35‰ < S < 36‰) is in agreement with the mean values during the summer, which

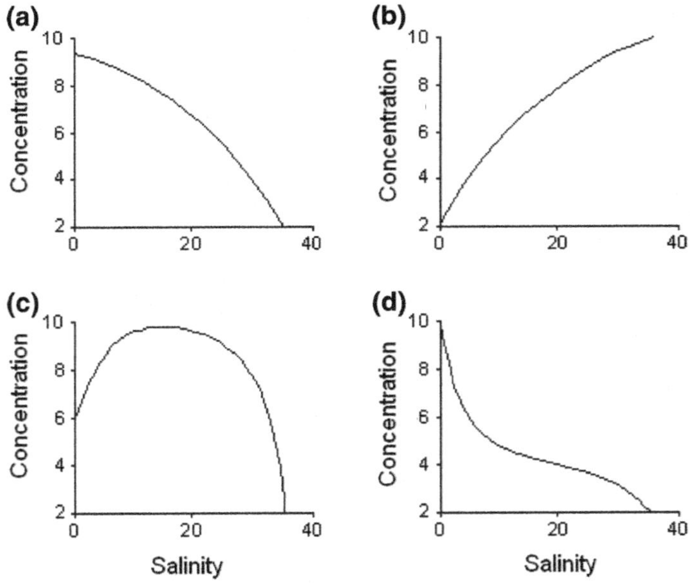

Fig. 6.17 Mixing diagrams schematically showing the sources and sinks of conservative or non-conservative properties' concentrations along an estuary: **a** source at the TRZ and sink in the MZ; **b** source in the coastal ocean and sink at the estuary head; **c** source at MZ and sinks at the head and in the coastal ocean; **d** and source at the TRZ and sink at the MZ

were calculated by Castro and Miranda (1998) for the continental shelf waters offshore of Itajaí (Santa Catarina, Brazil). This water with a higher density is able to be advected into the estuary by barotropic and baroclinic gradient pressure forces, generating high vertical stratification in the MZ during seasons with high fresh water discharge.

The T-S Diagram can be altered to become a *mixing diagram*, by substituting the temperature (heat concentration), in the ordinate axis, by an alternative property's concentration. According to the immediate purpose of the investigation, the correlation of a given property with salinity may be performed in several ways: with instantaneous concentration values during high and low tide, or with mean concentration values during tidal cycles. Among the numerous correlations which can be expected, some possibilities to identify concentrations of sources or sinks of the property are shown schematically in Fig. 6.17a–d.

If the property concentration is not at steady-state and the salinity at the coastal ocean has short temporal variations, these correlations may be more complex than those depicted in the figure.

References

Anderson, F. E. 1979. Laboratory Manual for Introductory Oceanography. Minnesotta, Burges Publishing., 141 p.

Bowden, K. F. 1967a. Circulation and Diffusion. In: Lauff, G. H. (ed.). Estuaries. American Association for the Advancement of Science, Washington, D. C., pp. 15–36. (Publication, 83).

Bowden, K. F. 1967b. Stability effects on turbulent mixing in tidal currents. Phys. Fluid. 10 (suppl), S278–S280.

Bradley, P. M. Kjerfve, B. & Morris, J. T. 1990. Rediversion Salinity Change in the Cooper River, South Carolina: Ecological Implications. Estuaries, vol. 13, No. 4, pp: 373–379.

Brown, W. S. & Arellano, E. 1980. "The Application of a Segmented Tidal Mixing Model to the Great Bay Estuary, N. H.". Estuaries, 3(4):248–257.

Castro, B. M. de & Miranda, L. B. 1998. Physical Oceanography of the Western Atlantic Continental Shelf Located Between 4°N and 34°S—Coastal Segment (4,W). In: Robinson, A. R. & Brink, K. H. (eds.). The Sea: The Global Coastal Ocean-Regional Studies and Synthesis. New York, Wiley, vol. 11, pp. 209–251.

Dyer, K. R. & Taylor, P. A. 1973. A Simple, Segmented Prism Model of Tidal Mixing in Well-mixed Estuaries. Estuar. Coast. Mar. Sci., 1:411–418.

Geyer, W. R. 1997. Influence of Wind on Dynamics and Flushing of Shallow Estuaries. Estuar. Coast. Shelf Sci., 44:713–722.

Hansen, D. V. & Rattray Jr., M. 1966. New Dimensions in Estuary Classification. Limnol. Oceanogr., 11(3):319–325.

Hughes, P. 1958. Tidal Mixing in the Narrows of the Mersey Estuary. Geophys. J. R. Astr. Soc., 1:271–283.

Ketchum, B. H. 1950. Hydrographic Factors Involved in the Dispersion of Pollutants Introduced into Tidal Waters. J. Boston Soc. Civ. Engrs., 37:296–314.

Ketchum, B. H. 1951. The Exchanges of Fresh and Salt Waters in Tidal Estuaries. J. Mar. Res., 10 (1):18–38.

Ketchum, B. H. 1953. Circulation in Estuaries. In: Johnson, J. W. (ed.). Proceedings of Third Conference on Coastal Engineering, Council on Wave Research. Cambridge, The Engineering Foundation, pp. 65–76.

Ketchum, B. H. 1955. Distribution of Coliform Bacteria and other Pollutants in Tidal Estuaries. Sew. Ind. Wastes, 27:1288–1296.

Ketchum, B. H. Redfield, A. C. & Ayres, J. C. 1951. The Oceanography of the New York Bight. Pap. Phys. Oceanogr. Met. MIT-WHOI, 7(1):1–46.

MacDonald, D. G. 2006. Estimating an Estuarine Mixing and Exchange Ratio from Boundary Data with Application to MT. Hope Bay (Massachusetts/Rhode Island), Est. coast., Shelf Sci., (70):326–332.

Maximon, L. C. & Morgan, G. W. 1955. A Theory of Tidal Mixing in a Vertically Homogeneous Estuary. J. Mar. Res., 14:157–175.

Miranda, L. B. 1984. Cinemática e Dinâmica de Estuários. São Paulo, BSP., 360 p.

Miranda, L. B. & Castro, B. M. de. 1993. Condições Oceanográficas no Canal de Bertioga. Relatório Técnico. São Paulo, Fundação de Estudos e Pesquisas Aquáticas, pp. 1–97.

Miranda, L. B. Bérgamo, A. L. & Castro, B. M. 2005. Interactions of river discharge and tidal modulation in a tropical estuary, NE, Brazil. Ocean Dynamics, vol. 55, pp. 430–440.

Miranda, L. B. Bérgamo, A. L. & Silva, C. A. R. 2006. Dynamics of a Tropical Estuary: Curimataú River, NE Brazil. J. of Coastal Research, vol. SI 39, pp. 697–701.

Nelson, A. W. & Lerseth, R. J. 1972. A Study of Dispersion Capability on San Francisco Bay-Delta Waters. Sacramento, California Department of Waters Resources.

Officer, C. B. 1976. Physical Oceanography of Estuaries (and Associated Coastal Waters). New York, Wiley. 465 p.

Officer, C. B. 1978. Some Simplified Tidal Mixing and Circulation Flux Effects in Estuaries. In: Kjerfve, B. (ed.). Estuarine Transport Processes. Columbia, University of South Carolina Press, pp. 75–93. (The Belle W. Baruch Library in Marine Science, 7).

Officer, C. B. & Kester, D. R. 1991. On Estimating the Non-advective Tidal Exchanges and Advective Gravitational Circulation Exchanges in an Estuary. Estuar. Coast. Shelf Sci., 32:99–103.

Pilson, M. E. Q. 1985. On the Residence Time of Water in Narragansett Bay. Estuaries, 8(1):2–14.

Schettini, C. A. F.; Kuroshima, K. N.; Pereira, J. Fo.; Rörig, L. R. & Resgalla Jr., C. 1998. Oceanographic and Ecological Aspects of the Itajaí-açu River Plume During High Discharge Period. Anais. Acad. Bras. Ci., 70(2):335–351.

Wood, T. 1979. A Modification of Existing Simple Segmented Tidal Prism Models of Mixing in Estuaries. Estuar. Coast. Mar. Sci., 8:339–347.

Quoted References

Fischer, H. B. List, E. J.; Koh, R. C. Y.; Imberger, J. & Brooks, N. H. 1979. Mixing in Inland and Coastal Waters. New York, Academic Press. 483 p.

Chapter 7
Hydrodynamic Formulation: Mass and Salt Conservation Equations

When hydrographic properties and motions in an estuary have spatial and temporal variation, they are termed as non-uniform and unsteady, as opposed to uniform and in steady-state. In the previous chapter, salinity in the estuary was, by hypothesis, in steady-state conditions in longitudinal segments during complete tidal cycles and at high and low tidal conditions. However, estuaries are dynamic systems, and salinity and current velocity vary in time and space from almost at rest (slack water) to speeds of up to several meters per second, in estuaries forced by macro-tides. In observational data analysis, it is usual to simulate steady-state conditions of hydrographic properties and circulation by calculating mean values during a time interval of one or more tidal cycles, under the assumption that the river discharge remains constant during this time. Tidal co-oscillation is the main driving force of the non-steady-state condition, however, in some situations other forces may also be important, such as the abnormal storm surge due to wind shear stress acting on the continental shelf.

The mathematical development of this chapter starts with the equation of mass conservation, also named the continuity equation, which complements the equation of motion, which will be studied in the next chapter. In practical applications it will be necessary to assume as given the estuary geometry, the river discharge and the initial and boundary conditions. For investigation of any hydrographic property, it will also be necessary to use the corresponding conservation equation; in the case of non-conservative properties, sources and sinks must also be specified.

For the application of the fundamental principles of Fluid Mechanics, it is necessary to consider an infinitesimally small fluid sample. Usually this sample is referred as a *material element,* or more often as a *volume element*. Another assumption is that all variables which will represent physical properties (scalar or vectorial, such as hydrographic properties or current velocity, respectively) are continuous functions of space and time; in this way the mathematical rules of the differential and integral calculus can be applied.

L. Bruner de Miranda et al., *Fundamentals of Estuarine Physical Oceanography*, Ocean Engineering & Oceanography 8, DOI 10.1007/978-981-10-3041-3_7

7.1 State of a Volume Element

When the fluid is in motion, its properties (scalar or vector) are functions of space (x, y, z) and time (t). Then, any property of the fluid, generically indicated by P, is expressed by the function:

$$P = P(x, y, z, t) = P(\vec{r}, t), \tag{7.1}$$

where $\vec{r} = \vec{r}(x, y, z)$ indicating the position vector of a small volume, δV, and x, y and z being its coordinates in space.

Conceptually, this volume of fluid presents the characteristics of properties associated with this elementary volume within a given water flow. According to Symon (1957) and Gill (1982), if this element is at the position $\vec{r}$ at the instant of time t, its position in the space may be generically indicated by the vector position $\vec{r} = \vec{r}[x(t), y(t), z(t)] = \vec{r}(t)$. Then, a generic property of the volume element is expressed by the following functional relationship:

$$P = P[x(t), y(t), z(t), t] = P[\vec{r}(t), t]. \tag{7.2}$$

From this expression it follows that the total rate of variation (dP/dt) of the property is,

$$\frac{dP}{dt} = \frac{\partial P}{\partial t} + \frac{\partial P}{\partial x}\frac{dx}{dt} + \frac{\partial P}{\partial y}\frac{dy}{dt} + \frac{\partial P}{\partial z}\frac{dz}{dt}, \tag{7.3}$$

or

$$\frac{dP}{dt} = \frac{\partial P}{\partial t} + \frac{d\vec{r}}{dt} \bullet \nabla P, \tag{7.4}$$

where the symbol ∇ is the *nabla* operator, $\nabla = (\frac{\partial}{\partial x})\vec{i} + (\frac{\partial}{\partial y})\vec{j} + (\frac{\partial}{\partial z})\vec{k}$, and the dot (•) indicates the scalar product. Equation (7.4) indicates that: (i) the rate at which the property, P, is changing with time at a fixed point in space is the partial derivative with respect to time (∂P/∂t), which is itself a function of x, y, z, and; (ii) the rate at which the property, P, is changing with respect to a point moving along with the fluid. Another component $(\frac{d\vec{r}}{dt})$ is the variation of the volume element's position in space,

$$\frac{d\vec{r}}{dt} = \vec{v} = \vec{v}(u, v, w) = \vec{v}(x, y, z, t), \tag{7.5}$$

where u = u(x, y, z, t), v = v(x, y, z, t) and w = w(x, y, z, t) are the velocity components of the velocity vector in the coordinate axes Ox, Oy and Oz, respectively, and are functions of time (t). Then, the equation that defines the state of a volume element, δV, of fluid is given by,

$$\frac{dP}{dt} = \frac{\partial P}{\partial t} + \vec{v} \bullet \nabla P. \tag{7.6}$$

Thus, the total time variation of the property, P, is composed of the local variation ($\partial P/\partial t$) and the variation due to the advection, which depends on the fluid velocity and the property gradient ($\vec{v} \bullet \nabla P$). When the local variation is zero ($\partial P/\partial t = 0$), the spatial property variation is considered to be in *steady-state*, and when it has no spatial variation it is *uniform.*

To simplify the notation of equations, it is useful to define the total derivative operator, d/dt, as:

$$\frac{d}{dt} = \frac{\partial}{\partial t} + \vec{v} \bullet \nabla. \tag{7.7}$$

This definition is very convenient, as may be seen considering the salt conservation equation of seawater, simply equating P = S. In fact, if the molecular and turbulent diffusion are neglected, the volume element in motion will retain the same concentration of its dissociate components, and its mass will remain constant during the motion. Mathematically,

$$\frac{dS}{dt} = \frac{\partial S}{\partial t} + \vec{v} \bullet \nabla S = 0. \tag{7.8a}$$

This equation indicates that during the motion there is an equilibrium between the local ($\partial S/\partial t$) and advective ($\vec{v} \bullet \nabla S$) variation. In the steady-state of the salinity field $\vec{v} \bullet \nabla S = 0$, or,

$$u\frac{\partial S}{\partial x} + v\frac{\partial S}{\partial y} + w\frac{\partial S}{\partial z} = 0. \tag{7.8b}$$

7.2 Mass and Salt Conservation Equations

Let us consider the fluid motion in a laminar flow regime, which usually holds for slow motions. Even if the volume element has a constant mass, its volume may vary due to the pressure acting on its surface during the motion. As density is defined by the ratio of mass by volume ($\rho = m/V$, $[\rho] = [ML^{-3}]$), it follows that density, being a dependant property, may vary with element volume changes. The equation relating the fluid's density with its motion (velocity) may be defined by the mass conservation principle, which is traditionally named continuity equation since it follows the conservation laws, and it is related to the density and velocity of a continuous medium.

The continuity equation is of fundamental importance to studies related to fluid motion, and its deduction may be obtained with different theoretical developments (Symon 1957; Brand 1959; Kinsman 1965; Neumann and Pierson 1966; Gill 1982, and others). A straightforward Eulerian formulation may be made using Gauss' divergence theorem, equating the local density time rate $(\partial\rho/\partial t)$, integrated in a differential volume element, V, enclosed by its area A,

$$\int_V (\frac{\partial \rho}{\partial t})dV, \tag{7.9a}$$

with the mass transport into the volume, V, through the closed surface area, A, which is expressed by:

$$-\int_A \rho\vec{v} \bullet \vec{n}dA = -\int_V (\nabla \bullet \rho\vec{v})dV, \tag{7.9b}$$

where $\vec{n}$ is the unity vector orthogonal to the closed surface, oriented outward of the volume V. Hence, equating the mass transport $[MT^{-1}]$ expressed by Eqs. 7.9a with the corresponding mass transport on the right-hand-side of Eq. 7.9b,

$$\int_V (\frac{\partial \rho}{\partial t})dV = -\int_V (\nabla \bullet \rho\vec{v})dV, \tag{7.9c}$$

it follows the mass conservation property inside the control volume V, or the continuity equation,

$$\frac{\partial \rho}{\partial t} + \nabla \bullet \rho\vec{v} = 0. \tag{7.10}$$

This equation indicates the following physical principle: the local density variation inside a volume element is due only to the divergence operator of the mass flux $(\rho\vec{v})$, $([\rho\vec{v}] = [ML^{-2}T^{-1}])$ through a closed surface of the fluid element. Using the divergent operator, the second term of this equation may be written as:

$$\nabla \bullet \rho\vec{v} = \vec{v} \bullet \nabla\rho + \rho\nabla \bullet \vec{v}, \tag{7.11}$$

combining expressions (7.11) and (7.10),

$$\frac{\partial \rho}{\partial t} + \vec{v} \bullet \nabla\rho + \rho\nabla \bullet \vec{v} = 0, \tag{7.12}$$

and, taking into account the total derivative (7.7), the continuity equation is reduced to the following expression:

$$\frac{d\rho}{dt} + \rho\nabla \bullet \vec{v} = 0, \tag{7.13a}$$

or

$$\frac{1}{\rho}\frac{d\rho}{dt} + \nabla \bullet \vec{v} = 0, \tag{7.13b}$$

which expresses the relationship between the time variation of fluid density and its velocity.

Equations (7.10) and (7.13b) are different mathematical expressions of the continuity equation. The first term of Eq. (7.13b) is the relative change of the total density variation, and the second is the divergent operator of the velocity field. At this point we should remember that the divergent operator may be positive, negative or zero, indicating the divergent, convergent or non-divergent fields, respectively.

The continuity equation in the differential form (7.13b) is valid for fluids in laminar motion with only one component, such as pure water. In the field of Physical Oceanography, seawater is considered a solution with two components (pure water + salt) and the mass of a volume element may vary due to salt diffusion through its geometric boundaries. Thus, when the continuity equation is applied to seawater, unless this diffusion process is negligible, it must be compensated by the introduction of a parcel which takes this into account to preserve the mass conservation principle.

To demonstrate that the salt diffusion may be disregarded in coastal and estuarine water masses which have non-constant ionic composition, Csanady (1982) presented the following development to the quantitative determination of the relative density time rate and the divergence of the velocity field of Eq. (7.13b). For a typical summer day, it is estimated that the time taken to heat the surface layer of the estuarine water mass by 1.0 °C is three hours. Thus, the temperature time rate increase is: $dT/dt = 1.0 \times 10^{-4}$ °C s^{-1}. Adopting a typical value of 1.0×10^{-4} °C^{-1} to the thermal expansion coefficient, the relative rate at which the density is changing with time at a fixed point in space (first term of Eq. (7.13b) is estimated to be 1.0×10^{-8} s^{-1}.

To estimate the molecular salt diffusion on the local density time rate, a value of 1.0×10^{-9} m^2 s^{-1} was adopted for the kinematic salt molecular diffusion coefficient (D). And, as the salt molecular diffusion obeys the Fickian law,

$$\frac{dS}{dt} = D\nabla^2 S, \tag{7.14}$$

the total salt variation (dS/dt) may be calculated, for the most unfavorable condition ($\nabla^2 S = 1$), as equal to 1.010^{-9} s^{-1}. Using a mean value for the saline contraction coefficient (β) of $7.5 \times 10^{-4} \approx 10^{-3}$, the estimated value of the kinematic molecular salt diffusion coefficient is 1.0×10^{-12} m^2 s^{-1}. These results indicate that the influences of the local heating and salt diffusion on the relative local density variation are by a magnitude of less than or equal to 1.0×10^{-8} s. For an estuary with a length of 10 km (1.0×10^4 m), a longitudinal density variation of 10 kg m^{-3} between its mouth and its head, and a velocity variation of 1.0 m s^{-1}, it follows that the relative density time variation (first term of the Eq. 7.13b) is less than or equal to 1.0×10^{-6} s^{-1}.

Let us now estimate the order of magnitude of the second parcel of Eq. (7.13b), representing the divergence of the velocity field. Observational data of estuaries indicate that the u-velocity component may vary from 0 to 1.0 m s^{-1} over distances of up to 1.0×10^4 m, and its divergence value is estimated in 1.0×10^{-4} s^{-1}. Comparing this value with the estimated value with the estimated value for the first parcel (1.0×10^{-6} s^{-1}), the conclusion is that the influence of the velocity divergence is predominant, even in the extreme conditions of the above example. Then, for practicality, the continuity equation is reduced to the simple expression in the Cartesian coordinate system:

$$\nabla \bullet \vec{v} = \frac{\partial u}{\partial x} + \frac{\partial v}{\partial y} + \frac{\partial w}{\partial z} = 0. \tag{7.15}$$

This mass conservation Eq. (7.15) may also be obtained from Eq. (7.13b) under the hypothesis that the fluid density is a constant (ρ = const.) or its relative value doesn't change during motion ($\frac{1}{\rho}\frac{\partial \rho}{\partial t} = 0$), which corresponds to the behavior of incompressible fluids. Thus, for practicality, estuarine water mass is considered to be an incompressible fluid. In some texts of Hydrodynamics, Eq. (7.13a, 7.13b) are named conservation of mass, and the expression continuity equation is usually used for Eq. (7.15).

As the motion regime in an estuary is transitional, changing from laminar to turbulent, the continuity equation must be adapted to take into account the turbulent flow. This may be accomplished by eliminating the random (or turbulent) small scale velocity fluctuations, dividing the velocity into two terms which are uncorrelated with one another: a mean time ($\langle \vec{v} \rangle$), and a turbulence velocity value ($\langle \vec{v'} \rangle$). The mean value $\langle \vec{v} \rangle$, is calculated from a time interval Δt which is long enough (generally a few minutes) to eliminate the turbulent fluctuations $\vec{v'}$, but short enough that the larger-scale variations do not affect the mean value. That is, the average value of the turbulent fluctuations should equal zero ($\langle \vec{v'} \rangle = 0$). Substituting this instantaneous value into Eq. (7.15), gives,

$$\nabla \bullet (\langle \vec{v} \rangle + \vec{v'}) = 0, \tag{7.16}$$

and calculating its mean time value for the time interval (Δt),

$$\frac{1}{\Delta t}\int_{0}^{\Delta t} \nabla \bullet (\langle \vec{v} \rangle + \langle \vec{v}' \rangle) dt = 0. \tag{7.17}$$

As the divergence is calculated as spatial derivatives of vector velocity, which is assumed to be a continuous function, according to the Schwartz's theorem it is possible to change the order of the derivative and integration operations. Taking into account that $\langle\langle \vec{v} \rangle\rangle = \langle \vec{v} \rangle$ and $\langle \vec{v'} \rangle = 0$, it follows that the expression of the continuity equation for a turbulent fluid flow is,

$$\nabla \bullet (\langle \vec{v} \rangle) = \nabla \bullet \vec{v} = 0, \tag{7.18}$$

where, to simplify the notation, the time mean value $(\langle \vec{v} \rangle)$ is substituted by $\vec{v}$ $(\langle \vec{v} \rangle = \vec{v} = u\vec{i} + v\vec{j} + w\vec{k})$. This vector now has u, v and w components, which are time mean values of a relatively short time interval (Δt). Then, the continuity equation, for a transitional or turbulent flow in the Cartesian frame of reference (Oxyz), is formally expressed by a similar equation which holds for laminar fluid flow (Eq. 7.15),

$$\nabla \bullet \vec{v} = \frac{\partial u}{\partial x} + \frac{\partial v}{\partial y} + \frac{\partial w}{\partial z} = 0. \tag{7.19}$$

When this expression of the continuity equation is integrated with respect to a geometric volume, such as for an estuary, at the free surface and bottom layers there will be sources and sinks of mass (evaporation-precipitation balance, snow, condensation on the surface, and bottom spring water), which must be adequately specified.

Now, let us apply the principle of mass conservation to other properties that are used to characterize the state of a water mass, such as its salt content. In practice, the principle of continuity is most often used together with the principle of conservation of salt to study the flow of relatively enclosed bodies of water, such as estuaries. By conservative properties we mean concentrations, such as salinity, that are altered locally, except at the boundaries, by diffusion and advection only.

The vector which characterizes the advective salt flux $(\vec{S})$, expressed by mass of salt per area and time $([\vec{S}] = [ML^{-2}T^{-1}])$, generated by a laminar motion, $\vec{v}$, is expressed by $\vec{S} = \rho S \vec{v}$. Substituting in Eq. (7.10) the density, ρ, with the scalar quantity, ρS, which has the same dimension as density ($[\rho S] = [ML^{-3}]$), but physically represents the concentration of mass of salt dissociated in seawater, it follows that:

$$\frac{\partial(\rho S)}{\partial t} + \nabla \bullet \rho S \vec{v} = 0, \quad (7.20)$$

or

$$\rho(\frac{\partial S}{\partial t} + \vec{v} \bullet \nabla S) + S(\frac{\partial \rho}{\partial t} + \nabla \bullet \rho \vec{v}) = 0. \quad (7.21)$$

These equations are the analytical expressions of the principle of conservation of salt, only due to the advection. As the expression between the parentheses of the second parcel of Eq. (7.21) is the continuity Eq. (7.10), and is equal to zero, then

$$\frac{\partial S}{\partial t} + \vec{v} \bullet \nabla S = 0, \quad (7.22)$$

or, in the scalar notation

$$\frac{\partial S}{\partial t} + u\frac{\partial S}{\partial x} + v\frac{\partial S}{\partial y} + w\frac{\partial S}{\partial z} = \frac{dS}{dt} = 0. \quad (7.23)$$

As stated previously, $\frac{dS}{dt}$ is the total time variation of the salinity. This differential equation is the principle of the conservation of salt, under the action of advection for a small volume of seawater, with the assumption that the molecular diffusion has been disregarded. However, as estuaries usually have a turbulent flow regime, the salt flux due to the turbulent diffusion is much higher. Thus, the influence of turbulent motion on the salt balance of estuarine waters must also be taken into account. Consider a cubic volume with surface area units normal to the coordinate axis, for an estuarine water mass without free surface. The salt conservation equation in the differential form is rigorously written as (Sverdrup et al. 1942; Pritchard 1958; Cameron and Pritchard 1963, and others):

$$\frac{\partial S}{\partial t} + u\frac{\partial S}{\partial x} + v\frac{\partial S}{\partial y} + w\frac{\partial S}{\partial z} = \frac{\partial}{\partial x}(K_x\frac{\partial S}{\partial x}) + \frac{\partial}{\partial y}(K_y\frac{\partial S}{\partial y}) + \frac{\partial}{\partial z}(K_z\frac{\partial S}{\partial z}). \quad (7.24)$$

In this equation, K_x, K_y and K_z are the kinematic[1] coefficients of turbulent diffusion of salt in the horizontal (Ox and Oy) and vertical (Oz) axes, respectively, which in general are functions of the spatial and temporal scales of the estuarine processes, with dimensions $[L^2T^{-1}]$. Equation (7.24) indicates that the local salinity variation ($\frac{\partial S}{\partial t}$) is dependent on the advection (velocity components u, v and w in the

[1]The dynamic coefficients of eddy diffusion (dispersion), A_x, A_y and A_z, which have dimensions of $[ML^{-1}T^{-1}]$, is obtained from the product of density, ρ, by the corresponding kinematic coefficient; $A_x = \rho K_x$, $A_y = \rho K_y$ and $A_z = \rho K_z$.

left-hand-side terms), and turbulent diffusion (terms in the right-hand-side) simulated by the Fickian law.

Another expression for the salt conservation equation may be obtained when combined with the continuity Eq. (7.19):

$$\frac{\partial S}{\partial t}+\frac{\partial(uS)}{\partial x}+\frac{\partial(vS)}{\partial y}+\frac{\partial(wS)}{\partial z}=\frac{\partial}{\partial x}(K_x\frac{\partial S}{\partial x})+\frac{\partial}{\partial y}(K_y\frac{\partial S}{\partial y})+\frac{\partial}{\partial z}(K_z\frac{\partial S}{\partial z}). \tag{7.25}$$

It is implicit in this equation that S = S(x, y, z, t) represents the average salinity obtained from a time interval that is long enough to eliminate the turbulent variations (S'), but short enough for this mean value not to be affected by long- term variations. In the same way, the velocity components u = u(x, y, z, t), v = v(x, y, z, t) and w = w(x, y, z, t) in the Eqs. (7.24) and (7.25) also represent mean values. If advection alone is responsible for the mixing process in steady-state conditions, these equations are reduce to their vector formulation $\vec{v}\bullet\nabla S=0$, or $\nabla\bullet S\vec{v}=0$.

The kinematics eddy diffusion coefficients K_x, K_y and K_z, with dimension $[L^2T^{-1}]$, are parameterized by cross correlations of the velocity turbulent fluctuations (u′, v′, w′) and S′, with expressions similar to those of the turbulent or eddy kinematic viscosity coefficients developed by Osborne Reynolds in 1894 (Pritchard 1954; Bowden 1963; Lacombe 1965, and others):

$$K_x=-\frac{\langle u'S'\rangle}{\frac{\partial S}{\partial x}};K_y=-\frac{\langle v'S'\rangle}{\frac{\partial S}{\partial y}};K_z=-\frac{\langle w'S'\rangle}{\frac{\partial S}{\partial z}}. \tag{7.26}$$

It should be noted that the numerators of Eq. (7.26), multiplied by the density $(\rho\langle u'S'\rangle,\ \rho\langle v'S'\rangle,\ \rho\langle w'S'\rangle)$, have dimensions of the salt fluxes generated by turbulent or eddy diffusion. Salinity and current velocity measurements in the James River estuary (Virginia, USA) taken over several tidal cycles in a cross section, gave the following results (Pritchard 1954):

- The horizontal advective (ρSu) and the vertical non-advective $(\rho\langle w'S'\rangle)$ fluxes of salt were the most important factors in maintaining the salt balance.
- The mean vertical advective (ρSw) and the horizontal non-advective $(\rho\langle u'S'\rangle)$ fluxes were of secondary importance, but still significant and small, respectively.
- In addition, the vertical non-advective flux $(\rho\langle w'S'\rangle)$ of salt is partly related to the magnitude of the oscillatory tidal currents, and is dependent on the vertical salinity stratification.

Pritchard's work confirmed the hypothesis that the mixing process in an estuary is mainly related to the tidal forcing, and suggested the possibility of calculating the turbulent diffusion terms using a modified version of Eq. (7.26) for a laterally homogeneous estuary, taking into account its width variation.

A conservation equation, similar to (7.24), may also be used in the mathematical simulation of a conservative concentration of a property dissociated in an estuary. If C = C(x, y, z, t) denotes the property's concentration, $[C] = [ML^{-3}]$, the conservation equation is:

$$\frac{\partial C}{\partial t} + u\frac{\partial C}{\partial x} + v\frac{\partial C}{\partial y} + w\frac{\partial C}{\partial z} = \frac{\partial}{\partial x}(K_{xC}\frac{\partial C}{\partial x}) + \frac{\partial}{\partial y}(K_{yC}\frac{\partial C}{\partial y}) + \frac{\partial}{\partial z}(K_{zC}\frac{\partial C}{\partial z}), \tag{7.27}$$

where K_{xC}, K_{yC} and K_{zC} are the kinematic eddy diffusion coefficients of the property, whose theoretical determinations are given by similar expressions as presented in Eqs. (7.26).

The quantity C may also represent the concentration of suspended cohesive or non-cohesive sediments, transported by velocities along the bottom which usually are very low in comparison to the velocities in the upper layer. For non-conservative substances, such as nutrients, dissolved oxygen, domestic effluents and radioactive substances, an additional term must be included in Eq. (7.27) to analytically represent sources and/or sinks. If the property has a first order exponential decay, its mathematical simulation is given by $\frac{\partial C}{\partial t} = -k_C C$, where k_C is a proportionality coefficient with dimension $[T^{-1}]$. In any case, it is important to remember that the velocity components of the advective terms (u, v, w) and the solution of Eq. (7.27), C = C(x, y, z, t), represent average values for a time interval Δt, which must be long enough to eliminate the turbulent fluctuations.

The partial differential Eqs. (7.24) or (7.25) and (7.27), which have the salinity, S = S(x, y, z, t), and concentration, C = C(x, y, z, t) fields as unknowns, are named as Eulerian formulations. Mathematically, the solutions may be obtained if the turbulent diffusion coefficients of these properties and the velocity field ($\vec{v}$) are known quantities. The solutions of these equations are also dependent on the initial and boundary conditions and the estuary geometry.

7.3 Integral Formulas: Mass and Salt Conservation Equations

7.3.1 Volume Integration

When the solution to an estuarine physics problem for a water body doesn't require detailed knowledge of the interior domain, a simple solution may be obtained by applying the continuity and salt conservation equations (or any other conservative property) integrated with respect to the volume domain.

To start, let us integrate the differential expression of the continuity Eq. (7.19) with respect to a small volume element (ΔV) limited by a closed continuous surface (ΔA). Under the assumption that all regularity conditions necessary for the application of Gauss's divergence theorem are met, the volume integral may be transformed into an integral in the area,

$$\iiint\limits_{\Delta V}(\nabla \bullet \vec{v})dV = \iint\limits_{\Delta A}(\vec{v} \bullet \vec{n})dA = 0. \qquad (7.28)$$

In this equation, the unitary vector $\vec{n}$ ($|\vec{n}| = 1$) is normal to the area ΔA and it is oriented from the interior of the volume ΔV to the exterior of the closed surface. This equation may be generalized to a finite volume, V, of the estuarine water mass, limited by an area A, then

$$\iiint\limits_{V}(\nabla \bullet \vec{v})dV = \iint\limits_{A}\vec{v} \bullet dA = 0. \qquad (7.29)$$

The integral over the area in Eq. (7.29) is volume transport $[L^3T^{-1}]$, through the geometric limits of volume V, enclosed by the area, A, and according to these conservation equations, are equal to zero. In the SI system of units this transport is calculated in $m^3\ s^{-1}$.

To obtain the integrated form of the salt conservation Eq. (7.24), for a differential water volume forced only by the advective process, the salinity and density fields in the volume, V, of the estuarine water mass must be, by hypothesis, stationary fields representing mean values during complete tidal cycles, and the salt conservation equation is reduced to the simplest differential expression:

$$\nabla \bullet \rho S\vec{v} = 0. \qquad (7.30)$$

Integrating this equation in the geometric volume, V, of the estuarine mixing zone (MZ), and applying the Gauss theorem, it follows that:

$$\iiint\limits_{V}(\nabla \bullet \rho S\vec{v})dV = \iint\limits_{A}\rho S\vec{v} \bullet \vec{n}dA = 0. \qquad (7.31)$$

The surface integral in the second member of this equation physically represents the advective salt transport $[MT^{-1}]$ (kg s^{-1} in the SI system of units) through the surface area A, enclosed by the geometric volume of the estuarine water mass. As this salt transport is equal to zero, the mass of salt entering the volume, V, is counterbalanced by an equal value exiting, according to the principle of conservation of salt.

Examples of the practical application of the conservation principles of mass and salt (Eqs. 7.19 and 7.24) applied on a relatively small scale are presented according to Officer (1976) and Team course (2001). Let us take a water volume, V, bounded by two vertical transverse sections, where areas A_1 and A_2 have uniform mean salinities S_1 and S_2, respectively. Water enters the channel through A_1 and exits through A_2, with mean velocities $\vec{v_1} = u_1\vec{n}$ and $\vec{v_2} = u_2\vec{n}$, respectively (Fig. 7.1). Two sources of input or output water will be considered: the fresh river discharge

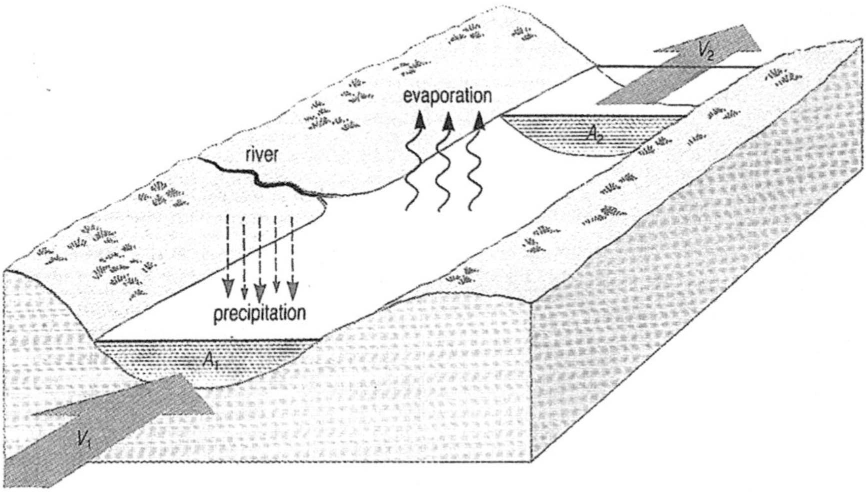

Fig. 7.1 Schematic diagram of a stationary estuarine water body bounded by vertical transverse sections, A_1 and A_2, and by the free surface (A_{su}) and the bottom. S_1, S_2, V_1 and V_2 are mean salinities and velocities, respectively (according the Team Course 2001)

(Q_f) and the input or outflow of fresh water through its free surface (A_{su}) by precipitation (P) and evaporation (E_v). Sources of bottom spring water and run-off will be disregarded.

Denoting $\bar{r}$ as the mean value of the difference evaporation to precipitation (E-P) per unit of time, the product $\vec{r}A_{su}$ is the volume transport of fresh water through the surface layer. Then, if $E_v > P$ or $E_v < P$, it follows that $\bar{r} > 0$ or $\bar{r} < 0$, respectively, and the volume transport across the surface area A_{su} is exiting or entering the system, respectively; when there is a counterbalance of precipitation and evaporation, the transport across the free surface is null ($\bar{r}A_{su} = 0$). However, it should be noted that usually $Q_f \gg \bar{r}A_{su}$, with the exception of estuaries in dry regions where hypersaline (or negative) estuaries are formed.

Under the assumption of steady-state conditions for the inflow and outflow of fresh water, we may apply the integrated continuity Eq. (7.29). Taking into account the particular geometry (Fig. 7.1) and the kinematic boundary condition ($\vec{v} \bullet \vec{n} = 0$), the scalar product $\vec{v} \bullet \vec{n} = 0$ is only different from zero on the transverse sections A_1 and A_2. Thus, applying the integrated formulation of continuity equation, the total volume of water entering this portion of the channel may be equal to the total volume leaving, resulting in the following expression:

$$-V_1A_1 + V_2A_2 + (Q_f + \bar{r}A_{su}) = 0. \tag{7.32}$$

Taking into account the geometric characteristics, the velocity and the salinity fields, applying Eq. (7.31) will give the following balance of the salt transport through the closed surface:

$$-\rho_1 S_1 V_1 A_1 + \rho_2 S_2 V_2 A_2 = 0, \tag{7.33}$$

because the salt transport through the bottom and the free surface are equal to zero. Under the assumption that the density variation may be disregarded in the salt balance ($\rho_1 \approx \rho_2$) this equation may be rewritten as:

$$-S_1 V_1 A_1 + S_2 V_2 A_2 = 0. \tag{7.34}$$

This approximation isn't restrictive, because salinity is a parameter that can be measured, and the density may be calculated with the equation of state of seawater. Using SI units for velocity, area, and $\bar{r}$, and psu units ($S \times 10^{-3}$) for salinity, the parcels of Eqs. (7.32) and (7.34) for the volume and salt transports are calculated in $m^3\ s^{-1}$ and $kg\ s^{-1}$, respectively.

As salinities S_1 and S_2 at the transverse sections A_1 and A_2, respectively, and the volume transports are known, Eqs. (7.32) and (7.34) may be solved for the velocities averages velocities in the transverse sections (V_1 and V_2), and for the volume transports through the cross sections A_1 and A_2 (V_1A_1 and V_2A_2), and the results are

$$V_1 = \frac{(\overline{r}A_{su} + Q_f)S_2}{A_1(S_2 - S_1)}, \quad \text{and} \quad T_{V1} = V_1 A_1 = \frac{(\overline{r}A_{su} + Q_f)S_2}{(S_2 - S_1)}, \tag{7.35}$$

and

$$V_2 = \frac{(\overline{r}A_{su} + Q_f)S_1}{A_2(S_2 - S_1)}, \quad \text{and} \quad T_{V2} = V_2 A_2 = \frac{(\overline{r}A_{su} + Q_f)S_1}{(S_2 - S_1)}, \tag{7.36}$$

With all quantities in the second member of these equations in units of the SI system, the velocity components (V_1 and V2) and the volume transports (T_{V1} and T_{V2}) are expressed in $m\ s^{-1}$ and $m^3\ s^{-1}$, respectively.

In extreme conditions where $Q_f = 0$, with evaporation is greater than precipitation ($E_v > P$, $\bar{r} > 0$) and $S_2 > S_1$, analysis of the solutions (7.35 and 7.36) indicates that the velocity directions ($V_1 > 0$ and $V_2 > 0$) are in agreement with those indicated in Fig. (7.1). For this ideal system, the flow is from the regions of low salinity towards the high salinity regions, in agreement with the salinity gradient direction.

Let us now consider the opposite process, that is, the precipitation rate exceeds the evaporation ($\bar{r} < 0$), which corresponds with $P > E_v$, and seawater is diluted by fresh water. Also, with $S_2 > S_1$ it follows from Eqs. (7.35) and (7.36) that the flow is in the opposite direction to the former condition ($V_1 < 0$ and $V_2 < 0$) and opposite to the salinity gradient.

Finally, with $Q_f \rightarrow 0$ and $\bar{r} \rightarrow 0$, the residual flow and volume transport through sections A_1 and A_2 are equal to zero. Hence, the difference of $P - E_v$ determines the driving motions in water bodies at coastlines, such as choked and hyper-saline coastal lagoons.

With H_0 denoting the mean depth of the water column of the closed water body shown in Fig. 7.1, the time interval (Δt) required for its interior volume of water to be completely removed from a choked coastal lagoon may be estimated by:

$$\Delta t = \frac{H_0 A_{su}}{V_2 A_2} = \frac{H_0 (S_2 - S_1)}{\bar{r} S_1}. \tag{7.37}$$

With the variables in this equation expressed in the SI units, the time interval Δt is calculated in seconds, and usually this quantity is converted in hours or days.

Consider now a similar problem, but for a salt wedge estuary in steady-state condition. The dynamics of this estuary is dominated by the river discharge, and the vertical salt distribution is generated by the entrainment. The continuity and the salt conservation equations integrated with respect to the volume (Eqs. 7.29 and 7.31) may be applied. Because the mean flow is one-dimensional in these equations, it will be considered along the longitudinal axis (Ox), oriented down-estuary (Fig. 7.2). This figure indicates the upper and lower salt-wedge transverse sections A_2 and A_1, and their mean velocity values are indicated by u_s and u_i, respectively. The mean salinities in these upper and lower sections are also considered as known, and are indicated by S_s and S_i, respectively. In Chap. 3 (Sect. 3.2) we have seen that for this estuarine type, the following inequality holds: $S_i \gg S_s$.

Hence, the integrated equations of continuity (Eq. 7.29) and the corresponding principle of salt conservation (Eq. 7.31) may be applied in the calculation of the intensity of the velocities u_s and u_i and the associated volume transports. Taking into account the MZ geometry and the kinematic boundary condition ($\vec{v} \bullet \vec{n} \neq 0$

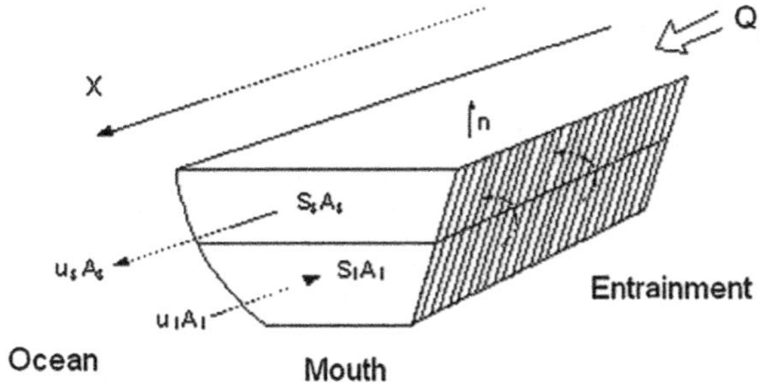

Fig. 7.2 Schematic diagram of a bidirectional motion through a vertical section localized at the mouth of a salt wedge estuary. The index of the quantities $A_{s,i}$, $u_{s,i}$ and $S_{s,i}$ indicate the areas of the *upper* (s) and *lower* (i) sections, and the corresponding mean velocity and salinity values, respectively. The unit vector, $\vec{n}$, (not shown) is normal to the closed surface oriented positively outward of the volume

only in the transverse sections A_s and A_i), from the conservation equations we have the following relationships:

$$\iint_A \vec{v} \bullet \vec{n}.dA = u_sA_s - u_iA_i - Q_f = 0, \tag{7.38}$$

and

$$\iint_A \rho S\vec{v} \bullet \vec{n}.dA = \rho_s S_s u_s A_s - \rho_i S_i u_i A_i = 0. \tag{7.39}$$

Disregarding the density variations ($\rho_s \approx \rho_i$) the equation system (7.38 and 7.39) is reduced to:

$$u_sA_s - u_iA_i = Q_f, \tag{7.40}$$

and

$$S_s u_s A_s = S_i u_i A_i. \tag{7.41}$$

If the mean salinity values (S_s and S_i), the area of the vertical sections (A_s and A_i), and the river discharges are all known, this equation system has only two unknowns, u_s and u_i, and the solutions are:

$$u_s = \frac{S_iQ_f}{A_s(S_i - S_s)} = \frac{Q_f}{A_s(1 - \frac{S_s}{S_i})}, \tag{7.42}$$

and

$$u_l = \frac{S_sQ_f}{A_i(S_i - S_s)} = \frac{Q_f}{A_i(\frac{S_i}{S_s} - 1)}. \tag{7.43}$$

With these results, it is also possible to calculate the transport of volumes (u_sA_s and u_iA_i) and salt ($u_sS_sA_s$ and $u_iS_i\,A_i$) in the upper and lower layers, respectively.

This practical application of the principles of continuity and conservation of salt integrated with respect to the volume exemplify how it is possible to calculate the mean velocities in transverse sections and the corresponding values of the volume and salt transports of a salt wedge estuary, when its geometry and scalar properties (salinity and river discharge) are known. In relation to Eqs. (7.42 and 7.43), which are used to calculate the velocities u_s and u_i, it is possible to observe that, even if A_s and A_i have the same areas, the velocity of the upper layer is always higher than the lower layer velocity ($u_s > u_i$) because $S_s \ll S_i$. Hence, this result is in agreement with the salt wedge estuary dynamics.

A numerical application of Eqs. (7.42 and 7.43) is presented to theoretically estimate the vertical velocity profiles at the mouth of the Fraser River estuary

(British Columbia, Canada). This estuary is classified as salt wedge (or type 4, according the Stratification-circulation diagram). The Fraser river estuary is a typical example of salt wedge estuary in a region of meso-tides. The following data were estimated from the article of Geyer and Farmer (1989): mean river discharge $Q_f = 3000\ m^3\ s^{-1}$, geometry at the upper and lower sections $A_s = 3750\ m^2$ and $A_i = 4500\ m^2$, and salinities $S_s = 14.0°/_{oo}$ and $S_i = 30.0°/_{oo}$, respectively, representing mean values at the upper and lower sections of the halocline, respectively. The estimated vertical profile of salinity and the theoretical simulations of the vertical velocity profile are presented in Fig. 7.3; the mean velocities at the upper and lower vertical sections are $u_s \approx 1.5\ m\ s^{-1}$ and $u_i \approx -0.6\ m\ s^{-1}$, respectively. The discontinuity of the vertical salinity profile at depth z = 5 m, generated similar characteristics in the velocity profile, because theoretical equations don't include dissipative forces due to the internal friction and at the bottom.

From the results of the velocity, the volume transport was calculated and its landward and seaward values were $Q_s = 5525\ m^3\ s^{-1}$ and $Q_i = -2525\ m^3\ s^{-1}$, respectively. Hence, the volume transport is in balance with the river discharge. The increase in the volume transport seaward, in comparison with the river discharge (Q_f), clearly indicates the influence on the upper transport, forced by the entrainment of seawater into the layer above the halocline.

We leave it to the reader to demonstrate the following dot marks:

- Solutions (7.42) and (7.43) identically satisfy the principles of mass and salt conservations;
- The mean speed at the mouth transverse section is calculated by: $\frac{Q_f}{(A_s + A_i)} = u_f$;
- The salt transports may be determined by $\rho_i u_i S_i A_i$ and $\rho_s u_s S_s A_s$, landward and seaward, respectively.

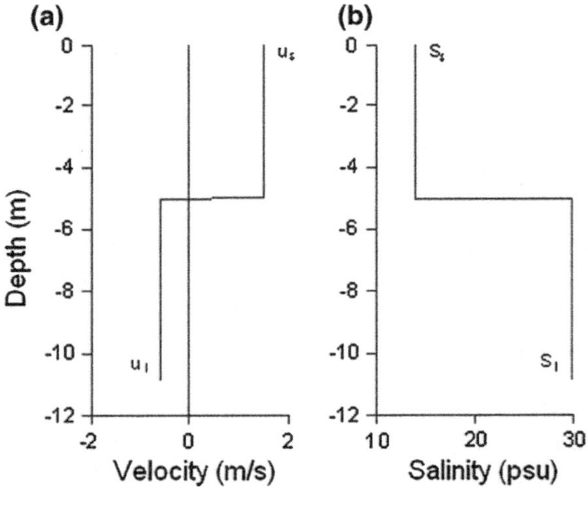

Fig. 7.3 Vertical theoretical velocity profiles of a salt wedge estuary (**a**). Estimates of salinity profile from experimental data (**b**), obtained near the estuary mouth

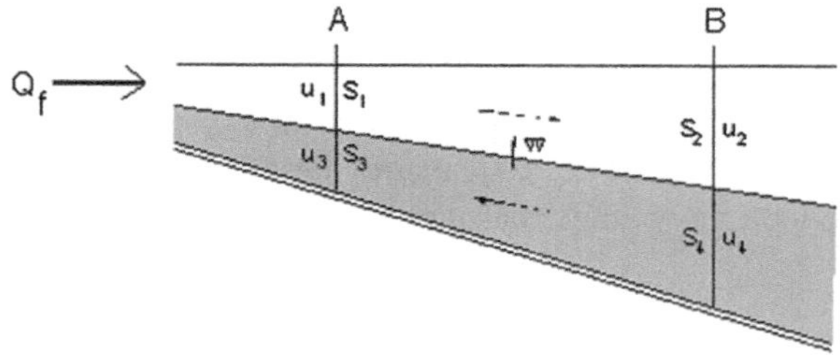

Fig. 7.4 Schematic diagram of bidirectional motion and salt transport through vertical sections, A and B, of a highly stratified estuary. The indexes 1–2 and 3–4 indicate physical properties in the *upper* and *lower* layers, respectively, bounded by the halocline (adapted from Defant 1961)

The classical Knudsen hydrographic theorem was presented at the beginning of the 19th century, stating relationships between a known salinity field and the velocity under stationary conditions. Let us assume that the estuary is highly stratified and its geometry and salinity are known. Under these conditions, the mean longitudinal motion, in relation to the Ox axis, is bidirectional in two layers separated by a sharp halocline; seaward and landward motions are in the surface and lower layers, respectively (Fig. 7.4).

Applying the continuity and salt conservation Eqs. (7.29 and 7.31) to the volume between the transverse sections (A, B), taking into account the channel geometry and the areas of the upper (A_1 + A_3) and lower (A_2 + A_4) layers, and knowing the salinities S_1 and S_3 (at A_1 and A_3), and S_2 and S_4 (at A_2 and A_4), the following volume and salt transport balances may be written:

$$-u_1A_1 + u_3A_3 = -Q_f, \tag{7.44}$$

$$u_2A_2 - u_4A_4 = Q_f, \tag{7.45}$$

$$-u_1A_1 + u_3A_3 + u_2A_2 - u_4A_4 = 0, \tag{7.46}$$

and

$$-S_1u_1A_1 + S_3u_3A_3 + S_2u_2A_2 - S_4u_4A_4 = 0, \tag{7.47}$$

with the approximation $\rho_1 \approx \rho_2 \approx \rho_3 \approx \rho_4$.

As the net salt transport across the transversal section A (sub-sections A_1 and A_3) and section B (sub-sections A_2 and A_4) must be equal to zero, the following equalities may be written from Eq. (7.47):

$$S_1u_1A_1 = S_3u_3A_3, \tag{7.48a}$$

and

$$S_2u_2A_2 = S_4u_4A_4. \tag{7.48b}$$

Equations (7.44), (7.45), (7.48a and 7.48b) form a system of four equations with four unknowns u_1, u_2, u_3 and u_4. Calculating these unknowns and multiplying by areas we then obtain the volume transports:

$$u_1 = \frac{S_3Q_f}{A_1(S_3 - S_1)}, \rightarrow u_1A_1 = \frac{S_3Q_f}{(S_3 - S_1)}, \tag{7.49a}$$

$$u_3 = \frac{S_1Q_f}{A_3(S_3 - S_1)}, \rightarrow u_3A_3 = \frac{S_1Q_f}{(S_3 - S_1)}, \tag{7.49b}$$

$$u_2 = \frac{S_4Q_f}{A_2(S_4 - S_2)}, \rightarrow u_2A_2 = \frac{S_4Q_f}{(S_4 - S_2)}, \tag{7.49c}$$

and

$$u_4 = \frac{S_2Q_f}{A_4(S_4 - S_2)}, \rightarrow u_4A_4 = \frac{S_2Q_f}{(S_4 - S_2)}. \tag{7.49d}$$

Then, with knowledge hydrologic and hydrographic data it is possible to calculate the velocity components (u_i, i = 1, 2, 3, 4), the volume (u_iA_i, i = 1, 2, 3, 4), and salt transports ($u_iA_iS_i$, i = 1, 2, 3, 4) across the upper and lower sections shown in Fig. 7.4. As $S_3 > S_1$ and $S_4 > S_2$, it follows from these equations that the velocity and volume transport modules are positive. As the flow direction has already been taken into account in the water column stratification ($S_1 < S_3$), from Eqs. (7.49a and 7.49b) it follows that $u_1 > u_3$, and the velocity in the upper layer is higher than the lower layer velocity. Also, if $A_2 = A_3$, from Eqs. (7.49c and 7.49d) it follows that $u_2 > u_4$. These theoretical inequalities, between the mean speeds in the upper and lower layers separated by the halocline, may be verified experimentally.

Although the Knudsen hydrographic theorem only takes into account the advective process, it is a good approximation for highly stratified and salt wedge estuaries, because vertical mixing due to turbulent diffusion is suppressed by the entrainment. This theorem has been applied by Scandinavian oceanographers in studies of the circulation in fjord type estuaries, and some examples may be found in Defant (1961) and Dyer (1973). To estimate the areas A_i (i = 1, 3) and A_j(j = 2, 4) usually the interface between the upper and lower layers is taken as the mean depth of the halocline.

According to Geyer (2010), let us make as an exercise the following simplification of the original Knudsen hydrographic theorem, displacing the cross-section areas at the positions A and B (Fig. 7.4) towards the estuary head and mouth, respectively, thus, at the new section A position the velocity and salinity have the following values: (i) $u_1 = u_f$ and $S_1 = S_f = 0$, and there will no more the quantities S_3 and u_3; ii) at the section B, now located at the estuary mouth, its properties above and below the halocline will remains with the same previous notations. Thus, applying the volume and salt transport conservations equations the following expressions of two equations with the unknowns Q_2 (or u_2) and Q_4 (or u_4) are written as:

$$Q_2 = Q_f + Q_4, \quad \text{and} \quad u_2A_2S_2 = u_4A_aS_4, \rightarrow Q_2S_2 = Q_4S_4, \tag{7.50a}$$

with the simplification $\rho_2 \approx \rho_4$. Solving this system of equations we find the following expressions to calculate volume transports and velocities and at the estuary head (A) and at position B:

$$Q_2 = u_2A_2 = \frac{S_4}{S_4 - S_2}Q_f, \quad \text{and} \quad , Q_4 = u_4A_4 = \frac{S_2}{S_4 - S_2}Q_f, \tag{7.50b}$$

or

$$u_2 = \frac{1}{A_2}\frac{S_4}{(S_4 - S_2)}Q_f, \quad \text{and} \quad , u_4 = \frac{1}{A_4}\frac{S_2}{(S_4 - S_2)}Q_f. \tag{7.50c}$$

We leave to the reader to demonstrate that these solutions satisfy the volume and salt transport conservation.

To establish the horizontal continuity of the flow, as indicated in Fig. (7.4), an upward mean velocity ($\overline{w}$), generated by entrainment, is necessary across the halocline. Thus, if A_h indicates the horizontal area of the halocline, the associate entrained volume transport is calculated by ($\overline{w}A_h$), which is generate by the volume transport convergence on the lower layer, and may be calculated by: $\overline{w}A_h = u_4A_4 - u_3A_3$. Then, from the volume transports calculated by Eqs. (7.52) and (7.54a, 7.54b), it follows that:

$$\overline{w}A_h = Q_f\left(\frac{S_2}{S_4 - S_2} - \frac{S_1}{S_3 - S_1}\right), \tag{7.51a}$$

and

$$\overline{w} = \frac{Q_f}{A_h}\left(\frac{S_2}{S_4 - S_2} - \frac{S_1}{S_3 - S_1}\right). \tag{7.51b}$$

In the presented theory, only the principle of mass conservation (continuity) and salt conservation in its integrated formulation were used, enabling the solutions for velocity field and transports at the boundaries of the estuary only. The driving forces were the river discharge input and the evaporation-precipitation rate, but the dissipative force (friction) and the turbulent diffusion were taken as negligible. Obtaining a solution for the inner circulation and property distributions in natural estuaries requires a complete set of differential equations, including the equations of motion, which will be presented in Chap. 8.

7.3.2 *Bi-Dimensional Formulation: Vertical Integration*

Estuaries are transitional water bodies with free surface and morphologic characteristics which may vary from a simply geometry, such as a channel, to complex system with a net of interconnected channels. The tridimensional equations of continuity and salt conservation have already been presented (Eqs. 7.19 and 7.24). Under the assumptions that the turbulent coefficients of salt diffusion and the velocity components are known, Eq. (7.24) may be solved to calculate the salinity field, S = S(x, y, z, t). However, its analytical solution is extremely difficult, perhaps even impossible, particularly for complex geometries.

Coastal plain estuaries which have a longitudinal channel geometry, low river discharge and high tidal amplitude are practically well-mixed (type 1 or C), and variations in velocity and property concentrations mainly occur in the transverse sections (plane Oxy). However, when estuaries are forced by moderate or high river discharge, variations in property concentrations may occur mainly in the Oxz plane, such as in partially mixed and salt wedge estuaries (types 2 and 4, or A and B). With these particular geometries and driving forces, the conservation equations may be simplified to two dimensions.

Let us now present the deduction of the two-dimensional continuity equation from its three-dimensional formulation (Eq. 7.19), which is often used in problems related to well-mixed estuaries. Properties variations in these estuaries are mainly in the Ox and Oy directions, oriented according to the reference system in Fig. 7.5.

To eliminate variations in the Oz direction, it is sufficient to integrate the continuity equation using the local depth $z = -H_0(x, y)$ and the ordinate of the free surface $z = \eta(x, y, t)$ as limits, disregarding the large-scale temporal depth variations due to erosion and sedimentation,

$$\int_{-H_0}^{\eta} \left(\frac{\partial u}{\partial x}\right)dz + \int_{-H_0}^{\eta} \left(\frac{\partial v}{\partial y}\right)dz + w|_{\eta} - w|_{-H_0} = 0. \tag{7.52}$$

In this equation, $w|_{\eta} = w(x, y, \eta, t)$ and $w|_{-H0} = w(x, y, -H_0, t)$ are values of the vertical velocity component at the surface and on the bottom, respectively. As its integration limits are functions of x, y and t, it is necessary simplify the equation to a more convenient expression for practical applications, using the Leibnitz rule of an integral derivation[2] (Severi 1956, p. 354):

[2]When the estuary bottom is plane (H_0 = const.), and due to the very long tidal wave, the tidal elevation may be considered uniform along the estuary, $\eta = \eta(t)$, and it is possible to change the order of the integral operator and the derivative. In these conditions $w|_{-H0} = 0$ and $w|_{\eta} = d\eta/dt = \partial\eta/\partial t$ are the kinematic boundary conditions.

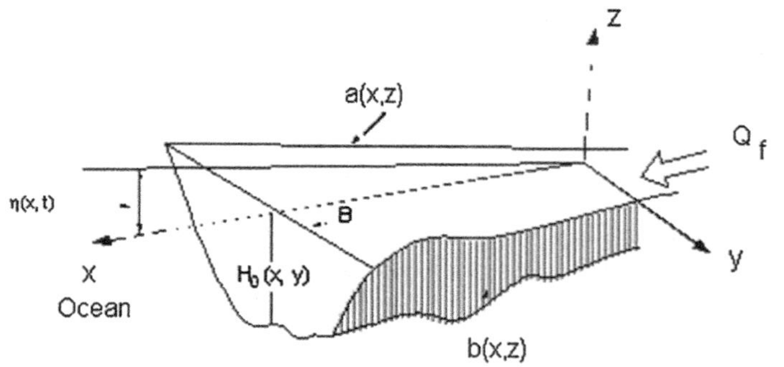

Fig. 7.5 Geometric limits of an estuary. The coordinates on the surface and bottom are indicated by $z = \eta(x, y, t)$ and $z = H_0(x, y)$, and from the right to *left*, a(x, z) and b(x, z) are lateral boundaries

$$\int_{-H_0}^{\eta}\left(\frac{\partial u}{\partial x}\right)dz = \frac{\partial}{\partial x}\left(\int_{-H_0}^{\eta} u dz\right) + u|_{-H_0}\frac{\partial(-H_0)}{\partial x} - u|_{\eta}\frac{\partial \eta}{\partial x}, \tag{7.53a}$$

and

$$\int_{-H_0}^{\eta}\left(\frac{\partial v}{\partial y}\right)dz = \frac{\partial}{\partial y}\left(\int_{-H_0}^{\eta} v dz\right) + v|_{-H_0}\frac{\partial(-H_0)}{\partial y} - v_{\eta}\frac{\partial \eta}{\partial y}, \tag{7.53b}$$

where $u|_\eta = u(x, y, \eta, t)$, $v|_\eta = v(x, y, \eta, t)$, $u|_{-H0} = u(x, y, -H_0, t)$ and $v|_{-H0} = v(x, y, -H_0, t)$ are values of velocity horizontal components in the free surface ($z = \eta$) and on the bottom ($z = -H_0$), respectively.

By substituting expressions (7.53a, 7.53b) into Eq. (7.52), and taking into account the vertical velocity components generated by the bottom topography and the sea-surface, because $H_0 = H_o(x, y)$ and $\eta = \eta(x, y, t)$, the following kinematic boundary conditions must be imposed:

$$w|_{-H_0} = u|_{-H_0}\frac{\partial(-H_0)}{\partial x} + v|_{-H_0}\frac{\partial(-H_0)}{\partial y}, \tag{7.54a}$$

and

$$w|_{\eta} = u|_{\eta}\frac{\partial \eta}{\partial x} + v|_{\eta}\frac{\partial \eta}{\partial y} + \frac{\partial \eta}{\partial t}. \tag{7.54b}$$

The final result is the expression,

$$\frac{\partial}{\partial x}(\int_{-H_0}^{\eta} udz) + \frac{\partial}{\partial y}(\int_{-H_0}^{\eta} vdz) + \frac{\partial \eta}{\partial t} = 0. \tag{7.55}$$

In this equation, the integrands u and v are, by hypothesis, independent of the depth. Finalizing the integration, the continuity equation in two dimensions for an estuary may be given by[3]:

$$\frac{\partial(uh)}{\partial x} + \frac{\partial(vh)}{\partial y} + \frac{\partial \eta}{\partial t} = 0, \text{or } \nabla_H \bullet h\vec{v} = -\frac{\partial \eta}{\partial t}, \tag{7.56a}$$

or, when the longitudinal depth variation ($\partial h/\partial x$) may be disregarded,

$$h\frac{\partial u}{\partial x} + \frac{\partial(vh)}{\partial y} + \frac{\partial \eta}{\partial t} = 0. \tag{7.56b}$$

where,

$$\int_{-H_0}^{\eta} dz = \eta + H_0 = h(x, y, t). \tag{7.56c}$$

The quantity $h(x, y, t) = \eta(x, y, t) + H_0(x, y)$ is the thickness of the water column and holds the identity $\partial h/\partial t = \partial \eta/\partial t$.

The bi-dimensional continuity Eq. (7.56a) has the following physical interpretation: the divergence ($\nabla_H \bullet h\vec{v} > 0$) or the convergence ($\nabla_H \bullet h\vec{v} < 0$) must be compensated by a decrease ($\partial h/\partial t < 0$) or an increase ($\partial h/\partial t > 0$) of the thickness of water layer, respectively, where the vector $h\vec{v}$ is the volume transport vertically integrated by width unity. As a result of the vertical integration, each term of the continuity equation has dimension of velocity [LT^{-1}], and the coordinate z was substituted by a geometric characteristic of the estuary (the depth, h).

By analogy, the vertically integrated deduction of the salt conservation equation results from the integration of Eq. (7.25) in the following limits: $z = -H_0(x, y)$ and $z = \eta(x, y, t)$ at the bottom and the free surface, respectively. Then, we have the following expression:

$$\int_{-H_0}^{\eta} (\frac{dS}{dt})dz = \int_{-H_0}^{\eta} [\frac{\partial}{\partial x}(K_x\frac{\partial S}{\partial x}) + \frac{\partial}{\partial y}(K_y\frac{\partial S}{\partial y})]dz + \int_{-H_0}^{\eta} \frac{\partial}{\partial z}(K_z\frac{\partial S}{\partial z})dz, \tag{7.57}$$

[3]This operation is equivalent to the *Mean Value Theorem*.

and the integrand of the term on the left-hand-side is the total derivative of the salinity (local + advective variations). By completing the integration of the last term on the right-hand-side of the equation, we have:

$$\int_{-H_0}^{\eta} \frac{\partial}{\partial z}(K_z \frac{\partial S}{\partial z})dz = K_z \frac{\partial S}{\partial z}|_{\eta} - K_z \frac{\partial S}{\partial z}|_{-H_0}. \quad (7.58)$$

As K_z is the kinematic turbulent diffusion coefficient of salt $[K_z] = [L^2T^{-1}]$, the terms of this equation have dimensions of velocity $[LT^{-1}]$, which may be interpreted physically as salt flux per density unit, through the surface ($z = \eta$) and the bottom ($z = -H_0$). As these salt fluxes must be zero within the estuary's geometric boundaries, the last parcel in the right-hand-side of Eq. (7.57) is equal to zero.

The vertical integration of the total derivative in the left-hand-side of the Eq. (7.57) is given by:

$$\int_{-H_0}^{\eta} \frac{dS}{dt}dz = \int_{-H_0}^{\eta} \frac{\partial S}{\partial t}dz + \int_{-H_0}^{\eta} \frac{\partial (uS)}{\partial x}dz + \int_{-H_0}^{\eta} \frac{\partial (vS)}{\partial y}dz + \int_{-H_0}^{\eta} \frac{\partial (wS)}{\partial z}dz. \quad (7.59)$$

Let us again apply the Leibnitz integration rule to the first three terms of the right-hand-side of this equation:

$$\int_{-H_0}^{\eta} \frac{\partial S}{\partial t}dz = \frac{\partial}{\partial t}(\int_{-H_0}^{\eta} Sdz) - S|_{\eta} \frac{\partial \eta}{\partial t}, \quad (7.60a)$$

$$\int_{-H_0}^{\eta} \frac{\partial (uS)}{\partial x}dz = \frac{\partial}{\partial x}(\int_{-H_0}^{\eta} uSdz) - uS|_{\eta} \frac{\partial \eta}{\partial x} + uS|_{-H_0} \frac{\partial (-H_0)}{\partial x}, \quad (7.60b)$$

and

$$\int_{-H_0}^{\eta} \frac{\partial (vS)}{\partial y}dz = \frac{\partial}{\partial y}(\int_{-H_0}^{\eta} vSdz) - vS|_{\eta} \frac{\partial \eta}{\partial y} + vS|_{-H_0} \frac{\partial (-H_0)}{\partial y}. \quad (7.60c)$$

Finally, integrating the last term of the right-hand-side of Eq. (7.59) gives,

$$\int_{-H_0}^{\eta} \frac{\partial (wS)}{\partial z}dz = wS|_{\eta} - wS|_{-H_0}. \quad (7.61)$$

In these equations, the quantities $S|\eta = S(x, y, \eta, t)$ and $S|_{-H0} = S(x, y, -H_0, t)$ are the salinity values at the surface and bottom, respectively.

At this stage the first two terms of the right-hand-side of Eq. (7.57), relating to the lateral influence of the turbulent diffusion, are still missing from the vertical integration of Eq. (7.57). Applying the Leibnitz rule to these terms, and they may be rewritten as:

$$\int_{-H_0}^{\eta} \frac{\partial}{\partial x}(K_x \frac{\partial S}{\partial x})dz = \frac{\partial}{\partial x}(\int_{-H_0}^{\eta} K_x \frac{\partial S}{\partial x}dz) - K_x \frac{\partial S}{\partial x}|_\eta \frac{\partial \eta}{\partial x} + K_x \frac{\partial S}{\partial x}|_{-H_0} \frac{\partial(-H_0)}{\partial x}, \tag{7.62a}$$

and

$$\int_{-H_0}^{\eta} \frac{\partial}{\partial y}(K_y \frac{\partial S}{\partial y})dz = \frac{\partial}{\partial y}(\int_{-H_0}^{\eta} K_y \frac{\partial S}{\partial y}dz) - K_y \frac{\partial S}{\partial y}|_\eta \frac{\partial \eta}{\partial y} + K_y \frac{\partial S}{\partial y}|_{-H_0} \frac{\partial(-H_0)}{\partial y}. \tag{7.62b}$$

Finally, substituting Eqs. (7.60a, b, c) and (7.62a, b) into Eq. (7.57), and taking into account that:

- The diffusive salt flux in the estuary boundaries are zero;

$$K_x \frac{\partial S}{\partial x}|_{-H_0\eta} = K_y \frac{\partial S}{\partial y}|_{-H_0\eta} = K_z \frac{\partial S}{\partial z}|_{-H_0\eta} = 0, \tag{7.63a}$$

- The kinematic boundary conditions indicated in Eqs. (7.54a, b) are valid when multiplied by $S|_{-Ho}$ and $S|\eta$:

$$wS|_{-H_0} = S|_{-H_0}[u|_{-H_0} \frac{\partial(-H_0)}{\partial x} + v|_{-H_0} \frac{\partial(-H_0)}{\partial y}], \tag{7.63b}$$

and

$$wS|_\eta = S|_\eta[u|_\eta \frac{\partial \eta}{\partial x} + v|_\eta \frac{\partial \eta}{\partial y} + \frac{\partial \eta}{\partial t}]. \tag{7.63c}$$

Then, the following vertically integrated formulation of the continuity equation has been obtained:

$$\frac{\partial}{\partial t}\left(\int_{-H_0}^{\eta} S dz\right) + \frac{\partial}{\partial x}\left[\int_{-H_0}^{\eta} (uS) dz\right] + \frac{\partial}{\partial y}\left[\int_{-H_0}^{\eta} (vS) dz\right]$$
$$= \frac{\partial}{\partial x}\left[\int_{-H_0}^{\eta} \left(K_x \frac{\partial S}{\partial x}\right) dz\right] + \frac{\partial}{\partial y}\left[\int_{-H_0}^{\eta} \left(K_y \frac{\partial S}{\partial y}\right) dz\right]. \tag{7.64}$$

Imposing the conditions that the estuary is well-mixed (vertically homogeneous), the quantities u, S, uS, vS, $K_x \frac{\partial S}{\partial x}$ and $K_y \frac{\partial S}{\partial y}$ are independent of depth, and, taking into account that the integral of the differential dz in the limits $z = -H_0$ and $z = \eta$ is equal to the local depth, $h = H_0 + \eta$, the integration of Eq. (7.64), yields the expression for the salt conservation equation:

$$\frac{\partial(Sh)}{\partial t} + \frac{\partial(uSh)}{\partial x} + \frac{\partial(vSh)}{\partial y} = \frac{\partial}{\partial x}\left(K_x h \frac{\partial S}{\partial x}\right) + \frac{\partial}{\partial y}\left(K_y h \frac{\partial S}{\partial y}\right). \tag{7.65}$$

Some simplifications may be made in this equation if, for instance, the tidal oscillation is much less than the estuary depth ($\eta \ll H_0$), $h(x, y) = H_0(x, y)$. Then the continuity Eq. (7.56a) and the salt conservation (7.65) may be rewritten as:

$$\frac{\partial(uH_0)}{\partial x} + \frac{\partial(vH_0)}{\partial y} + \frac{\partial \eta}{\partial t} = 0, \tag{7.66}$$

and

$$\frac{\partial(SH_0)}{\partial t} + \frac{\partial(uSH_0)}{\partial x} + \frac{\partial(vSH_0)}{\partial y} = \frac{\partial}{\partial x}\left(K_x H_0 \frac{\partial S}{\partial x}\right) + \frac{\partial}{\partial y}\left(K_y H_0 \frac{\partial S}{\partial y}\right). \tag{7.67}$$

Analysis of Eq. (7.66) indicates that when there is divergence ($\nabla_H \bullet H_0 \vec{v} > 0$) or convergence ($\nabla_H \bullet H_0 \vec{v} < 0$) of the volume transport along the depth axis, they must be compensated by negative or positive of the vertical velocity component on the surface ($\partial\eta/\partial t < 0$ or $\partial\eta/\partial t > 0$, respectively).

A simple salt conservation equation may be obtained from Eq. (7.67) by separating derived variables, such as $S \times h$, $S \times uh$ and $S \times vh$, and combining these variables with the continuity Eq. (7.56a) to give,

$$\frac{\partial S}{\partial t} + u\frac{\partial S}{\partial x} + v\frac{\partial S}{\partial y} = \frac{1}{h}\left[\frac{\partial}{\partial x}\left(K_x h \frac{\partial S}{\partial x}\right) + \frac{\partial}{\partial y}\left(K_y h \frac{\partial S}{\partial y}\right)\right] + \frac{\partial S}{\partial z}\Big|_{z=\eta}(P - E_V). \tag{7.68a}$$

This equation may be further simplified when h = const.,

$$\frac{\partial S}{\partial t} + u\frac{\partial S}{\partial x} + v\frac{\partial S}{\partial y} = [\frac{\partial}{\partial x}(K_x\frac{\partial S}{\partial x}) + \frac{\partial}{\partial y}(K_y\frac{\partial S}{\partial y})] + \frac{\partial S}{\partial z}|_{z=\eta}(P - E_V). \quad (7.68b)$$

In Eqs. (7.68a, b), the last term of the right-hand-side was introduced to simulate the salinity time variation due to fresh water exchanges at the free surface through precipitation (P) and evaporation (E_v) rates, $[P] = [E_v] = [LT^{-1}]$. When $P > E_V$ or $P < E_v$, there will be a fresh water source or sink at the surface ($z = \eta$), respectively; when $P = E_v$, there will be no fresh water interchanges at the free surface.

Pritchard (1954) used Eq. (7.68b) to study the salt balance in the James river coastal plain estuary (Virginia, USA) under steady-state conditions ($\partial S/\partial t = 0$) and with $P = E_v$. Based on a time series over several tidal cycles of salinity and current velocity, it was observed that the horizontal flux due to advection and the vertical non-advective salt flux were the most important factors in maintaining a simplified salt balance equation, such as:

$$u\frac{\partial S}{\partial x} + v\frac{\partial S}{\partial y} = \frac{\partial}{\partial y}(K_y\frac{\partial S}{\partial y}). \quad (7.69)$$

Equations (7.65) and (7.68a, 7.68b) are the physical-mathematical formulation of the Eulerian description of the bi-dimensional salinity field variation $S = S(x, y, t)$ in the water column. If the estuary geometry, the velocity field, the kinematic coefficients K_x and K_y, and the initial and boundary conditions are known, these equations may be integrated to calculate the mean salinity in the water column.

7.3.3 Bi-Dimensional Formulation: Lateral Integration

Let us now consider a second bi-dimensional model, under the assumption that the estuarine water mass is laterally homogeneous, which is generally the case of narrow partially-mixed estuaries; its circulation, salinity and others properties are independent on the Oy axis. In these conditions, the conservation equations of mass (7.19) and salt (7.24 or 7.25) must be integrated along the lateral direction, from $y = a(x, z)$ to $y = b(x, z)$ coordinates, and their mean values calculated. The difference being $b(x, z) - a(x, z) = B(x, z)$ will indicate the estuary width (Fig. 7.5). The dependence on the Oz direction will take into account the time variation of the lateral coordinates, because z is dependent on the time variation of the free surface, $\eta = \eta(x, t)$.

By analogy with the mathematical development used for the vertically homogeneous estuary, it is necessary to laterally integrate the tri-dimensional equations, calculate their means and then reduce them, applying the Leibnitz rule, with consideration that the integration limits of the estuary boundaries, $\underline{a}$ and $\underline{b}$ are functions of x and z, $a = a(x, z)$ and $b = b(x, z)$. Following this procedure, it is necessary to apply the boundary conditions in order to eliminate influences that are topographically generated by the bottom and margins of the estuarine channel, and impose the condition that salt advection and diffusion through its boundaries must be zero.

Beginning with the continuity Eq. (7.19), it follows that:

$$\int_a^b \frac{\partial u}{\partial x} dy + \int_a^b \frac{\partial v}{\partial y} dy + \int_a^b \frac{\partial w}{\partial z} dy = 0. \tag{7.70}$$

Because the integration of the second parcel is immediate, applying the Leibnitz rule, this equation is rewritten as:

$$\frac{\partial}{\partial x}\left(\int_a^b u dy\right) - u|_b \frac{\partial b}{\partial x} + u|_a \frac{\partial a}{\partial x} + v|_b - v|_a + \frac{\partial}{\partial z}\left(\int_a^b w dy\right) - w|_b \frac{\partial b}{\partial z} + w|_a \frac{\partial a}{\partial z} = 0, \tag{7.71}$$

where $u|_b = u(x, b, z, t)$, $v|_b = v(x, b, z, t)$, $u|_a = u(x, a, z, t)$ and $v|_a = v(x, a, z, t)$.

Due to variations in the margin geometry of the estuarine channel, the coordinates $a = a(x, z)$ and $b = b(x, z)$ may induce, due to topographic influences, transversal components of the velocity, which is necessary the imposition of the following boundary conditions:

$$v|_a = u|_a \frac{\partial a}{\partial x} + w|_a \frac{\partial a}{\partial z}, \tag{7.72a}$$

and

$$v|_b = u|_b \frac{\partial b}{\partial x} + w|_b \frac{\partial b}{\partial z}. \tag{7.72b}$$

For an estuarine channel with a uniform rectangular transversal section, the adherence principle states that at the margins $v|_a = v|_b = 0$.

Finally, with the hypothesis of lateral uniformity, or imposing the conditions that the velocity components are independent of the variable, y, it follows that the expression of the laterally integrated continuity equation is:

$$\frac{\partial(uB)}{\partial x} + \frac{\partial(wB)}{\partial z} = 0, \tag{7.73a}$$

and

$$\int_a^b dy = b(x, z, t) - a(x, z, t) = B(x, z, t). \tag{7.73b}$$

The continuity Eq. (7.73a) for a laterally homogeneous estuary indicates that the vector $\vec{v} = uB\vec{i} + wB\vec{k}$ is non-divergent ($\nabla_V \bullet B\vec{v} = 0$), and the associated current function, $\Psi = \Psi(x, z)$, is defined by,

$$\frac{\partial\psi(x,z)}{\partial x} = w(x,z)B \text{ and } \frac{\partial\psi(x,z)}{\partial z} = -u(x,z)B, \qquad (7.73c)$$

and satisfy identically the continuity equation; its dimension is equivalent to the volume transport $[\Psi(x, z)] = [L^3T^{-1}]$.

Applying an analogous procedure, the salt conservation Eq. (7.25) will be laterally integrated from $a = a(x, z)$ to $b = b(x, z)$,

$$\int_a^b \frac{dS}{dt} = \int_a^b \left[\frac{\partial}{\partial x}\left(K_x\frac{\partial S}{\partial x}\right) + \frac{\partial}{\partial y}\left(K_y\frac{\partial S}{\partial y}\right)\right]dy + \int_a^b \left(\frac{\partial}{\partial z}\left(K_z\frac{\partial S}{\partial z}\right)\right)dy, \qquad (7.74)$$

applying the Leibnitz rule and the following boundary conditions:

$$vS|_a = S|_a\left[u|_a\frac{\partial a}{\partial x} + w|_a\frac{\partial a}{\partial z}\right], \qquad (7.75a)$$

and

$$vS|_b = S|_b\left[u|_a\frac{\partial b}{\partial x} + w|_a\frac{\partial b}{\partial z}\right]. \qquad (7.75b)$$

Imposing the lateral homogeneity condition and that the salt flux, per density unity, at its geometric boundaries is zero,

$$K_x\frac{\partial S}{\partial x}|_{a,b} = K_z\frac{\partial S}{\partial z}|_{a,b} = 0, \qquad (7.75c)$$

it follows that the expression of the salt conservation equation for laterally homogeneous estuaries is:

$$\frac{\partial(BS)}{\partial t} + \frac{\partial(uBS)}{\partial x} + \frac{\partial(wBS)}{\partial z} = \frac{\partial}{\partial x}\left(BK_x\frac{\partial S}{\partial x}\right) + \frac{\partial}{\partial z}\left(BK_z\frac{\partial S}{\partial z}\right). \qquad (7.76)$$

Combining this result with the continuity Eq. (7.73a), it follows that the most usual salt conservation equation is:

$$\frac{\partial S}{\partial t} + u\frac{\partial S}{\partial x} + w\frac{\partial S}{\partial z} = \frac{1}{B}\left[\frac{\partial}{\partial x}\left(BK_x\frac{\partial S}{\partial x}\right) + \frac{\partial}{\partial z}\left(BK_z\frac{\partial S}{\partial z}\right)\right]. \qquad (7.77)$$

This equation is the Eulerian formulation of the bi-dimensional salinity variation S = S(x, z, t) of a laterally homogeneous and partially-mixed estuary (type 2 or B). If the estuary geometry, the velocity field, the turbulent diffusion coefficients and the initial and boundary conditions are known, this equation may be solved for the salinity field distribution.

Equation (7.77) may be further simplified according to the estuary characteristics; for a steady-state well-mixed estuary (type 1 or C) with a constant cross-sectional area A, a width B, and a constant kinematic turbulent diffusion coefficient (K_x), forced by the river discharge, the Eq. (7.77) is simplified to:

$$u_f \frac{dS}{dx} = K_x\left(\frac{d^2S}{dx^2}\right). \tag{7.78a}$$

Imposing the following boundary conditions: (i) $S(x)|_{x=0}=S_0$, and; (ii) $S|_{x\to\infty} = 0$, which indicate the salinity at the estuary mouth and head, respectively, the solution to this equation is:

$$S(x) = S_0 \exp\left(-\frac{u_f}{K_x}x\right), \tag{7.78b}$$

and the salinity decreases exponentially from the head down its mouth.

Another example of simplifying the Eq. (7.77) is presented for a highly stratified estuary, such as a salt wedge (type 4 or A). In this estuary, the dominant mixing process is advection, and the salinity increase across the halocline is due to entrainment. Then, the salt conservation equation is simplified to:

$$u\frac{\partial S}{\partial x} + w\frac{\partial S}{\partial z} = 0. \tag{7.79a}$$

However, with the exception of salt wedge estuaries, in the layer over the halocline, the diffusion term may still be important and holds the expression,

$$\frac{1}{B}\frac{\partial}{\partial z}\left(BK_z\frac{\partial S}{\partial z}\right), \tag{7.79b}$$

which may be further simplified if the width (B) is constant.

In coastal plain estuaries forced by macro or hyper-tides and with moderate river discharge, there will be random velocity fluctuations generated by internal turbulence and friction at the estuary boundaries. The vertical mixing of the upper and lower layers is enhanced, and the halocline is partially eroded. Thus, the salinity increase in the upper layer, increasing the seaward salt transport, while the salinity in the lower layer decreases landward, and the estuary becomes partially mixed (type 2 or B). If the estuary is laterally homogeneous and in steady-state (according to mean values over tidal cycles), the most important terms of the salt conservation Eq. (7.77) are the horizontal and vertical advection and the vertical diffusion. Then, the principle of salt conservation is reduced to:

$$u\frac{\partial S}{\partial x}+w\frac{\partial S}{\partial z}=\frac{1}{B}\frac{\partial}{\partial z}(BK_z\frac{\partial S}{\partial z}), \qquad (7.80)$$

because the non-advective horizontal term is small and may be disregarded (Pritchard 1954, 1955).

Although the two-dimensional equations in the planes (Ox, y) or (Ox, z) are simplifications of the tridimensional equation, their steady-state solutions for natural estuarine systems, from analytical and time dependent numerical methods, have some complexity. However, analytical solutions were obtained for simple geometries and steady-state conditions in the classical articles of Pritchard and Kent (1956), Rattray and Hansen (1962), Hansen and Rattray (1965), Fisher et al. (1972), Officer (1977), among others. A non-steady-state numerical solution using the natural geometry of the Potomac river estuary may be found in Blumberg (1975).

As a practical example, under steady-state conditions and with a constant width (B = const.), Eq. (7.80) will be used to calculate the salinity profile S = S(z) of a partially mixed estuary, with the following quantities known: the vertical velocity profile u = u(z) or u = u(Z), the mean longitudinal salinity gradient $(\partial S/\partial X)\approx\overline{S}_x$), and w = 0. From some algebraic rearrangement, the equation is reduced to:

$$\frac{\partial^2 S}{\partial z^2}=\frac{u(z)}{K_z}\overline{S}_x, \text{ or } \frac{d^2S}{dZ^2}=\frac{h^2}{K_z}u(Z)\overline{S}_x,$$

where Z = z/h ($0\le Z\le 1$), and with the non-dimensional depth (Z) used in the latter expression. The general solution of this second order ordinary differential equation is,

$$S(Z)=\frac{u(Z)h^2}{2K_z}\overline{S}_xZ^2+C_1Z+C_2.$$

The non-dimensional constants C_1 and C_2 are calculated using the following boundary conditions: at the surface $S(0)=S_s$, and at the bottom $S(1)=S_b$. In the general solution, they are given by:

$$C_1=S_s \text{ and } C_2=(S_b-S_s)-\frac{u(Z)h^2}{2K_z}\overline{S}_x,$$

where the difference S_b-S_s for partially mixed estuaries may vary from just a few to values to over twenty psu, for weakly and high stratified conditions, respectively.

Substituting the constants C_1 and C_2 into the general solution, the final vertical salinity profile is:

$$S(Z) = [\frac{u(Z)h^2}{2K_z}\overline{S}_x]Z^2 + [(S_b - S_s) - \frac{u(Z)h^2}{2K_z}\overline{S}_x]Z + S_s,$$

which satisfies the boundary conditions, and may be reduced to a final solution if the vertical velocity profile u = u(Z) has previously been calculated.

7.3.4 One-Dimensional Formulation: Integration in an Area

Estuaries that are long, narrow and shallow, with accentuated tidal forcing, and are vertically non-stratified and laterally homogeneous, can be studied as one-dimensional system, and the continuity and salt conservation equations may be simplified to be applied to these well-mixed estuaries (type 1, C). In these estuaries, the longitudinal variation is prevalent and the velocity, salinity and the concentration of properties may be taken as functions of the longitudinal distance and time, i.e., u = u(x, t), S = S(x, t) and C = C(x, t), respectively. The one-dimensional deduction of these equations are not trivial, because the integrated mean values of Eqs. (7.19), (7.25) and (7.27) must be along the transverse plane (Oyz), orthogonal to the axis Ox, as illustrated in Fig. 7.6.

Beginning with the integration of the continuity equation,[4] let us indicate by A = A(x, t) the area of the transverse section limited by the closed line c (Fig. 7.6), which varies both along the estuary and with time

$$\iint_A \frac{\partial u}{\partial x} dydz + \iint_A \frac{\partial v}{\partial y} dydz + \iint_A \frac{\partial w}{\partial z} dydz = 0, \tag{7.81}$$

where dydz = dA is a small elementary area.

The first term of Eq. (7.81) must be transformed by applying the Leibnitz derivation rule of a double integral, resulting in (Pritchard 1958; Okubo 1964):

$$\frac{\partial}{\partial x,t}(\iint_A u dydz) = \iint_A \frac{\partial u}{\partial x,t} dydz + \frac{1}{c}(\frac{\partial A}{\partial x,t}) \oint_c u dl. \tag{7.82}$$

In this equation, c is the length of the continuous curve limiting the area, A, depicted by the closed line going through in the positive sense of direction (Fig. 7.6), and dl is the differential arch element.

The remaining terms (second and third) of Eq. (7.82) may be adequately reduce using the Green's formula (Severi 1956, p. 369), which transform the surface integral into a line integral,

---

[4]This type of estuary is usually shallow and the influence of the gravitational circulation may be disregarded and the baroclinic bumping landward is negligible.

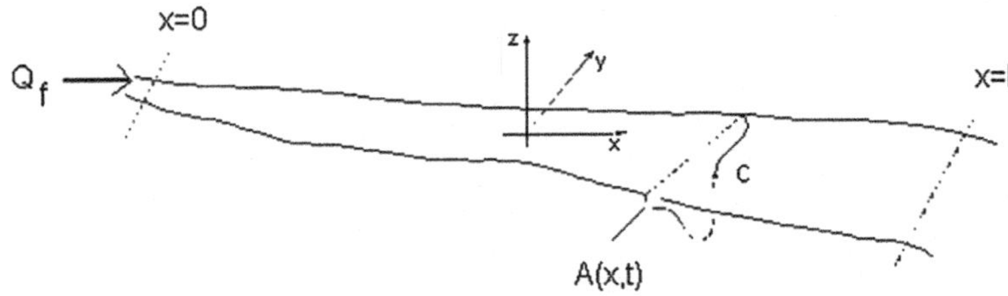

Fig. 7.6 Estuary approximated by one-dimensional model with uniform property distributions in the transversal section A = A(x, t). The boundary of the area, A, is the closed line circulating in the positive orientation (area is located at left)

$$\iint_A \frac{\partial v}{\partial y} dydz = \oint_c vdz, \tag{7.83}$$

and

$$\iint_A \frac{\partial w}{\partial z} dydz = -\oint_c wdy, \tag{7.84}$$

where, dy and dz are differential elements of the contour line (c).

Substituting Eqs. (7.82), (7.83) and (7.84) into the Eq. (7.81) we have the following expression:

$$\frac{\partial}{\partial x}\left(\iint_A udydz\right) - \frac{1}{c}\left(\frac{\partial A}{\partial x}\right)\oint_c ud\ell + \oint_c vdz - \oint_c wdy = 0. \tag{7.85}$$

As the u-velocity component is uniform in the transversal section A, this equation may be rewritten as:

$$\frac{\partial (uA)}{\partial x} - \frac{\partial A}{\partial x} u|_A + \oint_c vdz - \oint_c wdy = 0, \tag{7.86}$$

where the following identities have been taken into account: $\int_A dydz = A$, and $\frac{1}{c}\oint_c d\ell = 1$, and $u|_A = u$, is the velocity mean value in the area A.

Now, for the physical interpretation of the sum of the two last terms of Eq. (7.86), we may use a *consequence* the Green theorem, which states that (Severi 1956, p. 369):

$$A = \frac{1}{2}\left(\oint_c y dz - \oint_c z dy\right). \tag{7.87}$$

As $A = A(x, t)$, the local variation $\partial A/\partial t$ is formulated by:

$$\frac{\partial A}{\partial t} = \frac{1}{2}\left(\oint_c v dz - \oint_c w dy\right), \text{ or } 2\frac{\partial A}{\partial t} = \left(\oint_c v dz - \oint_c w dy\right), \tag{7.88}$$

and the functions being integrated, v and w are given by

$$v = \frac{dy}{dt}; w = \frac{dz}{dt}. \tag{7.89}$$

The second term of Eq. (7.86) may be rewritten as:

$$\frac{\partial A}{\partial x} u|_A = \frac{\partial A}{\partial x}\frac{dx}{dt} = \frac{\partial A}{\partial t}, \tag{7.90}$$

and substituting Eqs. (7.88) and (7.90) into the Eq. (7.86), the result is,

$$\frac{\partial (uA)}{\partial x} - \frac{\partial A}{\partial t} + 2\frac{\partial A}{\partial t} = 0, \tag{7.91}$$

and the analytical expression of the one-dimensional principle of continuity is reduced to (Pritchard 1958):

$$\frac{\partial (uA)}{\partial x} + \frac{\partial A}{\partial t} = 0. \tag{7.92a}$$

For a wide shallow estuary ($B \gg H_0$), $A = B(H_0 + \eta) \approx B\eta$, this equation is reduced to

$$\frac{1}{B}\frac{\partial (uA)}{\partial x} + \frac{\partial \eta}{\partial t} = 0. \tag{7.92b}$$

As the area, A, is a known geometric property in this equation, it may be solved using the mean velocity at the transverse section $u = u(x, t)$, as well as to the volume transport, $u(x, t)A = T_V(x, t)$; for convenience, this volume transport may be also denoted by Q ($uA = T_V = Q$). Then, in order to satisfy the mass conservation principle, if in a longitudinal location, x, the volume transport increases ($\partial T_V/\partial t > 0$) or decreases ($\partial T_V/\partial t < 0$), this must be compensated by a time decrease ($\partial A/\partial t < 0$) or increase ($\partial A/\partial t > 0$) of the cross-section area, respectively. However, in steady-state condition, the continuity equation is reduced to $T_V = Q_f = uA = \text{const}$, or $u = u_f = Q_f/A$.

Similar to the mass continuity, the one-dimensional salt conservation equation is obtained by multiplying the tridimensional expression (7.25) by the differential area dydz, developing the integral terms using the Leibnitz rule and the Green theorem, it follows that:

$$\begin{aligned}&\frac{\partial}{\partial t}(\iint_A \mathrm{Sdydz}) - \frac{\partial A}{\partial t}S|_A + \frac{\partial}{\partial x}(\iint_A \mathrm{uSdydz}) - \frac{\partial A}{\partial x}uS|_A\\&+\oint_c \mathrm{vSdz} - \oint_c \mathrm{wSdy} = \frac{\partial}{\partial x}[\iint_A (K_x\frac{\partial S}{\partial x})\mathrm{dydz}]\\&-\frac{1}{c}\frac{\partial A}{\partial x}\oint_c (K_x\frac{\partial S}{\partial x})\mathrm{dl} + \oint_c (K_y\frac{\partial S}{\partial y})\mathrm{dz} - \oint_c (K_z\frac{\partial S}{\partial z})\mathrm{dy}.\end{aligned} \tag{7.93}$$

Taking into account Eqs. (7.88) and (7.90), and that the salinity, S, and the u-velocity component are uniform in the area, A, the terms on the left-hand-side of Eq. (7.93) simplifies to:

$$\frac{\partial(AS)}{\partial t} + \frac{\partial(uAS)}{\partial x}, \tag{7.94}$$

because $\iint_A \mathrm{dydz} = A$, and the sum of the remaining terms are equal to zero,

$$-\frac{\partial A}{\partial t}S|_A - \frac{\partial A}{\partial x}uS|_A + S|_A(\oint_c \mathrm{vdz} - \oint_c \mathrm{wdy}) = 0. \tag{7.95}$$

In the first term on right-hand-side of the Eq. (7.93), salinity and the longitudinal salt diffusive term, $K_x(\partial S/\partial x)$, are uniform in the area A. As the line integrals are equal to zero, these terms are reduced to:

$$\frac{\partial}{\partial x}(AK_x\frac{\partial S}{\partial x}), \tag{7.96}$$

and it follows that the one-dimensional salt conservation equation (Pritchard 1958) is:

$$\frac{\partial(AS)}{\partial t} + \frac{\partial(uAS)}{\partial x} = \frac{\partial}{\partial x}(AK_x\frac{\partial S}{\partial x}). \tag{7.97}$$

Finally, developing the products indicate in the left-hand-side of the differential equation, and combining the result with the one-dimensional continuity Eq. (7.92a, 7.92b), the salt conservation equation may be simplified to,

$$\frac{\partial S}{\partial t} + u\frac{\partial S}{\partial x} = \frac{1}{A}[\frac{\partial}{\partial x}(AK_x\frac{\partial S}{\partial x})]. \quad (7.98)$$

Under steady-state conditions and for A = const., this equation may be further reduced to $u_f S = K_x(\frac{dS}{dx})$ or $\rho(u_f S) = \rho[K_x(\frac{dS}{dx})]$, which states that the downstream advective salt flux driven by the river velocity ($\rho u_f S$) will counteract the upstream diffusive flux driven by all other mechanism, $\rho K_x(\partial S/\partial x)$. This equation has been used to estimate the kinematic (or dynamic) eddy diffusion coefficient, which may be approximated by $K_x = u_f S/(\partial S/\partial x)$. However, coefficients calculated with this equation and used in analytical and numerical models are subjected to interpretation and must be validated with observational results of velocity and salinity profiles.

We have seen that when the geometry of a transverse section (A), the uniform longitudinal velocity component (u), the kinematic eddy diffusion coefficient (K_x) and the initial and boundary conditions are known, it is possible to solve the differential Eq. (7.98) for S = S(x, t) or S = S(x). The following statements, according to Okubo (1964) and Fischer et al. (1979), should be noted: (i) Time derivative describes the change of water property per tidal cycle, and A may indicate the cross-sectional area at mean time interval of the tidal cycle; (ii) Because of the cross-sectional area divergence, the effective mean flow velocity tends to decrease towards the sea; (iii) On the other hand, the local fresh water inflow through the sides of the estuarine may counteract, to some extent, the cross-section divergence of the mean flow.

Similar expressions to Eqs. (7.97) and (7.98) with a few adaptations, may be transformed into one-dimensional conservation equations for analysis of the one dimensional concentration, C = C(x, t), diffusion of pollutants, based on the time and space averaged equations formulated by:

$$\frac{\partial C}{\partial t} + u\frac{\partial C}{\partial x} = \frac{1}{A}[\frac{\partial}{\partial x}(AK_{xC}\frac{\partial C}{\partial x})] \pm (\text{sources or sink terms}), \quad (7.99a)$$

or

$$A\frac{\partial C}{\partial t} + Q_f\frac{\partial C}{\partial x} = [\frac{\partial}{\partial x}(AK_{xC}\frac{\partial C}{\partial x})] \pm (\text{sources or sink terms}). \quad (7.99b)$$

In these equations, C, is a non-dimensional concentration $[C] = [MM^{-1}]$ and K_{xC} $[K_{xC}] = [L^2T^{-1}]$ is the kinematic eddy-diffusion coefficient.

For the Eqs. (7.99a, 7.99b) the same restrictions hold as those indicated for Eq. (7.98). A rigorous deduction of this equation, describing the space and time variation of the pollutant diffusion in a one-dimensional estuary was developed by Okubo (1964). The conditions for which it is appropriate to use for pollutant diffusion in a one-dimensional estuary under steady-state condition are: (i) after a critical initial time period; (ii) when, due to tidal fluctuations of cross-sectional areas, the property concentration and density are sufficiently small compared with the respective mean value; (iii) when tidal mean velocity weighted by cross-sectional area is used for velocity (u), instead a simple tidal mean velocity, and; (iv) when an effective eddy diffusivity is defined, including non-homogeneity within the cross-section, and the river discharge into the estuary is small.

7.4 Simplifyed Forms of the Continuity Equation

In Chap. 2, a simplified one-dimensional continuity equation was presented for an estuary with width (B) and depth (H_0) constants (Eq. 2.17, Chap. 2). Let us derive another equation in which the condition that the width B $\neq$ const., starting with the bi-dimensional expression of the continuity (7.56a), performing its lateral integration from y = a(x, z) to y = b(x, z), the result is:

$$\int_a^b \frac{\partial(uh)}{\partial x} dy + \int_a^b \frac{\partial(vh)}{\partial y} dy + \int_a^b \frac{\partial h}{\partial t} dy = 0. \tag{7.100}$$

As $h = H_0 + \eta(x, t)$, in this equation it has been taken into account that $\partial\eta/\partial t = \partial h/\partial t$. Applying the Leibnitz rule to the first term, calculating the integral in the second term, and with the boundary conditions $vh|_{x=a} = vh|_{x=b} = 0$, it follows that:

$$\frac{\partial}{\partial x}[\int_a^b (uh) dy] + \int_a^b (\frac{\partial h}{\partial t}) dy = 0. \tag{7.101}$$

Because an estuary's length is generally much less than one quarter of the co-oscillating tidal wave length, the local depth (h) may be considered as independent on x and y, and the Eq. (7.101) may be rewritten as:

$$\frac{\partial}{\partial x}[\int_a^b (uh) dy] + \frac{\partial h}{\partial t} \int_a^b dy = 0, \tag{7.102a}$$

and

$$\frac{\partial}{\partial x}[\int_a^b (uh) dy] + B \frac{\partial h}{\partial t} = 0. \tag{7.102b}$$

As the u-velocity component is independent of the depth (z) and the water column height (h), in the first term it may be substituted by an integral (as in Eq. 7.56c); then, this equation may be re-written as:

$$\frac{\partial}{\partial x}[u\int_{a}^{b}(\int_{-H_0}^{\eta} dz)dy] + B(\frac{\partial h}{\partial t}) = 0, \quad (7.103a)$$

or

$$\frac{\partial Q}{\partial x} + B(\frac{\partial \eta}{\partial t}) = 0, \quad (7.103b)$$

where the volume transport is denoted by $Q = Q(x, t)$ and is calculated by

$$u[\int_{a}^{b}(\int_{-H_0}^{\eta} udz)dy] = u(x, t)A = Q(x, t) \quad (7.103c)$$

By the longitudinal integration of equation of Eq. (7.103b), from the estuary head ($x = 0$) down to the mouth ($x = L$), it follows that,

$$Q_f - Q_L(t) = \frac{\partial \eta}{\partial t}(\int_{0}^{L} Bdx) = (\frac{\partial \eta}{\partial t})A_{su}, \quad (7.104)$$

where Q_f and $Q_L(t)$ are the river discharge and the time variation of the volume transport at the estuary mouth, respectively. The integration of the estuary width (B) along its longitudinal length (L) may be identified as the surface area of the estuary (A_{su}), which is dependent on the tidal height, $A_{su} = A_{su}(\eta)$. If the river discharge (Q_f) is disregarded, this solution may be simplified as:

$$Q_L(t) = -(\frac{\partial \eta}{\partial t})A_{su}. \quad (7.105)$$

This equation may be used to estimate the volume transport through the mouth of a hyper-saline estuary or coastal lagoon in arid regions, and indicates that the temporal variation of the volume transport is proportional to the surface area and the tidal oscillation. As this surface area may be determined by the hypsometric characteristics of the coastal system (Fig. 7.7), with leveling techniques and aerial pictures, and the tide may be predicted, it is possible to calculate approximately the volume transport, $Q_L = Q_L(t)$.

It should be noted that during the tidal flood ($\partial\eta/\partial t > 0$) and ebb ($\partial\eta/\partial t < 0$) tides, the volume transport is negative $Q_L < 0$ and positive $Q_L > 0$, respectively. Then, the volume transport balance, $T_C = CQ_L(t)$, associated with any known conservative property concentration (C), may be used to estimate the importation ($T_C < 0$) or exportation ($T_C > 0$) of substances. With the concentration in units of property per volume [prL^{-3}], the property transport is calculated in [prT^{-1}].

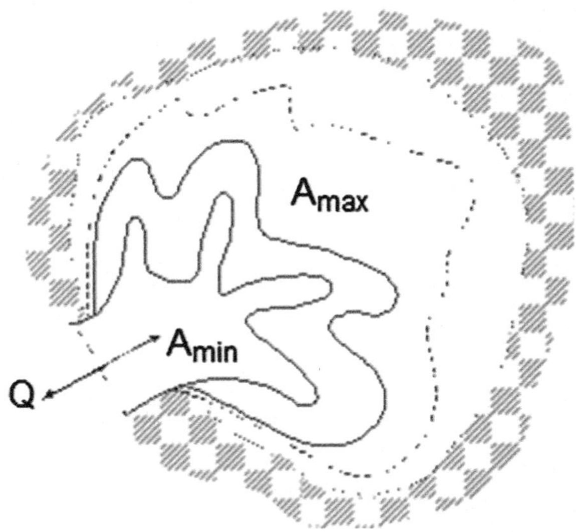

Fig. 7.7 Schematic diagram of a coastal lagoon showing the inflow and outflow of volume transport (Q_L) and the associate maximum and minimum flooding areas, which may be calculate by hypsometric techniques

7.5 Application of the One-Dimensional Continuity Equation

To exemplify an application of the one-dimensional continuity Eq. (7.92a, b), let us investigate its analytical solution for a classical estuarine channel. Under the assumption that in this estuary the velocity is uniform in the cross-sectional area, the equation may be integrated from a section located at the estuary head (x = 0) down to a longitudinal position x (Fig. 7.6):

$$\int_0^x \frac{\partial(uA)}{\partial x}dx + \int_0^x \frac{\partial A}{\partial t}dx = 0, \qquad (7.106a)$$

or

$$(uA)|_x - (uA)|_{x=0} + \int_0^x \frac{\partial A}{\partial t}dx = 0, \qquad (7.106b)$$

where $(uA)|_{x=0} = Q_f$. As the integration limits are independent on the time,

$$(uA)|_x + \frac{\partial}{\partial t}(\int_0^x Adx) = Q_f, \qquad (7.107a)$$

or

$$(uA)|_x + \frac{dV}{dt} = Q_f. \tag{7.107b}$$

In this equation, V is the volume of the estuarine water mass between the cross-sections located from x = 0 and the arbitrary position x, and dV/dt is its time variation. Solving this equation for the cross-section mean velocity, u = u(x, t), gives,

$$u(x,t) = \frac{1}{A}(Q_f - \frac{dV}{dt}), \tag{7.108}$$

where u = u(x, t) is the non-steady state u-velocity component generated by the river discharge and the barotropic gradient pressure force due to tidal oscillation, and its influence in the volume time variation (dV/dt). From this result, it is possible to compare the current intensities during the ebb (u_E) and flood (u_F) tides. In effect, because the river discharge may be taken as constant during the tidal period $u_f = Q_f/A$, and the ebb velocity (u_E) usually is higher than the flood (u_F), because the volume decrease (dV/dt < 0) of the mixing zone (MZ). During the flood, dV/dt > 0, and its intensity $u_F < u_E$. Under steady-state condition dV/dt = 0 then $u_f = Q_f/A$.

Taking into account that the mean free surface $\overline{A}_{su}$ is occupied by the MZ, between the estuary head (x = 0) and the estuary mouth at longitudinal position x, the geometric volume of the estuarine water mass is calculated by $V(t) = \overline{A}_{su}\eta(x,t)$. Thus, the Eq. (7.108) may be rewritten as:

$$u(x,t) = \frac{1}{A}(Q_f - \overline{A}_{su}\frac{\partial\eta}{\partial t}), \tag{7.109}$$

which confirms that this velocity is generated by the river discharge and the barotropic gradient pressure force. With this equation, the volume transport Q, $[Q] = [L^3T^{-1}]$ may also be calculated by,

$$Q(x,t) = u(x,t)A = Q_f - \overline{A}_{su}\frac{\partial\eta}{\partial t} = Q_f - \frac{dV}{dt}, \tag{7.110}$$

which has similarities with Eq. (7.104).

In estuaries, the time-rate of the volume stored in the space between a cross-section and its head may be approximate by sine functions of time (t) and the angular frequency (ω). Then, the integral of the Eq. (7.107b) during a complete tidal cycle is given by (Pritchard 1958):

$$\int_0^T (uA)dt = \int_0^T Q_f dt - \int_0^T \frac{dV}{dt}dt. \tag{7.111}$$

As the river discharge Q_f is constant, and the last integral on the right-hand-side is equal to zero, as it is supposed that during a complete tidal cycle there will be no time variation in the stored volume ($\frac{dV}{dt} = 0$) in the mixing zone (MZ), then,

$$\int_0^T (uA)dt = Q_f T = R, \tag{7.112}$$

This result confirms the fresh water volume conservation during this time interval, and that the net volume transport is equal to R.

7.6 Application of the One-Dimensional Salt Conservation Equation

The tidal prism models where semi-empirical developed without the formalism of the principles of mass and salt conservation presented in this chapter. Additionally, the basic ideas of Ketchum's methods were described, taking into account the physics of the continuum in the classical article of Arons and Stommel (1951), which will be presented in Chap. 10.

In this topic, the mass and salt conservation equations will be applied with the same hypothesis as the discrete models: one-dimensional, steady-state and well-mixed estuaries. Let us consider a simple geometry (A = const.), with the kinematic effective (eddy) diffusion coefficient (K_H) in the MZ taken as constant. In these conditions, because ∂A/∂t = 0, the continuity Eq. (7.92a, b) is simplified to ($\frac{\partial uA}{\partial x} = 0$), and uA = Q_f is independent of the longitudinal position. In turn, the salt conservation Eq. (7.98) is simplified to:

$$u_f \frac{\partial S}{\partial x} = K_H(\frac{\partial^2 S}{\partial x^2}). \tag{7.113}$$

As the river discharge and the estuary geometry are known, the solution S = S (x) of Eq. (7.113) is dependent only on the boundary conditions and a known K_H coefficient. As the salinity field is uniform in the cross section area, the problem has been reduced to one dimension, and the differential Eq. (7.113) is an ordinary second order differential equation with constants coefficients. Rearranging its terms the equation to be solved is:

$$\frac{d^2S}{dx^2} - \frac{u_f}{K_H}\frac{dS}{dx} = 0, \tag{7.114a}$$

or, reducing its order with a first integration from the river tidal zone (x = 0, where S = 0), and the longitudinal position x in the mixing zone (MZ), gives

$$u_f S = K_H \frac{dS}{dx} + A_1, \tag{7.114b}$$

where A_1 is an integration constant. Let us solve this equation in relation to a local coordinate system with Ox oriented positively seaward from x = −L (boundary TRZ/MZ) to x = 0 (estuary mouth) as shown in Fig. 7.8; with this orientation of the u-velocity component generated by the river discharge, (u_f) is positive.

At x = −L, the salinity and the product $K_H \frac{dS}{dx}$ are equal to zero and the integration constant $A_1 = 0$. Multiplying both members of the remaining equation by the density, it follows that the seaward advective salt flux counter balances the eddy diffusive salt flux due to the tidal forcing.

The simplifying hypothesis of this problem may limit its practical application because some influences have not been taken in consideration (mainly bottom and lateral friction, baroclinic forcing and vertical mixing). The solution, however, is very interesting because it demonstrates the physical nature of the estuary and its relationship with the main concepts presented in Chap. 6.

Returning to Eq. (7.114b), its solution with $A_1 = 0$ simplifies to

$$S(x) = S_0 \exp\left(\frac{u_f}{K_H} x\right). \tag{7.115}$$

This solution was determined with the following boundary conditions: the salinity at the mouth (x = 0) is equal to the salinity at coastal region $S(0) = S_0$ (the only salt source), and the estuary is long enough that for $L \to -\infty$ the salinity decreases to zero $S(-\infty) = 0$. As $x < 0$ and $u_f > 0$, it is easy to show that this solution identically satisfies these boundary conditions.

The equation for the longitudinal distribution of a solute discharged into a steady-state well-mixed estuary forced by tides and river discharge, with transverse sections that vary in the longitudinal direction A = A(x), was obtained in the classical work of Maximon and Morgan (1955). Their solution, when particularized for a constant transversal section (A = const.), reduces identically to the solution (7.115).

With this analytical solution for S = S(x), it is possible to apply the Eq. (6.11) (Chap. 6) to calculate the longitudinal variation of the fresh water fraction, f = f(x), which is necessary to theoretically calculate the fresh water volume stored in the estuary. Thus, this expression becomes,

$$f(x) = 1 - \exp\left(\frac{u_f}{K_H} x\right), \tag{7.116}$$

and for $x \to -\infty$, this solution converges to one $f(-\infty) \to 1$.

In these solutions (7.115 and 7.116), the longitudinal kinematic eddy diffusion coefficient (K_H) still remains an unknown physical quantity, which may be theoretically calculated by correlation of the turbulent fluctuations of the u-velocity and salinity (Eq. 7.26). However, its order of magnitude may be estimated if the

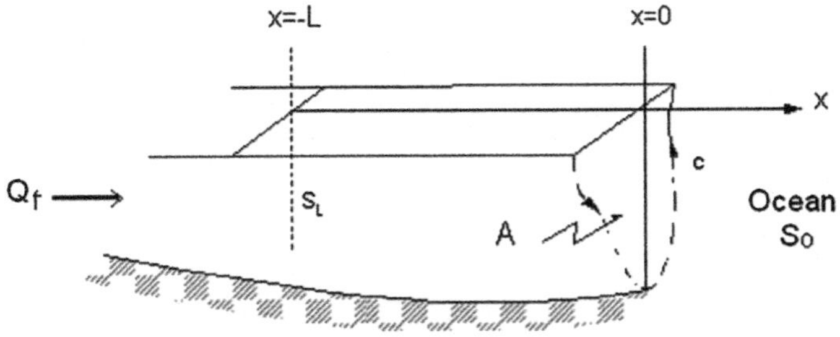

Fig. 7.8 One-dimensional model indicating the landward (x = −L) and seaward (x = 0) longitudinal coordinates of the mixing zone (MZ) between its head and mouth, respectively

longitudinal distance between the estuary mouth and head (mixing zone length) is known, along with the physical quantities such as Q_f, A and S_0. For example, if $Q_f = 100\ m^3\ s^{-1}$, $A = 10^3\ m^2$, $S_0 = 35°/_{oo}$ and $S(-L) = 1$, Eq. (7.116) may easily be solved for the coefficient K_H, and some results are presented in Table 7.1.

Although this table only shows the coefficient dependence on the mixing zone length, it is also directly dependent on the river discharge, fresh water fraction, and the estuary geometry (transverse cross-section area and length). As the transverse cross-section area usually varies along the estuarine channel, even for a constant river discharge, this coefficient will be dependent on the longitudinal distance, $K_H = K_H(x)$. This dependence was first investigated by Stommel (1953b), solving the Eq. (7.115) by finite differences around a longitudinal position, x,

$$K_H(x) = \frac{u_f S(x)}{\Delta S}\Delta x = \frac{Q_f S(x)}{A(x)\Delta S}\Delta x = \frac{Q_f[f(x)-1]}{A(x)\Delta f}\Delta x. \tag{7.117}$$

In the last term, the salinity $S = S(x)$ was changed by the fresh water fraction, as $S(x) = S_0[1-f(x)]$ and $\Delta S(x) = -S_0\Delta f(x)$. In this equation, the salinity and the fresh water fraction are mean values at the cross-sections separated by a distance Δx, and ΔS and Δf are finite intervals of salinity and the fresh water fraction, respectively.

Equation (7.117) is a simple and direct method to determine the K_H coefficient using known quantities that may be obtained experimentally. With all variables in the SI system of units, the kinematic coefficient, K_H, is calculated in $m^2\ s^{-1}$. For well-mixed estuaries with mean dimensions and moderate river discharge, we may,

Table 7.1 Estimates of the longitudinal kinematic eddy diffusion coefficient (K_H), based on the mixing zone length (−L) and the following data: $Q_f = 100\ m^3\ s^{-1}$, $A = 10^3\ m^2$ and $S_0 = 35°/_{oo}$

Mixing zone length L (m)	Coefficient K_H ($m^2\ s^{-1}$)
36	1
360	10
3600	100
36,000	1000

as a first approximation, adopt K_H values between 3.0×10^2 m^2 s^{-1} and 5.0×10^2 m^2 s^{-1}. Results obtained by various scientists who applied this method are compared by Officer (1978).

With the theoretical solutions of salinity, S = S(x), and the fresh water fraction f = f(x), it is possible to calculate the following fundamental physical quantities, defined in Chap. 6: fresh water volume V_f stored in the MZ, flushing time (t_q) and flushing rate (F). To exemplify this procedure, let us take $L = -10^4$ m, and the same values as we used previously: $Q_f = 100$ m^3 s^{-1}, $A = 10^3$ m^2, and $S_0 = 35.0‰$. With these values, it is easy to show that:

$$S(x) = 35\exp(3.6 \times 10^{-4}x), \quad (7.118a)$$

$$f(x) = 1 - \frac{S(x)}{35} = 1 - \exp(3.6 \times 10^{-4}x), \quad (7.118b)$$

$$V_f = A\int_{-L}^{0} f(x)dx = A\{L + 2.8 \times 10^3[\exp(-3.6 \times 10^{-4}L) - 1]\}, \quad (7.118c)$$

and

$$F = \frac{S_0}{S_0 - \overline{S}}Q_f = \frac{35}{35 - 9.5}Q_f. \quad (7.118d)$$

Calculating these analytical solutions with the numerical values given above gives the following results: $V_f = 7.28 \times 10^6$ m^3, $t_q = 20.2$ h and $F = 137$ m^2 s^{-1}. As this theory is applied to well-mixed estuaries, the flux rate (F) will be almost constant due to the dominance of the diffusion in the mixing process. The longitudinal variations of salinity and the fresh water fraction in this steady-state and one-dimensional model are presented in Figs. 7.9a, b, respectively.

Applying the conservation Eq. (7.114a) to a conservative property, with the same simplifying conditions as this example, and with its source located at the estuary mouth, it is easy to verify that the solution is similar to the Eq. (7.115), with the following substitutions: at the estuary mouth the salinity S_0 by the maximum concentration C_0, and K_H substituted by the corresponding eddy diffusion coefficient of the property K_{HC}. Then, the longitudinal distribution of the property concentration, C = C(x), decreases exponentially landward, following the exponential salinity decrease. This result has already been used qualitatively in Chap. 6, in a semi-empirical development (Eq. 6.83).

As a complementary exercise at the end of this topic, let us present another solution of the salt conservation Eq. (7.113), but in terms of the fresh water fraction (f). After integration of this equation, and considering the integration constant equal to zero, it may be rewritten as:

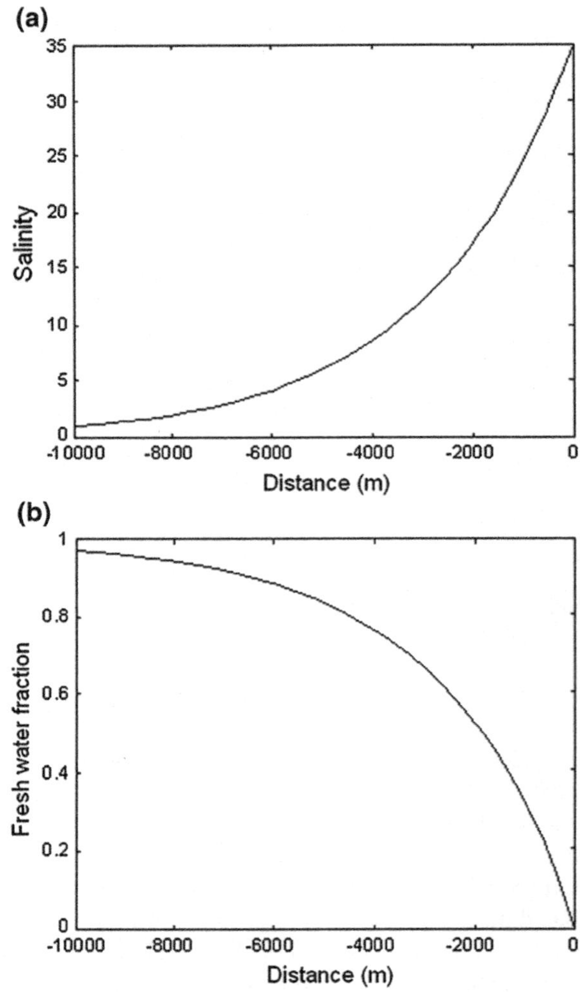

Fig. 7.9 Simulation of longitudinal variations of salinity and fresh water fraction of a one dimensional and steady-state well-mixed estuary, with length $L = -10^4$ m, and kinematic coefficient of eddy diffusion, $K_H = 280\ m^2\ s^{-1}$

$$u_f S - K_x \frac{dS}{dx} = 0, \tag{7.119}$$

and its first term may be associated with the mass conservation equation, for $u_f = Q_f/A$. Taking into account that the fresh water fraction $f = f(x)$ defined in the Eq. 6.10 (Chap. 6), it follow that the salinity may be expressed in terms of the fresh water fraction $S = S_0(1 - f)$, and

$$\frac{dS}{dx} = -S_0 \frac{df}{dx}, \tag{7.120}$$

where $0 \leq f \leq 1$. Combining Eqs. (7.119) and (7.120) and taking into account the mass conservation equation, it follows:

$$(\frac{Q_f}{A})S + S_0 K_H (\frac{df}{dx}) = 0. \tag{7.121}$$

Dividing both members of this equation by S_0, and knowing that the ratio S/S_0 is equal to $(1 - f)$, the salt conservation equation in terms of the fresh water fraction is rewritten as (Officer 1978):

$$Q_f = fQ_f - K_H A(\frac{df}{dx}). \tag{7.122}$$

Thus, the fresh water fraction, $f = f(x)$, may be calculated solving the following differential equation:

$$\frac{df}{dx} = (f - 1)\frac{Q_f}{K_H A}, \tag{7.123a}$$

or

$$\frac{df}{dx} = (f - 1)\frac{u_f}{K_H}. \tag{7.123b}$$

This equation satisfies identically the longitudinal variation of the fresh water fraction: at the estuary mouth ($f \to 0$) and $\frac{df}{dx} < 0$, thus $f = f(x)$ decreases seaward. Otherwise, at the estuary head $\frac{df}{dx} = 0$, and $f = f(x)|_{x=1} \to$ $f = \text{const.} = 1$. An exponential solution of $f = f(x)$ has already been presented in Eq. (7.118b).

7.7 Steady-State Concentration Distribution of a Non-conservative Substance

A study of the longitudinal distribution of the concentration of a non-conservative property (tracer), $C = C(x)$, in a well-mixed estuary under steady-state conditions is now presented as a particular solution of Eq. (7.99a). By hypothesis, the concentration is almost invariable from one tidal cycle to the next, and the river and effluent discharges are also constant during this time interval. As the concentration of a non-conservative substance will be considered (for conservative, the longitudinal variation follows the salinity very closely), its concentration may decrease with time even without the influence of tidal diffusion, and it is necessary to add a term into the equation to take that influence into account. In this case, the decrease with time may be expressed according to the first order exponential decay,

$$C(t) = c_0 \exp(-kt), \tag{7.124}$$

where c_0 is the initial concentration, $C(0) = c_0$, and k, $[k] = [T^{-1}]$ is a coefficient of proportionality, and $t = k^{-1}$ is the time required for the concentration to decay from the initial value c_0 to c_0/e.

With these definitions, the conservation equation for calculation of the steady-state longitudinal distribution, $C = C(x)$, of a non-conservative substance, undergoing a first order concentration decay, in a one-dimensional estuarine channel is reduced to the following second order differential equation:

$$\frac{d^2C}{dx^2} - \frac{u_f}{K_{HC}}\frac{dC}{dx} + \frac{k}{K_{HC}}C = 0, \tag{7.125}$$

relative to the same referential system presented in Fig. 7.8. The non-dimensional coefficient K_{HC} is a tracer's kinematic eddy (effective) horizontal diffusion. Even though this particular example has limited potential to be directly applied to an estuary, its solution is of interest in showing the approximate forms of solutions that can be expected, and it is appropriate to view this equation as a postulate (or empirical) model, subject to verification (Fischer et al. 1979). This equation has as general solution $C(x) = e^{mx}$, and the characteristic second grade root of this equation has two solutions in the real numeric field:

$$C(x) = c_0 \exp[\frac{u_f x}{2K_{HC}}(1 \pm \sqrt{1+\Psi})], \tag{7.126}$$

where $\Psi = 4K_{HC}k/u_f^2$ is a non-dimensional quantity, and the x variable is negative, according to the reference system used. Because the velocity generated by the river discharge ($u_f > 0$) in well-mixed estuaries is usually low, and the quantity Ψ is inversely proportional to u_f, the value of Ψ may be very large.

In order to exemplify the application of the solution presented in Eq. (7.126), let us consider the input of a tracer in an estuary with a mass transport W, in kg s^{-1}, in the longitudinal position $x = x_w$ ($x_w < 0$). Then, the initial condition $C(0) = c_0$ is satisfied, and the equation's solution must be separated into two terms (Stommel 1953b, Fischer et al. 1979):

- Landward from the effluent throw position x_w ($x \leq - x_w$)

$$C_{ac}(x) = c_0 \exp[\frac{u_f}{2K_{HC}}(1 + \sqrt{1+\psi})(x - x_w)]. \tag{7.127a}$$

- Seaward from the effluent throw position ($x \leq - x_w$)

$$C_{ac}(x) = c_0 \exp[\frac{u_f}{2K_{HC}}(1 + \sqrt{1+\psi})(x - x_w)]. \tag{7.127b}$$

In these solutions, it is easy to see that for $x = x_w$ the concentration at this point, $C(x_w)$, is equal to c_0, $C(x_w) = c_0$ and, according to the simplifying hypothesis, the initial concentration is determined by the equation,

$$c_0 = \frac{W}{Q_f} f_0, \tag{7.128}$$

where f_0 is the fresh water fraction at the effluent's launching position, and W/Q_f is the initial concentration due to dilution by the river discharge (Eq. 6.78, Chap. 6), without the influence of diffusion or concentration decay.

Let us exemplify numeric results using the above equations under the assumption that the tracer input is made into the estuary whose longitudinal salinity distribution, $S = S(x)$, and fresh water fraction, $f = f(x)$, were theoretically obtained and presented in the Eqs. (7.118a, b), which were calculated with $A = 10^3\ m^2$ and $Q_f = 100\ m^3\ s^{-1}$. For the tracer discharge point, we adopt the following positions: $x_W = -3000$ m and -5000 m, where, according to the Eq. (7.118b), $f(x)|_{x=0} = f_0 = 0.65$ and $f_0 = 0.83$, respectively. Assuming of a tracer transport of 20.0 kg s^{-1}, $K_{xC} = 60\ m^2\ s^{-1}$ and $k = 2.3 \times 10^{-4}\ s^{-1}$, the longitudinal landward (C_{ac}) and seaward (C_{ab}) distributions from the discharge positions are given by:

$$C_{ac}(x) = 0.13.\exp[2.94 \times 10^{-3}(x + 3 \times 10^3)], \tag{7.129a}$$

$$C_{ab}(x) = 0.13.\exp[-1.55 \times 10^{-3}(x + 3 \times 10^3)], \tag{7.129b}$$

and

$$C_{ac}(x) = 0.17.\exp[2.94 \times 10^{-3}(x + 5 \times 10^3)], \tag{7.130a}$$

$$C_{ab}(x) = 0.17.\exp[-1.55 \times 10^{-3}(x + 5 \times 10^3)]. \tag{7.130b}$$

In these solutions, the initial concentrations $C(x_w) = c_0 = 0.13$ kg m^{-3} and 0.17 kg m^{-3}, were calculated with Eq. (7.128); however, the primary phase of the tracer decay has not been taken into account. As may be observed, due to the seaward decrease of the fresh water fraction, the displacement of the tracer input in this direction reduces the initial tracer concentration (c_0).

Distributions of concentrations, $C = C(x)$, for the discharge positions considered in this exercise, are presented in Fig. 7.10, along with the corresponding salinity $S = S(x)$ and fresh water fraction $f = f(x)$ variations, representing the longitudinal behavior of a conservative property above and below the input position, respectively.

The analytical solution of this problem of a conservative substance is related to the semi-empirical solution of Ketchum (1955), shown in Fig. (6.14, Chap. 6), and the analytical solution of Stommel (1953a) may be similarly interpreted. If salinity

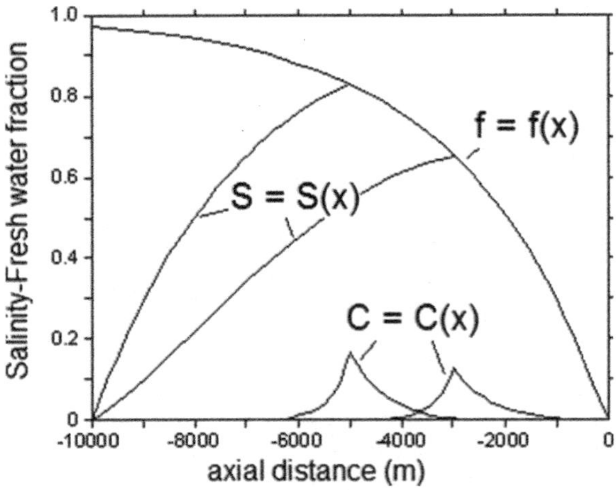

Fig. 7.10 Longitudinal distributions of the concentration, C = C(x), of a non-conservative tracer undergoing a first order decay, with the fresh water fraction, f = f(x) and salinity S = S(x), simulating conservatives properties. The one dimensional solutions were obtained landward and seaward of the tracer input position at −3000 m and −5000 m. (adapted from Stommel 1953a, b). The estuary mouth is located at x = 0

is introduced at the locations x = −6 km and −3 km, upstream of the outfall they will follow the S = S(x) curves, as shown in Fig. 7.10; however, the remainder of the solution downstream of the outfall their concentrations will follow the fresh water concentration curves, f = f(x). In this simple case it is clear that no matter what the location of the outfall, the concentration is every-where reduced upstream of the outfall if the location of the outfall is moved towards the sea. This solution also indicates that, up to the intersection of the curve S = S(x) with fresh water fraction curve, f = f(x) indicates the maximum input concentration of the effluent.

Solutions for the concentration distribution, C = C(x), for a non-conservative tracer discharged into the estuary by outfalls located at $x = -3.0 \times 10^3$ m and $x = -5.0 \times 10^3$ m, are also shown in the Fig. 7.10. A remarkable feature of these distributions is that the peak concentration, even at the outfalls, are very much reduced in comparison to the conservative tracer. As shown in the figure, the tracer concentration extends both upstream and downstream of the outfall and, unlike a conservative tracer, its concentration can be reduced at a point bellow by an upstream displacement of the outfall (Stommel 1953b; Ketchum, *unpublished report*). These theoretical and practical results indicate the following influence of the river discharge on the tracer longitudinal distribution: an increase in river discharge causes an increase in the fresh water fraction in the mixing zone, reducing the landward displacement of the tracer, and its initial concentration at the discharge position, as well as the seaward concentration of a conservative substance.

References

Arons, A. B. & Stommel, H. 1951. "A Mixing-Length Theory of Tidal Flushing". Trans. Am. Geophys.Un., 32(3):419–421.

Blumberg, A. F. 1975. A Numerical Investigation into the Dynamics of Estuarine Circulation. Tech. Rept. Chesapeake Bay Institute, The Johns Hopkins University. n. 91. 110 p. + Apêndices.

Bowden, K. F. 1963. The Mixing Processes in a Tidal Estuary. J. Air Wat. Pollut., 7:343 356.

Brand, L. 1959. Análisis Vectorial. Compania Editorial Continental, S.A., 333 p.

Cameron, W. M. & Pritchard, D. W. 1963. "Estuaries". In: Hill, M. N. (ed.). The Sea. Ideas and Observations on Progress in the Study of the Seas. New York, Interscience, pp. 306–324.

Csanady, G. T. 1982. Circulation in the Coastal Ocean. D. Reidel Publishing, Dordrecht, 279 p. (Environmental Fluid Mechanics).

Defant, A. 1961. Physical Oceanography. Oxford, Pergamon Press, vol. 1. 729 p.

Dyer, K. R. 1973. Estuaries: A Physical Introduction. London, Wiley. 140 p.

Fisher, J. S.; Ditmars, J. D. & Ippen, A. T. 1972. Mathematical Simulation of Tidal Time Averages of Salinity and Velocity Profiles in Estuaries. Massachusetts Institute of Technology, Mass., Rept. MITSG 72–11, 157 p.

Fischer, H. B.; List, E. J.; Koh, R. C. Y.; Imberger, J. & Brooks, N. H. 1979. Mixing in Inland and Coastal Waters. New York, Academic Press. 483 p.

Geyer, W.R. 2010. Estuarine salinity structure and circulation. In: ed. Valle-Levinson A. Contemporary Issues in Estuarine Physics. Cambridge University Press, pp. 12–26.

Geyer, W. R. & Farmer, D. M. 1989. Tide-Induced Variations of the Dynamics of a Salt Wedge Estuary. J. Phys. Oceanogr., v.19, pp. 1060–1072.

Gill, A. E. 1982. Atmospheric-Ocean Dynamics. New York, Academic Press. 662 p.

Hansen, D. V. & Rattray Jr., M. 1965. Gravitational Circulation in Sraits and Estuaries. J. Mar. Res., 23(1):102–122.

Ketchum, B. H. 1955. Distribution of Coliform Bacteria and other Pollutants in Tidal Estuaries. Sew. Ind. Wastes, 27:1288–1296.

Kinsman, B. 1965. Wind Waves-Their Generation and Propagation on the Ocean Surface. New Jersey, Prentice-Hall. 676 p.

Lacombe, H. 1965. Cours d'Océanographie Physique (Théories de la circulation générale. Houles et vagues). Paris, Gauthiers-Villars. 392 p.

Maximon, L. C. & Morgan, G. W. 1955. A Theory of Tidal Mixing in a Vertically Homogeneous Estuary. J. Mar. Res., 14:157–175.

Neumann, G. & Pierson, W. J. Jr. 1966. Principles of Physical Oceanography. London, Prentice-Hall. 545 p.

Officer, C. B. 1976. Physical Oceanography of Estuaries (and Associated Coastal Waters). New York, Wiley. 465 p.

Officer, C. B. 1977. Longitudinal Circulation and Mixing Relations in Estuaries. Estuaries, Geophysics, and the Environment. Washington, D. C., National Academy of Sciences, pp. 13–21.

Officer, C. B. 1978. Some Simplified Tidal Mixing and Circulation Flux Effects in Estuaries. In: Kjerfve, B. (ed.). Estuarine Transport Processes. Columbia, University of South Carolina Press, pp. 75–93 (The Belle W. Baruch Library in Marine Science, 7).

Okubo, A. 1964. Equations Describing the Diffusion of an Introduced Pollutant in a One-dimensional Estuary. In: Yoshida, K. (ed.). Studies on Oceanography. Tokyo, University of Tokyo Press, pp. 216–226.

Pritchard, D.W. 1954. A Study of Salt Balance in a Coastal Plain Estuary. J. Mar. Res., 13 (1):133 144.

Pritchard, D. W. 1955. Estuarine Circulation Patterns. Proc. Am. Soc. Civ. Eng., 81:717:1–11.

Pritchard, D. W. 1958. The Equations of Mass Continuity and Salt Continuity in Estuaries. J. Mar. Res., 17:412–423.

Pritchard, D. W & Kent, R. E. 1956. A Method of Determining Mean Longitudinal Velocities in a Coastal Plain Estuary. J. Mar. Res., 15(1):81–91.

Rattray Jr., M. & Hansen, D. V. 1962. A Similarity Solution for Circulation in an Estuary. J. Mar. Res., 20(2):121–133.

Severi, F. 1956. Lecciones de Análisis. Barcelona, Labor, v. 2. 429 p.

Stommel, H. 1953a. The Role of Density Currents in Estuaries. Proc. Minnesota International Hydraulics Convention, Minneapolis, University of Minnesota, pp. 305–312.

Stommel, H. 1953b. Computation of Pollution in a Vertically Mixed Estuary. Sew. Ind. Wastes, 25 (9):1065–1071.

Svedrup, H. U.; Johnson, M. W. & Fleming, R. H. 1942. The Oceans, their Physics, Chemistry and General Biology. New Jersey, Prentice-Hall. 1042 p.

Symon, K. R. 1957. Mechanics. Addison-Wesley Publishing Company, Inc., sec. ed. 358 p.

Team Course. 2001. Ocean Circulation. Oxford, Open University Course Team/Pergamon Press. 2nd. Edition 286 p.

Chapter 8
Hydrodynamic Formulation: Equations of Motion and Applications

Estuarine hydro- and thermodynamics processes control a variety of physical phenomena which are described by the water level, field of motion (tidal currents and gravitational circulation), and hydrographic properties (salinity, temperature, pressure). These phenomena include the flushing and residence times, as well as pollutant diffusion, erosion, sediment transport and sedimentation. Marine life is also strongly influenced by these processes, not only physiological processes of the individual organisms, their aggregation, stability, population growth and expansion of populations.

In this chapter, let us assume that the equation of motion of a geophysical fluid is known. The deduction of the terms of this equation, which formulates the generating and dissipative forces of the motion, may be found in several books (Sverdrup et al. 1942; Lacombe 1965; Neumann and Pierson 1966; Pedloski 1979, among others). However, the estuarine water body is enclosed by a particular geometry, and special attention will be given to the simplifications of this equation to their analytical and numerical solutions for the motion and mixing processes.

8.1 Equations of Motion

To formulate a problem related to the conservation equations of mass, salt, or any property (Chap. 7) of an estuary, it is necessary to associated to them with the equation of motion to theoretically calculate the estuarine circulation. The equation of motion is named the Navier-Stokes equation, and has been adapted to geophysical fluids. The horizontal components of this equation, in relation to a Cartesian reference system, Oxyz, are:

L. Bruner de Miranda et al., *Fundamentals of Estuarine Physical Oceanography*, Ocean Engineering & Oceanography 8, DOI 10.1007/978-981-10-3041-3_8

$$\frac{\partial u}{\partial t}+\frac{\partial(uu)}{\partial x}+\frac{\partial(uv)}{\partial y}+\frac{\partial(uw)}{\partial z}\pm f_0 v=-\frac{1}{\rho}\left(\frac{\partial p}{\partial x}\right)+\frac{1}{\rho}\left(\frac{\partial \tau_{xx}}{\partial x}+\frac{\partial \tau_{xy}}{\partial y}+\frac{\partial \tau_{xz}}{\partial z}\right), \quad (8.1)$$

and

$$\frac{\partial v}{\partial t}+\frac{\partial(vu)}{\partial x}+\frac{\partial(vv)}{\partial y}+\frac{\partial(vw)}{\partial z}\pm f_0 u=-\frac{1}{\rho}\left(\frac{\partial p}{\partial y}\right)+\frac{1}{\rho}\left(\frac{\partial \tau_{yx}}{\partial x}+\frac{\partial \tau_{yy}}{\partial y}+\frac{\partial \tau_{yz}}{\partial z}\right), \quad (8.2)$$

with the Oz axis oriented in the direction contrary to the gravity acceleration ($\vec{g}$). The notations u, v and w indicate the velocity components according to the Ox, Oy and Oz axis, respectively.

The terms $\partial u/\partial t$, $\partial v/\partial t$ and $\partial w/\partial t$ are the local accelerations components and the non-linear terms [$\partial(uu)/\partial x$, $\partial(uv)/\partial y$ and $\partial(uw)/\partial z$, and [$\partial(vu)/\partial x$, $\partial(vv)/\partial y$ and $\partial(vw)/\partial z$ are advective accelerations, introduced according to the Euler's expansion. When the local and advective accelerations are zero, the motion is said to be in *steady-state*. The resulting acceleration or motion attenuations are due to the forces acting in the second member of Eqs. 8.1 and 8.2; the pressure gradient force components, per mass unity, first terms on the right-hand-side, are calculated in a fixed point in space, while the fluid elements themselves move through it. The motion attenuation is due to the frictional shear stresses forces, and its formulations are presented in the last three terms of the of equation components.

As the estuarine water mass is a geophysical fluid, its motions are deflected by the Coriolis acceleration defined by the parameter $f_0 = 2\Omega\sin(\theta_0)$, where θ_0 is the geographical latitude and $\Omega = 7.29 \times 10^{-5}\ s^{-1}$ is the modulus of the angular velocity of the Earth. The latitude varies between the intervals $0 \leq \theta_0 \leq 90\ °$ and $-90 \leq \theta_0 \leq 0\ °$, in the northern and southern hemispheres, where $f_0 \geq 0$ and $f_0 \leq 0$, respectively. In the southern hemisphere ($-f_0 v$), in wide and highly stratified estuaries forced by diurnal tides, the dynamical influence of the Coriolis acceleration in the longitudinal component of the equation of motion may produce a lateral ascending slope of the halocline to the left of the motion; the layer over this interfaced is deeper and the motion may be more intense. In the article of Chant (2010), the importance of the dynamical balance between friction and the Coriolis acceleration (Ekman forcing) has been emphasize that, in fact the Earth's rotation, can be an important contributor to the structure not only for lateral, but also along channels flows, even in estuaries that are significantly narrower than an internal Rossby radius, which is inversely proportional to f_0.

The first term to the right of the second member of Eqs. 8.1 and 8.2 is the gradient pressure force per unit of mass. As shown by Kinsman (1965) in the deduction of this simple expression, which has the pressure (p) as the only unknown, there are two basic simplifications in a cubic volume element: (i) the density is approximate by a constant value (ρ = const.), and (ii) the pressure variation, from one side of the cubic volume to the opposite one, varies linearly through the distances dx, dy and dz. The quantities τ_{xx}, τ_{xy}, τ_{xz}, τ_{yx}, τ_{yy} and τ_{yz} are the Reynolds turbulent shear due to the internal dissipative frictional forces, expressed in force per unit area, with dimension of $[ML^{-1}T^{-2}]$. The frictional force, such as wind shear, acting on the air-sea interaction layer, is responsible for accelerating the water motion as it transfers some of its energy to the water, being important mainly in wide estuaries and coastal lagoons. All terms of Eqs. (8.1) and (8.2) have the dimension of acceleration, or force per unit of mass, but sometimes they will be referred generically in this text as *forces*.

Similar to the open ocean, in an estuary the vertical velocity component (w) is very low in comparison with the horizontal components u and v ($w \ll u, v$), Then, in the Oz component of equation of motion, the local and advective accelerations of vertical motions have very low orders of magnitude compared with the other terms, and may be disregarded. As the vertical frictional shear stress τ_{zx}, τ_{zy}, and τ_{zz} are also parameterized as functions of the shear of the vertical velocity component (w), they may also be disregarded. The vertical Coriolis component is another term of the equation of motion which is very small in comparison to the gravity acceleration, and may also be disregarded. Then, the vertical component of the equation of the motion used to solve problems related to ocean and estuarine circulation is reduced to the classical hydrostatic equation,

$$\frac{1}{\rho}\frac{\partial p}{\partial z} = -g, \tag{8.3}$$

where g is the modulus of the gravity acceleration ($g \approx 9.8\ m\ s^{-2}$).

This approximation, disregarding the vertical accelerations components (local and advective), but taking into account the vertical velocity component (w) in the advective accelerations of Eqs. (8.1 and 8.2), may be justified if $|w| \leq O(uH_0/L)$, where L is the estuary length. Retaining the w-velocity component in the equation of motion simplifies its solution, and it is possible to calculate the w-component, closing the tridimensional character of gravitational estuarine circulation (Stommel 1953a). As shown in Chap. 2 (Eq. 2.10a, b), when the hydrostatic equilibrium is assumed (Eq. 8.3), the gradient pressure force has three components: barometric, barotropic and baroclinic. Although the tidal oscillation is one of the main forcing of estuarine circulation, the hydrostatic approximation is valid because the wave length of the diurnal or semidiurnal components is usually greater than 500 km.

The components of the Reynolds frictional stress are generated by the turbulent velocity variations. Then, the dependent variables u, v, w, p and ρ of these components are mean values calculated over a time interval that is large enough to eliminate the turbulent fluctuations of velocity, but short enough for these mean values to still be functions of time. In the classical work of Osborne Reynolds, published in 1884, the frictional stress components were defined as functions of the turbulent velocity fluctuations u′, v′ and w′, and parameterized by the eddy dynamic viscosity coefficients, A_{xx}, A_{xy}, A_{xz}, A_{yx}, A_{yy} and A_{yz}, which are several orders of magnitude higher than the corresponding molecular coefficients.

This parameterization was made semi-empirically and by analogy to the molecular shear stress, which is characteristic of the laminar motions and are expressed by:

$$\tau_{xx} = -\rho < u'u' > = -A_{xx}\frac{\partial u}{\partial x}, \tag{8.4}$$

$$\tau_{xy} = -\rho < u'v' > = -A_{xy}\frac{\partial u}{\partial y}, \tag{8.5}$$

$$\tau_{xz} = -\rho < u'w' > = -A_{xz}\frac{\partial u}{\partial z}, \tag{8.6}$$

$$\tau_{yx} = -\rho < v'u' > = -A_{yx}\frac{\partial v}{\partial x}, \tag{8.7}$$

$$\tau_{yy} = -\rho < v'v' > = -A_{yy}\frac{\partial v}{\partial y}, \tag{8.8}$$

$$\tau_{yz} = -\rho < v'w' > = -A_{yz}\frac{\partial v}{\partial z}. \tag{8.9}$$

The dynamic coefficients of viscosity have as dimensions $[ML^{-1}T^{-1}]$ (expressed in kg m^{-1} s^{-1} in SI units). According to the notations used, the symbol $< >$ indicates a time mean value, $< > = (1/\Delta t)\int () \, dt$, and the integral is calculated between the limits 0 and Δt, and the finite value of this quantity is the correlation of the turbulent velocity fluctuations, that multiplied by the density has the dimension of shear stress (force per unit of area $[ML^{-1}T^{-2}]$), which in SI units is expressed in *Pascal* (Pa = 1 N m^{-2}). For these eddy viscosity coefficients, the following simplifications are usually assumed in analytical solutions of the equation of motion:

$$A_{xx} = A_{xy} = A_x; \; A_{yx} = A_{yy} = A_y, \tag{8.10}$$

and

$$A_{xz} = A_{yz} = A_z. \tag{8.11}$$

As these coefficients are defined by random velocity fluctuations and parameterized as functions of the mean velocity (u, v) gradients, they become dependent on the motion characteristic rather than the estuary's physical properties. These coefficients also are dependent on the spatial scale of the motion, and there is no simple method for its determinations unless the random velocity fluctuations are accurately measured. To date, this has remained difficult to solve. In turn, the vertical eddy viscosity coefficient (A_z) is also dependent on the vertical stability of the water column; if there is stable stratification, turbulent mixing requires some of the tidal energy to be used to raise the potential energy of the water column, but most of the vertical mixing energy is extracted from the bottom and internal shear forced by tidal currents (Fischer et al. 1979). Hence, it is natural to find expressions of this eddy coefficient in function of the Richardson number (Ri) (Eq. 2.34, Chap. 2), such as the Munk and Anderson (1948) classical expression.

Numerical values of the eddy kinematic viscosity coefficients ($N_z = A_z/\rho$, is the kinematic version of A_z, and $[N_z] = [L^2T^{-1}]$, were calculated from tidal current measurements off Red Wharf Bay (England) by Bowden and Fairbairn (1952); at a given depth this coefficient varied during the tidal period and was related to the amplitudes and phases of the semi-diurnal tidal constituents. To overcome this difficulty due to its variability, the frictional *mean kinematic eddy viscosity coefficient* was defined, and used in the steady-state analytical models as in the articles of Hansen and Rattray (1965), Officer (1977) and others, with values ranging from $N_z = 4.1 \times 10^{-3}\ m^2\ s^{-1}$ to $9.0 \times 10^{-5}\ m^2\ s^{-1}$, at the surface and mid depths, respectively. The kinematic eddy diffusion coefficient has also been calculated by Pritchard (1956) and Bowden (1963) for the Mersey estuary (UK), with values ranging from $K_z = 5.0 \times 10^{-2}\ m^2\ s^{-1}$ to $K_z = 7.1 \times 10^{-3}\ m^2\ s^{-1}$ at the surface and mid-depths, respectively.

With the parameterizations presented above (Eqs. 8.4 to 8.9) and the approximations of equalities (8.10) and (8.11), the analytical simulations of the eddy shear stress, responsible for internal energy dissipation, may be incorporated in the horizontal equations of motion, which are expressed as:

$$\frac{\partial u}{\partial t} + \frac{\partial (uu)}{\partial x} + \frac{\partial (uv)}{\partial y} + \frac{\partial (uw)}{\partial z} - f_0 v$$
$$= -\frac{1}{\rho}\frac{\partial p}{\partial x} + \frac{1}{\rho}[\frac{\partial}{\partial x}(A_x \frac{\partial u}{\partial x}) + \frac{\partial}{\partial y}(A_y \frac{\partial u}{\partial y}) + \frac{\partial}{\partial z}(A_z \frac{\partial u}{\partial z})], \tag{8.12}$$

and

$$\frac{\partial v}{\partial t} + \frac{\partial (vu)}{\partial x} + \frac{\partial (vv)}{\partial y} + \frac{\partial (vw)}{\partial z} + f_0 u$$
$$= -\frac{1}{\rho}\frac{\partial p}{\partial y} + \frac{1}{\rho}[\frac{\partial}{\partial x}(A_x \frac{\partial v}{\partial x}) + \frac{\partial}{\partial y}(A_y \frac{\partial v}{\partial y}) + \frac{\partial}{\partial z}(A_z \frac{\partial v}{\partial z})]. \tag{8.13}$$

There is no clear evidence supporting the simplifications required to obtain the analytical expressions of Eqs. (8.12 and 8.13), but the hypothesis for the simplest form of the third component (Eq. 8.3) of the general equation of motion was justified. Others simplifications will be made to these equations according to the relative importance of their terms and estuary geometry, enabling relative simple analytic and numerical solutions. Because the dissipative terms $A_x(\partial u/\partial x)$, $A_x(\partial v/\partial y)$, $A_y(\partial u/\partial y)$ and $A_y(\partial v/\partial y)$ generally have orders of magnitude much less than $A_z(\partial u/\partial z)$ and $A_z(\partial v/\partial z)$, they may be disregarded in analytical solutions. With these simplifications the horizontal components of the general equation of motion are reduced to:

$$\frac{\partial u}{\partial t}+\frac{\partial(uu)}{\partial x}+\frac{\partial(uv)}{\partial y}+\frac{\partial(uw)}{\partial z}-f_0 v=-\frac{1}{\rho}\frac{\partial p}{\partial x}+\frac{1}{\rho}[\frac{\partial}{\partial z}(A_z\frac{\partial u}{\partial z})], \tag{8.14}$$

and

$$\frac{\partial v}{\partial t}+\frac{\partial(vu)}{\partial x}+\frac{\partial(vv)}{\partial y}+\frac{\partial(vw)}{\partial z}+f_0 u=-\frac{1}{\rho}\frac{\partial p}{\partial y}+\frac{1}{\rho}[\frac{\partial}{\partial z}(A_z\frac{\partial v}{\partial z})]. \tag{8.15}$$

To solve some theoretical problems of estuarine dynamics using a simplified formula, while taking into account some of the physical characteristics in natural estuarine systems, a mean depth value for the viscosity coefficient A_z (or N_z) may be used. In these conditions, the equations are simplified to:

$$\frac{\partial u}{\partial t}+\frac{\partial(uu)}{\partial x}+\frac{\partial(uv)}{\partial y}+\frac{\partial(uw)}{\partial z}-f_0 v=-\frac{1}{\rho}\frac{\partial p}{\partial x}+N_z(\frac{\partial^2 u}{\partial z^2}), \tag{8.16}$$

and

$$\frac{\partial v}{\partial t}+\frac{\partial(vu)}{\partial x}+\frac{\partial(vv)}{\partial y}+\frac{\partial(vw)}{\partial z}+f_0 u=-\frac{1}{\rho}\frac{\partial p}{\partial y}+N_z(\frac{\partial^2 v}{\partial z^2}), \tag{8.17}$$

where $N_z = A_z/\rho$ is the mean kinematic eddy viscosity coefficient in the water column. This coefficient has dimensions of $[L^2T^{-1}]$ ($m^2\ s^{-1}$ SI units).

As previously shown (Eq. 2.10a, b Chap. 2), the first term in the right-hand-side of these equations are the barotropic and baroclinic pressure gradient pressure force, generated by inclinations of the free surface and mass (density or salinity) longitudinal distribution in the estuarine water body, respectively. Disregarding the barometric component of the gradient pressure force, the Ox and Oy components are:

$$-\frac{1}{\rho}(\frac{\partial p}{\partial x})=-g(\frac{\partial \eta}{\partial x})-\frac{g}{\rho}(\int_z^{\eta}\frac{\partial \rho}{\partial x}dz), \tag{8.18a}$$

and

$$-\frac{1}{\rho}(\frac{\partial p}{\partial y})=-g(\frac{\partial \eta}{\partial y})-\frac{g}{\rho}(\int_z^{\eta}\frac{\partial \rho}{\partial y}dz), \tag{8.18b}$$

where $\eta = \eta(x, y, t)$. Substituting in the baroclinic component the longitudinal the density gradient by the corresponding salinity gradient using the linear equation of state of seawater $\rho(S) = \rho_0(1 + \beta S)$, it follows:

$$-\frac{1}{\rho}(\frac{\partial p}{\partial x}) = -g(\frac{\partial \eta}{\partial x}) - \beta g(\int_{z}^{\eta} \frac{\partial S}{\partial x} dz), \quad (8.19a)$$

and

$$-\frac{1}{\rho}(\frac{\partial p}{\partial y}) = -g(\frac{\partial \eta}{\partial y}) - \beta g(\int_{z}^{\eta} \frac{\partial S}{\partial y} dz), \quad (8.19b)$$

with the ratio ρ/ρ_0 approximate by one (1) in the last expression. As shown in the Chap. 2, the barotropic component (first terms on the right-hand-side of the equations) is dependent on the time oscillation and the free surface inclination in the longitudinal ($\partial\eta/\partial x$) and lateral ($\partial\eta/\partial y$) directions, but is independent on the depth (z). The barotropic circulation in the longitudinal direction (tidal currents) predominates in the majority of estuaries. As the longitudinal component varies periodically, the seaward ($\partial\eta/\partial x > 0$) and the landward ($\partial\eta/\partial x < 0$) intensities are dependent on the tidal height; at high and low tidal heights, the water motion tends to zero (slack water). The vertical mixing process is also dependent on the tidal forcing and on the vertical stratification of the estuarine water mass. As demonstrated by Pritchard (1954), it is possible to estimate the vertical salt eddy diffusion coefficient (K_z) if the magnitude of the tidal oscillatory circulation is known.

In the second term on the right-hand-side of Eqs. (8.18a, b and 8.19a, b), the baroclinic pressure gradient intensity is expressed by the longitudinal (transversal) density (salinity) gradient, as a function of the depth; it is always oriented landward, because $\partial\rho/\partial x$ and $\partial S/\partial x$) are positive. As changes in the salinity field occur during the tidal cycle, this component presents complex time variations during the tidal cycle. Although the intensity of this component is one order of magnitude less than the barotropic pressure gradient, and it is an important component of the estuarine circulation. The increase of the u-velocity component with depth due to the baroclinic gradient pressure and the bottom shear stress, is one of the forces responsible for the vertical velocity shear ($\partial u/\partial z$), and the vertical mixing variation due to the river discharge and/or the tidal forcing intensities may be investigated with the estuarine Richardson number (Eq. 2.36, Chap. 2). The transversal components of the baroclinic pressure gradient (Eqs. 8.18b and 8.19b), will be used later in the presentation of a simple model related to the secondary circulation.

During the ebb tide, the barotropic component of the gradient pressure force acts in the opposite direction to the baroclinic component, and the intensity of the u-velocity component decreases with depth; however, during the flood tide, these components are oriented landward, resulting in the longitudinal velocity increasing with depth. In partially mixed estuaries, due to the simultaneous action of the barotropic and baroclinic components and the upward motion due to the vertical mixing, the time mean vertical velocity profile during one or more tidal cycles, simulating nearly steady-state circulation, presents a characteristic bi-directional motion which is seaward and landward in the upper and lower layers, respectively.

The equations of motion (8.14) and (8.15), associated with the continuity Eq. (7.19), were simplified and reduced to a set of equations in the coordinates system Oxz. This equation system was solved by Ianniello (1979) to derive analytical solutions for the longitudinal and cross-channel residual currents induced in narrow tidal channels with variable breadth and depth. The system was forced by first order non-linear tidal forcing with the following simplifications: sources of river input, as well as wind stress were omitted, and the ratio of the tidal amplitude (η_0) to the local depth (H_0) was supposed to be much less than one ($\ll 1$). Solutions for channels with exponentially decreasing breadth and depth profiles, in comparison to the solutions for a channel with breadth and depth constants, indicate significant differences in the residual currents.

The intensity of the baroclinic component is dependent on the longitudinal density (salinity) gradient and the depth (last term on the right-hand-side of Eqs. (8.18a and 8.19a). To demonstrate the importance of this component, let us examine Fig. 8.1, which presents vertical profiles of the isopicnals, $\rho - (1/2\rho)$, ρ, and $\rho + (1/2\rho)$ in an estuary with two distinct vertical stratifications: well-mixed (vertical isopicnals) and highly stratified, which may be generated by low and high river discharges, respectively. Using the method of finite differences to calculate the longitudinal density gradient in the vicinity of the point M, we may observe an inequality in the gradients $\Delta\rho/\Delta x < \Delta\rho/\Delta x'$, for $\Delta x' < \Delta x$. Then, the longitudinal density gradient and, consequently, the intensity of the baroclinic component are also dependent on the vertical stratification of the water column, increasing with higher vertical stratification, and decreasing when the estuary is almost vertically homogeneous.

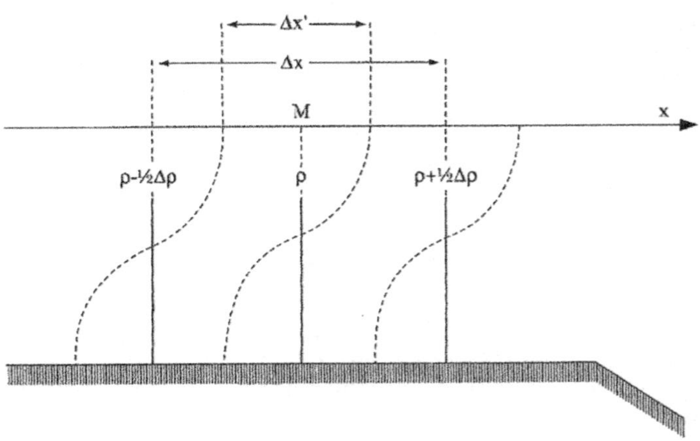

Fig. 8.1 Different vertical density stratification to calculate by finite differences the longitudinal density gradient and the intensity of the baroclinic pressure gradient force in the vicinity of the point M

The previously introduced Eqs. (8.1–8.3) and (8.12 and 8.13) associated with the equations of mass conservation (Eq. 7.19, Chap. 7) and salt conservation (Eq. 7.25, Chap. 7), and a simplified equation of state of seawater (Eq. 4.11, Chap. 4) constitute an Eulerian equation system with six partial differential equations having u, v, w, p, ρ and η as unknowns, and the initial and boundary conditions are correctly specified, these unknowns may theoretically be calculated in the space and time (x, y, z, t).

As the objective of a tridimensional model is in generally to theoretically calculate sea-surface elevations, η = η(x, y, t), velocity components, u = u(x, y, z, t), v = v(x, y, z, t), w = w(x, y, z, t) and salinity S = S(x, y, z, t) fields, this equation system will be a closed one (number of equations is equal to the number of unknowns) if the coefficients of eddy viscosity and diffusion are also known. However, the following quantities are also necessary:

- Estuary geometry;
- River discharge;
- Salinity of the coastal region;
- Amplitude and tidal phase at the estuary mouth;
- Wind field at the free surface.

The components of the equation of motion (Eqs. 8.12 and 8.13) are the starting point for reducing the tri-dimensional formulation to the bi- and one-dimensional formulations, using the theoretical procedure already made (Chap. 7) to the equations (mass conservation) and salt conservation, these simplifications are mainly necessary for analytic solutions.

8.2 Boundary and Integral Conditions

To guarantee a unique solution to the equation of motion, the determinate initial, upper and lower boundary conditions and, in some cases, specified conditions expressed as integrals (boundary integral conditions) must be specified. The initial conditions states values to the unknowns at the initial instant of time (t = 0), as for instance, the velocity field is at rest at a given spatial position u(x, y, z, 0) = 0; v(x, y, z, 0) = 0, w(x, y, z, 0) = 0, and the boundary conditions indicate specified values at the boundaries of the space occupied by the estuarine water mass.

To simplify the presentation of boundary and integral boundary conditions, let us start with the equation of motion (8.16). To establish the lower and upper boundary conditions, both members of the equation must be multiplied by the differential, dz, whose terms integrate from the bottom ($z = -H_0$) to the surface (z = η),

$$\int_{-H_0}^{\eta}\left[\frac{\partial u}{\partial t}+\frac{\partial(uu)}{\partial x}+\frac{\partial(uv)}{\partial y}+\frac{\partial(uw)}{\partial z}-f_0 v\right]dz=\int_{-H_0}^{\eta}\left(-\frac{1}{\rho}\frac{\partial p}{\partial x}\right)dz+\int_{-H_0}^{\eta}\left[\frac{\partial}{\partial z}\left(N_z\frac{\partial u}{\partial z}\right)\right]dz. \tag{8.20}$$

Taking into account that the gradient pressure force has two components, the barotropic and baroclinic (Eq. 8.18a), the first term on the right-hand-side may be written as,

$$-g(H_0+\eta)\frac{\partial\eta}{\partial x}-\frac{g}{\rho}\int_{-H_0}^{\eta}\left(\int_{z}^{\eta}\frac{\partial\rho}{\partial x}dz\right)dz\approx-(gH_0)\frac{\partial\eta}{\partial x}-\frac{g}{\rho}\int_{-H_0}^{\eta}\left(\int_{z}^{\eta}\frac{\partial\rho}{\partial x}dz\right)dz, \tag{8.21}$$

taking into account that $\eta \ll H_0$, and that the barotropic component is independent of depth, $\frac{\partial}{\partial z}\left(\frac{\partial\eta}{\partial x}\right)=\frac{\partial}{\partial x}\left(\frac{\partial\eta}{\partial z}\right)=0$. If the longitudinal density gradient is also independent of depth (taken, for example, as a depth mean value), it follows that:

$$-gH_0\frac{\partial\eta}{\partial x}-\frac{1}{2}\frac{g}{\rho}\frac{\partial\rho}{\partial x}H_0^2, \tag{8.22}$$

Although this relationship doesn't characterize a boundary condition, it indicates that the barotropic and baroclinic components may be considered, as a first approximation, to be proportional to H_0 and H_0^2, respectively.

The integral of the shear stress due to the eddy viscosity of the fluid (last term of the right-hand-side of Eq. 8.20) may be physically interpreted as the difference of the longitudinal components of the wind stress (τ_{Wx}) and the bottom friction (τ_{Bx}), and it is simplified to

$$\int_{-H_0}^{\eta}\left[\frac{\partial}{\partial z}\left(N_z\left(\frac{\partial u}{\partial z}\right)\right)\right]dz=N_z\left(\frac{\partial u}{\partial z}\right)\Big|_{z=\eta}-N_z\left(\frac{\partial u}{\partial z}\right)\Big|_{z=-H_0}=-\frac{1}{\rho}(\tau_{Wx}-\tau_{Bx}). \tag{8.23}$$

The quantities τ_{Wx} and τ_{Bx} have dimensions of force per unit of area ($ML^{-1}T^{-2}$). The wind component (τ_{Wx}) acts on the estuarine free surface and may accelerate or non-accelerate the estuarine circulation due to changes in the wind direction, and the bottom stress τ_{Bx} attenuates the motion due to the frictional forces on the bottom; thus, these stresses are the upper and lower boundary conditions, respectively.

The wind stress (τ_{Wx}) must be taken into account in the theoretical simulation of estuaries which have a large free surface and an energetic wind field. However, this forcing has also been taken into account in studies of the influence of the wind stress on the dynamics of shallow and small estuaries as, for example, in those investigated by Geyer (1997). The imposition of the upper boundary condition is analytically formulated by,

$$\tau_{Wx} = \rho N_z \frac{\partial u}{\partial z}|_{z=\eta} = A_z \frac{\partial u}{\partial z}|_{z=\eta}, \tag{8.24}$$

where $A_z = \rho N_z$, $[A_z] = [ML^{-1}T^{-1}]$ is the dynamic eddy viscosity coefficient. With the orientation axis according to Fig. 8.1 and the wind stress is acting in the Ox axis direction, $\tau_{Wx} > 0$; when the wind forcing may be disregarded, this boundary condition is reduced to a null vertical velocity shear, $(\partial u/\partial z|_{z=\eta}) = 0$.

When this boundary condition is applied ($\tau_{Wx} \neq 0$), it will appear explicitly as a forcing term in the solution of the equation of motion, and the wind speed (U) must be taken into account for its determination using the semi-empirical relationship:

$$\tau_{Wx} = \rho_{ar} C_D |U| U_V. \tag{8.25}$$

The quantities $|U|$ and U_V are the wind intensity modulus and its component in the direction of the Ox axis, calculated by

$$U_V = U \cos(\vartheta), \tag{8.26}$$

according the methodology presented in Chap. 5 (Eq. 5.6).

The ocean's free surface responds to the wind shear stress by generating a motion intensity that is approximately 1.3% of the wind velocity, measured at a height of 10 m above the sea-level. This difference between the wind velocity and that generated at the ocean surface is due to the density differences of the seawater and the air. If the wind intensity is less than 6 m s^{-1}, a typical value of the drag coefficient (C_D) is 1.1×10^{-3}; however, for higher wind intensities this coefficient increases almost linearly, according to the equation (Smith 1980; Gill 1982)

$$C_D = (0.61 + 0.063|U|) \times 10^{-3}, \tag{8.27}$$

with the wind intensity in m s^{-1}. Alternative equations for calculating the C_D coefficient may be found in the classical Roll's (1965) and Charnock's (1981) books and in Liu et al. (1979).

The bottom boundary condition, denoted by τ_{Bx}, is a shear stress which dissipates energy and is formulated by

$$\tau_{Bx} = \rho N_z \frac{\partial u}{\partial z}|_{z=-H_0} = A_z \frac{\partial u}{\partial z}|_{z=-H_0}. \tag{8.28}$$

Depending to the problem where this boundary condition will be applied, it may assume one of the following characteristics for the main estuary axis:

- Maximum friction, when $u(x, y, z, t)|_{z=-H0} = 0$;
- Minimum friction, when $\tau_{Bx} = 0$, or $\partial u/\partial z|_{z=-H0} = 0$; or
- Moderated friction, when $\tau_{Bx} \neq 0$.

The first condition is often referred as a simplified condition or without slippery, and is also known as imposition of the adherence principle. In the second case, as the dynamic eddy viscosity coefficient is different from zero ($A_z \neq 0$), this implies that the vertical velocity shear must be zero at the bottom.

The final characteristic of a moderate bottom friction is usually the best condition to apply to minimize the deviation between theoretical and experimental results of vertical velocity profiles. Investigations of wide channels with free surface have indicated that the bottom shear stress may be parameterized as a function of the square of the velocity. For a partially-mixed estuary (type 2 or B), this parameterization is given by

$$\tau_{Bx} = \frac{\rho g u}{C_y^2}(u^2 + v^2)^{1/2}, \tag{8.29}$$

where C_y is the Chézy coefficient and its dimension $[C_y] = [L^{1/2}T^{-1}]$. This coefficient was obtained by the French engineer Antoine Chézy in 1769, to analytically describe the uniform and one-dimensional velocity in a channel. This equation of motion may be theoretically obtained from the balance of gravity and frictional forces. However, in the Chézy's pioneer investigation, the analytical formulation was obtained from experimental results in channels with different free surface slopes, and observational velocity data obtained in the Seine river (Paris, France). The uniform velocity (u) is expressed by:

$$u = C_y\sqrt{R_H(\frac{\partial \eta}{\partial x})}, \tag{8.30a}$$

and the quantities in this equation are:

- The Chézy coefficient calculated by: $C_y = \sqrt{\frac{2g}{f_r}}$, where f_r is a non-dimensional parameter ($f_r = 1$ in SI units). Its dimension is $[C_y] = [L^{1/2}T^{-1}]$.
- R_H is the hydraulic radius, defined by the ratio of the transversal section area of the channel (A) and the *wet perimeter* (P_W): $R_H = A/P_w$. The perimeter, P_W, is defined as the contour length of the cross section in contact with the water, which effectively exerts frictional resistance to the motion. According to this definition $[R_H] = [L]$.

In 1867, almost one century after the Chézy coefficient had been used in Hydraulics, the French engineer Philippe Gauckler, introduced the known Gauckler formula, to calculate the u-velocity in channels, which was later re-developed by the Irish engineer Robert Manning in 1890, and it is also named Manning's formula:

$$u = \frac{g^{1/2}(R_H)^{2/3}}{n} (\frac{\partial \eta}{\partial x})^{1/2}. \quad (8.30b)$$

In this equation, u has the same meaning as in Eq. (8.30a), and n is the Manning number which simulates the energy dissipation by *f*rictional forces, and its dimension is $[n] = [L^{1/6}]$. Comparing Eqs. (8.30a and b), it follows that the relationship with the Chézy coefficient is: $n = \frac{g^{1/2} R_H^{1/6}}{C_y}$. Values of the Chézy coefficient and the Manning numbers vary according to the frictional forces at the bottom of the channel and, according to the literature, have orders of magnitude of 60.0 $m^{1/2}s^{-1}$ and 0.03 $m^{1/6}$, respectively. The low values of the Manning's friction coefficient ($n = 0.015$ — 0.020 $m^{1/6}$) found in the Fly river estuary (Papua New Guinea) and the South Alligator river (Australia) by King and Wolanski (1996) were attributed to be due the drag reduction by the suspended clay and/or to fluid mud on the bottom, thus relaxing the non-slip condition for the bulk of the water column.

As the motion in estuaries is predominantly longitudinal, the bottom shear stress (Eq. 8.29) simplifies to:

$$\tau_{Bx} = \frac{\rho g \mid u \mid u}{C_y^2} = \rho k \mid u \mid u, \quad (8.31)$$

where $k = g / C_y^2$ is a non-dimensional coefficient named *roughness coefficient*. This expression is formally similar to the previously presented analytical expression of the shear stress as function of a reference velocity ($\tau = \rho C_D U_r^2$, Eq. 5.14, Chap. 5). Because, in the theoretical deduction of the Chézy equation, the frictional force per unit area was taken as proportional to the square of the velocity, which is uniform in the cross sectional area, the quantity $|u|$ in the above equation is the velocity modulus.

Experimental results indicate that the Chézy coefficient depends not only on the bottom slope, but also on the shear at the bottom and on the channel geometry. Therefore, care should be taken when applying theoretical models to estuarine channels and simulating the energy dissipation with Eq. (8.31). The Chézy coefficient variations may be used to calibrate the model results against experimental data.

As the rough or slippery bottom characteristic of estuarine channels is also related to tidal currents intensities, another analytical expression which may be applied under moderate bottom friction was formulated in the classical article of Bowden (1953) and applied by Prandle (1982):

$$\tau_{Bx} = \rho N_z(\frac{\partial u}{\partial z})|_{z=-H_0} = \rho\,(\frac{4}{\pi})kU_0u|_{z=-H_0}, \tag{8.32}$$

where U_0 is the velocity amplitude generated by the tide, and k is a non-dimensional coefficient. As this equation is related to a bottom boundary condition, the velocity at the bottom is theoretically calculated by the general equation of motion.

Let us now present the integral boundary condition used in bi-dimensional models in the Oxz space. This boundary condition is obtained from the continuity equation, and thus represents an alternative formulation of this principle to transform the set of equations to be solved as a hydrodynamic closed system (the number of equations is equal to the number of unknowns). As a starting point, the continuity equation is integrated in a transversal section (Eq. 7.92a, Chap. 7); taking into account that u-velocity component is uniform in the cross-section area, this equation may be written as:

$$\frac{\partial}{\partial x}[\iint_A u(x,z,t)dA] + \frac{\partial A}{\partial t} = 0, \tag{8.33}$$

where:

$$[\iint_A u(x,z,t)dA] = u(x,z,t)A. \tag{8.34}$$

Integrating Eq. (8.33) from a position at the estuary head, x = 0, to a generic seaward position, x, and with a river discharge Q_f (Fig. 8.2), the result is (Pritchard 1971):

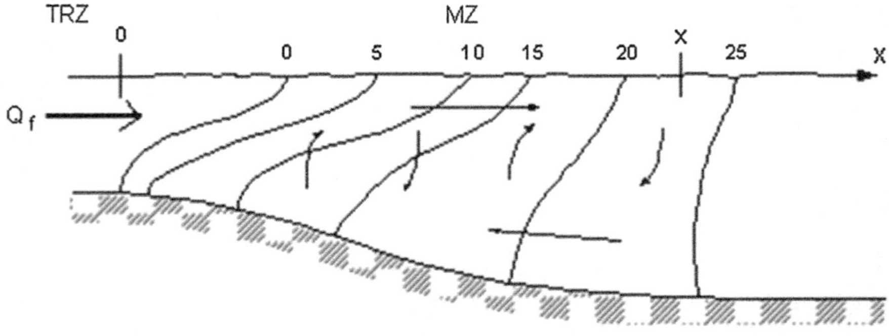

Fig. 8.2 Schematic diagram of a partially mixed estuary. The river input and the mixing zones are indicated by TRZ and MZ, respectively. Zero (0) and x indicates the initial and final integration interval

$$\iint_A u(x,z,t)dA - Q_f + \frac{\partial}{\partial t}\left(\int_0^x A dx\right) = 0. \quad (8.35)$$

As the result of Eq. (8.35) is equal to the one obtained in Chap. 7 (Eq. 7.107a, b) and, as its last term is the water volume landward of the cross-section located in the position x, it follows that:

$$\iint_A u(x,z,t)dA + \frac{\partial V}{\partial t} = Q_f. \quad (8.36)$$

Under steady-state condition, $\partial V/\partial t = 0$, and taking into account that the integrand in the first term in the left-hand-side is constant in the cross-section area (laterally homogeneous estuary), this equation is reduced to:

$$B\int_{-H_0}^{\eta} u(x,z)dz = Q_f \quad (8.37)$$

where the estuary width, $B = \int_a^b dy = (b - a)$. Under the assumption that $B = \text{const.}$, $A = B(H_0 + \eta) \approx BH_0$, and $Q_f = u_f BH_0$, the equation is reduced to:

$$\frac{1}{H_0}\int_{-H_0}^{0} u(x,z)dz = \frac{Q_f}{A} = u_f. \quad (8.38)$$

Hence, we have an alternative relationship to the continuity equation. As it is applied by an integral formulation, it is referred to as an integral boundary condition. In particular, if, in a given problem, it is necessary to impose a null volume transport, this boundary condition is reduced to

$$\int_{-H_0}^{0} u(x,z)dz = 0. \quad (8.39)$$

The equations of motion, continuity and salt conservation, associated with an equation of state of seawater and the appropriate initial and boundary conditions, form a system of equations which includes the main forcing and the energy dissipation that control the circulation and mixing processes of an estuarine system.

8.3 Bi-Dimensional Formulations: Vertical and Lateral Integration

In Chap. 7, we presented the bi-dimensional formulations of the mass (continuity) and salt conservation equations used in solutions of vertical and laterally homogeneous estuaries, which have (x, y, t) and (x, z, t) as independent variables, respectively. Similarly, the main principles related to the deduction of the bi-dimensional circulation, starting with the tridimensional equations of motion, will be presented in this here.

8.3.1 Vertical Integration

To obtain the vertically integrated equation of motion, which can be applied in studies of vertically homogeneous estuaries, let us start with the Eqs. (8.14 and 8.15), where the lateral momentum exchanges are considered negligible. The vertical integration of these equations in the water column from $z = -H_0(x, y)$ and $z = \eta(x, y, t)$ are

$$\int_{-H_0}^{\eta} \frac{\partial u}{\partial t} dz + \int_{-H_0}^{\eta} \frac{\partial (uu)}{\partial x} dz + \int_{-H_0}^{\eta} \frac{\partial (uv)}{\partial y} dz + \int_{-H_0}^{\eta} \frac{\partial (uw)}{\partial z} dz - f_0 \int_{-H_0}^{\eta} v dz$$
$$= - \int_{-H_0}^{\eta} \frac{1}{\rho} \frac{\partial p}{\partial x} dz + \int_{-H_0}^{\eta} \frac{1}{\rho} \frac{\partial}{\partial z} (A_z \frac{\partial u}{\partial z}) dz, \qquad (8.40)$$

and

$$\int_{-H_0}^{\eta} \frac{\partial v}{\partial t} dz + \int_{-H_0}^{\eta} \frac{\partial (vu)}{\partial x} dz + \int_{-H_0}^{\eta} \frac{\partial (vv)}{\partial y} dz + \int_{-H_0}^{\eta} \frac{\partial (vw)}{\partial z} dz + f_0 \int_{-H_0}^{\eta} u dz$$
$$= - \int_{-H_0}^{\eta} \frac{1}{\rho} \frac{\partial p}{\partial y} dz + \int_{-H_0}^{\eta} \frac{1}{\rho} \frac{\partial}{\partial z} (A_z \frac{\partial v}{\partial z}) dz. \qquad (8.41)$$

Applying the Leibnitz rule of derivation of an integral to the terms in the left-hand-side of Eq. (8.40), and applying the upper and lower boundary conditions to the last term of the right-hand-side of this equation it follows that:

$$\begin{aligned}
&\frac{\partial}{\partial t}(\int_{-H_0}^{\eta} udz) + \frac{\partial}{\partial x}[\int_{-H_0}^{\eta} (uu)dz] + \frac{\partial}{\partial y}[\int_{-H_0}^{\eta} (uv)dz] \\
&- u|_{\eta}\frac{\partial \eta}{\partial t} + u|_{-H_0}\frac{\partial(-H_0)}{\partial t} - (uu)|_{\eta}\frac{\partial \eta}{\partial x} + (uu)|_{-H_0}\frac{\partial(-H_0)}{\partial x} \\
&- (uv)|_{\eta}\frac{\partial \eta}{\partial y} + (uv)|_{-H_0}\frac{\partial(-H_0)}{\partial y} + (uw)|_{\eta} - (uw)|_{-H_0} - f_0\int_{-H_0}^{\eta} vdz \\
&= -\int_{-H_0}^{\eta}(\frac{1}{\rho}\frac{\partial p}{\partial x})dz + \frac{1}{\rho}(\tau_{Wx} - \tau_{Bx}).
\end{aligned} \tag{8.42}$$

Factoring the terms of the left-hand-side of the equation, which becomes zero due to the surface and bottom boundary conditions ($u|_{-H0} = u|\eta = 0$),

$$u|_{-H_0}[\frac{\partial(-H_0)}{\partial t} + u|_{-H_0}\frac{\partial(-H_0)}{\partial x} + v|_{-H_0}\frac{\partial(-H_0)}{\partial y} - w|_{-H_0}] = 0, \tag{8.43}$$

and

$$u|_{\eta}[\frac{\partial \eta}{\partial t} + u|_{\eta}\frac{\partial \eta}{\partial x} + v|_{\eta}\frac{\partial \eta}{\partial y} - w|_{\eta}] = 0. \tag{8.44}$$

Thus, the Ox component of the equation of motion is reduced to the following expression:

$$\begin{aligned}
&\frac{\partial}{\partial t}(\int_{-H_0}^{\eta} udz) + \frac{\partial}{\partial x}[\int_{-H_0}^{\eta} (uu)dz] + \frac{\partial}{\partial y}[\int_{-H_0}^{\eta} (uv)dz] - f_0\int_{-H_0}^{\eta} vdz \\
&= -\int_{-H_0}^{\eta}\frac{1}{\rho}\frac{\partial p}{\partial x}dz + \frac{1}{\rho}(\tau_{Wx} - \tau_{Bx}).
\end{aligned} \tag{8.45}$$

As the estuary is vertically homogeneous, the integration may be completed, and the longitudinal component of the general equation of motion is simplified to:

$$\frac{\partial(uh)}{\partial t} + \frac{\partial(uuh)}{\partial x} + \frac{\partial(uvh)}{\partial y} - f_0 vh = -\frac{h}{\rho}\frac{\partial p}{\partial x} + \frac{1}{\rho}(\tau_{Wx} - \tau_{Bx}). \tag{8.46}$$

With a similar procedure, it follows that the equation to calculate the transversal vertically integrated velocity component (v) is:

$$\frac{\partial(vh)}{\partial t}+\frac{\partial(uvh)}{\partial x}+\frac{\partial(vvh)}{\partial y}+f_0uh=-\frac{h}{\rho}\frac{\partial p}{\partial y}+\frac{1}{\rho}(\tau_{Wy}-\tau_{By}). \quad (8.47)$$

When combined with the corresponding bi-dimensional form of the continuity Eq. 7.56a (Chap. 7), Eqs. (8.46) and (8.47) may be transformed into expressions that are more convenient for practical applications. Thus, they are simplified to:

$$\frac{\partial u}{\partial t}+u\frac{\partial u}{\partial x}+v\frac{\partial u}{\partial y}-f_0v=-\frac{1}{\rho}\frac{\partial p}{\partial x}+\frac{1}{\rho h}(\tau_{Wx}-\tau_{Bx}), \quad (8.48)$$

and

$$\frac{\partial v}{\partial t}+u\frac{\partial v}{\partial x}+v\frac{\partial v}{\partial y}+f_0u=-\frac{1}{\rho}\frac{\partial p}{\partial y}+\frac{1}{\rho h}(\tau_{Wy}-\tau_{By}). \quad (8.49)$$

It should be noted that in the longitudinal component (Eq. 8.48), the gradient pressure force has two mainly components: the barotropic and baroclinic pressure gradients (Eq. 8.18a, b or 8.19a, b). As the baroclinic component is dependent on the mass or density (salinity) distribution, for a complete closed hydrodynamic system it is necessary to associate these equations with the vertically integrated salt conservation (Eq. 7.68a, b, Chap. 7).

$$\frac{\partial S}{\partial t}+u\frac{\partial S}{\partial x}+v\frac{\partial S}{\partial y}=\frac{1}{h}\left[\frac{\partial}{\partial x}\left(K_xh\frac{\partial S}{\partial x}\right)+\frac{\partial}{\partial y}\left(K_yh\frac{\partial S}{\partial y}\right)\right]+\frac{\partial S}{\partial z}\Big|_{z=0}(P-E_V), \quad (8.50)$$

and a simplified equation of state of seawater,

$$\rho(S)=\rho_0(1+\beta S), \quad (8.51)$$

where ρ_0 is a density constant, and β is the mean coefficient of saline contraction in the estuarine MZ.

Equations (8.48 to 8.51) are analytical formulations of the main processes that control the circulation and the mixing processes in a well-mixed estuary. If the initial, boundary and integral boundary conditions, the eddy viscosity and diffusion coefficients are known, these equations, combined with the continuity Eq. (7.56a, Chap. 7), may be solved by numerical or analytical methods (under simplifying conditions) to obtain the following unknowns: the free surface elevation, $\eta=\eta(x, y, t)$, the velocity components $u = u(x, y, t)$ and $v = v(x, y, t)$, and the salinity $S = S(x, y, t)$.

The pioneering work of Ferraz (1975), investigating the tidal heights and tidal currents in the Northern channel of the Amazon river, between the estuarine mouth and the Macapá (Amapá, Brazil), is an example of this equation system being solved numerically, approximated by the barotropic condition (ρ = cte).

The following studies of other Brazilian estuarine systems should also be sited:

- Results of the circulation of the estuarine region of the Patos Lagoon during the summer season, obtained from time series analysis of the wind stress, tidal heights and river discharges, combined with simulation of the barotropic mean vertical velocity, were presented by Möller et al. (1996). In this study, it was observed that wind stress oriented towards NE and SW, forced by the passages of meteorological systems frontal zones, generated ebbing and flooding oscillatory motions with nodal points at the central region of the lagoon.
- Intensive field observations and a vertically-averaged two-dimensional hydrodynamic model were used in dynamical studies of the well-mixed Fly River estuary (Papua New Guinea) by Wolanski et al. (1997). Tidal currents were successfully reproduced by the model, which was coupled with a non-steady state dispersive model to calculate the longitudinal distribution of salinity in the estuary. The main conclusions were that: (i) the river discharge, the spring-neap tidal cycle, the wind and the presence of fluid mud are all important in determining the salinity distribution in the estuary; (ii) the water slope, necessary to evacuate the fresh water input in the estuary, was at its largest and lowest at spring and neap tides, respectively.
- The tidally generated barotropic circulation in the Santos-São Vicente Bay (São Paulo, Brazil) and in estuarine channels of São Vicente, Santos and Bertioga were investigated by Harari and Camargo (1998).

8.3.2 *Cross-Section Integration*

Relatively narrow partially mixed estuaries (type 2 or B) forced by meso-tides may be approximated by a laterally homogeneous condition, and the tridimensional equations may be reduced to two-dimensions in the Oxz plane. In these estuaries, both tidal and gravitational circulations are important. Using a mathematical development similar to that used in the deduction of the bi-dimensional equations of continuity (Eq. 7.73a) and salt conservation (Eq. 7.77), it is necessary to laterally integrate the tri-dimensional equations of motion (8.14) between the coordinates $a = a(x, z)$ and $b = b(x, z)$ located at the estuary margins (Fig. 7.5, Chap. 7). Applying the Leibnitz rule to the derivation of an integral, and factorizing the common parcels $u|_a$ and $u|_b$, and the kinematic boundary conditions (Eqs. 7.72a and b, Chap. 7) to the u-velocity component, yields

$$u|_a(u|_a\frac{\partial a}{\partial x} - v|_a + w|_a\frac{\partial a}{\partial z}) = 0, \tag{8.52}$$

and

$$u|_b(u|_b\frac{\partial b}{\partial x}-v|_b+w|_b\frac{\partial b}{\partial z})=0. \quad (8.53)$$

Thus, the equation of motion is reduced to

$$\frac{\partial}{\partial t}(\int_a^b udy)+\frac{\partial}{\partial x}[\int_a^b (uu)dy]+\frac{\partial}{\partial z}[\int_a^b (uw)dy] = -\int_a^b (\frac{1}{\rho}\frac{\partial p}{\partial x})dy+\frac{\partial}{\partial z}[\int_a^b \frac{1}{\rho}(A_z\frac{\partial u}{\partial z})dy]. \quad (8.54)$$

As this model doesn't take into account the Oy direction, the component of the equation of motion in the Oy direction and the Coriolis acceleration were omitted in this equation. Applying the condition of transverse homogeneity, the indicated integration may be calculated, resulting in:

$$\frac{\partial(uB)}{\partial t}+\frac{\partial(uuB)}{\partial x}+\frac{\partial(uwB)}{\partial z}=-\frac{1}{\rho}(B\frac{\partial p}{\partial x})+\frac{1}{\rho}\frac{\partial}{\partial z}(A_zB\frac{\partial u}{\partial z}), \quad (8.55)$$

where B = b − a is the estuarine channel width.

Combining the Eq. (8.55) with the laterally integrated continuity equation (Eq. 7.73a, Chap. 7)

$$\frac{\partial(uB)}{\partial x}+\frac{\partial(wB)}{\partial z}=0, \quad (8.56)$$

it follows that the laterally integrated equation of motion is,

$$\frac{\partial u}{\partial t}+u\frac{\partial u}{\partial x}+w\frac{\partial u}{\partial z}=-\frac{1}{\rho}\frac{\partial p}{\partial x}+\frac{1}{B}\frac{\partial}{\partial z}(BN_z\frac{\partial u}{\partial z}), \quad (8.57a)$$

or, in terms of the barotropic and baroclinic pressure gradient,

$$\frac{\partial u}{\partial t}+u\frac{\partial u}{\partial x}+w\frac{\partial u}{\partial z}=-g\frac{\partial \eta}{\partial x}-\frac{g}{\rho}\int_z^\eta \frac{\partial \rho}{\partial x}dz+\frac{1}{B}\frac{\partial}{\partial z}(BN_z\frac{\partial u}{\partial z}). \quad (8.57b)$$

In these equations, N_z is the kinematic eddy viscosity coefficient and has the dimension $[N_z] = [L^2T^{-1}]$. As the baroclinic component of the gradient pressure force is dependent on the density (Eq. 8.57b), which also depends on the salinity distribution, it is necessary to associate the equation of motion with laterally integrated salt conservation equation and a linear equation of state of sea water,

$$\frac{\partial S}{\partial t}+u\frac{\partial S}{\partial x}+w\frac{\partial S}{\partial z}=\frac{1}{B}[\frac{\partial}{\partial x}(BK_x\frac{\partial S}{\partial x})+\frac{\partial}{\partial z}(BK_z\frac{\partial S}{\partial z})]+\frac{\partial S}{\partial z}|_{z=0}(P-E_v),\quad (8.58)$$

and

$$\rho=\rho_0(1+\beta S).\quad (8.59)$$

This system of Eqs. (8.56)–(8.59) constitutes a physical-mathematical formulation of the main processes that control the motions and mixing processes in a laterally homogeneous estuary. Under the assumption that the eddy viscosity coefficients are known, this is a closed hydrodynamic equation system. Imposing the initial and boundary conditions, it may be solved by analytical methods (Rattray and Hansen 1962; Fisher et al. 1972; Hamilton and Rattray 1978) and by numerical methods as a function of the non-dimensional depth (Blumberg 1975; Paiva and Rosman 1993, and others). From this solution, it is possible to calculate transverse mean values of the following quantities:

- The longitudinal and vertical velocity components u = u(x, z, t) and w = w(x, z, t);
- The salinity field S = S(x, z, t).

8.4 One-Dimensional Formulation

In relatively shallow and straits estuarine systems with low fresh water discharge and forced by medium and high tidal heights, hydrographic properties are almost uniformly distributed in transverse sections; such estuaries are generally considered to be well-mixed estuaries (types 1 or C). Under these conditions, as previously seen in the development of the one-dimensional equations of mass and salt conservation, the tri-dimensional equations of motion (Eq. 8.14) may be reduced to a simple formulation with only a single spatial variable and time (x, t) in the longitudinal direction (Ox).

The mathematical procedure is the same procedure used previously in the one-dimensional formulation of the mass and salt conservation equations. As before, we must integrate of the tri-dimensional equation in the area normal to the longitudinal axis (Ox), under the assumption that all properties are uniform in this area. The mathematical development follow that presented by Pritchard (1958), where we find an analysis of the main physical and mathematical concepts evolved in the deduction of these equations, formerly used without the application of the correct mathematical formalism.

Multiplying the Ox-longitudinal component of the equation of motion (Eq. 8.14) by a differential element dA = dxdz and integrating all terms in the cross-sectional area A = A(x, t) yields,

$$\begin{aligned}&\iint_A \left(\frac{\partial u}{\partial t}\right)dydz + \iint_A \left(\frac{\partial uu}{\partial x}\right)dydz + \iint_A \left(\frac{\partial uv}{\partial y}\right)dydz\\&+ \iint_A \left(\frac{\partial uw}{\partial z}\right)dydz - f_0 \iint_A vdydz\\&= - \iint_A \left(-\frac{1}{\rho}\frac{\partial p}{\partial x}\right)dydz + \iint_A \frac{1}{\rho}\frac{\partial}{\partial z}\left(A_z\frac{\partial u}{\partial z}\right)dydz.\end{aligned} \tag{8.60}$$

The first and second terms on the left-hand-side of this equation may be reduced by applying the Leibnitz rule of the derivation of a double integral, as used in the Eq. (7.82) (Chap. 7). The third and fourth terms of the left-hand-side and the last term on the right-hand-side of this equation may be transformed by applying the Green's formula and its consequences (as applied in the Chap. 7.) in the closed line (c) integrals around the area A. Then, the equation may be rewritten as

$$\begin{aligned}&\frac{\partial}{\partial t}\left(\iint_A udA\right) + \frac{\partial}{\partial x}\left[\iint_A (uu)dA\right] - \frac{1}{c}\frac{\partial A}{\partial t}\oint_c udl\\&-\frac{1}{c}\frac{\partial A}{\partial x}\oint_c (uu)dl + \oint_c (uv)dy - \oint_c (uw)dz\\&= - \iint_A \left(-\frac{1}{\rho}\frac{\partial p}{\partial x}\right)dA - \oint_c \left(\frac{A_z}{\rho}\frac{\partial u}{\partial z}\right)dz.\end{aligned} \tag{8.61}$$

In this equation, the volume transport generated by the Coriolis acceleration is null and has been neglected because the model doesn't solve the lateral dimension. This equation may be further simplified, taking into account the following equalities:

$$\frac{\partial A}{\partial x}(uu)|_a = \frac{\partial A}{\partial t}u|_a, \tag{8.62}$$

and

$$\oint_c (uv)dz - \oint_c (uw)dy = u|_a\left(\oint_c vdz - \oint_c wdy\right) = u|_a\left(2\frac{\partial A}{\partial t}\right). \tag{8.63}$$

Then

$$\frac{\partial}{\partial t}\left(\iint_A udA\right)+\frac{\partial}{\partial x}\left[\iint_A (uu)dA\right]=-\iint_A \left(\frac{1}{\rho}\frac{\partial p}{\partial x}\right)dA-\oint_c \left(\frac{A_z}{\rho}\frac{\partial u}{\partial z}\right)dy. \tag{8.64}$$

Imposing the condition that the physical properties are constant in the transversal section (A), this equation is simplified to

$$\frac{\partial (uA)}{\partial t}+\frac{\partial (uuA)}{\partial x}=-\frac{A}{\rho}\frac{\partial p}{\partial x}-\oint_c \left(\frac{A_z}{\rho}\frac{\partial u}{\partial z}\right)dy. \tag{8.65}$$

As the line integral in the last term of the right-hand-side of this equation must be calculated around the perimeter, c, of the area A (Fig. 7.8, Chap. 7), it physically represents the wind stress acting on the estuary's free surface, plus the frictional stress acting in the wet perimeter (P_m) of the transversal section A. Then, the following equality holds:

$$-\oint_c \left(\frac{A_z}{\rho}\frac{\partial u}{\partial z}\right)dy=\frac{1}{\rho}(\tau_{Wx}B-\tau_{Bx}P_m). \tag{8.66}$$

Combining the results of Eqs. (8.65) and (8.66), we have:

$$\frac{\partial (uA)}{\partial t}+\frac{\partial (uuA)}{\partial x}=-\frac{A}{\rho}\frac{\partial p}{\partial x}+\frac{1}{\rho}(\tau_{Wx}B-\tau_{Bx}P_m). \tag{8.67}$$

Finally, this equation may be simplified taking into account the one-dimensional expression of the continuity equation (Eq. 7.92, Chap. 7),

$$\frac{\partial (uA)}{\partial x}+\frac{\partial A}{\partial t}=0, \tag{8.68}$$

and the one-dimensional equation of motion takes the following expression:

$$\frac{\partial u}{\partial t}+u\frac{\partial u}{\partial x}=-\frac{1}{\rho}\frac{\partial p}{\partial x}+\frac{1}{\rho A}(\tau_{Wx}B-\tau_{Bx}P_m), \tag{8.69a}$$

or

$$\frac{\partial u}{\partial t}+u\frac{\partial u}{\partial x}=-g\frac{\partial \eta}{\partial x}-\frac{g}{\rho}\int_{-H_0}^{\eta}\frac{\partial \rho}{\partial x}dz+\frac{1}{\rho A}(\tau_{Wx}B-\tau_{Bx}P_m). \tag{8.69b}$$

As the mean value of the baroclinic gradient pressure force is dependent on the longitudinal mass distribution in the estuary, and hence on the salinity distribution, it is necessary to associate the last equation with the salt conservation equation (Eq. 7.98, Chap. 7), in which was introduced in the right-hand-side the term to take into account the salinity variations at the surface due to precipitation and evaporation.

$$\frac{\partial S}{\partial t} + u\frac{\partial S}{\partial x} = \frac{1}{A}\frac{\partial}{\partial x}(K_x A\frac{\partial S}{\partial x}) + \frac{\partial S}{\partial z}|_{z=0}(P - E_v) \tag{8.70}$$

and an equation of state of seawater or one of its simplified expressions, such as the linear expression,

$$\rho(S) = \rho_0(1 + \beta S). \tag{8.71}$$

The equation system, 8.68–8.71, is the physical-mathematical formulation of the main processes that control the circulation and mixing in estuaries which have geometries and physical characteristics that approximate a one-dimensional model. When the initial conditions, the boundary and integral boundary conditions, and the values of the eddy viscosity, diffusion coefficients and the $P - E_v$ rates are known, these equations may be solved by analytical and numerical models (Ippen and Harleman 1961; Harleman and Ippen 1967; Thatcher and Harleman 1972, and others), enabling determination of the unknowns:

- Surface elevation $\eta = \eta(x, t)$.
- The u-velocity component, $u = u(x, t)$, and
- Salinity $S = S(x, t)$.

In the last term of the one-dimensional equation of motion (8.69a, b), as previously indicated, the wind stress may be calculated by semi-empirical equations when the wind speed is known (Eq. 8.25). As such, the bottom frictional stress formulated in Eq. 8.29 is reduced to

$$\tau_{Bx} = \frac{g}{C_y^2}\rho|u|u, \tag{8.72}$$

To introduce the volume transport (Q) in this equation and the wet perimeter (P_m), as an artifact, the term on the right-hand-side must be multiplied and divided by the square of the cross-section area (A^2), and the result of which is multiplied by the wet perimeter, P_m. Then, the bottom stress is calculated by

$$\tau_{Bx}P_m = \frac{P_M \rho g}{C_y^2 A}|Q|Q, \tag{8.73}$$

where Q is the volume transport $[Q] = [L^3T^{-1}]$, and the ratio A/P_m is the hydraulic radius ($R_H = A/P_m$), which substituted in the above equation gives,

$$\tau_{Bx} = \frac{g}{C_y^2}\rho\frac{|Q|Q}{P_m R_H}. \tag{8.74}$$

In these conditions, combining the parametric formulation of the longitudinal wind shear stress (Eq. 8.25), the bottom shear (Eq. 8.74), and the decomposition of the gradient pressure force in its barotropic and baroclinic components (Eq. 8.18a), it follows that another expression for the one-dimensional equation of motion (Eq. 8.67) with the volume transport (uA = Q) as an unknown, is:

$$\frac{\partial Q}{\partial t}+u\frac{\partial Q}{\partial x}+Q\frac{\partial u}{\partial x}=-gA\frac{\partial \eta}{\partial x}-\frac{gAh}{\rho}\frac{\partial \rho}{\partial x}+\frac{\rho_{ar}}{\rho}C_D|U|U_W B-\frac{g|Q|Q}{C_y^2 A R_H}, \quad (8.75)$$

where the longitudinal density gradient ($\partial\rho/\partial x$) is independent of depth.

In the case of a steady-state well-mixed estuary (type 1 or C) which has a small river discharge, and where the wind-stress may be disregarded, Eq. (8.75) is simplified to the following dynamical equilibrium:

$$g\frac{\partial \eta}{\partial x}+\frac{g|Q|Q}{A^2C_y^2R_H}|=0. \quad (8.76)$$

This result expresses the balance between the barotropic component of the gradient pressure force and the bottom dissipative shear stress. This simplification holds for shallow estuaries because the high intensity of frictional forces reduces the importance of the local and advective accelerations terms.

The equation of motion (8.76) has two unknowns, $\eta = \eta(x)$ and $Q = Q(x)$, and to solve this equation it must be associated with the following continuity equation, which may be obtained from its one-dimensional formulation (Eq. 7.108, Chap. 7), integrated between the longitudinal positions x and x + δx, which yields:

$$\frac{dV}{dt}=Q|_x-Q|_{x+\delta x}, \quad (8.77)$$

which is simplified with a null fresh water discharge ($Q_f = 0$).

An implicit bi-dimensional numerical model developed by Wolanski et al. (1980), based on the solution of Eqs. (8.76 and 8.77), was applied by Kjerfve et al. (1991) in studies of the circulation of the complex system of interconnected estuarine channels in the North Inlet (South Caroline, USA). This is a well-mixed estuary, consisting of inundated channels with depths varying from 2 m and 5 m. Due to the medium semi-diurnal tidal heights, the banks of the channel are vegetated with Spartina-alterniflora. The channel system was subdivided in 75 cells, delimitated by transverse sections. Simulations of the mean u-velocity component at cross-sectional areas were compared with direct velocity measurements from experimental data for several cells, and, with adjustments of the Manning's (Chézy's) coefficient, the results were very close.

8.5 Simplifyed Formulation and Application

The velocity generated by the fresh water discharge ($u_f = Q_f/A$) may be calculated with the application of the continuity Eq. (7.92b, Chap. 7) under steady conditions. However, to calculate deviations of the constant value u_f, it is necessary to take into account the energy dissipation due to the internal and bottom friction, which is dependent of the eddy viscosity coefficient, and the bottom stress. Due to these frictional forces, the velocity generated by the fresh water discharge will present a vertical shear ($\partial u_f/\partial z \neq 0$) due to the energy dissipation. The basic equations which formulate this behavior are the bi-dimensional hydrodynamic equations in the Oxz plane (Eq. 8.57b) under steady-state conditions, with the gradient pressure force decomposed in the barotropic and baroclinic components (Eq. 8.18a), and the continuity equation formulated as integral boundary condition, expressed in Eq. (8.38).

The theoretical development will follow the articles of Prandle (1985, 2004, 2009), focusing on the investigations of the characteristics of the vertical velocity profile forced by the river discharge and the wind stress, considering bottom boundary conditions.

8.5.1 *Velocity Generate by the River Discharge*

To investigate the velocity structure generated by the river discharge, let us consider the density constant (ρ = const.), eliminating the local and advective acceleration and the wind stress on the free surface. For the bottom boundary conditions, the following hypothesis will be taken into account: maximum shear or non-slip conditions, and moderate friction (slippery bottom). To proceed with this investigation, the Oxz referential (Fig. 8.3) will be adopted, with the origin of the vertical axis at the estuary bottom and with the following simplifying hypothesis:

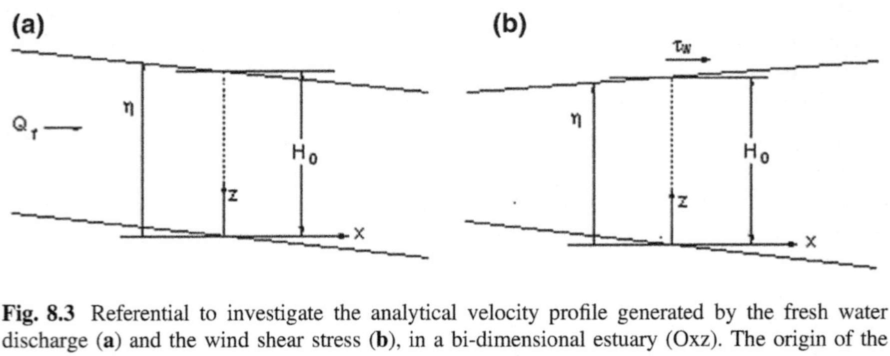

Fig. 8.3 Referential to investigate the analytical velocity profile generated by the fresh water discharge (**a**) and the wind shear stress (**b**), in a bi-dimensional estuary (Oxz). The origin of the vertical axis is located on the bottom (according to Prandle 1985)

- Constant transversal section area (B and H_0 = const.) and kinematic eddy viscosity coefficient constant (N_z = const.).
- Steady-state and uniform motion.

Under these simplifications, the linear steady-state equation of motion is forced by the barotropic component of the gradient pressure force, and Eq. (8.57b) is simplified to the following ordinary equation:

$$-g\frac{d\eta}{dx}+N_z\frac{d^2u}{dz^2}=0;\ \text{or}\ -g\eta_Q+N_z\frac{d^2u}{dz^2}=0. \tag{8.78}$$

In this equation, $d\eta/dx = \eta_Q$ is the steady-state slope of the free surface generated by the river discharge (Q_f), and indicates a balance, per unit of mass, between the barotropic component and the dissipating energy due to the vertical velocity shear. As there are two unknowns in this equation, u = u(z) and η_Q, its solution will only be obtainable using the integral boundary condition,

$$\frac{1}{H_0}\int_0^{H_0}u(z)dz=\frac{Q_f}{BH_0}=u_f. \tag{8.79}$$

As Eq. (8.78) is a second order differential equation with constant coefficients, it must be integrated two times in relation to the independent variable, z, and the result is the general solution:

$$u(z)=\frac{g}{2N_z}\eta_Q z^2+A_1z+A_2, \tag{8.80}$$

where A_1 and A_2 are integration constants with the dimensions $[A_1] = [T^{-1}]$ and $[A_2] = [LT^{-1}]$.

Initially, under the assumption of maximum bottom friction, these constants may be calculated with the following boundary conditions,

- Null wind stress on the surface, $\tau_{Wx} = \rho N_z(du/dz)|_{z=H0} = 0$, and
- Maximum friction on the bottom $u(0) = u|_{z=0} = 0$.

With these boundary conditions applied to the general solution (Eq. 8.80), it follows the analytical expressions for the A_1 and A_2,

$$A_1=-\frac{g\eta_Q}{N_z}H_0;\ \text{and}\ A_2=0, \tag{8.81}$$

and the particular solution of Eq. (8.80) is

$$u_Q(z)=\frac{g}{N_z}\eta_Q(\frac{z^2}{2}-H_0z), \tag{8.82}$$

where $u_Q = u_Q(z)$ is the velocity component generated by the river discharge. Factoring the second member of this equation by H_0^2 and introducing the non-dimensional depth, $Z = z/|H_0|$, which is more convenient for practical applications, yields the solution

$$u_Q(Z) = \frac{gH_0^2}{N_z}\eta_Q\left(\frac{Z^2}{2} - Z\right). \quad (8.83)$$

As this solution still has the unknown, η_Q, it may be determined applying the integral boundary condition (8.79), which written as a function of the non-dimensional depth, taking into account that $u(z) = u_Q(z)$, is

$$\frac{1}{H_0}\int_0^{H_0} u_Q(z)dz = \int_0^1 u_Q(Z)dZ = u_f. \quad (8.84)$$

Substituting the analytical expression, $u_Q = u_Q(Z)$, of Eq. (8.83) into Eq. 8.84, completing the integration and solving the equation for the second unknown, η_Q, its expression as a function of known quantities is

$$\eta_Q = \frac{d\eta}{dx} = -3\frac{N_z Q_f}{gBH_0^3} = -3\frac{N_z u_f}{gH_0^2}. \quad (8.85)$$

The steady-state free surface slope is directly proportional to the river discharge (Q_f) (or its velocity u_f) and inversely proportional to the third (or second) power of the channel depth. Because η_Q ($d\eta_Q/dx$) is negative, the free surface slopes downward and varies linearly with the longitudinal distance (Fig. 8.3a). Substituting the solution (8.85) into Eq. (8.83), the vertical velocity profile[1] is given by

$$u_Q(Z) = \frac{d\eta}{dx} = \frac{Q_f}{BH_0}\left(-\frac{3}{2}Z^2 + 3Z\right) = 3u_f\left(-\frac{1}{2}Z^2 + Z\right). \quad (8.86)$$

This solution agrees with the integral boundary condition (Eq. 8.84) and indicates that when the internal and bottom friction are taken into account, the velocity generated by the river discharge has a parabolic profile and the vertical velocity shear ($\partial u_Q/\partial z$) increases linearly from $u_f = 0$ at the surface to $3u_f$, at the bottom.

To investigate the influence of the wind stress ($\tau_{Wx} \neq 0$), a second particular solution for Eq. (8.80) may be obtained, maintaining the maximum friction at the bottom. Let us consider the following upper boundary condition:

[1]This velocity profile is equal to the one obtained by Officer (1978) with the Oz axis also oriented against the gravity acceleration, but with its origin at the free surface.

- Wind stress on the surface, $\tau_{Wx} = -\rho N_z(du/dz)|_{z=H0.}$

As du/dz < 0, the wind stress is being applied in the orientation indicated in Fig. 8.3b. Applying this condition, the values of the integration constants, A_1 and A_2, are:

$$A_1 = (-\frac{\tau_{Wx}}{\rho N_z} - \frac{g\eta_Q H_0}{N_z});\text{ and } A_2 = 0. \tag{8.87}$$

Combining these results with the general solution (8.80), the vertical velocity profile $u_{\tau W} = u_{\tau W}(z)$, forced by the wind stress (τ_W), the slope of the sea surface (η_Q), is calculated by

$$u_{\tau W}(z) = \frac{g\eta_Q}{2N_z}z^2 + (-\frac{\tau_W}{\rho N_z} - \frac{g\eta_Q H_0}{N_z})z, \tag{8.88a}$$

and as function of the non-dimensional depth ($Z = z/|H_0|$),

$$u_{\tau W}(Z) = \frac{g\eta_Q H_0^2}{2N_z}Z^2 - (\frac{\tau_W H_0}{\rho N_z} + \frac{g\eta_Q H_0^2}{N_z})Z. \tag{8.88b}$$

As seen in the previous solution, the surface slope generated by the river discharge (η_Q) is still unknown, which may be determined by applying the integral boundary condition (8.84), in which the function to be integrated is $u_{\tau W}(z)$. Then, solving for η_Q,

$$\eta_Q = \frac{3N_z}{gH_0^2}(u_f + \frac{\tau_{Wx}H_0}{\rho N_z}) \tag{8.89}$$

To obtain the free surface slope due to only the wind stress ($\eta_{\tau W}$), it is sufficient to eliminate the river velocity ($u_f = 0$) from this equation,

$$\eta_{\tau w} = 3\frac{\tau_{Wx}}{\rho g H_0}. \tag{8.90}$$

This result indicates that the free surface slopes upward in the seaward direction, as indicated in Fig. 8.3b, and is directly proportional to the wind stress and inversely proportional to the channel depth. Combining this result with the Eq. (8.88b), the final result for the vertical velocity profile is directly proportional to the wind stress and the depth, and is given by,

$$u_{\tau_W}(Z) = \frac{H_0 \tau_{Wx}}{\rho N_z}(1.5Z^2 - Z). \tag{8.91}$$

The theoretical profile of Eq. (8.86) was superimposed onto the experimental results of Prandle (1982, 1985) as the dashed line in Fig. (8.4a). Analysis of this figure indicates that the maximum friction at the bottom, $u_Q(0) = 0$, doesn't correctly simulate experimental data; therefore, it is apparent that a moderate condition for the bottom friction is more adequate for theoretical simulation of the vertical profile generated by the river discharge. In Prandle's articles, this problem has been solved using a bottom boundary condition expressed by Eq. (8.32),

- $\tau_{Bx} = \rho N_z (\frac{\partial u}{\partial Z})|_{z=-H_0} = \rho\,(\frac{4}{\pi}) k U_0 u|_{z=-H_0}$,

where the bottom friction is directly proportional to the velocity amplitude generated by the tide (U_0),[2] and to the slip bottom velocity ($u|_{z=0} \neq 0$). Comparing this expression of τ_{Bx} with Eqs. (8.31) or (8.72), verifies that they are formally similar, and the constant 4 k/π may be considered proportional to g / C_y^2.

As demonstrate by Bowden (1953), the coefficient $4kU_0/\pi$ may be applied when $U_0 \gg u$, and this condition is generally holds to partially mixed estuaries (type 2 or B). In the expression τ_{Bx}, the kinematic eddy viscosity coefficient, N_z, was taken as directly proportional to the velocity generated by the tide velocity amplitude (U_0) and the depth (H_0), expressed as $N_z = kU_0H_0$, where the coefficient, k, known as *bed friction coefficient*, was taken as constant and equal to $k = 1.5 \times 10^{-3}$ (Bowden 1967b), or $k = 2.5 \times 10^{-3}$ according to Ianniello (1977) and Prandle (1982).

With this new expression for the bottom boundary condition (τ_{Bx}), the integration constants A_1 and A_2 are,

$$A_1 = -\frac{g\eta_Q H_0}{N_z};\ A_2 = -\frac{g\eta_Q \pi H_0}{4kU_0}. \tag{8.92}$$

Substituting these constants into the general solution (Eq. 8.80), it follows that the new expressions for the velocity generated by the river discharge are,

$$u_Q(z) = \frac{g\eta_Q}{2N_z}(z^2 - 2H_0 z - \frac{\pi N_z H_0}{2kU_0}), \tag{8.93}$$

or in terms of the non-dimensional depth Z,

$$u_Q(Z) = \frac{g\eta_Q}{2N_z} H_0^2 (Z^2 - 2Z - \frac{\pi N_z}{2kH_0U_0}). \tag{8.94}$$

---

[2]For a frictionless estuary the order of magnitude of U_0 may be estimated by Eq. (2.24a, b, Chap. 2).

These solutions, similar to Eqs. (8.82) and (8.83), are still functions of the unknown, η_Q, which is the free surface slope generate by the river discharge. As with the previous solution, this unknown may be calculated by applying the integral boundary condition (Eq. 8.84). Then, substituting the condition into the expression $u_Q = u_Q(Z)$ (Eq. 8.94), completing the integration and solving the result to the unknown $d\eta_Q/dx = \eta_Q$, yields

$$\eta_Q = -\frac{u_f}{g(\frac{H_0^2}{3N_z} + \frac{\pi H_0}{4kU_0})} = -0.89\frac{kU_0u_f}{gH_0} = -0.89\frac{kU_0Q_f}{gBH_0^2}. \tag{8.95}$$

In this equation, it has been taken into account that $N_z = kU_0H_0$ and $u_f = Q_f/BH_0$. Similar to Eq. (8.85), all quantities in the last member have positive values, and the free surface slope, η_Q, is constant and negative, decreasing in the seaward direction, and is directly proportional to the river discharge (Q_f) and the tidal amplitude velocity (U_0), but it is inversely proportional to the estuary width (B) and the square of its depth (H_0).

Combining the Eqs. (8.95) and (8.93) and reducing the result to the simplest expression gives us the following equation to calculate the velocity generated by the river discharge:

$$\begin{aligned} u_Q(z) &= \frac{u_f}{\frac{1}{3} + (\frac{\pi N_z}{4kH_0U_0})}\left(-\frac{z^2}{2H_0^2} + \frac{z}{H_0} + \frac{\pi N_z}{4kH_0U_0}\right) \\ &= 0.89u_f\left(-\frac{z^2}{2H_0^2} + \frac{z}{H_0} + \frac{\pi}{4}\right), \end{aligned} \tag{8.96}$$

or, introducing the dimensionless depth, Z,

$$u_Q(Z) = 0.89u_f\left(-\frac{Z^2}{2} + Z + \frac{\pi}{4}\right). \tag{8.97}$$

The vertical relative velocity profile (u_Q/u_f) generated by the river discharge is presented in Fig. 8.4a in comparison with experimental results. We may observe that the theoretical solution calculated with moderate bottom friction (Eq. 8.92) reproduces the vertical velocity profile more accurately than maximum bottom friction.

8.5.2 *Velocity Generate by the Wind Stress*

As another example, let us investigate the influence of the wind stress forcing (τ_{wx}) as a surface boundary condition, under the assumption that the fresh water discharge may be disregarded. Adopting the same simplifying hypothesis and solving

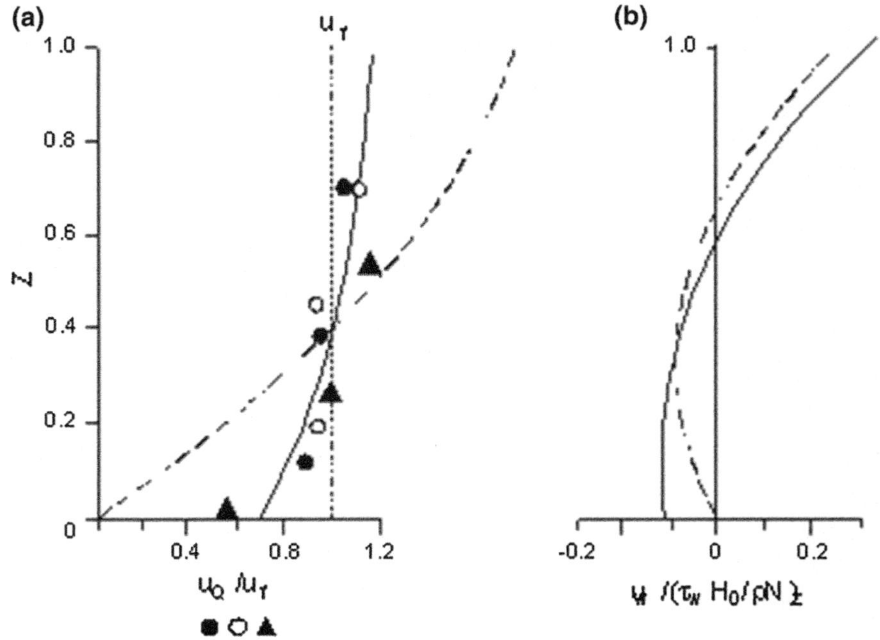

Fig. 8.4 Vertical velocity profiles generated by the river discharge (**a**), in comparison with experimental data measured at three depths (*filled circle*, *open circel*, *filled triangle*), and by the wind stress (**b**), according to Prandle (1985). Continuous and dashed lines are profiles calculated with moderate and maximum bottom friction, respectively

the problem with two alternate bottom boundary conditions (maximum and moderate friction), these solutions are similar to the former. Then, the constants of the integration of the general solution will be calculated with the following conditions:

- Wind stress on the surface: $\tau_{Wx} = -\rho N_z(\partial u/\partial z)|_{z=H0}$.
- Maximum friction at the bottom: $u(0) = u|_{z=0} = 0$, or
- Moderate friction at the bottom: $\tau_{Bx} = \rho N_z(\partial u/\partial z)|_{z=0} = \rho(4/\pi)kU_0u|_{z=0}$.

With the first two boundary conditions applied to the general solution (Eq. 8.80), the integration constants A_1 and A_2 are:

$$A_1 = \frac{\tau_{Wx} - \rho g \eta_W H_0}{\rho N_z}; \text{ and } A_2 = 0, \tag{8.98}$$

where $\eta_W = d\eta_w/dx$ is the free surface slope generated by the wind. Substituting these constants into the general solution, it follows that the vertical velocity profile, $u_w = u_w(z)$, generate by steady-state wind is expressed by:

$$u_W(z) = \frac{g\eta_W}{2N_z}(z^2 - 2H_0z) + \frac{\tau_{Wx}}{\rho N_z}z. \quad (8.99)$$

As a function of the non-dimensional depth, $Z = z/H_0$, this velocity component is given by

$$u_W(Z) = \frac{g\eta_W H_0^2}{2N_z}(Z^2 - 2Z) + \frac{\tau_{Wx}H_0}{\rho N_z}Z. \quad (8.100)$$

This result is dependent on the second unknown, η_W, which may be calculated by applying the integral boundary condition. Under steady-state conditions and with river discharge equal to zero ($Q_f = 0$), the net volume transport during a tidal cycle is also zero; this boundary condition is equivalent to a mean velocity in the water column ($\bar{u}_w$) equal to zero, which is analytically expressed by:

$$\bar{u}_w = \frac{1}{H_0}\int_0^{H_0} u_W(z)dz = \int_0^1 u_W(Z)dZ = 0. \quad (8.101)$$

As the function being integrated, $u_w = u_w(z)$ or $u_w = u_w(Z)$, is known (Eqs. 8.94 and 8.95), this equation may be integrated and the final result resolved for η_w is given by:

$$\eta_w = \frac{d\eta_w}{dx} = 1.5\frac{\tau_w}{\rho g H_0}. \quad (8.102)$$

This result indicates that the free surface displacement, $\eta_w = \eta_w(x)$, varies linearly with the longitudinal distance, and is directly dependent on the wind stress and inversely dependent on the local depth. The only quantity which may be positive or negative is the wind stress (τ_w), and the free surface slope may be positive or negative when the wind direction is seaward or landward, respectively. Combining the solutions of Eqs. (8.100) and (8.102), the analytical expression to calculate the vertical velocity profile forced by the wind is:

$$u_w(Z) = \frac{\tau_W H_0}{\rho N_z}(0.75Z^2 - 0.5Z), \quad (8.103)$$

where Z is the non-dimensional depth. As the factor, $\tau_w H_0/\rho N_z$, has the dimension of velocity [LT^{-1}], the ratio $u_w(Z)/(\tau_w H_0/\rho N_z)$ is a non-dimensional quantity and

$$\frac{u(Z)}{(\frac{\tau_W H_0}{\rho N_z})} = (0.75Z^2 - 0.5Z), \quad (8.104)$$

This non-dimensional vertical velocity profile generated by the wind stress is presented in Fig. 8.4b as a dashed-line profile.

Lets us now solve the problem applying the bottom boundary condition of moderate friction formulated by: $\tau_{Bx} = \rho(4/\pi)kU_0u|_{z=0}$. Applying this condition to the general solution (8.80) yields the following expressions for the integration constants A_1 and A_2:

$$A_1 = \frac{g\eta_w H_0}{N_z} + \frac{\tau_W}{\rho N_z}, \tag{8.105}$$

and

$$A_2 = \frac{\pi g\eta_W H_0}{4kU_0} + \frac{\pi\tau_W}{4\rho\rho k_0}. \tag{8.106}$$

Substituting these constants into the general solution (8.80), simplifying the result and expressing the final solution as a function of the non-dimensional depth, gives the solution:

$$u_W(Z) = \frac{g\eta_W H_0^2}{N_z}\left(\frac{1}{2}Z^2 - Z - \frac{\pi}{4}\right) + \frac{\tau_W H_0}{\rho N_z}\left(Z + \frac{\pi}{4}\right). \tag{8.107}$$

This solution is still a function of the surface slope (η_W), but this dependence may be eliminated by applying the integral boundary condition (8.101) which, when reduced to the simplest expression yields,

$$\eta_W = \frac{d\eta}{dx} = 1.149\frac{\tau_W}{\rho g H_0} \approx 1.15\frac{\tau_W}{\rho g H_0}. \tag{8.108}$$

This result has the same analytical expression as the surface slope presented in (8.102), but with a lower coefficient of proportionality, indicating that this slope is, to some extent, independent on the bottom friction characteristics.

Substituting the expression (8.108) into Eq. (8.107) and simplifying the result, the steady-state vertical velocity profile generated by the wind stress is given by (Prandle 1982, 1985):

$$u_W(Z) = \frac{\tau_W H_0}{\rho N_z}(0.574Z^2 - 0.149Z - 0.117), \tag{8.109}$$

and the non-dimensional vertical velocity profile is

$$\frac{u_W(Z)}{\frac{\tau_W H_0}{\rho N_z}} = (0.574Z^2 - 0.149Z - 0.117). \tag{8.110}$$

This velocity profile, for the moderate friction at the bottom, is presented in Fig. 8.4b (continuous line) in comparison with the profile for maximum bottom friction (dashed line). This figure indicates that the changes from maximum to moderate bottom friction produced significant changes in the u-velocity component of the bottom layers.

So far we have demonstrated simplified analytical solutions of the equation of motion for one-dimensional motions, which is the starting point for solutions of the salt conservation equation. Let us complete this topic with a simple solution of the following problem: with a known steady-state velocity profile u = u(Z), calculate the corresponding salinity profile S = S(Z).

The appropriate salt conservation equation to be used will be a simplified expression of Eq. (8.58), i.e., steady-state, vertical velocity component much less than the longitudinal ($w \ll u$), B = const. and ($P - E_v = 0$) and, in terms of the non-dimensional depth,

$$u(Z)\frac{dS}{dx} = \frac{K_z}{H_0^2}\left(\frac{d^2S}{dZ^2}\right) \quad \text{or} \quad \frac{d^2S}{dZ^2} = S_X\frac{H_0^2}{K_z}u(Z). \tag{8.111}$$

Adopting for the vertical velocity profile the one generate by the fresh water discharge (Eq. 8.86), i.e., $u(Z) = u_f(-\frac{3}{2}Z^2 + 3Z)$, the salt conservation equation is given by,

$$\frac{d^2S}{dZ^2} = S_X\frac{H_0^2}{K_z}\left[3u_f\left(-\frac{Z^2}{2} + Z\right)\right], \tag{8.112a}$$

or

$$\frac{d^2S}{dZ^2} = \frac{3S_xH_0^2u_f}{K_z}\left(-\frac{Z^2}{2} + Z\right) \tag{8.112b}$$

This differential equation will be solved with the following known boundary conditions: $S(1) = S_S$ and $S(0) = S_F$ are surface and bottom salinities, respectively. With two successive integrations,

$$S(Z) = 3\iint \frac{S_xH_0^2u_f}{K_z}\left(-\frac{Z^2}{2} + Z\right)dZdZ, \tag{8.113a}$$

its general solution is

$$S(Z) = \frac{S_xH_0^2u_f}{2K_Z}\left(-\frac{Z^4}{6} + Z^3\right) + C_1Z + C_2. \tag{8.113b}$$

Applying the boundary conditions at depths Z = 0 and the surface Z = 1, the dimensionless integration constants C_1 and C_2 are calculated by,

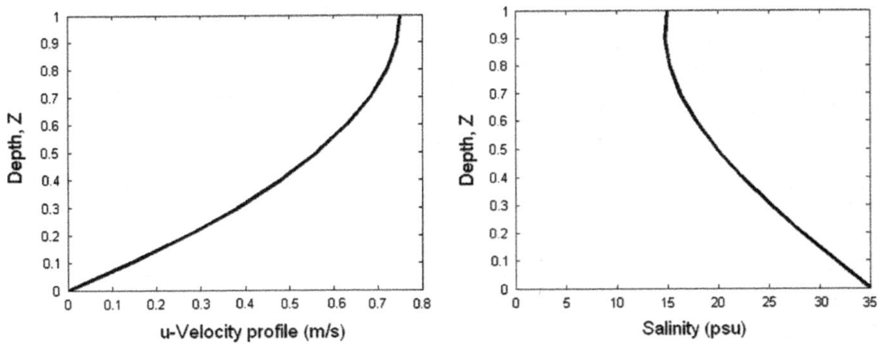

Fig. 8.5 Steady-state theoretical vertical profiles of the u-velocity component (*left*) and salinity (*right*)

$$C_1 = -\frac{5}{12}\frac{S_x H_0^2 u_f}{K_z} - (S_F - S_S), \text{ and } C_2 = S_S. \quad (8.114)$$

Substituting C_1 and C_2 in the general solution (8.113b), and simplifying the resulting expression, the steady-state salinity vertical profile is given by

$$S(Z) = \frac{1}{12}\frac{S_x H_0^2 u_f}{K_z}(-Z^4 + 6Z^3 - 5Z) - (S_F - S_S)Z + S_F, \quad (8.115)$$

which identically satisfies the boundary conditions at free surface and at the bottom.

The theoretical profiles of the vertical u-velocity component and the salinity, calculated with Eqs. (8.86) and (8.115) are shown in Fig. 8.5. The following parameters values were used in the calculations: $S_{mouth} = 35.0°/_{oo}$, $S_{head} = 1.0°/_{oo}$, estuary mixing length $L = 10^4$ m, and $S_x = 3.4 \times 10^{-5}$ m^{-1}, $H_0 = 10.0$ m, $u_f = 0.5$ m s^{-1}, $K_z = 5.0 \times 10^{-3}$ m^2 s^{-1}, and surface and bottom salinities $S_S = 30°/_{oo}$ and $S_F = 35°/_{oo}$, respectively, indicating a high stratified partially mixed estuary.

The salinity profile in this figure indicates the advective influence of the river discharge generating a low vertical salinity gradient in the upper layer, and a highly stratified condition towards the bottom layers.

8.6 Shallow Water Tidal Current and Phase Velocity

The topic presented in Chap. 2, related to the progressive wave propagation in a one-dimensional channel, will be now implemented. The analytical treatment of the velocity and the associated free surface oscillation of a frictionless fluid with constant density (ρ) will be presented according Pugh (1987) and Franco (1988).

In this development, the following simplifications were applied: (i) channel with constant width (B) and infinite length; (ii) small wave amplitude $\eta(x, t) \ll H_0$; (iii) the wave height is comparable with the water depth, $\eta(x, t) \approx H_0$, and (iv) energy dissipation due to friction is neglected.

Taking into account these simplifications, the main characteristics of the wave propagation is formulated by the equations of continuity (8.68) and motion (8.69a), reducing these equation system into the following analytical expressions:

$$(H_0 + \eta)\frac{\partial u}{\partial x} + u\frac{\partial \eta}{\partial x} = -\frac{\partial \eta}{\partial t}, \tag{8.116a}$$

and

$$\frac{\partial u}{\partial t} + u\frac{\partial u}{\partial x} = -g\frac{\partial \eta}{\partial x}. \tag{8.116b}$$

The simplification of the continuity Eq. (8.116a) was obtained taking into account that the transverse section has an area $A = B(H_0 + \eta)$ and that H_0 is the mean channel depth.[3]

In the De Saint Venant's pioneer work, published in 1871, the following general solution was obtained for the tidal velocity generated by the barotropic influence,

$$u\,(x,t) = C_1\sqrt{g[H_0 + \eta\,(x,t)]} - C_2, \tag{8.117}$$

where C_1 and C_2 are constants to be determined. Adopting to this solution the initial conditions, $\eta(x, 0) = 0$ and $u(x, 0) = 0$ at the initial instant of time ($t = 0$), the barotropic tidal influence on the velocity is null, and the relationship between C_1 and C_2 is:

$$C_2 = C_1\sqrt{gH_0}. \tag{8.118}$$

In comparison to Eq. (2.20, Chap. 2), the barotropic value of the tidal current in shallow water now has the proportionality coefficient C_1.

Combining Eqs. (8.117) and (8.118), the tidal velocity in shallow water is calculated by

$$u\,(x,t) = C_1[\sqrt{g[H_0 + \eta\,(x,t)]} - \sqrt{gH_0}]. \tag{8.119}$$

This is an intermediate solution, which is dependent on the constant C_1 and the tidal oscillation, $\eta = \eta(x, t)$. To calculate C_1, it is necessary to derive this equation in relation to x and t, and combine the result with the continuity Eq. (8.116a),

[3]If the advective acceleration is disregarded and, $\eta \ll H_0$, this system of equations is reduced to the Eqs. (2.14) and (2.17) (Chap. 2).

taking into account the analytical expression (8.119). Following this procedure, according to Franco (1988), yields:

$$\frac{\partial u}{\partial x} = \frac{gC_1}{2\sqrt{g(H_0+\eta)}}\frac{\partial \eta}{\partial x}, \tag{8.120a}$$

$$\frac{\partial u}{\partial t} = \frac{gC_1}{2\sqrt{g(H_0+\eta)}}\frac{\partial \eta}{\partial t}, \tag{8.120b}$$

and combining this result with Eq. (8.116a)

$$\frac{\partial \eta}{\partial t} = [-\frac{3C_1}{2}\sqrt{g(H_0+\eta)} + C_1\sqrt{gH_0}\frac{\partial \eta}{\partial x}]. \tag{8.121}$$

Combining this result with Eq. (8.120b), we have:

$$\frac{\partial u}{\partial t} = [-\frac{3gC_1^2}{4} + \frac{gC_1^2\sqrt{gH_0}}{2\sqrt{g(H_0+\eta)}}]\frac{\partial \eta}{\partial x}. \tag{8.122}$$

Substituting the expressions of u, ∂u/∂x and ∂u/∂t in the equation of motion (8.116b) and simplifying the result, gives

$$(-\frac{gC_1^2}{4} + g)\frac{\partial \eta}{\partial x} = 0. \tag{8.123}$$

As $\partial\eta/\partial x \neq 0$, with exception of high and low water, it follows from this equation that $C_1 = \pm 2$. Using the positive value ($C_1 = 2$) in order to obtain a result in the real numeric field, the final solution of Eq. (8.119) is the De Saint Venant solution for the tidal velocity

$$u(x,t) = 2[\sqrt{g(H_0+\eta)} - \sqrt{gH_0}]. \tag{8.124}$$

From this equation, it follows that the velocity is zero at the equilibrium tide ($\eta = 0$) However, during the flood tide ($\eta > 0$) and, $g(H_0 + \eta)^{1/2} > (gH_0)^{1/2}$ and there will be a gradual increase in the velocity, reaching its highest value at the flood tide. During the ebb tide, the motion occurs in the opposite direction, and the highest velocity will be at the ebb tide.

If $C_1 = 2$ is also substituted into the Eq. (8.121) the result for the tidal phase velocity (c_0) is:

$$-\frac{dx}{dt} = c_0 = 3\sqrt{g(H_0+\eta)} - 2\sqrt{gH_0}. \tag{8.125}$$

Then, c_0 is dependent on the tidal height, $\eta = \eta(x, t)$, and only when $\eta \ll H_0$ this solution is equal to that of Eq. (2.20, Chap. 2), where this quantity is dependent only on the channel's mean water depth (H_0).

Equation (8.125) was applied by Franco (1988, 2009) to investigate the diurnal inequality observed in the North Channel of the Amazon river (Amazon, Brazil). The occurrence of this phenomenon was also investigate by Kjerfve and Ferreira (1993), who's studies were based on time-series of currents measured in the Mearin river (Maranhão, Brazil), illustrated in Chap. 2 (Fig. 2.7). In the figure, it can be observed that the free surface elevation $\eta = \eta(x, t)$ is positive during the flood and negative during the ebb tide, generating high and low phase velocities in the flood and ebb tides, respectively. Further analysis, using Eq. (8.125), indicates high and low tidal velocities during the flood and ebb tides, respectively. This phenomenon causes the tidal bore in estuaries in the Brazilian states Pará and Amazon, during the April and May full moon.

As the wave amplitude is comparable with the total water depth the wave celerity may be approximate by (Pugh 1987; Lessa 1996):

$$c_0 = \sqrt{g\,(H_0 + 1.5\eta} \qquad (8.126)$$

and the wave speed decreases as the water depth decreases, which causes the crest of the wave to move closer to the through ahead, resulting in a distorted tidal wave shape with a steep face and a gentle back-slope due to bottom friction and the associated non-linearity.

8.7 Periodic Stratification Tidal Generate: Potential Energy Anomaly

The energy resulting from cyclic tidal pumping at different time scales, mainly semi-diurnal, diurnal and fortnightly, and the influence of this energy on the vertical stability of an estuary, may be calculated by the *potential energy anomaly* (Φ), defined by Simpson et al. (1990),

$$\Phi = \frac{1}{h}\int_{-h}^{\eta} g\,(\bar{\rho} - \rho)\,z\,dz, \qquad (8.127)$$

where h is the depth, $\rho = \rho(z)$, and $\bar{\rho}$ is the depth mean density in the water column, $[\bar{\rho} = \frac{1}{h}\int_{-h}^{0} \rho\,(z)\,dz]$. This physical quantity has dimension $[\Phi] = [ML^2T^{-1}/L^3]$ and is calculated in J/m^3 in SI units. Physically, this quantity is the energy per volume unit required to generate vertical mixing processes. In other words, according to Prandle (2009), it represents the amount of energy required to mix the water column to a uniform density and hence, it is inversely the effectiveness of the vertical mixing of stratified estuarine water mass. When the estuary is well-mixed $\Phi \rightarrow 0$, and Φ increases with vertical stratification, this process is named *Strain-Induced Periodic Stratification* (SIPS).

Under the assumption that the depth is either independent of the tidal oscillation, or is higher than the tidal height, the expression of the time variation ($\partial\Phi/\partial t$) may be calculated by

$$\frac{\partial \Phi}{\partial t} = \frac{g}{h} \int_{-h}^{0} \left(\frac{\partial \bar{\rho}}{\partial t} - \frac{\partial \rho}{\partial t}\right) zdz, \tag{8.128}$$

with the approximation η=0.

For a one-dimensional fluid motion in the Ox direction, it is possible to verify in the continuity equation (Eq. 7.12, Chap. 7) that for a non-divergent field of motion ($\nabla \bullet \vec{v} = 0$), the local variation of density ($\partial\rho/\partial t$) is related to the advective term with the *advection of density* equation,

$$\frac{\partial \rho}{\partial t} = -u\frac{\partial \rho}{\partial x} \text{ and } \frac{\partial \bar{\rho}}{\partial t} = -\bar{u}\frac{\partial \bar{\rho}}{\partial x}. \tag{8.129}$$

In this equation, $\bar{u}$ is the depth-mean value in the water column, and the longitudinal density gradient is independent of the depth. Combining this result with Eq. (8.128), it follows that the expression of the time variation of the potential energy ($\partial\Phi/\partial t$) as function of the u-velocity component, u = u(z), is

$$\frac{\partial \Phi}{\partial t} = \frac{g}{h}\frac{\partial \rho}{\partial x} \int_{-h}^{0} [u(z) - \bar{u})]\, zdz. \tag{8.130}$$

In this equation, the vertical velocity profile u = u(z) may be based in experimental data or theoretically calculated. In terms of the non-dimensional depth ($Z = z/|h|$), this equation is expressed by

$$\frac{\partial \Phi}{\partial t} = gh\frac{\partial \rho}{\partial x} \int_{-1}^{0} [u(Z) - \bar{u})]ZdZ. \tag{8.131}$$

The time variation of the potential energy anomaly ($\partial\Phi/\partial t$) is used to investigate the tidal mixing processes; the calculated values of this physical quantity depend on the vertical salinity stratification. Forecast criterion on the SIPS occurrence using theoretical velocity profiles and variability characteristics of the potential energy anomaly are presented in the Simpson's et al. article.

An alternative expression to calculate the potential energy anomaly (Φ) was suggested by Prandle (2004), taking into account the steady-state vertical salinity profile calculated by Officer (1976),

$$S(z) = \overline{S} + \rho \frac{gS_x^2}{N_z K_z} \frac{h^2}{10^4} (-83z^5 + 224z^4 - 62z^3 - 146z^2 + 33), \tag{8.132}$$

and, using the definition of the potential energy anomaly as a function of the salinity, we have:

$$\Phi_E = \frac{1}{h} \int_0^h [S(z) - \overline{S}]\, g\,(z - h) dz, \tag{8.133}$$

and substituting (8.132) into the Eq. (8.133), we can obtain its time-averaged value:

$$\Phi_E = \frac{7.10^{-4} \rho g^2 S_x^2 h^6}{N_z K_z}. \tag{8.134}$$

Tidal cycle simulations of the potential energy anomaly (Φ_E) were analysed by Prandle (2004) and, for some experiments, the results indicated reasonable agreement with observational data, indicating a complete vertical mixing for a period of approximately one-third of the tidal cycle, following the maximum flood current.

Nearly steady-state vertical profiles of the u-velocity component calculated from hourly profiles sampled during three complete tidal cycles in the Piaçaguera estuarine channel, located in the upper reaches of the (Santos-São Vicente Estuary System (Fig. 1.5, Chap. 1) were used to estimate the time variation of the potential energy anomaly according to the equation (8.130). From the observations made during neap and spring tidal cycle in June/July, 2001, the channel was classified as partially mixed and weakly stratified (type 2a). These measurements used to calculate the time variation of the potential energy anomaly (Eq. 8.130). Adopting the value of 1.2×10^{-4} kg m^{-4} for the longitudinal density gradient, and with theoretical simulations of the steady-state u-velocity profiles using the Hansen and Rattray (1966) analytical model, the results indicated the potential energy anomaly rate decreased from 1.5×10^{-3} J m^{-3} s^{-1} to 0.56 J m^{-3} s^{-1}, for the neap to the spring tide transitional period (Miranda et al. 2012).

As stated by Simpson et al. (1990), the level of stratification in water column is crucial in controlling the intensity of vertical mixing and hence the vertical fluxes of water properties as heat, salt, momentum, and nutrients elements. The latter may be of critical importance in limiting biological productivity. By inhibiting vertical displacement, stratification also serves to influence the degree of light exposure experienced by marine organisms. Phytoplankton located in shallow layer above a strong pycnocline receive a much more generous input of light energy than in an environment where vertical mixing is complete and these organisms are regularly displaced over the full water column depth. In the example presented above, on the maintenance of the partially mixed conditions in the Piaçaguera channel, during a fortnightly transitional period, the vertical stability of the water column may be helpful to the living organisms on the shallow upper layer above the picnocline.

References

Blumberg, A. F. 1975. A Numerical Investigation into the Dynamics of Estuarine Circulation. Tech. Rept. Chesapeake Bay Institute, The Johns Hopkins University. n. 91. 110 p. + Apêndices.

Bowden, K. F. 1953. Note on Wind Drift in a Channel in the Presence of Tidal Currents. Proc. R. Soc. Lond., A219, pp. 426–446.

Bowden, K. F. 1963. The Mixing Processes in a Tidal Estuary. J. Air Wat. Pollut., 7:343 356.

Bowden, K. F. 1967b. Stability effects on turbulent mixing in tidal currents. Phys. Fluid. 10 (suppl), S278-S280.

Bowden, K. F. & Fairbairn, L.A. 1952. A Determination of the Frictional Forces in a Tidal Current. Proc. Royal.Soc. Lond. A, 214. p. 371–392.

Chant, R.J. 2010. Estuary secondary circulation. In: ed. Valle-Levinson A. Contemporary Issues in Estuarine Physics. Cambridge University Press, pp. 100–124.

Charnock, H. 1981. Air-sea Interaction. In: Warren, B. A. & Wunsch, C. (eds.). Evolution of Physical Oceanography. Cambridge, MIT Press, pp. 482–503.

Ferraz, L. A. de C. 1975. Tidal and Current Prediction for the Amazon's North Channel using a Hydrodynamical-Numerical Model. M. Sc. Dissertation. California, Naval Post-graduate School. 85 p.

Franco, A. S. 1988. Tides: Fundamentals, Analysis and Prediction. São Paulo, Fundação Centro Tecnológico de Hidráulica. 249 p.

Franco, A. S. 2009. Marés – Fundamentos, Análise e Previsão. Diretoria de Hidrografia e Navegação. Rio de Janeiro, 2nd ed., 344 p.

Fischer, H. B.; List, E. J.; Koh, R. C. Y.; Imberger, J. & Brooks, N. H. 1979. Mixing in Inland and Coastal Waters. New York, Academic Press. 483 p.

Fisher, J. S.; Ditmars, J. D. & Ippen, A. T. 1972. Mathematical Simulation of Tidal Time Averages of Salinity and Velocity Profiles in Estuaries. Massachusetts Institute of Technology, Mass., Rept. MITSG 72–11, 157 p.

Geyer, W. R. 1997. Influence of Wind on Dynamics and Flushing of Shallow Estuaries. Estuar. Coast. Shelf Sci., 44:713–722.

Gill, A. E. 1982. Atmospheric-Ocean Dynamics. New York, Academic Press. 662 p.

Hamilton, P. & Rattray Jr., M. 1978. Theoretical Aspects of Estuarine Circulation. In: Kjerfve B. (ed.). Estuarine Transport Processes. Columbia, University of South Carolina, pp. 37–73. (Belle W. Baruch Library in Marine Science, 7).

Hansen, D. V. & Rattray Jr., M. 1965. Gravitational Circulation in Sraits and Estuaries. J. Mar. Res., 23(1):102–122.

Hansen, D. V. & Rattray Jr., M. 1966. New Dimensions in Estuary Classification. Limnol. Oceanogr., 11(3):319–325.

Harari, J. & Camargo, R. 1998. Modelagem Numérica da Região Costeira de Santos (SP): Circulação de Maré. Rev. Bras. Oceanogr., São Paulo, 46(2):135–156.

Harleman, D. R. F. & Ippen, A. T. 1967. Two-Dimensional Aspects of Salinity Intrusion in Estuaries: Analysis and Velocity Distributions. Committee on Tidal Hydraulics. Tech. Bull., Corps of Engineers, U. S. Army, n. 13.

Ianniello, J. P. 1977. Tidally induced residual currents in estuaries of constant breadth and depth. Journal of Marine Research, 35(4), 755–786.

Ianniello, J. P. 1979. Tidally induced residual currents in estuaries of variable breadth and depth. J. Phys. Oceanogr., 9, 962–974.

Ippen, A. T. & Harleman, D. R. F. 1961. One-Dimensional Analysis of Salinity Intrusion in Estuaries. Committee on Tidal Hydraulics. Tech. Bull. Corps of Engineers U. S. Army, n. 5. 120 p.

Kinsman, B. 1965. Wind Waves-Their Generation and Propagation on the Ocean Surface. New Jersey, Prentice-Hall. 676 p.

King B. & Wolanski, E. 1996. Bottom Friction Reduction in Turbid Estuaries. In: ed. Pattiaratchi C., Coastal and Estuarine: Mixing in Estuaries and Coastal Seas. American Geophysical Union, pp. 325:337.

Kjerfve, B. & Ferreira, H. O. 1993. Tidal Bores: First Ever Measurements. Ciência Cult., São Paulo, 45(2):135–137.

Kjerfve, B.; Miranda, L. B. & Wolanski, E. 1991. Modelling Water Circulation in an Estuary and Intertidal Salt Marsh System. Neth. J. Sea Res., 28(3):141–147.

Lacombe, H. 1965. Cours d'Océanographie Physique (Théories de la circulation générale. Houles et vagues). Paris, Gauthiers-Villars. 392 p.

Lessa, G. 1996. Tidal Dynamics and Sediment Transport in a Shallow Macrotidal Estuary. In: ed. Pattiaratchi C., Coastal and Estuarine: Mixing in Estuaries and Coastal Seas. American Geophysical Union, pp. 338:360.

Liu, W. T.; Katsaros, K. B. & Businger, J. A. 1979. Bulk Parametrisation of Air-Sea Exchanges of Heat and Water Vapour Including Molecular Constraints at the Interface. J. Atmos. Sci., 36:1722–1735.

Miranda, L. B.; Dalle Olle, E.; Bérgamo, A.L.; Silva, L.S. & Andutta, F.P. 2012. Circulation and salt intrusion in the Piaçaguera Channel, Santos (SP). Braz. J. Oceanography, 60(1):11–23.

Möller, O. O. Jr., 1996. Hydrodynamique de la Lagune dos Patos (30°S, Brésil). Mesures et Modélisation. Tese de Doutorado. Bordeaux, L'Université de Bordeaux I. École Doctorale des Sciences de la Terre et de la Mer. 204 p.

Munk, W. H. & Anderson E. R. Anderson 1948. Notes on the theory of the thermocline, J. Mar. Res., 3, 276–295, 1948

Neumann, G. & Pierson, W. J. Jr. 1966. Principles of Physical Oceanography. London, Prentice-Hall. 545 p.

Officer, C. B. 1976. Physical Oceanography of Estuaries (and Associated Coastal Waters). New York, Wiley. 465 p.

Officer, C. B. 1977. Longitudinal Circulation and Mixing Relations in Estuaries. Estuaries, Geophysics, and the Environment. Washington, D. C., National Academy of Sciences, pp. 13–21.

Officer, C. B. 1978. Some Simplified Tidal Mixing and Circulation Flux Effects in Estuaries. In: Kjerfve, B. (ed.). Estuarine Transport Processes. Columbia, University of South Carolina Press, pp. 75–93. (The Belle W. Baruch Library in Marine Science, 7)

Paiva, A. M. & Rosman, P. C. C. 1993. Modelagem Numérica de Circulação em Estuários Estratificados. X Simpósio Brasileiro de Recursos Hídricos – I Simpósio de Recursos Hídricos do Cone Sul. ABRH, Rio de Janeiro, vol. 5, pp. 485–494.

Pedloski, J. 1979. Geophysical Fluid Dynamics. 2. ed., New York, Springer-Verlag. 624 p.

Prandle, D. 1982. The Vertical Structure of Tidal Currents and other Oscillatory Flows. Continent. Shelf Res., 1(2):191–207.

Prandle, D. 1985. On Salinity Regimes and the Vertical Structure of Residual Flows in Narrow Tidal Estuaries. Estuar. Coast. Shelf Sci., 20:615–635.

Prandle, D. 2004. Saline Intrusion in Partially Mixed Estuaries. Est. Coast. Shelf Sci. (59):385–397.

Prandle, D. 2009. Estuaries: Dynamics, Mixing, Sedimentation and Morphology. Cambridge University Press, N.Y., 236 p.

Pritchard, D. W. 1954. A Study of Salt Balance in a Coastal Plain Estuary. J. Mar. Res., 13 (1):133 144.

Pritchard, D.W. 1956. The Dynamic Structure of a Coastal Plain Estuary. J. Mar. Res., 15(1):33–42.

Pritchard, D. W. 1958. The Equations of Mass Continuity and Salt Continuity in Estuaries. J. Mar. Res., 17:412–423.

Pritchard, D. W. 1971. Two-Dimensional Models. In: Ward Jr., G. H. & Espey Jr., W. H. (eds.). Estuarine Modelling: An Assessment Capabilities and Limitations for Resource Management and Pollution Control. Austin, Tracor, pp. 22–33. (Water Pollution Control Research Series).

Pugh, D. T. 1987. Tides, Surges and Mean Sea-Level, Wiley, New York.

Rattray Jr., M. & Hansen, D. V. 1962. A Similarity Solution for Circulation in an Estuary. J. Mar. Res., 20(2):121–133.
Roll, H. U. 1965. Physics of the Marine Atmosphere. New York, Academic Press. 497 p. (International Geophysics Series, 7).
Simpson, J. H.; Brown, J.; Mattews, J. & Allen, G. 1990. Tidal Straining, Density Currents, and Stirring in the Control of Estuarine Stratification. Estuaries, 13(2), p. 125–132.
Smith, S. D. 1980. Wind Stress and Heat Flux over the Ocean in Gale Force Winds. J. Phys. Oceanogr., 10:709–726.
Stommel, H. 1953a. The Role of Density Currents in Estuaries. Proc. Minnesota International Hydraulics Convention, Minneapolis, University of Minnesota, pp. 305–312.
Svedrup, H. U.; Johnson, M. W. & Fleming, R. H. 1942. The Oceans, their Physics, Chemistry and General Biology. New Jersey, Prentice-Hall. 1042 p.
Thatcher, M. L. & Harleman, D. R. F. 1972. Prediction of Unsteady Salinity Intrusion in Estuaries: Mathematical Model and User's Manual. Massachusetts Institute of Technology, Mass., Rep. MITSG 72–21. 193 p.
Wolanski, E.; Jones, M. & Bunt, J. S. 1980. Hydrodynamics of a Tidal Creek-Mangrove Swamp System. Aust. J. Mar. Freshwat. Res., 31:431–450.
Wolanski, E.; King, B. & Galloway, D. 1997. Salinity Intrusion in The Fly River Estuary, Papua New Guinea. J. Coast. Res., 13(4):983–994.

Chapter 9
Circulation and Mixing in Steady-State Models: Salt Wedge Estuary

Now that we have laid in the previous chapters the basic estuarine hydrodynamic framework, let us present in the following chapters practical applications of the analytical and numerical studies on the circulation in estuaries and its influence in the distributions of properties concentration.

Among the characteristics to be analytically studied in the field of estuary kinematics and dynamics, there are some which may be approximated using a simple geometry and steady-state conditions, where the estuary can be theoretically treated as a one or two-dimensional system. Natural estuarine channels usually don't have uniform transverse sections that may expand and contract in an irregular manner; however, a common characteristic is that a channel's length is much greater than its width. With the aim of applying the concepts developed in the equations of preceding chapters, steady-state analytical solutions for salt wedge, partially mixed and vertical and laterally well-mixed estuaries will be presented in the present and following chapters. With some approximations, these estuaries may have their circulation and salinity stratification simulated with relatively simple analytical models. Although these solutions will only simulate steady-state conditions, and residual motions and salinity stratifications will be obtained, they are of great practical importance because: (i) their solutions may indicate if the estuary is flushing out or not undesirable substances that are discharged into estuaries; and, (ii) may be used to validate non-steady state numeric solutions.

The general kinematic and dynamic characteristics of estuaries classified as salt wedge (types A or 4) by Pritchard (1955), Hansen and Rattray (1966) were presented in Chap. 3. They were studied in laboratory experiments, combining one and two-dimensional models by several investigators such as Farmer and Morgan (1953), Sanders et al. (1953). The water masses in the upper layers of salt wedge estuaries have very low salinities, and their seaward velocities are much higher than the compensating landward motion below a sharp picnocline; in other words, as stated by Geyer and Farmer (1989), a salt-wedge occurs in an estuary when the river discharge is adequate to maintain a strong gradient between fresh and salt water against the mixing tendency of tide and wind-induced turbulence.

L. Bruner de Miranda et al., *Fundamentals of Estuarine Physical Oceanography*, Ocean Engineering & Oceanography 8, DOI 10.1007/978-981-10-3041-3_9

Due to the continuous seaward motion in the surface layer, the velocity shear at the picnocline interface between fresh and salt water produces an entrapment of some salt water from the wedge into the upper fresh water layer. In this situation, there is little or no mixing of fresh water into the salt wedge. The salt water volume in the upper layer subsequently increases seaward, and a slow upstream movement of water in the salt wedge occurs to compensate for the upward loss into the fresh water.

The water mass in a characteristic salt-wedge has low stratification, with salinity very close to the one of the coastal water and a sharp halocline is between the transition of the lower (salt-wedge) and upper layers. The circulation continuity is provided by the *entrainment* phenomenon generating slow ascending vertical motions across the picnocline due to oscillating internal waves. This type of estuary is usually dominated by the fresh water discharge, and eddy diffusion may only be important in the surface layer above the halocline. In steady-state conditions and with lateral homogeneity, the dominant terms in the salt-balance equation are the vertical and longitudinal advection, and the diffusive longitudinal term can be disregarded. In the upper layer, the eddy diffusion term may also be taken into account under the influence of strong winds.

In these estuaries, the upper layer above the halocline has its velocity mainly forced by the fresh water discharge. A classical example is the South Pass in the delta of the Mississippi river (Mississippi, USA), which maintains nearly steady-state conditions characteristics over several tidal cycles; it may however, be significantly influenced by tidal motions, causing considerable variation in the vertical structure of salinity and velocity within a tidal cycle (Wright 1970). Another example is the seaward reaches of the Itajai-açu river (Santa Catarina, Brazil), which is forced by micro-tides and has been classified as a salt wedge estuary in conditions where river discharges around 300 $m^3\ s^{-1}$, and the saline wedge is displaced landward up to 18 km from its mouth. However, when the river discharge reaches values up to 1000 $m^3\ s^{-1}$, the seawater is completely evacuated through its mouth (Döbereiner 1985, quoted in Schettini (2002)). The salt-wedge extension in the estuarine plume was empirically correlated with the river discharge, presenting an exponential decay with the increase in river discharge (Schettini and Truccolo 1999).

Under the assumption of nearly steady-state conditions, the landward salt-wedge propagation varies mainly at a seasonal time scale, forced by the river discharge. The theory which will be developed in this chapter can't be generalized for all salt wedge estuaries, and holds only for an *arrested* salt wedge estuaries. According to classical authors Farmer and Morgan (1953), Schijf and Schonfeld (1963) (quoted in Geyer and Farmer (1989)); the designation *arrested salt-wedge* for this estuary type refers to a regime in which the pressure gradient force is balanced by inertial and frictional forces within the estuary, and its interfacial structure attains a quasi-steady-state configuration.

In the literature we find studies of salt wedge estuaries forced by meso-tides, for example, the Fraser river estuary (Vancouver, Canada). In this river, the salt-wedge varies along the estuary during the tidal cycle towards an equilibrium condition against the free surface slope variations. The advancing of the salt-wedge front position vs. time, provided by tracking with echo-sounding images for three sets of observations, indicated that the advance of salt-wedge intrusion length varied from 9 to 18 km for high and low river discharge, respectively (Geyer 1986). The interaction of the tidal flow with the density-driven motion of the salt-wedge, during different phases of the tide and river discharge has been clearly illustrated by Geyer and Farmer (1989), showing that the highly stratified vertical salinity structure, existing at high and low tides, becomes poorly stratified at the end of the flood tide, and the salt-wedge water remains under the strong picnocline at the estuary mouth.

In salt wedge estuaries, which will be analytically investigated in this chapter, the physical process of momentum exchanges in the fresh-salt water interface will be simulated by a shear named *interfacial stress*. This stress, which is force per unit of area, is mainly provided by the river input, and causes a seaward ascending inclination of the salt-wedge (Fig. 9.1). In this figure, we may observe that the longitudinal salinity gradients in the layers above and below the halocline are absent or very low, and in the theoretical treatment of the salt-wedge its dynamical consequences will be disregarded. This figure also indicates the displacement of the salt wedge front position due to the influence of the river discharge variation.

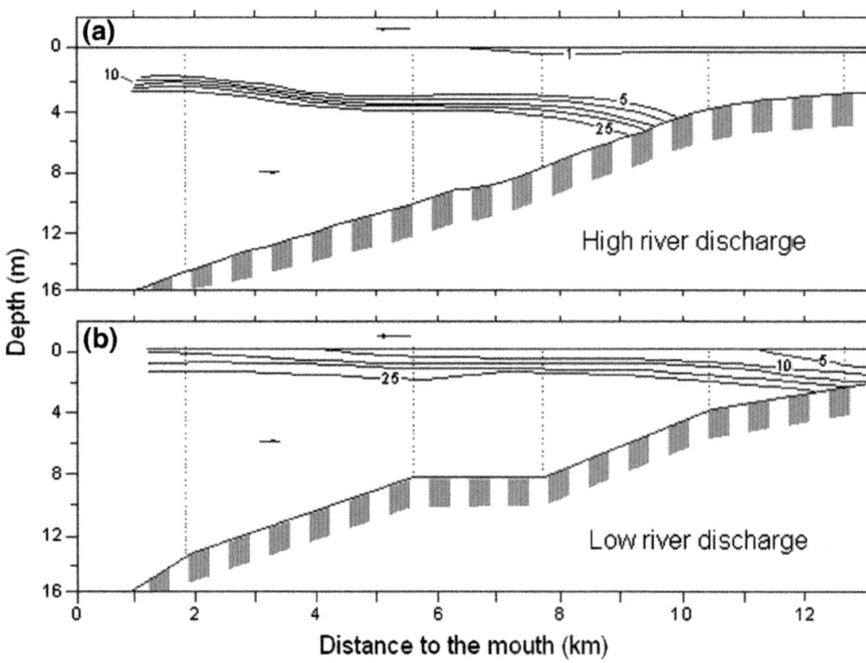

Fig. 9.1 Salinity stratification in a salt wedge estuary in conditions of high (**a**) and low (**b**) river discharge in the river Duwamish (Seattle, USA) (according to Dawson and Tilley 1972)

9.1 Hypothesis and Theoretical Formulation

A salt wedge estuary is schematically represented in Fig. 9.2, with the referential system and the adopted notation of properties and variables in the upper and lower layers indicated with indices 1 and 2, respectively. In this development, the estuary is assumed to be narrow and laterally homogeneous, and the Oz will be oriented in the gravity acceleration direction, which requires a signal change in the mathematical expression of the longitudinal component of the barotropic gradient pressure force (Eq. 8.18a,b, Chap. 8), as its previous orientation was against the gravity acceleration. It should be observed that ∂η/∂x is negative and the interface slopes in the landward direction.

Taking into account the Oxz axis orientation, the gradient pressure force has the following expression:

$$-\frac{1}{\rho}(\frac{\partial p}{\partial x}) = g(\frac{\partial \eta}{\partial x}) - \frac{g}{\rho}(\int_{z}^{\eta} \frac{\partial \rho}{\partial x} dz). \qquad (9.1)$$

The first theoretical investigations to calculate the vertical velocity profile and the salt intrusion length of the salt wedge estuary, using the continuity and motion equations in the upper and lower layers, were developed by Farmer and Morgan (1953), Sanders et al. (1953), followed by Shi-Igai and Sawamoto (1969). In these studies, the main results of which are described in this chapter, the motion attains a

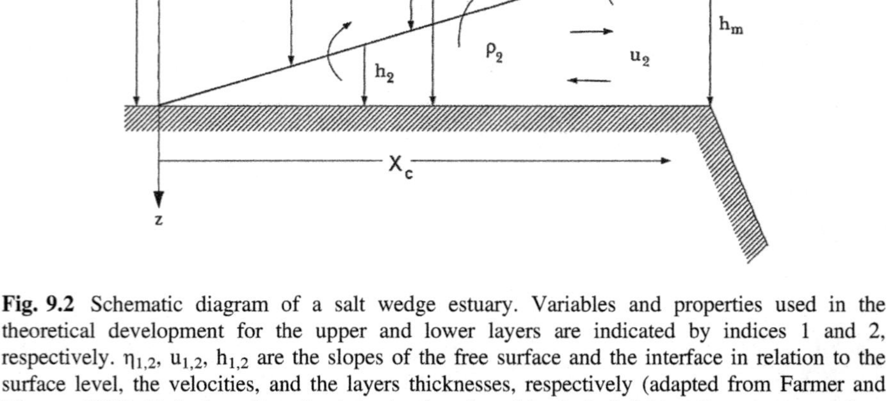

Fig. 9.2 Schematic diagram of a salt wedge estuary. Variables and properties used in the theoretical development for the upper and lower layers are indicated by indices 1 and 2, respectively. $\eta_{1,2}$, $u_{1,2}$, $h_{1,2}$ are the slopes of the free surface and the interface in relation to the surface level, the velocities, and the layers thicknesses, respectively (adapted from Farmer and Morgan 1953). X_c is the salt-wedge intrusion length and h_m is its height at the estuary mouth

quasi-steady condition, in which the baroclinic pressure gradient is balanced by inertial and frictional forces within the estuary. In these two-layer motions, the following simplifying hypotheses were adopted:

- Simple geometry: width (B) and depth (H_0) constants;
- No vertical mixing between the upper and lower layers;
- The u-velocity component in the upper layer is generated by the river discharge;
- The wind stress on the surface is disregarded;
- The interfacial shear stress $f_i[f_i] = [ML^{-1}T^{-2}]$, on the halocline is proportional to the square of the upper layer velocity, and the proportionality constant, k, which is non-dimensional,

$$f_i = k\rho_1 u_1^2, \tag{9.2}$$

- The velocity in the salt wedge, u_2, is much less than in the upper layer, $u_2 << u_1$;
- The longitudinal acceleration due to the advection, $u_2 \frac{\partial u_2}{\partial x}$, and the volume transport in the salt-wedge, Q_2, $[Q_2] = [L^3T^{-1}]$ will be disregarded ($Q_2 \ll Q_f$).

The comparison of the interfacial shear stress f_i (Eq. 9.2) with the bottom shear stress (τ_{Bx}) of a one-dimensional estuarine channel (Eq. 8.31, Chap. 8) indicates that the coefficient, k, corresponds to the ratio of the gravity acceleration to the square of the Chézy coefficient (g/C_y^2).

The theoretical development of this analytical model has the following objectives: (i) Calculate the vertical velocity profile u = u(x, z), the free surface and the halocline interface slopes $d\eta_1(x)/dx$ and $d\eta_2(x)/dx$, respectively (ii) Determination of the salt-wedge intrusion length, X_C, (Fig. 9.2) and the energy dissipation due to the interfacial shear stress and viscosity. To achieve this, the hydrodynamics formulation must take into account the mass and momentum conservation equations, which must be adequately simplified and solved in order to satisfy the specified boundary and integral boundary conditions.

9.2 Circulation and Salt-Wedge Intrusion

9.2.1 The Upper Layer

The one-dimensional equations of motion and the continuity (Eqs. 8.67 and 8.68, Chap. 8) are used to formulate the hydrodynamics of the upper layer which, according to the simplifying conditions, are:

$$\frac{\partial(uuA)}{\partial x} = -\frac{A}{\rho}\frac{\partial p}{\partial x} - \frac{1}{\rho}\tau_{Bx}P_m, \tag{9.3}$$

and

$$\frac{\partial(uA)}{\partial x} = 0. \tag{9.4}$$

The simplified equation of the state of seawater (Eq. 8.71, Chap. 8) will provide the hydrodynamic closure of this equation system,

$$\rho(S) = \rho_0(1 + \beta S). \tag{9.5}$$

The salinity in the upper, $S_1 \approx 0$, and lower layers, $S_2 = S_0$, will be taken as constants, and the salinity at the coastal region, S_0, is the only salt source for the estuarine water mass formation; from the equation of state of seawater Eq. (9.5), it follows that the density in these layers are constants and $\rho_1 < \rho_2$.

For the layer above the halocline, the last term on the right-hand-side of Eq. (9.3), representing the formulation of the bottom energy dissipation, will as a first approximation be substituted by the interfacial shear stress (Eq. 9.2). As the estuary width (B) is usually much greater than its depth ($B \gg H_0$), the wet perimeter, P_m, in Eq. (9.3) may be approximated by its width ($P_m = B$). As in Eqs. (9.3) and (9.4), the partial derivation may be changed to the total derivation, because x is only independent variable in these equations, and they may be rewritten as:

$$\frac{d(uuA)}{dx} = gA\frac{\partial \eta}{\partial x} - Bku^2, \tag{9.6}$$

and

$$\frac{d(uA)}{dx} = 0. \tag{9.7}$$

Under the assumption that k is a known coefficient, this equation system is closed and the two unknowns, $u = u(x)$ and $\eta = \eta(x)$, may be calculated.

Applying these equations to the upper layer (1) where $\eta_1(x) \le z \le \eta_2(x)$ and $h_1(x) = \eta_2(x) - \eta_1(x)$, and taking into account the particular geometry of the problem, the continuity equation is simplified to:

$$\frac{d[Bu_1(x)h_1(x)]}{dx} = 0. \tag{9.8}$$

In this equation, $h_1(x)$ is the thickness of the upper layer, and it is possible to calculate the uniform velocity field, $u_1 = u_1(x)$, integrating from the estuary head ($x = 0$) seaward up to a generic position x,

$$Bu_1(x)h_1(x) - Q_f = 0, \tag{9.9}$$

and

$$u_1(x) = \frac{Q_f}{Bh_1(x)}. \tag{9.10}$$

This last result indicates that if the halocline or picnocline depth $h_1(x)$ is known, the cross sectional mean velocity, $u_1 = u_1(x)$, in the upper layer may be calculated.

As B = const. Equation (9.8) may be rewritten as

$$\frac{d[u_1(x)h_1(x)]}{dx} = \frac{dQ_1(x)}{dx} = 0, \tag{9.11}$$

and the product of the mean velocity in the upper layer by the halocline depth, $Q_1(x)$, with dimension $[Q_1] = [L^2T^{-1}]$, is independent of the longitudinal distance (x), and this value is equal to Q_f/B, or

$$u_1(x)h_1(x) = \frac{Q_f}{B}; \text{or } u_1(x) = \frac{Q_f}{Bh_1(x)}. \tag{9.12}$$

With this procedure applied to the salt-wedge, it follows the trivial result due to the hypothesis that the volume transport in the lower layer is zero

$$Bu_2(x)h_2(x) = Q_2 = 0, \tag{9.13}$$

where $h_2(x)$ is the lower layer thickness. This result indicates that the mean depth velocity, $\bar{u}_2(x)$, is equal to zero. However, the salt-wedge presents a vertical velocity gradient (vertical shear), whose profile $u_2 = u_2(z)$ will be determined during this theoretical development.

Now, let us continue, applying the equation of motion (9.6) to the upper layer:

$$\frac{d[u_1(x)u_1(x)Bh_1(x)]}{dx} = gBh_1(x)\frac{d\eta_1(x)}{dx} - kBu_1^2(x). \tag{9.14}$$

Calculating the derivative of the first term of this equation, it follows that:

$$u_1(x)\frac{d[u_1(x)h_1(x)]}{dx} + u_1(x)h_1(x)\frac{du_1(x)}{dx} = gh_1(x)\frac{d\eta_1(x)}{dx} - ku_1^2(x), \tag{9.15}$$

and combining with Eq. (9.11), gives

$$u_1(x)h_1(x)\frac{du_1(x)}{dx} = gh_1(x)\frac{d\eta_1(x)}{dx} - ku_1^2(x). \tag{9.16}$$

The only unknown in this equation is the slope of the free surface $\eta_1 = \eta_1(x)$, because the velocity $u_1 = u_1(x)$ has already been determined by the Eq. (9.10). Hence, the unknown $\eta_1 = \eta_1(x)$ may be calculated by the following expression:

$$\eta_1(x) = \frac{k}{g}\int_0^x \frac{u_1^2(x)}{h_1(x)}dx + \frac{1}{2g}[u_1^2(x) - u_1^2(0)], \tag{9.17}$$

or, taking into account the $u_1(x)$ solution (Eq. 9.10), where $u_1(0) = Q_f/BH_0 = u_f$, the solution may also be expressed as:

$$\eta_1(x) = \frac{kQ_f^2}{gB^2}\int_0^x \frac{1}{h_1^3(x)}dx + \frac{1}{2g}[u_1^2(x) - u_f^2]. \tag{9.18}$$

As with Eqs. (9.10) and (9.18), it is possible to calculate $u_1(x)$ and $\eta_1(x)$, and thus the hydrodynamic problem for the upper layer of the salt wedge estuary is solved.

In the following development, let us calculate the relationship between the first derivatives of the free surface slope $(d\eta_1/dx)$ and that of the salt-wedge $(d\eta_2/dx)$, which will be used later to calculate the salt-wedge intrusion length. Thus, the first term of Eq. (9.16) may be combined with Eq. (9.10), resulting in:

$$u_1(x)h_1(x)\frac{d[u_1(x)]}{dx} = u_1(x)h_1(x)\frac{d}{dx}[\frac{Q_f}{Bh_1(x)}] = -\frac{[u_1(x)Q_f]}{Bh(x)_1}\frac{d[(h_1(x)]}{dx}, \tag{9.19}$$

or

$$-\frac{u_1(x)Q_f}{Bh_1(x)}[\frac{d\eta_2(x)}{dx} - \frac{d\eta_1(x)}{dx}] = -u_1^2(x)[\frac{d\eta_2(x)}{dx} - \frac{d\eta_1(x)}{dx}]. \tag{9.20}$$

Combining Eqs. (9.20) and (9.16) and rearranging its terms, it follows that:

$$[\frac{gh_1(x)}{u_1^2(x)} - 1][\frac{d\eta_1(x)}{dx} - \frac{d\eta_2(x)}{dx}] = k. \tag{9.21}$$

Taking into account that

$$[\frac{gh_1(x)}{u_1^2(x)}] \gg 1, \tag{9.22}$$

Equation (9.21) is reduced to the following relationship between the first derivatives of the sea surface slope $\eta_1(x)$ and $\eta_2(x)$:

$$\frac{gh_1(x)}{u_1^2(x)}\left[\frac{d\eta_1(x)}{dx}\right]+\frac{d\eta_2(x)}{dx}=k. \tag{9.23}$$

With Eq. (9.10), it is possible to calculate the volume transport per unit width of the cross-section (Q_1),

$$\frac{Q_f}{B}=Q_1, \tag{9.24}$$

which is independent of the longitudinal distance x (Eq. 9.11). Its introduction into the Eq. (9.23) is convenient, and to achieve this, it is necessary to multiply and divide the factor $d\eta_1(x)/dx$ by the square of the depth of the upper layer, $h_1^2(x)$. Then, according to Officer (1976) the result is:

$$\frac{gh_1^3(x)}{Q_1^2}\left(\frac{d\eta_1(x)}{dx}\right)+\frac{d\eta_2(x)}{dx}=k. \tag{9.25}$$

All terms on the left-hand-side of this equation are dimensionless.

9.2.2 *The Lower Layer (Salt-Wedge)*

According to the simplified physics adopted for this estuary, in the salt-wedge ($\eta_2 \le z \le \eta_1$) which has a thickness equal to $h_2(x)$, the velocity u_2 is much less than that of the upper layer ($u_2 \ll u_1$), and the advective acceleration may be disregarded. This layer characteristic has already been demonstrated in the model that used only the continuity and salt conservation equations (Eqs. 7.43 and 7.44), resulting in constant values of the longitudinal velocity component in the upper layer of the salt wedge estuary, and $u_1 > u_2$ (Fig. 7.3, Chap. 7). However, the motion direction in the halocline is reverted due to the entrainment and the bottom friction, and it is expected that this velocity component, although with low intensity, should present a vertical shear ($\partial u_2/\partial z \neq 0$). Thus, at any given longitudinal distance this velocity (u_2) is dependent on the depth. According to the longitudinal velocity component of the bi-dimensional equation of motion (equation, 8.57a, Chap. 8), the hydrodynamic equilibrium is reduced to the balance of the barotropic pressure gradient generated by the interface slopes η_1 and η_2 and the frictional force,

$$\frac{1}{\rho_2}\frac{\partial p_2}{\partial x}=\frac{\partial}{\partial z}\left[N_z\frac{\partial u_2(x,z)}{\partial z}\right], \tag{9.26}$$

This equation is a simplified formulation of the bi-dimensional equation of motion in the Oxz plane, whose gradient pressure force only has the barotropic component, because the density (ρ_2) at this layer is independent of the longitudinal distance. According to the linear equation of state (Eq. 9.5), knowing the salinity, the value of which may be obtained from experimental results, the density in the lower layer (ρ_2) is also known. Under the assumption that the kinematic eddy viscosity coefficient (N_z) is given, the only unknown in the Eq. (9.26) is the velocity in the salt-wedge, $u_2 = u_2(x, z)$, which may be calculate as follows.

The pressure p_2 at a depth z of the salt-wedge (Fig. 9.2) may be calculated by:

$$p_2(x,z) = g\rho_1(\eta_2 - \eta_1) + g\rho_2(z - \eta_2). \tag{9.27}$$

By derivation of this equation in relation to the longitudinal distance (x), it follows that the expression for the barotropic pressure gradient is:

$$\frac{\partial p_2}{\partial x} = -g\rho_1[\frac{\partial \eta_1(x)}{dx}] - g(\rho_2 - \rho_1)[\frac{d\eta_2(x)}{dx}], \tag{9.28}$$

which is independent of the depth and is dependent only on the slopes of the free surface and the salt-wedge.

Proceeding with the integration of Eq. (9.26) in the vertical direction of the salt-wedge, and taking into account that the first term is the barotropic pressure gradient yields,

$$\frac{\partial p_2}{\partial x}[\eta_3(x) - \eta_2(x)] = \rho_2 N_z[\frac{\partial u_2(x, z)}{\partial z}|_{z=\eta_3} - \frac{\partial u_2(x, z)}{\partial z}|_{z=\eta_2}]. \tag{9.29}$$

Remembering that $h_2(x) = \eta_3(x) - \eta_2(x)$ is the salt-wedge thickness, as indicated Fig. 9.2, and the two terms of its right member are the components of the shear stress acting at the bottom ($z = \eta_3$) and surface ($z = \eta_2$) of the salt wedge, this equation may be rewritten as:

$$h_2(x)\frac{\partial p_2}{\partial x} = \tau_{zx}|_{z=\eta_3} - \tau_{zx}|_{z=\eta_2}. \tag{9.30}$$

Combining this equation with Eq. (9.28) we have

$$-gh_2(x)[\rho_1\frac{d\eta_1(x)}{dx} + (\rho_2 - \rho_1)\frac{d\eta_2(x)}{dx}] = \tau_{zx}|_{z=\eta_3} - \tau_{zx}|_{z=\eta_2}. \tag{9.31}$$

In this equation, the shear stress on the superior interface of the salt-wedge is, by hypothesis, equal to the interfacial stress ($f_i = k\rho_1 u_1^2$). Then,

$$\tau_{zx}|_{z=\eta_3} = \tau_{zx}(\eta_2) = f_i = k\rho_1 u_1^2(x), \tag{9.32}$$

and the Eq. (9.31) may be rewritten as

$$-gh_2(x)[\rho_1\frac{d\eta_1(x)}{dx}+(\rho_2-\rho_1)\frac{d\eta_2(x)}{dx}]=\tau_{zx}|_{z=\eta_3}-k\rho_1u_1^2(x). \quad (9.33)$$

As the term on the left-hand side of Eq. (9.26) is the barotropic pressure gradient, the velocity may be approximated by the following quadratic expression (Officer 1976): $u_2(x,z)=a+bz+cz^2$. The coefficients of this expression may be determined by applying the boundary and integral boundary conditions, and one of these coefficients will be function of x. In this development, the Oz axis will have its origin at the bottom and will be oriented upward, against the gravity acceleration. The new ordinate will be denoted by ($\overline{z}$); it will be related to the orientation of the first vertical variable (z) orientation by the relation $\overline{z}=H_0-z$, and at the bottom $z = H_0$ and $\overline{z}=0$. With the introduction of this new variable, the vertical velocity profile in the salt-wedge will be given by

$$u_2(x,\,\overline{z})=a+b\overline{z}+c(\overline{z})^2, \quad (9.34)$$

and the coefficients a, b and c may be calculated with the following boundary and integral boundary conditions:

$$u_2(x,\,\overline{z})|_{\overline{z}=0}=0, \quad (9.35)$$

$$u_2(x,\,\overline{z})|_{\overline{z}=h_2}=u_1(x), \quad (9.36)$$

and

$$\int_0^{h_2}u_2(x,\,\overline{z})d\overline{z}=Q_2=0. \quad (9.37)$$

The latter condition is due to the hypothesis that the net volume transport in the salt-wedge is zero.

Applying the boundary condition (9.35), it follows immediately that a = 0, and for the remaining conditions, (9.36) and (9.37), the result is an algebraic system of two equations and two unknowns b and c,

$$u_1=bh_2+ch_2^2, \quad (9.38)$$

and

$$\frac{1}{2}bh_2^2+\frac{1}{3}ch_2^3=0. \quad (9.39)$$

This system of equations may be solved, giving the results: $b = -2u_1/h_2$ and $c=3u_1/h_2^2$, and the vertical velocity profile $u_2 = u_2(x,\ z)$ has the following expression:

$$u_2(x, \bar{z}) = -\frac{2u_1}{h_2}\bar{z} + \frac{3u_1}{h_2^2}\bar{z}^2, \quad (9.40)$$

or, returning to the z variable

$$u_2(x, z) = -\frac{2u_1}{h_2}(H_0 - z) + \frac{3u_1}{h_2^2}(H_0 - z)^2. \quad (9.41)$$

Analysis of these solutions indicates that the velocity is zero at $\bar{z} = (2/3)h_2$ and $z = H_0$ - $(2/3)h_2$, and there is a minimum point in this vertical velocity profile at depth $\bar{z} = (1/3)h_2$ or $z = H_0 - (1/3)h_2$. At this depth, the minimum velocity at the salt-wedge is $u_2 = -(1/3)u_f$.

9.2.3 *Vertical Velocity Profile*

The combined solutions of Eqs. (9.10) and (9.41), used to calculate the velocities $u_1 = u_1(x)$ and $u_2 = u_2(x,z)$ in the upper and lower layers of the halocline, respectively, are the theoretical solutions of the vertical velocity profile in the salt wedge estuary, which are driven by the fresh water discharge and the barotropic influences of the free surface slope and salt-wedge interface with the river discharge, respectively. The energy dissipating forces, which counteract the river discharge and baroclinic pressure gradient, are the vertical friction, due to the viscosity, and the interfacial and bottom shear stresses.

According to classical investigations cited in the article of Geyer and Farmer (1989), the designation *arrested salt wedge* for this estuary refers to a regime in which the baroclinic pressure gradient is balanced by inertial and frictional forces within the estuary, and its interfacial structure attains a quasi-steady configuration. A practical example of this theory will be presented at the end of this chapter.

9.2.4 *Salt-Wedge Intrusion Length*

Knowing the analytical expression of the vertical velocity profile in the salt wedge estuary (Eqs. 9.40 or 9.41), it is possible to calculate the frictional stresses, $(\tau_{zx}|_{z=\eta_3})$ and $(\tau_{zx}|_{z=\eta_2})$ at the depths $z = H_0$ (or $\bar{z} = 0$) and $z = \eta_2$ (or $\bar{z} = h_2$), respectively,

$$\tau_{zx}|_{z=\eta_3} = \tau_{zx}(\eta_3) = -\rho_2 N_z \frac{\partial u_2}{\partial \bar{z}}|_{\bar{z}=0} = \frac{2\rho_2 N_z u_1}{h_2}, \quad (9.42)$$

and

$$\tau_{zx}|_{z=\eta_2} = \tau_{zx}(\eta_2) = -\rho_2 N_z \frac{\partial u_2}{\partial \bar{z}}|_{\bar{Z}=h_2} = -\frac{4\rho_2 N_z u_1}{h_2}. \tag{9.43}$$

Combining these equations and taking into account that the first term on the right-hand-side of Eq. (9.43) is the interfacial frictional shear that may be approximated by Eq. 9.32, it follows that,

$$\tau_{zx}|_{\bar{z}=0} = -\frac{1}{2}\tau_{zx}|_{\bar{z}=h_2} = \frac{1}{2}f_i = \frac{1}{2}\rho_1 k u_1^2. \tag{9.44}$$

By subtracting Eqs. (9.42) and (9.43),

$$\tau_{zx}|_{z=\eta_3} - \tau_{zx}|_{z=\eta_2} = \frac{6\rho_2 N_z u_1}{h_2}. \tag{9.45}$$

In this equation, the quantity $\tau_{zx}|_{z=\eta_2} = \tau_{zx}(\eta_2)$ is equal to the interfacial shear stress (f_i), and the following relationship exists between the coefficients k and the kinematic eddy viscosity coefficient, N_z,

$$N_z = k(\frac{\rho_1 h_2 u_1}{4\rho_2}), \tag{9.46}$$

and substituting this result into Eq. (9.45),

$$\tau_{zx}|_{z=\eta_3} - \tau_{zx}|_{z=\eta_2} = \frac{3}{2}k\rho_1 u_1^2. \tag{9.47}$$

Finally, combining this result with Eq. (9.31) gives the following relationship of the derivatives of sea surface ($d\eta_1/dx$) and salt-wedge ($d\eta_2/dx$), slopes:

$$-gh_2[\rho_1 \frac{d\eta_1(x)}{dx} + (\rho_2 - \rho_1)\frac{d\eta_2(x)}{dx}] = \frac{3}{2}\rho_1 k u_1^2. \tag{9.48}$$

As an artifice, multiplying the term on the left-hand-side by the ratio h_1^2/h_1^2 and dividing both equation members by $\rho_1 u_1^2$ and using the approximation $\rho_1 \approx \rho_2$, yields

$$-\frac{gh_1^2 h_2}{Q_1^2}[\frac{d\eta_1(x)}{dx} + \delta\frac{d\eta_2(x)}{dx}] = \frac{3}{2}k, \tag{9.49}$$

where $(u_1 h_1)^2 = Q_1^2$ is the square value of the river discharge per unit width, and the quantity δ is defined by

$$\delta = \frac{\rho_2 - \rho_1}{\rho_2} = \frac{\Delta\rho}{\rho_2}. \tag{9.50}$$

Equations (9.25) and (9.49) are components of an algebraic system with two unknowns, $d\eta_1(x)/dx$ and $d\eta_2(x)/dx$. Then, for the second unknown the result is;

$$\frac{d\eta_2(x)}{dx}\left[\frac{h_2(x)}{h_1(x)} - \frac{gh_1^2(x)\delta h_2(x)}{Q_1^2}\right] = k\left[\frac{h_2(x)}{h_1(x)} + \frac{3}{2}\right]. \tag{9.51}$$

A trivial solution of this equation is to consider that the interfacial shear stress ($f_i = k\rho_1 u_1^2$) is equal to zero, which may be simulated with k = 0. However, for a salt-wedge occurrence ($k \neq 0$) the Eq. (9.51) may be solved for $d\eta_2/dx$ and integrated to calculate the unknown, $\eta_2 = \eta_2(x)$,

$$\eta_2(x) = \int_0^x \left\{\frac{k\left[\frac{h_2(x)}{h_1(x)} + \frac{3}{2}\right]}{\left[\frac{h_2(x)}{h_1(x)} - \frac{g'h_1^2(x)h_2(x)}{Q_1^2}\right]}\right\} dx. \tag{9.52}$$

As the main objective of this topic is to calculate the salt-wedge intrusion length, X_c, Eq. (9.51) will be used for this purpose. As the ordinate η_3 may be taken as a constant, let us apply the approximation,

$$\frac{d[h_2(x)]}{dx} = -\frac{d[\eta_2(x)]}{dx}, \tag{9.53}$$

and combining this with Eq. (9.51), factoring in the first term by the ratio h_2/h_1, and rearranging the terms, we have,

$$h_2(x)\left[1 - \frac{g'h_1^3(x)}{Q_1^2}\right]\frac{dh_2(x)}{dx} = -k\left[h_2(x) + \frac{3h_1(x)}{2}\right], \tag{9.54}$$

and analysis of the salt wedge estuary (Fig. 9.2) showed the following relationships:

$$H_0 = h_1 + h_2 + \eta_1 \text{ and } h_1 + h_2 \gg \eta_1. \tag{9.55}$$

Thus, Eq. (9.54) may be rewritten as a function of the non-dimensional salt-wedge height $H(x) = h_2(x)/H_0$, which varies in the interval $0 \leq H(x) < 1$, and the differential $dh_2(x)$ is

$$dh_2(x) = H_0 dH(x), \tag{9.56}$$

and combined with the relationship (9.55) the initial solution is:

$$H(x)[1 - \frac{g'H_0^3[1 - H(x)]^3}{Q_1^2}]\frac{dH(x)}{dx} = -\frac{k[3 - H(x)]}{2H_0}. \tag{9.57}$$

Considering the ratio

$$(\frac{Q_1^2}{g'H_0^3}) = \gamma, \tag{9.58}$$

which may be considered constant, because in the hypothesis of a steady-state condition the fresh water (Q_f) is also constant, Eq. (9.57) can be rewritten as,

$$H(x)\{\frac{[1 - H(x)]^3 - \gamma}{\gamma}\}\frac{dH(x)}{dx} = k\frac{[3 - H(x)]}{2H_0}. \tag{9.59}$$

This equation is an ordinary differential equation with separable variables which may be integrated from the landward limit of the salt-wedge, x = 0, up to a seaward longitudinal position, x,

$$(\frac{k\gamma}{2H_0})x = \int_0^H \frac{H(x)[1 - H(x)]^3 - \gamma H(x)}{[3 - H(x)]}dH. \tag{9.60}$$

As a case limit for this result, we may observe that for $x \to 0$, implies that $H(x) \to 0$, because by definition $H(x) = h_2(x)/H_0$, and $h_2(0) = 0$ at the interior limit of the salt-wedge (Fig. 9.2).

Developing the algebraic expression of the integrand in Eq. (9.60), and using the additive propriety of integrals yields:

$$\begin{aligned}(\frac{k\gamma}{2H_0})x = {} & (1 - \gamma)\int_0^H \{\frac{H(x)}{[3 - H(x)]}\} dH \\ & - 3\int_0^H \{\frac{H^2(x)}{[3 - H(x)]}\}dH + 3\int_0^H \{\frac{H^3(x))}{[3 - H(x)]}\} dH \\ & - \int_0^H \{\frac{H^4(x)}{[3 - H(x)]}\} dH.\end{aligned} \tag{9.61}$$

Taking into account the following algebraic equalities:

$$\frac{H^2(x)}{[3 - H(x)]} = -H(x) + 3\frac{H(x)}{[3 - H(x)]}, \tag{9.62}$$

$$\frac{H^3(x)}{[3 - H(x)]} = -H^2(x) - 3H(x) + 9\frac{H(x)}{[3 - H(x)]}, \tag{9.63}$$

and

$$\frac{H^4(x)}{[3 - H(x)]} = -H^3(x) - 3H^2(x) - 9H(x) + 27\frac{H(x)}{[3 - H(x)]}. \tag{9.64}$$

Substituting them into the integrands of the lasts three terms of the right-hand-side of expression (9.61) and simplifying the result, we have:

$$\frac{k\gamma}{2H_0}x = -(\gamma - 8)\int_0^H \{ \frac{H(x)}{[3 - H(x)]}\} dH + 3\int_0^H H(x)dH + \int_0^H H^3(x)dH. \tag{9.65}$$

The first term of the right-hand-side of this equation may be easily integrate remembering that its indefinite integral is given by (Granville et al. 1956),

$$\int \frac{H}{(3 - H)}dH = 3 - H - 3\ln(3 - H). \tag{9.66}$$

The integration of the second and third terms is immediate, and follow the relationship between the longitudinal distance, x, and the non-dimensional salt-wedge height:

$$\frac{k\gamma}{2H_0}x = \frac{3}{2}H^2(x) + \frac{1}{4}H^4(x) + (\gamma + 8)\{ 3\ln[\frac{3 - H(x)}{3}] + H(x)\} . \tag{9.67}$$

Equation (9.58), which defines the quantity γ, is a function of the river discharge, mass stratification and the estuary depth. As $Q_1 = u_1(0)H_0 = u_f H_0$ at the estuary head, γ may be expressed as:

$$\gamma = \frac{u_f^2}{g'H_0} = \frac{u_f^2}{g\frac{\Delta\rho}{\rho_2}H_0}. \tag{9.68}$$

This dimensionless number is equal to the square of the densimetric Froude number ($\gamma = F_m$), defined in the Chap. 2 (Eq. 2.39). This number has been investigated by Farmer and Morgan (1953), who simulated the circulation in salt wedge estuaries and observed that this number converges to 1 ($F_m \rightarrow 1$) in the transition of the fresh water flow to the salt water reservoir. In the salt wedge estuary, this number may be estimated using the following data: $g = 10\ ms^{-2}$, $u_f = 0.1\ ms^{-1}$, $\Delta\rho/\rho_2 = 3.0 \times 10^{-4}$ and $H_0 = 10$ m, resulting in $\gamma = 0.3$. As $g < 1 \rightarrow F_m < 1$, this indicates a subcritical vertical stratification which is characteristic of highly stratified estuaries. Usually the parameter $\gamma \ll 8$ and may be disregarded in the last term of Eq. (9.67), which may be simplified for praticality to:

$$\frac{k\gamma}{2H_0}x = \frac{3}{2}H^2(x) + \frac{1}{4}H^4(x) + 8\{\ 3\ln[\frac{3 - H(x)}{3}] + H(x)\}\ . \qquad (9.69)$$

To calculate the salt-wedge intrusion length (X_c), let us define its non-dimensional depth at the estuary mouth as $H_m = h_m/H_0$ (Fig. 9.2), which can be obtained with observational data. Then, if the depth $H \rightarrow H_m$ in the second member of Eq. (9.69), the generic distance x of the first member approaches X_c, and this may be calculated by:

$$\frac{k\gamma}{2H_0}X_c = \frac{3}{2}H_m^2 + \frac{1}{4}H_m^4 + 8\{\ 3\ln[\frac{3 - H_m}{3}] + H_m\}\ . \qquad (9.70)$$

Solving this equation for the salt-wedge intrusion length, X_c, it follows that:

$$X_c = 2\frac{g'H_0^2}{ku_f^2}[\frac{3}{2}H_m^2 + \frac{1}{4}H_m^4 + 8\{\ 3\ln[\frac{(3 - H_m)}{3}] + H_m\}\ . \qquad (9.71)$$

This result indicates that X_c is directly proportional to the square of the estuary depth (H_0^2), and inversely proportional to the coefficient of interfacial frictional shear (k) and the square of the velocity generated by the river discharge (u_f). Besides the seasonal variation of u_f, its input in the estuary may be the altered by utilization of river water in agriculture, industrial and for domestic use, interfering with the salt-wedge intrusion length. The estuarine channel depth (H_0) may decrease due to sedimentation processes and may be modified by dredging. Consequently, the theoretical results (Eq. 9.71) clearly indicate that human interference may have anomalous influences on this natural environment.

The salt-wedge configuration can be conveniently analysed through its non-dimensional formulation, which may be obtained by the ratio of Eqs. (9.69) and (9.70),

$$\frac{x}{X_c} = [\frac{H^2(x)}{H_m}].\{\ \frac{\frac{3}{2} + \frac{H^2(x)}{4} + \frac{8}{H^2(x)}[3\ln(\frac{3-H(x)}{3}) + H(x)]}{\frac{3}{2} + \frac{H_m^2}{4} + \frac{8}{H_m^2}[3\ln(\frac{3-H_m}{3}) + H_m]}\}\ . \qquad (9.72)$$

From this solution, we have the following limiting cases:

- When $H = H_m \rightarrow x/X_c = 1$; and
- For $H = 0 \rightarrow x/X_c = 0$.

These results are equivalent to the simplest analytical expressions obtained by Farmer and Morgan (1953); Officer (1976) for determination of the steady-state configuration of the salt-wedge. Analysis of Eq. (9.72) indicates that there will be similarities in the salt wedge configurations for different estuaries. These similarities are due to the fact that when the interfacial Froude number is less than one, the salt-wedge configuration is independent of the water mass salinity in the coastal sea and of the velocity generated by the river discharge. The non-dimensional salt-wedge configuration deduced by Farmer & Morgan (op. cit), with the notation adapted to that used in this chapter, is

$$\frac{x}{X_c} = (\frac{H(x)}{H_m})^2[3 - 2(\frac{H(x)}{H_m})]. \tag{9.73}$$

This analytical solution was compared to observational data of the South Pass of the Mississippi river delta (Mississippi, USA), and to laboratory experiments, with the results found to be in close agreement (Fig. 9.3).

The non-dimensional salt-wedge configuration of the South Pass (Mississippi river) was also simulated by Wright (1970), using the Eq. (9.73) and the following quantities: $H_0 = 11.5$ m, $h_m = 7.3$ m and $H_m = 0.63$. Taking h = 0.0; 0.05; 0.1; 0.2; 0.3; 0.4; 0.5 and 0.6, the following values were obtained for the non-dimensional ratio x/X_c = 0.0; 0.02; 0.08; 0.27; 0.49; 0.70; 0.88 and 1.0, respectively. The results of the correlation, H/H_m, as a function of the non-dimensional distance, x/X_c, are shown comparatively in Fig. 9.3 (black points), and are almost coincident with the classical results of Farmer and Morgan (1953), with only a small deviation near the estuary mouth. In this figure, it is also possible

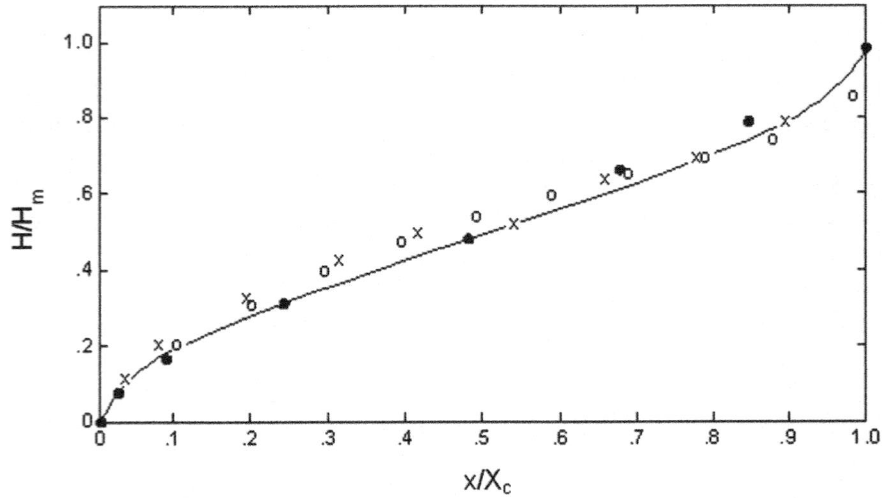

Fig. 9.3 Non-dimensional salt-wedge configuration. The continuous line is the theoretical result obtained with the Eq. (9.73). Observational data from the South Pass of the Mississippi river, and experimental laboratory data are indicated by o and x, respectively (according to Farmer and Morgan, 1953). Black points • were introduced to indicate theoretical results calculated with Equation (9.74)

to observe that for small values of the non-dimensional depth, that is, in the vicinity of the interior salt-wedge limit, the non-dimensional profile is approximately convex, gradually becoming linear in the medium portion of the wedge and, finally, slightly concave in the proximity of the estuary mouth.

This theoretical model has been generalized by Rattray and Mitsuda (1974) in order to include the bottom topography, with its declivity and the bottom friction. Also, in the theoretical development of the upper layer, a simplified equation of motion was used, which included the advective acceleration. To analytically formulate the motion in the lower layer (salt-wedge), several approximations from the classical articles also were used.

A theory of the density current in a stratified two-layer estuary flow with complete vertical mixing in the upper layer was developed by Prandle (1985). This theory was extended to the special case of a channel with a flat bed, constant breadth and depth. The theoretical result was further simplified, neglecting some undesirable effects, and an estimate of the salt-wedge intrusion length, L_{length}, was calculated by:

$$L_{length} = 0.26\frac{gH_0^2}{k\bar{u}}\frac{\Delta\rho}{\rho} = 0.26\frac{g'H_0^2}{k\bar{u}}. \tag{9.74a}$$

This result was compared with the following expression of the intrusion length, L_A, of an arrested salt wedge estuary given by G. H. Keulegan in 1949 (quoted in Ippen & Harleman, 1961), adding useful support to the above expression,

$$L_A = A\frac{g^{5/4}H_0^{9/4}}{\bar{u}^{5/2}}\left(\frac{\Delta\rho}{\rho}\right)^{3/4}. \tag{9.74b}$$

During the investigation of the dynamical interaction of the tidal flow with the estuarine circulation of the Fraser river salt wedge estuary, which has a characteristic two layer circulation, the *internal or densimetric Froude number*, G, (Chap. 2, Eq. 2.39), has been expressed by Geyer and Farmer (1989) as:

$$G^2 = (F_1)^2 + (F_2)^2, \tag{9.75}$$

where $(F_i)^2 = u_i^2/g'h_i$, (i = 1, 2), u_1 and u_2 are velocities in the upper and lower layers, respectively, g' is the reduced gravity, and h_1 and h_2 are the thicknesses of the upper and lower layers, respectively. For the simplified two-layer flow of a salt wedge estuary in a rectangular channel with a uniform depth-mean volume transport and a quasi-steady interface elevation, the momentum equations for the upper and lower layers where combined to form the following equation for density-driven shear flow (Geyer and Farmer, op. cit.):

$$\frac{\partial}{\partial t}(u_2 - u_1) = -[(1 - G^2)g'\frac{\partial \eta}{\partial x} + \frac{C_D|u_2|u_1}{h_2} + C_E(\frac{1}{h_1} + \frac{1}{h_2})|u_2 - u_1|(u_2 - u_1)]. \tag{9.76}$$

In this equation, $\eta = \eta(x)$ is the interface elevation, C_D and C_E are the bottom and interfacial drag coefficients, respectively; $C_E \ll C_D$ unless the interface is unstable. Since the interface slopes downward in the landward direction (Fig. 9.2), the first term on the right-hand-side of Eq. (9.76) will be positive or negative for subcritical and supercritical flows, respectively. The bottom drag term will be positive or negative, depending on the direction of the near-bottom flow. The magnitude of the interfacial drag term is difficult to ascertain, since it depends on the stability of the interface; however, its sign will always be such that it acts in opposition to the shear.

The solution of an *arrested salt wedge* is obtained when the left-hand-side of Eq. (9.76) vanishes and the baroclinic pressure gradient balances the drag terms. For this to occur, the flow must be subcritical, with the baroclinic pressure gradient balancing the drag of the landward deep flow.

Studies of the time dependent mixing in salt wedge estuary were presented by Partch and Smith (1978), analyzing measurements of salinity and velocity profiles, taken at short time intervals in comparison to the tidal period, as well as direct measurements of vertical turbulent salt flux and turbulent kinetic energy. Their results indicated that the turbulent mixing through the density interface is highly time dependent with the most intense mixing occurring at the maximum speed, and when the flow approaches critical conditions.

9.3 Theory and Experiment

Exemplifying the theory that has been developed, let us perform an analysis of the longitudinal salinity stratification presented in Fig. 9.1a. As previously indicated, this experimental result, which was observed during a period of high river discharge ($Q_f \approx 148\ m^3\ s^{-1}$) in the salt wedge estuary of the Duwamish river (Seattle, Washington, USA), was published by Dawson and Tilley (1972).

To adequate this experimental result to the presented theory, it is necessary to approximate the estuary with a simple geometry, for example: the bottom with a planel surface, with a mean depth of 10 m ($H_0 = 10$ m) and a constant width (B = 140 m, from hydrographic charts, Corps of Engineers 1973, quoted in Rattray and Mitsuda 1974). From Fig. 9.1a, it is possible to estimate the salt-wedge intrusion length as 10^4 m with a mean slope estimated as $d\eta_2/dx = 2.0 \times 10^{-4}$ (approximately 1.0 m for a length of 5000 m). The fresh water velocity at the estuary head is estimated as $0.10\ ms^{-1}$. Another quantity which may be estimated from the figure is the non-dimensional depth of the salt-wedge at the estuary mouth, calculated by the ratio $H_m = h_m/H_0 \approx 0.4$ ($h_m = 4$ m and $H_0 = 10.0$ m). With this value, using Eq. (9.73), it is possible to calculate the non-dimensional configuration of the salt wedge, which is similar to that presented in Fig. 9.3.

These results show that it is possible to have good agreement between values obtained theoretically and experimentally, such as the salt-wedge intrusion length, using a determined value for the interfacial friction coefficient, k. However, this doesn't represent proof of the hypothesis used in the theory, because k is a measure of the eddy shear at the salt and fresh water interface, and its value varies not only with different estuary conditions, but also from one estuary to the other (Farmer and Morgan 1953; Rattray and Mitsuda 1974).

Let us continue to theoretically calculate the vertical velocity profile at the landward position, x≈7.8 km, in the salt wedge estuary (Fig. 9.1a). At this position, the thicknesses of the upper and lower layers during high river discharge are approximately $h_1 = 4.0$ m, $h_2 = 6.0$ m. Then, according to Eq. (9.10), the velocity in the upper layer ($0 \leq z \leq 4$ m) is calculated by

$$u_1(x) = u_f = \frac{Q_f}{Bh_1(x)} = 0.26\,\mathrm{ms}^{-1}.$$

The lower layer (salt-wedge) is delimited by the depth interval (4 m $\leq$z 10 m), and the theoretical vertical velocity profile is calculated by Eq. (9.41). Using the values already determined for this profile, we have:

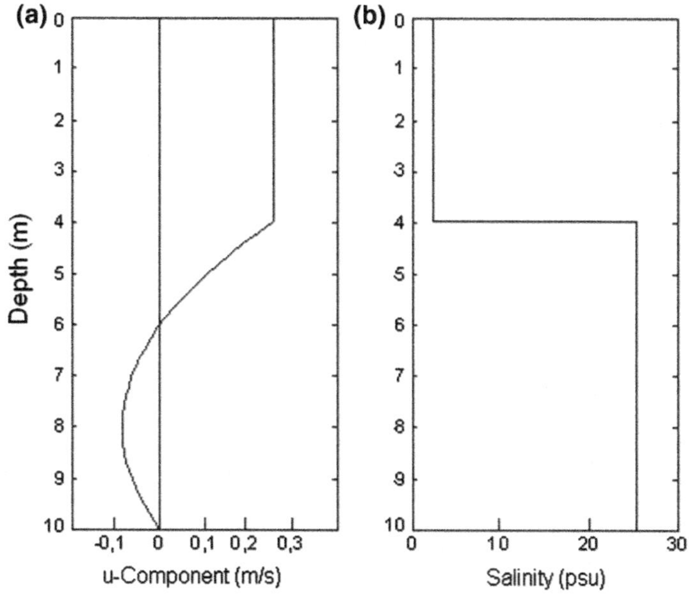

Fig. 9.4 **a** Theoretical vertical velocity profile in the salt wedge estuary of the Duwamish river. **b** Experimental vertical salinity profile. The physical quantities necessary to calculate these profiles were estimated from Fig. 9.1a

$$u_2(x,z) = -(\frac{0.52}{6})(10 - z) + (\frac{0.78}{36})(10 - z)^2,$$

or

$$u_2(x,z) = -8.7x10^{-2}(10 - z) + 2.2x10^{-2}(10 - z)^2.$$

Composing the profiles, $u_1(x)$ and $u_2(x, z)$, in the upper ($0 \leq z \leq 4$ m) and lower ($10 \text{ m} \leq z \leq 4$ m) layers, we obtain the theoretical velocity profile in the water column, as shown in Fig. 9.4a; this solution is represented graphically, with the vertical salinity profile (Fig. 9.4b estimated from Fig. 9.1a. Figure 9.4a indicate that above the halocline, the flow is seaward with constant velocity, $u_1(x) = 0.26 \, \text{ms}^{-1}$. In the salt-wedge, an accentuated decrease is observed in the velocity, $u_2(x, z)$, and at 6 m depth the velocity is zero. For greater depths, the motion is landward and reaches a velocity of $-0.09 \, \text{ms}^{-1}$. At the depth interval of the salt-wedge the vertical velocity shear is forced by the barotropic pressure gradient and the free surface slope, and due to the imposed boundary condition the velocity at the bottom is zero.

As the theoretical velocity profile has been obtained with simplifying hypothesis, it must be validated by comparison with experimental velocity profiles.

References

Dawson, W. A. & Tilley, L. J. 1972. Measurement of Salt Wedge Excursion Distance in the Duwamish River Estuary, Seattle, Washington, by Means of the Dissolved-Oxygen Gradient. Geological Survey Water-Supply. Washington, D. C., U. S. Department of Interior, Paper 1873-D, pp. D1–D27.

Farmer, H. G. & Morgan, G. W. 1953. The Salt Wedge. In: Johnson, J. W. (ed.). Proc. of Third Conference on Coastal Engineering. Council on Wave Research. Cambridge, The Engineering Foundation, pp. 54–64.

Geyer, W. R. 1986. The Advance of a Salt Wedge Front: Observations and a Dynamical Model. In: Dronkers, J & Van Leussen W. (eds.). Physical Processes in Estuaries. Berlin, Springer-Verlag, pp. 181–195.

Geyer, W. R. & Farmer, D. M. 1989. Tide-Induced Variations of the Dynamics of a Salt Wedge Estuary. J. Phys. Oceanogr., v.19, pp.1060–1072.

Granville, W. A.; Smith, P. F. & Longley, W. R. 1956. Elementos de Cálculo Diferencial e Integral. Trad. J. Abdelhay. 2 ed., Rio de Janeiro, Editora Científica. 695 p.

Hansen, D. V. & Rattray Jr., M. 1966. New Dimensions in Estuary Classification. Limnol. Oceanogr., 11(3):319–325.

Harleman, D. R. F. & Ippen, A. T. 1967. Two-Dimensional Aspects of Salinity Intrusion in Estuaries: Analysis and Velocity Distributions. Committee on Tidal Hydraulics. Tech. Bull., Corps of Engineers, U. S. Army, n. 13.

Keulegan, G. H. 1949. Interfacial Instability and Mixing in Stratified Flows. J. Res. U. S. Geol. Surv., 43:487–500.

Officer, C. B. 1976. Physical Oceanography of Estuaries (and Associated Coastal Waters). New York, Wiley. 465 p.
Partch, E.N. & Smith, J.D. 1978, Time Dependent Mixing in a Salt Wedge Estuary. Estuarine and Coastal Marine Science. 6, pp. 3–19.
Prandle, D. 1985. On salinity Regimes and the Vertical Structure of Residual Flows in Narrow Tidal Estuaries. Estuar. Coast. Shelf Sci., 20:615–635.
Pritchard, D. W. 1955. Estuarine Circulation Patterns. Proc. Am. Soc. Civ. Eng., 81:717:1–11.
Rattray Jr., M. & Mitsuda, E. 1974. Theoretical Analysis of Conditions in a Salt Wedge. Estuar. Coast. Mar. Sci., 2:375–394.
Sanders, J. L.; Maximon, L. C. & Morgan, G. W. 1953. On the Stationary Salt Wedge – a Two Layer Free Surface Flow. Tech. Rept., Brown University, n. 1. 44 p.
Schettini, C. A. F. 2002. Caracterização Física do Estuário do Rio Itajaí-açu, SC. Revista Brasileira Recursos Hídricos, 7(1):123–142.
Schettini & Truccolo. E. C. 1999. Dinâmica da Intrusão Salina no Estuário do Rio Itajaí-açu. In: Congresso Latino Americano de Ciências do Mar, 8, Trujillo, Peru, Resumenes ampliados, Tomo II, UNT/ALICMAR, p. 639–640.
Shi-Igai, H. & Sawamoto, M. 1969. Experimental and Theoretical Modeling of Saline Wedges. Proc. of the 13th Congress Internat. Assoc. Hydraulic Res., Kyoto. Science Council of Japan, 3 (C): 29–36.
Wright, L. D. 1970. Circulation, Effluent Diffusion and Sediment Transport, Mouth of South Pass, Mississippi River Delta. Baton Rouge, Louisiana State University Press. 56 p.

Quoted References

Corps of Engineers. 1973 (quoted in Rattray & Mitsuda, 1974. Theoretical Analysis of Conditions in a Salt-Wedge. Estuar. Coast. Mar. Sci., 2:375–394).
Döbereiner, C.E. 1985. Comportamento hidráulico e sedimentológico do estuário do rio Itajaí, SC. Rio de Janeiro, Instituto Nacional de Pesquisas Hidroviárias (INPH), Relatório 700/03, 34 p. (quoted in Schettini (2002), p. 132).
Ippen, A. T. & Harleman, D. R. F. 1961. One-Dimensional Analysis of Salinity Intrusion in Estuaries. Committee on Tidal Hydraulics. Tech. Bull. Corps of Engineers U. S. Army, n. 5. 120 p.
Schijf, J.B. & Schonfeld, 1963. Theoretical considerations on the motion of salt and fresh water. Proc. Minnesota Int. Hydraul. Conv. 5th Congress I.A.H.R., pp. 321–333. (quoted in Geyer & Farmer (1989), p. 1060).

Chapter 10
Circulation and Mixing in Steady-State Models: Well-Mixed Estuary

In this chapter, the analytic model of circulation and mixing in a well-mixed and laterally homogeneous estuary (Types 1 or D) will be presented. In contrast to that of the salt wedge estuary, the vertical salinity stratification of a well-mixed estuary is the complete opposite, being very weak. These conditions are characteristic of estuaries in regions of low river discharge, where the circulation and mixing processes are dominated by tidal forcing. As the vertical salinity (density) gradient is very low, it may be practically neglected, and in steady-state conditions, the fresh water discharge and tidal forcing remain constant during tidal cycles. In practice, these simplifying conditions are simulated by mean values during tidal cycles, resulting in a one-directional seaward circulation (Fig. 10.1).

In the mixing zone (MZ), the longitudinal salinity (density) gradient is much less intense than that observed in partially mixed estuaries, as indicated in Chap. 8 (Fig. 8.1). However, the integrated influence of the baroclinic pressure gradient associated with the internal friction is one of the processes responsible for the occurrence of the small vertical velocity shear. Eventually, in relatively deep estuaries, this integrated influence may generate weak gravitational circulation and landward motions in bottom layers.

10.1 Hydrodynamic Formulation and Hypothesys

Let us consider a laterally homogeneous, well-mixed estuary with the objective of introduce an analytical model to calculate the vertical velocity profile, $u = u(x, z)$, the free surface slope, $\partial\eta/\partial x$, and the longitudinal salinity variation $S = S(x, z)$, which are generated by the fresh water discharge, the gradient pressure force and the surface wind stress. The approach we will take to achieve this follows the articles of Arons and Stommel (1951), Maximon and Morgan (1955), Officer (1976, 1977) and Prandle (1985). By hypothesis, the estuary has a simple geometry with constant width (B) and depth (h), as schematically shown in Fig. 10.2. The Oxz

L. Bruner de Miranda et al., *Fundamentals of Estuarine Physical Oceanography*, Ocean Engineering & Oceanography 8, DOI 10.1007/978-981-10-3041-3_10

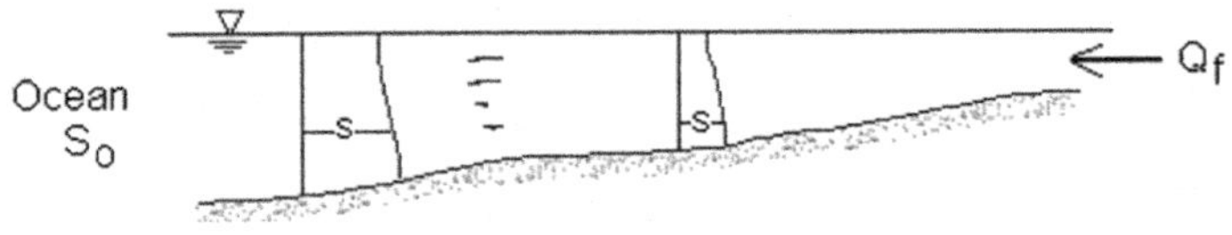

Fig. 10.1 Steady-state vertical salinity and velocity distribution in a well-mixed estuary, with small vertical salinity gradients which may be found in nature. S_0 is the salinity at the coastal ocean, which is a boundary condition to the salt conservation equation

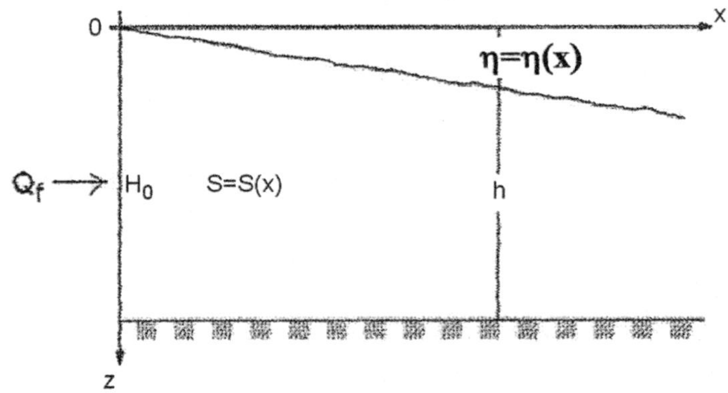

Fig. 10.2 Diagram of a well-mixed estuary and the coordinate system used in the theoretical development. S = S(x, z) or ρ = ρ(x, z), η = η(x) and H_0 indicate the longitudinal salinity or density distribution, the free surface slope and the depth at the estuary head, respectively

referential system will be used, with the vertical *a*xis (Oz) originating at the free surface and oriented in the direction of the gravity acceleration ($\vec{g}$), and the Ox axis oriented seaward. Then, according to the barotropic and baroclinic pressure gradients (Eq. 8.18 and Chap. 8), it follows that,

$$-\frac{1}{\rho}\frac{\partial p}{\partial x} = g\frac{\partial \eta}{\partial x} - \frac{g}{\rho}\int_{\eta}^{z}\frac{\partial \rho}{\partial x}dz. \qquad (10.1)$$

As the longitudinal density gradient is taken as independent of the depth, $\frac{\partial}{\partial z}\left(\frac{\partial \rho}{\partial x}\right) = 0$, the longitudinal density (salinity) gradient is substituted by the depth-mean value $\frac{\partial \bar{\rho}}{\partial x}\left(\frac{\partial \bar{S}}{\partial x}\right)$, this equation is simplified to

$$-\frac{1}{\rho}\frac{\partial p}{\partial x} = g\frac{\partial \eta}{\partial x} - \frac{g}{\rho}\frac{\partial \bar{\rho}}{\partial x}(z-\eta) = g\frac{\partial \eta}{\partial x} - \frac{g}{\rho}\frac{\partial \bar{\rho}}{\partial z}z. \qquad (10.2)$$

In the last term of the right-hand-side of this equation, the approximation $z - \eta \approx z$ was made, and the baroclinic pressure gradient increases linearly with depth. In some cases, although the density gradient is considered to be independent of the depth, it may vary along the longitudinal distance, $\partial\overline{\rho}/\partial x = g(x)$, where $g = g(x)$ is a function which remains to be determined.

To satisfy the mass continuity principle, the free surface must slope downward in the seaward direction ($\partial\eta/\partial x > 0$) generating the barotropic pressure gradient, which physically simulates the motion generated by the fresh water discharge.

The following simplifying assumption must be specified:

- Local, advective and Coriolis accelerations are disregarded;
- The lateral friction has low intensity in comparison to the energy dissipation due to the vertical eddy diffusion.
- The vertical kinematic eddy viscosity (N_z) and diffusivity (K_z) coefficients are supposed to be independent of the depth.

Taking into account these simplifications, the laterally integrated continuity and motion Eqs. (8.56 and 8.57b, Chap. 8), combined with the gradient pressure force (Eq. 10.2) may be simplified, and reduced to

$$\frac{\partial u}{\partial x} + \frac{\partial w}{\partial z} = 0, \tag{10.3}$$

and

$$g\frac{\partial \eta}{\partial x} - \frac{g}{\rho}\frac{\partial \overline{\rho}}{\partial x}z + N_z\frac{\partial^2 u}{\partial z^2} = 0. \tag{10.4}$$

It should be noted that in the equation of motion, a simple balance of the barotropic and baroclinic pressure gradients and the energy dissipation due to the friction is taken into account. These equations must be complemented with the one-dimensional salt conservation equation (Eq. 7.78a Chap. 7), in steady-state condition, along with the linear state of seawater equation

$$u_f\frac{\partial S}{\partial x} = K_x\left(\frac{\partial^2 S}{\partial z^2}\right), \tag{10.5}$$

$$\rho(S) = \rho_0(1 + \beta S), \tag{10.6}$$

where the saline contraction coefficient, β, may be approximated by a constant.

Under the assumption that the kinematic eddy viscosity (N_z) and diffusion (K_z) coefficients are constants and known, Eqs. (10.3) to (10.6) form a closed hydrodynamic system. Its solution, satisfying the boundary and integral boundary conditions, will give the following unknowns: $u = u(x, z)$, $\eta = \eta(x)$, $S = S(x, z)$ and $\rho = \rho(x)$.

As the coefficient, N_z, is known, Eq. (10.4) may be rewritten as

$$\frac{\partial^2 u}{\partial z^2} = -\frac{g}{N_z}\frac{\partial \eta}{\partial x} + \frac{g}{\rho N_z}\frac{\partial \overline{\rho}}{\partial x}z, \quad (10.7a)$$

or in terms of the longitudinal salinity gradient,

$$\frac{\partial^2 u}{\partial z^2} = -\frac{g}{N_z}\frac{\partial \eta}{\partial x} + \frac{\rho_0 g \beta}{\rho N_z}\frac{\partial \overline{S}}{\partial x}z. \quad (10.7b)$$

In Eq. (10.7b), the longitudinal density gradient has been substituted by the salinity gradient, using the simplified equation of state (10.6); this gradient may be calculated using steady-state salinity values, or theoretically estimated with the salt conservation Eq. (10.5).

The salt conservation equation was solved in Chap. 7 (Eq. 7.113) as a first approximation of the mixing process in a well-mixed estuary. In this solution, the salinity distribution presented an exponential decrease in the mixing zone, $S(x) = S_0 . \exp(u_f x/K_x)$, and its longitudinal salinity gradient ($\partial S/\partial x$) behaved similarly.

For the solution of the equation of motion (10.7b), it is convenient to rewrite the equation as

$$\frac{\partial^2 u}{\partial z^2} = \overline{a}\eta_x + \overline{c}S_x z, \quad (10.8)$$

where the following coefficient changes have been used: $\overline{a} = -g/N_z$ and $\overline{c} = \beta g/N_z$, both with dimensions $[\overline{a}, \overline{c}] = [L^{-1}T^{-1}]$, and the approximation $\rho_0/\rho \approx 1$. In this equation, the non-dimensional surface slope ($\partial \eta/\partial x$) and the longitudinal salinity gradient, $\frac{\partial \overline{S}}{\partial x} = S_x$, $[S_x] = [L^{-1}]$, are denoted by η_x and S_x, respectively.

The non-homogeneous second order differential equation with constant coefficients (Eq. 10.8) may be integrated twice

$$u(x, z) = \iint (\overline{a}\eta_x + \overline{c}S_x z)dzdz, \quad (10.9a)$$

with the general solution:

$$u(x, z) = \frac{1}{6}\overline{c}S_x z^3 + \frac{1}{2}\overline{a}\eta_x z^2 + C_1 z + C_2. \quad (10.9b)$$

In this equation, the integration constants, C_1 and C_2, have the dimensions $[C_1] = [T^{-1}]$ and $[C_2] = [LT^{-1}]$, and will be determined with the following boundary conditions:

- Wind stress at the surface, τ_{Wx};

$$\rho N_z \frac{\partial u}{\partial z}|_{z=0} = A_z \frac{\partial u}{\partial z}|_{z=0} = -\tau_{Wx}, \tag{10.10}$$

where A_z, is the dynamic eddy viscosity coefficient, and the wind is oriented in the seaward direction, and $\tau_{Wx} > 0$ because $(\partial u/\partial z < 0)$.

- Maximum friction (non-slippery) at the bottom ($z = H_0 - \eta \approx H_0$);

$$u(x, z)|_{z=H_0} = 0. \tag{10.11}$$

Another bottom boundary condition which may be applied is a slippery bottom, simulated by,

$$\rho N_z \frac{\partial u}{\partial z}|_{z=H_0} = A_z \frac{\partial u}{\partial z}|_{z=H_0} = \tau_{Bx}. \tag{10.12}$$

In this last boundary condition, τ_{Bx}, is the bottom frictional which is a dissipative energy source.

10.2 Solution with Maximum Bottom Friction

Applying the boundary conditions (10.10) and (10.11) to the general solution (Eq. 10.9b) the integration constants are expressed by,

$$C_1 = -\frac{\tau_{Wx}}{A_z}, \tag{10.13}$$

and

$$C_2 = -\frac{1}{6}\bar{c}S_x H_0^3 - \frac{1}{2}\bar{a}\eta_x H_0^2 + \frac{\tau_{Wx}}{A_z} H_0, \tag{10.14}$$

with dimensions $[C_1] = [T^{-1}]$ and $[C_2] = [LT^{-1}]$.

Substituting these constants into the general solution (10.9b) we have

$$u(x, z) = \frac{1}{6}\bar{c}S_x(z^3 - H_0^3) + \frac{1}{2}\bar{a}\eta_x(z^2 - H_0^2) - \frac{\tau_{Wx}}{A_z}(z - H_0). \tag{10.15}$$

It is easy to verify that this solution has the dimension of velocity $[u(x, z)] = [LT^{-1}]$, but it is still dependent on the free surface slope (η_x), which is an unknown and is calculated using the continuity equation in the form of an integral boundary condition, which is given by,

$$\int_0^{H_0} u(x,z)dz = \frac{Q_f}{B}, \text{ and } u_f = \frac{Q_f}{BH_0}. \tag{10.16}$$

Substituting the solution (10.15) into the integrand, and completing the integration, we have

$$\frac{Q_f}{B} = -\frac{1}{8}\bar{c}S_x H_0^4 - \frac{1}{3}\bar{a}\eta_x H_0^3 + \frac{\tau_{Wx}}{2A_z}H_0^2, \tag{10.17}$$

where the approximation $\eta + H_0 \approx H_0$ was made under the assumption that $\eta \ll H_0$. Solving expression (10.17) for the unknown, η_x, and taking into account the expressions $\bar{a} = -g/N_z$ and $\bar{c} = \beta g/N_z$, we have:

$$\frac{\partial \eta}{\partial x} = \eta_x = 3\frac{N_z u_f}{gH_0^2} + \frac{3}{8}\beta S_x H_0 - \frac{3}{2}\frac{\tau_{Wx}}{\rho g H_0}, \tag{10.18}$$

or

$$\eta_x = 3\frac{N_z u_f}{gH_0^2} + 0.375\frac{H_0}{\rho_0}\frac{\partial \rho}{\partial x} - 1.5\frac{\tau_{Wx}}{\rho g H_0}. \tag{10.19}$$

Without the wind stress forcing ($\tau_{Wx} = 0$) this result is close to that calculated by Geyer (2010). In this solution $u_f = Q_f/BH_0$, and it has been taken into account that $\beta S_x = (1/\rho_0)\partial\rho/\partial x)$, which was obtained using the linear equation of state (10.6). The first and second terms on the right-hand-side of this equation are proportional to the fresh water discharge and the longitudinal density (salinity) gradient, $\partial\rho/\partial x$, and always have positive contributions ($u_f > 0$ and $\partial\rho/\partial x > 0$). The wind forcing however, may change its direction ($\tau_{Wx} > 0$ or $\tau_{Wx} < 0$ seaward or landward, respectively). Hence, the steady-state free surface slope is controlled by the forcings: river discharge, longitudinal density (salinity) gradient and wind stress. In normal conditions, $\eta_x > 0$ and the free surface slope is seaward; however, in abnormal wind conditions, the wind stress may be higher than the other forces. With the following order of magnitude: $N_z = 0.05\ m^2\ s^{-1}$, $u_f = 0.1\ m\ s^{-1}$, $H_0 = 10$ m and $(1/\rho_0)\partial\rho/\partial x = 3 \times 10^{-6}\ m^{-1}$ (density variation of 30 kg m^{-3} for an estuary with length of 10^4 m), the right-hand-side terms of Eq. (10.19) have the following orders of magnitude:

$$O\left(\left|\frac{3N_z u_f}{gH_0^2}\right|\right) \approx 10^{-5}, \tag{10.20}$$

$$O\left(\left|0.375\frac{H_0}{\rho_0}\frac{\partial \rho}{\partial x}\right|\right) \approx 10^{-5}. \tag{10.21}$$

These results indicate that the river discharge and the longitudinal density (salinity) gradient are positive and have the same order of magnitude, but this situation may change for longer estuaries ($\gg 10^4$ m). However, due to particular strong seaward or landward winds, its forcing influence may be higher than the combined effects of the river discharge and the longitudinal density (salinity) gradient, generating an ascending free surface from the estuary head towards the mouth and a bidirectional circulation, as shown in Fig. 10.3.

Substituting Eq. (10.18) into solution (10.15) simplifying the result and rewriting it in terms of the non-dimensional depth ($Z = z/|H_0|$), yields the following expression for calculation of the vertical u-velocity profile (Officer 1976, 1977):

$$u(x, Z) = \left[\frac{\beta g S_x H_0^3}{48 N_z}\right](1 - 9Z^2 + 8Z^3) + \frac{3}{2}u_f(1 - Z^2) + \frac{\tau_{Wx} H_0}{4A_z}(1 - 4Z + 3Z^2), \quad (10.22)$$

where $A_z = \rho N_z$ is the dynamic viscosity coefficient $[A_z] = [ML^{-1}T^{-1}]$. Recalculating the numerical coefficients and rewriting the results as functions of the longitudinal density gradient, gives

$$u(x, Z) = \frac{gH_0^3}{\rho_0 N_z}\frac{\partial \rho}{\partial x}(0.167Z^3 - 0.188Z^2 + 0.0208) - u_f(1.5Z^2 - 1.5) + \frac{\tau_{Wx} H_0}{A_z}(0.75Z^2 - Z + 0.25). \quad (10.23)$$

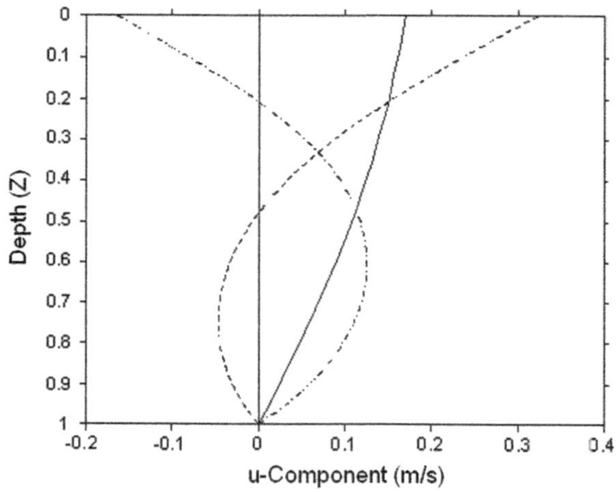

Fig. 10.3 Vertical u-velocity profiles of a well-mixed estuary. Under normal steady-state conditions, the u-velocity profile is one-directional (*black line*); strong wind flowing seaward and landward generating a bidirectional flow (*dashed* and *dash-point lines*, respectively)

The analysis of the relative importance of the forces on the right-hand-side of this equation may be made using the same values previously used to estimate the terms of Eq. (10.19), complemented with the wind stress, $\tau_{Wx} = 0.2\ \mathrm{N\ m^{-2}}$:

$$O\left(\left|\frac{gH_0^3}{N_z}\frac{1}{\rho_0}\frac{\partial\rho}{\partial x}\right|\right) \approx 5 \times 10^{-2}\ \mathrm{m\ s^{-1}}, \tag{10.24}$$

$$O(|u_f|) = 1 \times 10^{-1}\ \mathrm{m\,s^{-1}}, \tag{10.25}$$

$$O\left(\left|\frac{\tau_{Wx}H_0}{A_z}\right|\right) = 8 \times 10^{-2}\ \mathrm{m\,s^{-1}}. \tag{10.26}$$

These results indicate that these forces may have different orders of magnitude, to generate the one-directional u-velocity component of well-mixed estuaries. However, the estuary circulation may be substantially altered by abnormal weather events (wind intensity) and changes in the river discharge. Vertical u-velocity profiles were calculated with Eq. (10.23), using the order of magnitude values presented in (10.24) and (10.25), but with strong wind intensities oriented seaward and landward generating bidirectional circulation as shown in the Fig. (10.3). Another example, showing the influence of baroclinic pressure gradient force to generate bi-directional motion may be found in Fontes et al. (2015).

In normal conditions, the motion is seaward and its velocity decreases with depth, characterizing the classical steady-state circulation of a well-mixed estuary. For this type of estuary, the circulation parameter $\left(p_c = \frac{u_s}{u_f}\right)$ calculated in the analytical theory of Hansen and Rattray (1966) was 1.5; in the profile presented in Fig. 10.3, this parameter is equal to 1.7, which agrees closely to the theoretical value.

The dynamical influence of strong winds generating bi-directional motions forced by the barotropic pressure gradient, seaward and landward directions due to positive and negative free surface slopes, respectively, may be investigated with Eq. (10.23). For strong local wind generating a positive surface slope ($\partial\eta/\partial x > 0$), the velocity at the surface layer is positively composed with the river velocity, and may generate higher intensities than those generated by an opposing wind direction, as shown in Fig. 10.3.

Using Eq. (10.23) to calculate the surface and bottom velocities, the following results will be obtained:

$$u(x,0) = 2.08 \times 10^{-2}\left(\frac{gH_0^3}{N_z}\frac{1}{\rho}\frac{\partial\rho}{\partial x}\right) + 1.5u_f + 2.5 \times 10^{-1}\left(\frac{\tau_{Wx}H_0}{A_z}\right), \tag{10.27}$$

and

$$u(x, 1) = 0, \tag{10.28}$$

due to the maximum friction at the bottom.

If the first term on the left-hand-side of Eq. (10.27) is calculated from observational data and the surface wind stress is also known, Eq. (10.27) may be used to estimate the dynamic viscosity coefficient, A_z (or $N_z = A_z/\rho$); solving Eq. (10.27) for this coefficient yields the expression:

$$A_z = \frac{1}{[2u(x,0) - 3u_f]}\left[4.16 \times 10^{-2}\left(gH_0^3 \frac{\partial \rho}{\partial x}\right) + 5 \times 10^{-1}\tau_{Wx}H_0\right]. \tag{10.29}$$

Calculating the orders of magnitude of the expression between right brackets on the numerator of the right-hand-side, it is possible to verify that under normal conditions, the term of the longitudinal density gradient is one order of magnitude higher than that associated with the wind stress term, and the expression may be simplified to

$$A_z \approx \frac{4.16 \times 10^{-1} g\left(H_0^3 \frac{\partial \rho}{\partial x}\right)}{[2u(x,0) - 3u_f)]}. \tag{10.30}$$

In this result, it was assumed that $g \approx 10$ m s^{-2}. With $u(x, 0) = 0.2$ m s^{-1} and the same values previously used for the others variables, it follows that $A_z \approx 1.0 \times 10^2$ kg m^{-1}s^{-1}.

The occurrence of extreme points in the vertical velocity profile (Eq. 10.22) may be identified by a simple mathematical treatment, consisting of its derivation, and equating to zero the result, we have:

$$\frac{\partial u}{\partial Z} = \frac{\beta g S_x H_0^3}{48N_z}(-18Z + 24Z^2) - 3u_f Z + \frac{\tau_{Wx}H_0}{4N_z\rho}(-4 + 6Z) = 0. \tag{10.31}$$

Under the assumption that the wind stress may be disregarded, this equation is reduced to

$$\frac{\partial u}{\partial Z} = \frac{\beta g S_x H_0^3}{48N_z}(-18Z + 24Z^2) - 3u_f Z = 0. \tag{10.32}$$

As the derivative value is zero for $Z = 0$, and at this point Eq. (10.22) has a positive value, $u(x, 0) > 0$, it follows that it is of maximum value, and the vertical velocity profile may be calculated with Eq. (10.23), using the definition of the dynamic eddy viscosity coefficient $\rho_0 N_z = A_z$,

$$u(x, Z) = \frac{gH_0^3}{A_z}\frac{\partial \rho}{\partial x}(0.167Z^3 - 0.188Z^2 + 0.0208) - u_f(1.5Z^2 - 1.5). \tag{10.33}$$

In estuaries where the fresh water discharge is the main driving force and the influence of the baroclinic pressure gradient may be disregarded, but the vertical diffusion generated by the tidal energy is great enough for the estuary to be classified as well mixed, the vertical velocity profile in calm weather ($\tau_{wx} \approx 0$) is reduced to:

$$u(x, Z) = -u_f(1.5Z^2 - 1.5), \tag{10.34}$$

and its mean value in the water column is

$$\int_0^1 u(x, Z)dZ = u_f \int_0^1 (-1.5Z^2 + 1.5)dZ = u_f. \tag{10.35}$$

Thus, the free surface slope Eq. (10.19) is simplified to

$$\frac{\partial \eta}{\partial x} = \eta_x = 3u_f \frac{N_z}{gH_0^2}, \tag{10.36}$$

and, from the Eq. (10.34), the relative velocity is:

$$\frac{u(x, Z)}{u_f} = -1.5(Z^2 - 1), \tag{10.37}$$

which is presented in Fig. 10.4. The vertical velocity shear observed in this figure is only due to the vertical eddy diffusion (inner friction), because the baroclinic pressure gradient has been disregarded.

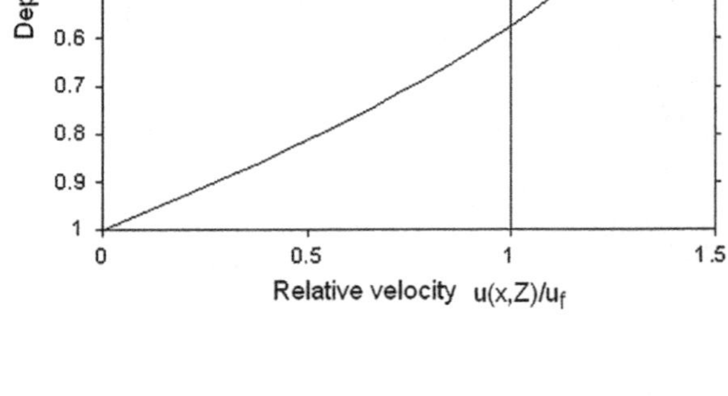

Fig. 10.4 Vertical relative velocity profile in a well-mixed estuary with the longitudinal density (salinity) and wind stress disregarded ($\partial\rho/\partial x = 0$ and $\tau_{Wx} = 0$), calculated with Eq. (10.37). The constant value for $u(x, Z)/u_f = 1$ is the relative value of the mean velocity in the water column

10.3 Vertical Velocity Profile: Moderate Bottom Friction

Let us now calculate the vertical profile of the u-velocity component with the bottom boundary condition, τ_{Bx}, simulated with a slippery bottom, according to formulation (10.12). However, in this case, the frictional bottom stress expressed by the semi-empirical relationship $\tau_{Bx} = \rho(4/\pi)kU_T u|_{z=Ho}$ (Eq. 8.32 and Chap. 8), which may be applied when the tidal velocity amplitude (U_T) is much higher than the u-velocity component ($U_T \gg u$), according to Bowden (1953). Taking into account the boundary condition at the free surface (10.10), the integration constants C_1 of the general solution (10.9b) is expressed by:

$$C_1 = -\frac{\tau_{Wx}}{\rho N_z} = -\frac{\tau_{Wx}}{A_z}, \tag{10.38}$$

which is equal to that obtained in (10.13). As the bottom shear stress will be simulated by $\tau_{Bx} = \rho(4/\pi)kU_T u|_{z=Ho}$, applying the bottom boundary condition (10.12) the expression for the integration constant C_2, is formulated by:

$$C_2 = \bar{c}S_x H_0^3\left(\frac{\pi}{8} - \frac{1}{6}\right) + \bar{a}\eta_x H_0^2\left(\frac{\pi}{4} - \frac{1}{2}\right) - \frac{\tau_{Wx}H_0}{A_z}\left(\frac{\pi}{4} - 1\right), \tag{10.39}$$

and its dimension is $[C_2] = [LT^{-1}]$.

Substituting the integration constants C_1 and C_2 into the general solution (Eq. 10.9b), reducing to the result to the simplest expression in terms of the non-dimensional depth (Z) and taking into account that $\bar{a} = -g/N_z$ and $\bar{c} = \beta g/N_z$, the final expression of the u-velocity profile is

$$\begin{aligned} u(x, Z) = & \frac{\beta g S_x H_0^2}{kU_T}\left[\frac{Z^3}{6} + \left(\frac{\pi}{8} - \frac{1}{6}\right)\right] \\ & - \frac{\eta_x g H_0}{kU_T}\left[\frac{Z^2}{2} + \left(\frac{\pi}{4} - \frac{1}{2}\right)\right] - \frac{\tau_{Wx}}{\rho k U_T}\left[Z + \left(\frac{\pi}{4} - 1\right)\right]. \end{aligned} \tag{10.40}$$

This expression is still a function of the free surface slope (η_x), to its determination as a function of a known physical quantity, it is necessary to apply the integral boundary condition expressed by,

$$\frac{1}{(H_0 + \eta)}\int_{\eta}^{H_0} u(x, z)dz = \int_0^1 u(x, Z)dZ = u_f, \tag{10.41}$$

where the approximation $H_0 + \eta \approx H_0$ will be made. From Eqs. (10.40) and (10.41), we have for η_x the following expression:

$$\eta_x = \beta S_x H_0 \frac{\left(\frac{\pi}{8} - \frac{1}{8}\right)}{\left(\frac{\pi}{4} - \frac{1}{3}\right)} - \frac{u_f k U_T}{g H_0}\left[\frac{1}{\left(\frac{\pi}{4} - \frac{1}{3}\right)}\right] - \frac{\tau_{Wx}}{\rho g H_0}\frac{\left(\frac{\pi}{4} - \frac{1}{2}\right)}{\left(\frac{\pi}{4} - \frac{1}{3}\right)}. \tag{10.42}$$

Reducing this equation to a more simple expression, it may be rewritten as

$$\eta_x = -2.212\frac{kU_T u_f}{gH_0} + 5.92\text{x}10^{-1}\beta S_x H_0 - 0.631\frac{\tau_{Wx}}{\rho g H_0} \tag{10.43}$$

or as a function of the longitudinal density gradient,

$$\eta_x = -2.212\frac{kU_T u_f}{gH_0} + 5.92\text{x}10^{-1}\frac{H_0}{\rho_0}\frac{\partial \rho}{\partial x} - 0.631\frac{\tau_{Wx}}{\rho g H_0}. \tag{10.44}$$

If the river discharge velocity (u_f), the longitudinal salinity (density) gradient and the wind stress are known, it is possible to calculate η_x with Eq. (10.43) or (10.44); the quantities U_T and H_0 are supposed to be known. Comparing Eqs. (10.44) and (10.19), it is possible to verify the following changes, which are due to the bottom boundary condition changes:

- For the river discharge velocity (u_f) and the wind stress (τ_{Wx}), the factors of these terms decreased from 3.0 to ≈ -2.2, and from 1.5 to ≈ 0.63, respectively;
- The factor of the longitudinal density gradient increased from ≈ 0.37 to ≈ 0.59.

To evaluate the importance of the forces producing the steady-state free surface slope (Eq. 10.44), the following orders of magnitude of its terms were calculated:

$$O\left(\left|2.212\frac{kU_T u_f}{gH_0}\right|\right) \approx 2.7 \times 10^{-6}, \tag{10.45}$$

$$O\left(\left|0.592\frac{H_0}{\rho_0}\frac{\partial \rho}{\partial x}\right|\right) \approx 3.5 \times 10^{-5}, \tag{10.46}$$

and

$$O\left(\left|0.631\frac{\tau_{Wx}}{\rho g H_0}\right|\right) \approx 6.3 \times 10^{-7}, \tag{10.47}$$

where the following numeric values were applied: $k = 2.5 \times 10^{-3}$, $U_T = 1.0\ \text{m s}^{-1}$, $H_0 = 20$ m ($N_z = 5 \times 10^{-2}\ \text{m}^2\ \text{s}^{-1}$), $u_f = 0.1\ \text{m s}^{-1}$, $(1/\rho_0)\partial\rho/\partial x = 3 \times 10^{-6}\ \text{m}^{-1}$ (density variation of 30 kg m^{-3} for a distance of 10 km) and $\tau_{Wx} = 0.2$ Pa.

These orders of magnitude (10.45) to (10.47) indicate that the influence of the longitudinal density gradient predominates and, although the river discharge velocity term is negative, the combination of these terms generates a positive free surface slope (downward towards the estuary mouth), as also observed in Eq. (10.19). The wind stress has the least influence, but intense landward wind forcing ($\tau_{Wx} < 0$) generates a positive free surface slope; or, intense seaward wind stress ($\tau_{Wx} > 0$) may generate negative slopes if its intensity is higher than the combined influences of the longitudinal density gradient and the river velocity input.

Substituting the equation of the free surface slope (10.44) into Eq. (10.40), numerically simplifying the result, and solving it as a function of the longitudinal density gradient, the analytic velocity profile has the following expression:

$$\begin{aligned} u(x, Z) = {} & \frac{gH_0^2}{kU_T} \frac{1}{\rho_0} \frac{\partial \rho}{\partial x} (0.167Z^3 - 0.296Z^2 + 0.058) \\ & + u_f(1.106Z^2 + 0.630) + \frac{\tau_{Wx}}{\rho k U_T} (0.316Z^2 - Z + 0.395). \end{aligned} \tag{10.48}$$

This solution has some similarities to Eq. (10.23) with the non-slippery bottom boundary condition, $u(x, z)|_{z=H0} = 0$.

Theoretical simulations of the steady-state u-velocity component with Eq. (10.48) under different intensities of the longitudinal density gradient, river discharge velocity and wind stress are presented in Fig. 10.5. To calculate these velocity profiles, the coefficient k and the quantities U_0 and H_0 were taken as $k = 2.5 \times 10^{-3}$, $U_T = 0.7\ \text{m s}^{-1}$ and $H_0 = 10$ m.

The intensity decreases in the profiles shown in Fig. 10.5a were due to the decrease in the longitudinal density gradient from 4.8×10^{-4} to 2.4×10^{-4} kg m^{-4}. At the bottom, we may observe the influence of the slippery bottom boundary condition generating a landward motion (very attenuated in the simulation with the weakest longitudinal density gradient).

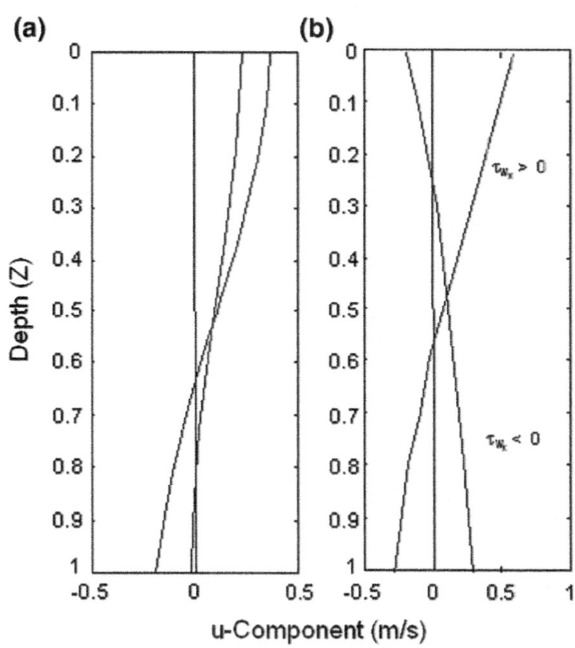

Fig. 10.5 Vertical velocity profiles simulated with the one-dimensional analytic model with the slippery bottom boundary condition (Eq. 10.49). **a** Normal wind conditions. **b** Profiles showing the reverted motions when local high intensity wind stress force the circulation seaward ($\tau_{Wx} > 0$) and landward ($\tau_{Wx} < 0$)

The profiles shown in Fig. 10.5b were calculated with the same longitudinal density gradient as Fig. 10.5a, to demonstrate only the influence of high wind stress acting seaward ($\tau_{Wx} = 1.0$ Pa) and landward ($\tau_{Wx} = -1.0$ Pa). For the seaward wind direction $\tau_{Wx} > 0$ the velocity in the layer deeper than $|Z| > 0.57$ is a relatively intense motion in both the seaward and landward directions (Fig. 10.5a), deviating from the classical estuarine circulation of a well-mixed estuary. For the landward wind direction ($\tau_{Wx} < 0$), the balance of forces that drive the well-mixed estuary circulation is suppressed, and the circulation changes to the opposite direction: landward above $|Z| < 0.25$ and seaward bellow $|Z| > 0.25$.

In the next topic, these theoretical results will be compared with nearly steady-state observational velocity profiles.

10.4 Theory and Observational Data

In the literature of estuarine dynamics, there is a great number of articles focusing on the comparison of theoretical values and observational data of steady-state estuarine circulation. To address this, we will compare experimental data from well-mixed (or almost well-mixed) estuaries with the solutions of the theoretical Eqs. (10.23) and (10.48). In this process, the nearly steady-state vertical velocity profiles were calculated from time-mean observational values, measured over one or more complete tidal cycles, which satisfies, as a first approximation, the hypothesis for steady-state estuarine conditions used in the theoretical development.

Under these hypotheses, the estuarine circulation is driven by the longitudinal density (salinity) gradient, by the river discharge, represented by the volume transport, Q_f (or the velocity u_f), and a constant wind stress. However, for most estuaries in this study, these data weren't simultaneously measured with the experimental velocity data used in the validation process. As such, the comparison of theory and experimental data must be taken only as a test of the theoretical development.

For the first comparison, consider the theoretical vertical velocity profile calculated by Eqs. (10.22) or (10.23), exemplified with observational data obtained in the estuary of the Cananéia-Iguape estuarine system (Fig. 1.5 and Chap. 1). Experimental data obtained during the spring tide indicated a weakly stratified vertical salinity profile (Fig. 10.6b), with salinities differing between the bottom and surface by less than 1.0‰ ($\delta S = S_f - S_s \approx 0.6$‰), with time-mean values calculated for a period of two semi-diurnal tidal cycles. According to the classification criteria of Pritchard (1955), this estuary may be classified as well-mixed. The analytical simulation of the u-velocity component presented by Miranda et al. (1995), and its comparative analysis with the experimental data (Fig. 10.6b) indicated good agreement between these results. In this figure, for $|Z| > 0.55$ it is also possible to observe the influence of the baroclinic pressure gradient, indicated by landward motions, because at the mooring station the estuary was relatively deep ($\approx$10 m), increasing the baroclinic component influence in the gradient pressure force.

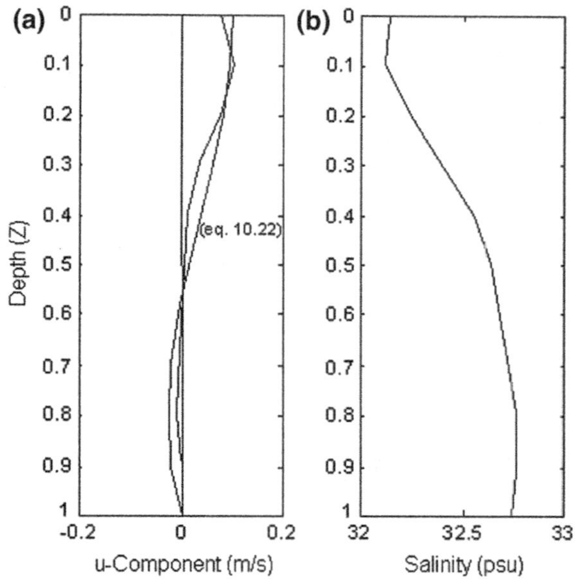

Fig. 10.6 **a** Steady-state time-mean vertical salinity profile, exhibiting a well mixed condition (type C or 1). **b** u-velocity profile from experimental data of the Cananéia estuarine channel, shown comparatively with the simulation with the steady-state one-dimensional analytical model with maximum friction at the bottom (Eq. 10.22). **b** According to Miranda and Castro (1996)

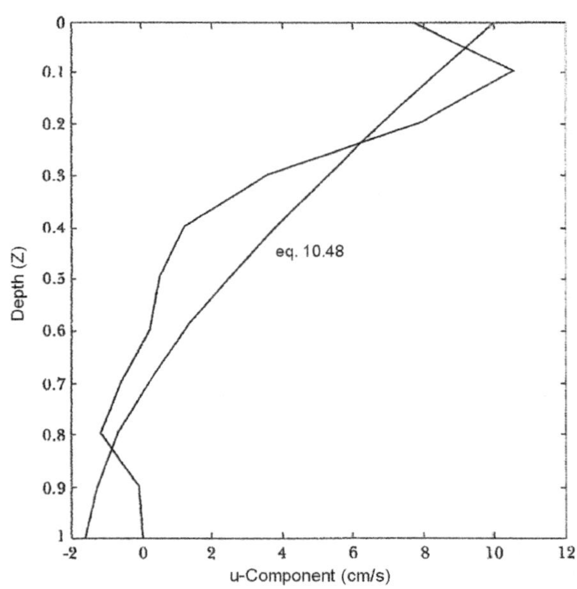

Fig. 10.7 Vertical mean u-velocity profile calculated with measurements in the Cananéia estuarine channel in comparison with the analytical profile calculated with the steady-state analytical model (Eq. 10.48) with moderate slippery bottom boundary condition

Let us use the same set of observational data measured in the same estuary system for comparison with the analytical model of Eq. (10.48), obtained with the moderate bottom friction conditions. As we have seen in (Eq. 10.40), the bottom friction may be adjusted with the tidal velocity amplitude (U_T), and the coefficient k. The best theoretical profiles in comparison to the experimental data (Fig. 10.7), was calculated with the following numeric values: $H_0 = 10.0$ m; $\partial\rho/\partial x = 4.0$ 10^{-4} kg m^{-4}; $U_T = 0.5$ m s^{-1}; $\rho_0 = 10^3$ kg m^{-3}, $k = 2.5 \times 10^{-3}$ and $\tau_{Wx} = 0.2$ Pa. The results presented in this figure, indicate a reasonable agreement between the theory and the observational data, and confirm that the main basic physical principles have been taken into account in the theoretical calculations.

10.5 Longitudinal Salinity Simulation

The longitudinal salinity distribution in an idealized well-mixed estuary, with the vertical eddy diffusion predominantly forced by tidal oscillation, was calculated analytically and compared with experimental data in the classical articles of Stommel (1951) and Arons and Stommel (1951), using the basic principle of physics of continuous medium. The development is similar to that presented in (Eq. 7.113 and Chap. 7), where the eddy diffusion coefficient was assumed to be constant. However, the basic difference is that in the Arons-Stommel's article, the current function and the tidal advective displacement due to the barotropic pressure gradient were taken into account. As in the previous simulations, the simplifying hypotheses for a laterally homogeneous estuary with simple geometry and uniform transverse section and depth are:

- The river discharge, the baroclinic pressure gradient and bottom friction are disregarded.
- The channel length is small in comparison to a quarter tidal wave length, thus the tidal elevation will be simultaneous and uniform over the entire channel.

In this simulation, the circulation and the salinity distribution analysis is made in relation to the longitudinal axis (Ox), oriented positively seaward with its origin ($x = 0$) at the TRZ/MZ interface. As in the previous solutions, the salinities at the estuary head and coastal ocean $S|_{x=0} = 0$, $S|_{x=L} = S_0$, respectively, are the boundary conditions.

According to the second hypothesis, the tidal height, $\eta = \eta(t)$, is expressed as:

$$\eta(t) = \eta_0 \cos(\omega t), \tag{10.49}$$

where $\omega = \pi/T_P$ is the angular frequency. The velocity generated by the barotropic pressure gradient has already been determined from the simplified expressions of the continuity and motion equations (Eqs. 2.17 and 2.22, Chap. 2),

$$u(x, t) = U_0 \sin(\omega t), \tag{10.50}$$

with the velocity amplitude calculated by,

$$U_0 = \frac{\eta_0 \omega x}{H_0}. \tag{10.51}$$

As the longitudinal velocity (Eq. 10.50) is known, the displacement (D_M) generated by the tidal current may be calculated by

$$D_M = \int u(x,t)dt = U_0 \int \sin(\omega t)dt = -\frac{U_0}{\omega}\cos(\omega t) = D_0 \cos(\omega t) + \text{const}. \tag{10.52}$$

The amplitude of the advective displacement, $D_0 = -U_0/\omega$, may be combined with the amplitude of barotropic velocity (Eq. 10.51),

$$D_0 = -\frac{U_0}{\omega} = -\frac{\eta_0 x}{H_0}. \tag{10.53}$$

Applying the one-dimensional salt conservation equation under steady-state conditions (Eq. 7.98 and Chap. 7) with the integration constant equal to zero,

$$u_f S = K_x \frac{dS}{dx}, \tag{10.54a}$$

To integrate this equation for an estuary with finite length it is convenient to introduce the non-dimensional longitudinal distance defined by $X = x/L$. As $x = 0$ and $x = L$ indicate the interior limits of the MZ and the estuary mouth, respectively, it follows that $X = 0$ and $X = 1$, indicate these positions, respectively. With the introduction of the longitudinal ordinate X, the differential Eq. (10.54a) is rewritten as

$$u_f S = \frac{K_x}{L}\frac{dS}{dX}. \tag{10.54b}$$

The kinematic longitudinal eddy diffusivity, K_x, is expressed in terms of the tidal velocity, U_0, and the modulus of a characteristic length (D_0), which has been taken as the total excursion of a particle due to the tides, and the following analytical expression may be assumed:

$$K_x = 2BU_0|D_0|, \tag{10.55}$$

where 2B is a non-dimensional coefficient. This form of diffusion coefficient regards the tides as a turbulent motion superposed upon the steady-state river flow through the estuary (Arons and Stommel 1951). Taking into account the expressions of U_0 and D_0 (Eqs. 10.51 and 10.53), the K_x coefficient may be expressed by

$$K_x = \frac{2B\eta_0^2\omega x^2}{H_0^2} = \frac{2B\eta_0^2\omega}{H_0^2}(LX)^2. \qquad (10.56)$$

Substituting of this expression of K_x into the salt conservation Eq. (10.54b), and integrating by separation of variables, yields the following expression:

$$\frac{dS}{S} = \frac{F}{X^2}dX, \qquad (10.57)$$

where the non-dimensional number F (*flushing number*) is defined by

$$F = \frac{u_f H_0^2}{2B\eta_0^2\omega L} = \frac{u_f H_0^2 T}{4\pi B\eta_0^2 L}. \qquad (10.58)$$

Integrating the differential Eq. (10.57), follows the general solution:

$$\ln(S) = -\frac{F}{X} + A_1, \qquad (10.59)$$

where A_1 is the integration constant, which may be easily calculated with the mouth boundary condition: for $X = 1 \rightarrow S(1) = S_0$ and $A_2 = \ln(S_0) + F$. Combining this integration constant with Eq. (10.59) the following expression is obtained:

$$\ln\left(\frac{S}{S_0}\right) = F\left(1 - \frac{1}{X}\right), \qquad (10.60)$$

or

$$\frac{S}{S_0} = \exp\left[F\left(1 - \frac{1}{X}\right)\right]. \qquad (10.61)$$

Parametric curves of the flushing number (F) calculated from the correlation of the relative salinity (S/S_0) versus the non-dimensional longitudinal distance (X), for an idealized estuary is shown in Fig. 10.8. According to Arons and Stommel (1951) these curves were developed for a much idealized situation, and it is somewhat surprising and encouraging to find that the empirical data from actual surveys can be plotted on these curves with such good agreement for the parameters. The figure is also very illustrative and presents the following characteristics: for $X \rightarrow 0$ and $X \rightarrow 1$, the curvatures bend towards minimum and maximum longitudinal gradients of the relative salinity values, and the longitudinal variation for the Alberni Inlet and the Raritan estuaries are close to the experimental data for $F = 0.3$ and $F = 0.8$, respectively. An attempt to calculate the proportionality factor between these results was unsuccessful, with the values being an order of magnitude different for these estuaries. Therefore, it appears that although the theoretical curves are in good agreement with the observational data, an a priori calculation of the

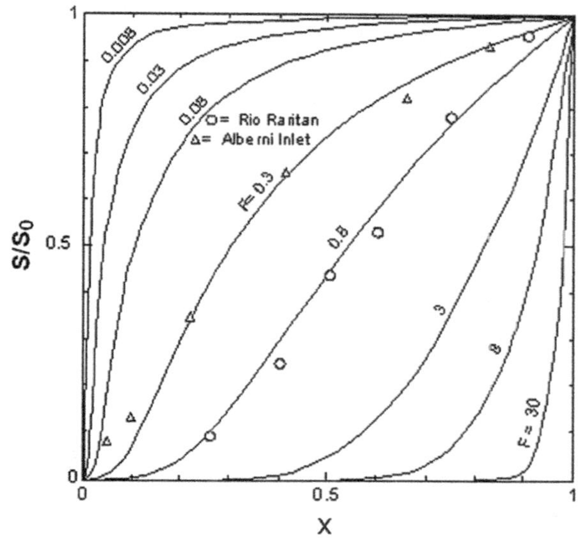

Fig. 10.8 Parametric curves of the flushing number (F) from the correlation of the relative salinity (S/S_0) versus the non-dimensional longitudinal distance (X). The curves were developed for an idealized estuary, shown comparatively with observational data of Alberni Inlet (*triangle*) and Raritan river estuary (*circle*). According to Arons and Stommel (1951)

flushing number, F, is not yet feasible; nevertheless, this number may be a convenient concept to characterize estuaries, just as the family of curves themselves is a convenient semi-empirical expression of the mean salinity longitudinal distribution (Arons and Stommel 1951).

10.6 Analytical Simulation

Let us present in this topic some examples of the determination of the steady-state u-velocity component and salinity vertical profiles of a well-mixed estuary, with constant width and uniform depth. The equations of continuity and motion (10.3) and (10.4) and salt conservation Eq. (10.5) will be solved with different boundary conditions, and kinematic eddy viscosity coefficient expressed by $N_z = \kappa U_T H_0$, with and $\kappa = 2.5 \times 10^{-3}$, and $U_T \gg u$, according to Bowden and Fairbain (1952), Bowden (1953) and Prandle (2009).

10.6.1 Basic Equations: Upper and Lower Boundary Conditions and Integral Boundary Condition

For an estuary under steady-state conditions, with width B = constant, referred to the Oxz reference system (with z = 0 and $z = H_0$ the surface and bottom, respectively)) and forced by the barotropic pressure gradient force, the equations of motion (8.57b, Chap. 8), continuity and salt conservation (7.73a, and 7.78a, Chap. 7), are approximate by,

$$g\frac{\partial \eta}{\partial x} + N_z\frac{\partial^2 u}{\partial z^2} = 0, \quad (10.62)$$

$$\frac{\partial u}{\partial x} + \frac{\partial w}{\partial z} = 0, \quad (10.63)$$

and

$$u(z, x)\frac{\partial S}{\partial x} = \frac{\partial}{\partial z}\left(K_z\frac{\partial S}{\partial z}\right). \quad (10.64)$$

In the continuity Eq. (10.63), the u- and w-velocity components may be expressed in terms of the *current function* $\Psi = \Psi(x, z)$, $[\Psi] = [L^2T^{-1}]$, with $u(x, z) = -\frac{\partial \psi}{\partial z}$, and $w(x, z) = \frac{\partial \psi}{\partial x}$. Thus the general solution of Eq. (10.62) is

$$u(x, z) = -\frac{\partial \psi}{\partial z} = -\frac{g\eta_x}{2N_z}z^2 + C_1 z + C_2. \quad (10.65)$$

In this solution, the surface slope due to the barotropic forcing $\left(\frac{\partial \eta}{\partial x} = \eta_x\right)$ may be positive, negative, and null during the flood, ebb or at high and low water, respectively. The integrations constants, with dimensions $[C_1] = [T^{-1}]$ and $[C_2] = [LT^{-1}]$ are calculated applying the following boundary conditions:

- Upper boundary condition.

The wind stress (τ_W) is applied on the free surface z = 0:

$$\rho N_z\left[\left(\frac{\partial u}{\partial z}\right)\right]_{z=0} = -\rho N_z\left(\frac{\partial^2 \psi}{\partial z^2}\right)\Big|_{z=0} = \tau_W, \quad (10.66a)$$

- Lower boundary condition.

This boundary condition is expressed according to the bottom characteristics; for maximum friction it is simulated by:

$$u(x, z)\big|_{z=H_0} = -\frac{\partial \psi}{\partial z}\Big|_{z=H_0} = 0, \quad (10.66b)$$

and for moderate bottom stress (τ_B):

$$-\rho N_z\left(\frac{\partial u}{\partial z}\right)\Big|_{z=H_0} = \frac{\partial^2\psi}{\partial z^2}\Big|_{z=H_0} = \tau_B. \tag{10.66c}$$

- Integral boundary condition is formulated by:

$$\frac{Q_f}{BH_0} = u_f = \frac{1}{H_0}\int_0^{H_0} u(x,z)dz = \int_0^1 u(x,\,Z)dZ. \tag{10.66d}$$

In the integral boundary condition, Z is the non-dimensional depth, and indicates the volume conservation of the river volume transport.

10.6.2 Barotropic Pressure Gradient, Wind Stress and Maximum Bottom Friction

Considering the surface slope (η_x) as known the integration constant C_1 of Eq. (10.65) is calculated applying the upper boundary condition (10.66a),

$$\rho N_z\left[-\left(\frac{g}{N_z}\right)\eta_x z + C_1\right]_{z=0} = \tau_W, \tag{10.67a}$$

and $C_1 = \frac{\tau_W}{\rho N_z}$.

The integration constant C_2 is calculated applying the bottom boundary condition (10.66b). After derivation of the stream function in relation to the depth and calculating the result for z = H_0, this constant is:

$$C_2 = \frac{g\eta_x H_0^2}{2Nz} - \frac{\tau_W H_0}{\rho N_z}. \tag{10.67b}$$

Substituting the expressions of C_1 and C_2 into the general solution (10.65), and rearranging its terms the solution for the u-velocity component is,

$$u(x,z) = -\frac{1}{2}\frac{g\eta_x}{N_z}(z^2 - H_0^2) + \frac{\tau_W}{\rho N_z}(z - H_0), \tag{10.68}$$

or, in function of the non-dimensional depth, ($Z = \frac{z}{|H_0|}$),

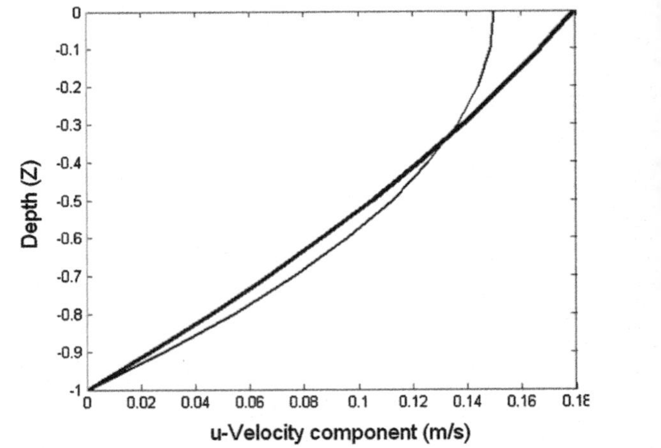

Fig. 10.9 Comparison of the u-velocity component of a steady-state theoretical vertical velocity profile (*dashed line*), and the observational data (*thin line*), forced by barotropic pressure gradient during spring tide in the Peruípe river Estuary (after Andutta et al. 2013)

$$u(x, Z) = -\frac{1}{2}\frac{gH_0^2}{N_z}\eta_x(Z^2 - 1) + \frac{H_0}{\rho N_z}\tau_W(Z - 1). \tag{10.69}$$

Taking into account that $N_z = \kappa U_T H_0$, this result may be rewritten as:

$$u(x, Z) = -\frac{1}{2}\frac{gH_0}{\kappa U_T}\left[\eta_x(Z^2 - 1) - \frac{2\tau_W}{\rho g H_0}(Z - 1)\right]. \tag{10.70}$$

This solution identically satisfies the surface and bottom boundary conditions, the flood and ebb circulation are generated by the barotropic forcing during ebb ($\partial\eta/\partial x > 0$) and flood ($\partial\eta/\partial x < 0$), and the wind stress may be seaward ($\tau_W > 0$) or landward ($\tau_W < 0$).

Figure 10.9 presents the u-vertical velocity profile using Eq. (10.69). This simulation were used the following numerical values: $H_0 = h = 9.6$ m; $N_z = \kappa U_T H_0 = 2.5 \times 10^{-3}$ $m^{.2}$ s^{-1}; $\rho = 1020.0$ kg m^{-3}; $g = 9.80$ m s^{-2}; $\frac{\partial\eta}{\partial X} \approx \frac{\Delta\eta}{\Delta X} = 10.0 \times 10^{-6}$ (ebb tide) and $\tau_W = 0$. This theoretical profile is shown in comparison to the nearly steady-state profile sampled in the Peruípe river estuary (Nova Viçosa, BA, Brazil), classified as well-mixed estuary during spring tide, according to the observational data of Andutta et al. (2013).

Under this advective influence the theoretical vertical salinity profile is calculated from the Eq. (10.64) written in terms of the non-dimensional depth (Z),

$$\frac{\partial^2 S}{\partial Z^2} = \frac{\overline{S_x}h^2}{K_z}u(x, Z), \tag{10.71}$$

where $\overline{S_x}$ is a known positive longitudinal salinity gradient and $u = u(x, Z)$ is given by the solution (10.69). Thus the vertical salinity profile $S = S(x, Z)$ is obtained by the integration of the following second order differential equation:

$$\frac{\partial^2 S}{\partial Z^2} = \frac{\overline{S_x} h^2}{K_z}\left[\frac{1}{2}\frac{gh^2\eta_x}{N_z}(Z^2-1) + \frac{\tau_w h}{\rho N_z}(Z-1)\right], \tag{10.72a}$$

or

$$\frac{\partial^2 S}{\partial Z^2} = \frac{\overline{S_x} g h^4 \eta_x}{2N_z K_z}(Z^2-1) + \frac{\overline{S_x} h^3 \tau_w}{\rho N_z K_z}(Z-1). \tag{10.72b}$$

In this equation the barotropic pressure gradient ($g\eta_x$) during the flood and ebb tide conditions are positive and negative, respectively. Its solution is obtained from two successive integrations, and the result is:

$$S(Z) = -\frac{\overline{S_x} g h^4 \eta_x}{2K_z N_z}\left(\frac{1}{12}Z^4 - \frac{1}{2}Z^2\right) + \frac{\overline{S_x} h^3 \tau_w}{\rho N_z K_z}\left(\frac{1}{6}Z^3 + \frac{1}{2}Z^2\right) + C_1 Z + C_2, \tag{10.73}$$

where C_1 and C_2 are constants of integration to be calculated with the imposition of the following boundary conditions:

$$S(Z)_{z=0} = S(0) = S_0, \tag{10.74a}$$

and

$$\frac{\rho K_z}{h}\frac{\partial S}{\partial Z}\Big|_{z=1} = 0. \tag{10.74b}$$

Applying the first condition it follows that $C_2 = S_0$, which is the salinity at the free surface. The second bottom boundary condition specify that the salt flux through the bottom is zero, and this will imply that $[C_1] = [ML^{-2}T^{-1}]$, but from (10.73) it must be non-dimensional quantity; then, as ($\rho K_z/h \neq 0$) the boundary condition to be applied must be reduced to $(\partial S/\partial Z)|_{z=1} = 0$ and the integration constant C_1 is expressed by:

$$C_1 = -\frac{1}{3}\frac{gh^4\overline{S_x}\eta_x}{N_z K_z} + \frac{1}{2\rho}\frac{\overline{S_x} h^3 \tau_w}{N_z K_z}. \tag{10.75}$$

Substituting the expressions C_1 and C_2 in the general solution (10.73) the final solution for the vertical salinity profile is

$$S(Z) = S_0 + \frac{gh^4\overline{S_x}\eta_x}{2K_z N_z}\left(\frac{1}{12}Z^4 - \frac{1}{2}Z^2\right) - \left(\frac{1}{3}\frac{gh^4\overline{S_x}\eta_x}{N_z K_z} - \frac{1}{2\rho}\frac{\overline{S_x} h^3 \tau_w}{N_z K_z}\right)Z, \tag{10.76a}$$

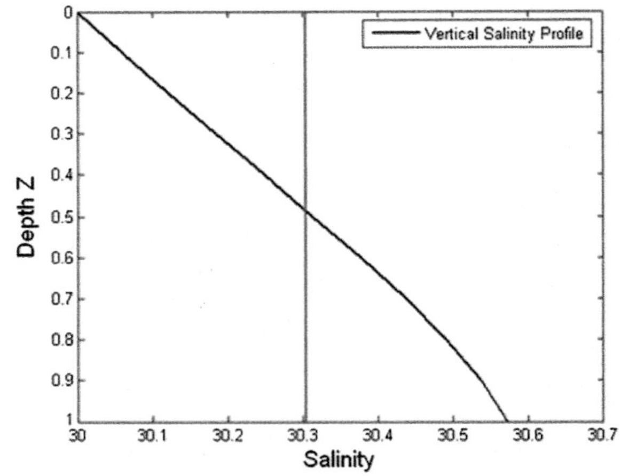

Fig. 10.10 Steady-state salinity profile of a well-mixed estuary forced by the barotropic gradient pressure force, calculated with upper boundary condition (S = 30‰) and no salt flux at the bottom, and $\tau_{w=0}$. The vertical line indicates the mean-depth vertical salinity S = 30.3‰

or

$$S(Z) = S_0 + \frac{gh^4\overline{S_x}\eta_x}{4K_zN_z}\left(\frac{1}{6}Z^4 - Z^2 - \frac{8}{3}Z\right) + \frac{1}{2\rho}\frac{\overline{S_x}h^3\tau_W}{N_zK_z})Z. \qquad (10.76b)$$

For no wind stress ($\tau_w = 0$) the solution simplify to:

$$S(Z) = S_0 + \frac{gh^4\overline{S_x}\eta_x}{4K_zN_z}\left(\frac{1}{6}Z^4 - Z^2 - \frac{4}{3}Z\right). \qquad (10.76c)$$

The vertical salinity profile S = S(Z), calculated with the following values: $h = H_0 = 9.6$ m; $S_0 = 30‰$; $K_z = 1.0 \times 10^{-2}\ m^2\ s^{-1}$; $N_z = 1.0 \times 10^{-3}\ m^2\ s^{-1}$; $\overline{S_x} = -3.5\times10^{-6}m^{-1}$, $|\eta_x| = 1.0 \times 10^{-4}$ and $\tau_w = 0$, is shown in Fig. 10.10.

10.6.3 Barotropic Pressure Gradient, Wind Stress, River Discharge and Maximum Bottom Friction

The general solution (10.65) was developed under the hypothesis that the barotropic forcing $\left(\frac{\partial\eta}{\partial X} = \eta_x\right)$ was known, but as the river discharge is a given property, the solution may be made using the integral boundary condition (10.66d) with the already known analytic profile (10.70) to find the surface slope, thus it follow:

$$u_f = \int_0^1 u(x, Z)dZ = -\frac{1}{2}\frac{gH_0^2}{N_z}\left\{\int_0^1 [\eta_x(Z^2 - 1) - \frac{1}{\rho g H_0}\tau_W(Z - 1)]\right\}dZ. \qquad (10.77)$$

Solving this equation to the free surface slope its analytical expression is:

$$\eta_x = -\frac{3N_z}{gH_0^2}u_f + \frac{3}{2}\frac{\tau_W}{\rho g H_0}. \tag{10.78}$$

Thus, η_x is directly proportional to the river velocity and, in normal conditions, it is usually higher than that generate by the wind stress (τ_W). Substituting this result into Eq. (10.70) the analytical expression of the u-velocity profile is:

$$u(x, Z) = \left[\frac{3}{2}u_f - \frac{3}{4}\frac{\tau_W H_0}{\rho N_z}\right](1 - Z^2) + \frac{\tau_W H_0}{\rho N_z}(1 + Z). \tag{10.79a}$$

This solution has two components or modes: the fresh water discharge and the wind stress. Taking into account that the kinematic eddy viscosity may be approximated by $N_z = \kappa U_T H_0$ this solution may be rewritten as:

$$u(x, Z) = \frac{3}{2}\left[(u_f - \frac{1}{2}\frac{\tau_W}{\rho\kappa U_T}\right](1 - Z^2) + \frac{\tau_W}{\rho\kappa U_T}(1 + Z). \tag{10.79b}$$

As an example, this profile is shown in Fig. 10.11 calculated with the following parameter values: $Q_f = 100\ m^3\ s^{-1}$, $A = 1000\ m^2$, $u_f = 0.1\ m\ s^{-1}$, $\rho = 1020.0$ kg m^{-3}, $U_T = 0.5\ m\ s^{-1}$, $\kappa = 5.0 \times 10^{-3}$, $H_0 = 10$ m, and the landward wind stress of $\tau_W = 0.3$ Pa.

When in the solution (10.79b) the wind stress is null ($\tau_W = 0$) the vertical velocity profile is simplified to:

$$u(x, Z) = \frac{3}{2}u_f(1 - Z^2), \tag{10.80}$$

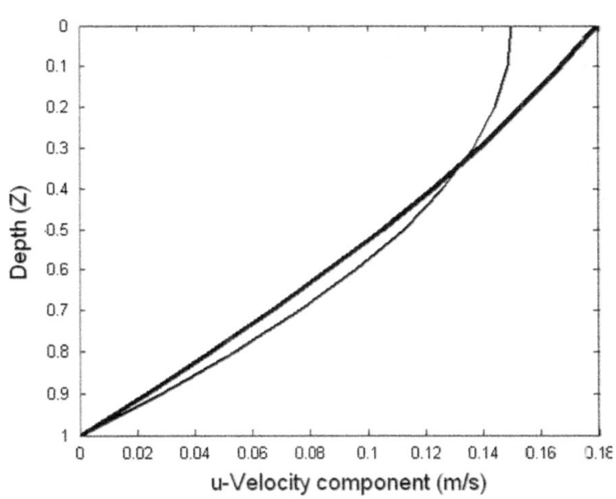

Fig. 10.11 Profile of the u-velocity of a well-mixed estuary calculated with the maximum friction at the bottom and forced by the barotropic pressure gradient, river discharge (*thick line*) and landward wind stress (*thin line*)

and the u-velocity profile forced only by the fresh water discharge has a parabolic configuration, and its value at the surface and bottom are $u(x,0) = \frac{3}{2}u_f$ and $u(x, 1) = 0$, respectively.

The analytical solution of the salt conservation Eq. (10.64), with the boundary conditions (10.74a), (10.74b), and the advective and diffusive influences of the vertical velocity profile of Eq. (10.80), is expressed by:

$$S(x, Z) = S(0) + \frac{3}{2}\frac{H_0^2\overline{S_x}u_f}{K_z}\left(-\frac{1}{12}Z^4 + \frac{1}{2}Z^2 + \frac{2}{3}Z\right). \quad (10.81)$$

10.6.4 Barotropic Pressure Gradient, River Discharge, Wind and Moderate Bottom Friction

With these forcing conditions the integration constants C_1 and C_2 of the general solution (10.65) are calculated imposing the following boundary conditions: wind stress at the free surface (10.66a) and moderate bottom friction (10.66c), with the last expressed by:

$$\left[-A_z\left(\frac{\partial u}{\partial z}\right)\right]_{z=H_0} = \tau_B = \rho C_B[u(x, z)]_{z=H_0}, \quad (10.82a)$$

where C_B, with dimension $[C_B] = [LT^{-1}]$, is a numeric parameter related to the bottom shear stress (τ_B). From this relationship it follow,

$$\left(\frac{\partial u}{\partial z}\right)|_{z=H_0} = \frac{C_B u(x, H_0)}{N_z}. \quad (10.82b)$$

In this boundary condition it will be assumed that $C_B = (4/\pi)kU_T$, where k is a non-dimensional coefficient and U_T is the tidal velocity amplitude.

Applying the upper boundary condition to the first integration of solution (10.65),

$$\rho N_z\left(\frac{\partial u}{\partial z}\right)_{z=0} = \tau_W = \rho N_z\left(\frac{g}{N_z}\eta_x z\right)|_{z=0} + \rho N_z C_1, \quad (10.83)$$

and knowing the integration constant $C_1 = \frac{\tau_W}{\rho N_z}$, its the partial solution is

$$u(x, z) = \frac{1}{2}\frac{g}{N_z}\eta_x z^2 + \frac{\tau_W}{\rho N_z}z + C_2. \quad (10.84)$$

Applying the bottom boundary condition (10.82b) the integration constant C_2 is expressed by,

$$C_2 = \frac{\tau_B}{\rho C_B} - \frac{1}{2}\frac{g\eta_x}{N_z} H_0^2 + \frac{\tau_w}{\rho N_z} H_0, \qquad (10.85a)$$

or

$$C_2 = \frac{g\eta_x}{C_B} H_0 + \frac{\tau_W}{\rho C_B} - \frac{g\eta_x H_0^2}{2N_z} + \frac{\tau_W}{\rho N_z} H_0, \qquad (10.85b)$$

with dimension $[C_2] = [LT^{-1}]$. Substituting this result in the Eq. (10.84) the u-velocity component is given by

$$u(x, z) = \frac{g\eta_x}{2N_z}(z^2 - H_0^2) + \frac{\tau_W}{\rho N_z}(z + H_0) + \frac{1}{C_B}\left(\frac{\tau_W}{\rho} - g\eta_x H_0\right). \qquad (10.86a)$$

or in function of the non-dimensional depth $Z = z/|H_0|$,

$$u(x, Z) = \frac{g\eta_x H_0^2}{2N_z}(Z^2 - 1) + \frac{\tau_W H_0}{\rho N_z}(Z + 1) + \frac{1}{C_B}\left(\frac{\tau_W}{\rho} - g\eta_x H_0\right). \qquad (10.86b)$$

We leave to the reader to verify that these solutions satisfy identically the surface and the bottom boundary conditions (10.82a), (10.82b). For a null wind stress ($\tau_W = 0$) the Eq. (10.86b) is reduced to:

$$u(x, Z) = \frac{g\eta_x H_0^2}{2N_z}(Z^2 - 1) - \frac{1}{C_B} g\eta_x H_0. \qquad (10.87a)$$

At the bottom ($Z = 1$), the velocity $u(x, 1) = u_B$ is calculated by

$$u(x, 1) = u_B = -\frac{1}{C_B} g\eta_x H_0, \qquad (10.87b)$$

and it is directly and inversely proportional to the sea surface slope and the bottom parameter C_B, respectively; and for $C_B \to \infty$ the bottom velocity is zero, $u(x, 1) = 0$ which corresponds to the maximum bottom friction.

The above solutions (10.87a), (10.87b) are in function of the surface slope η_x which may be eliminated by introducing the river discharge (Q_f) as an integral boundary condition (Eq. 10.66d):

$$u_f = \frac{g\eta_x H_0^2}{2N_z}\int_0^1 (Z^2 - 1)dZ + \frac{\tau_W H_0}{\rho N_z}\int_0^1 (Z + 1)dZ + \frac{1}{C_B}\int_0^1 \left(\frac{\tau_W}{\rho} - g\eta_x H_0\right)dZ, \qquad (10.88)$$

Solving the integrals the final result for the surface slope is,

$$\eta_x = \frac{-u_f + \frac{\tau_W}{\rho N_z}\left(\frac{H_0}{2} - \frac{N_z}{C_B}\right)}{\frac{gH_0^2}{N_z}\left(\frac{1}{3} - \frac{N_z}{H_0 C_B}\right)}. \quad (10.89a)$$

If the wind stress may be disregarded ($\tau_W = 0$) this result is further simplified:

$$\eta_x = -\frac{u_f}{\frac{gH_0^2}{N_z}\left(\frac{1}{3} - \frac{N_z}{H_0 C_B}\right)}. \quad (10.89b)$$

As the river discharge flows seaward ($u_f > 0$) and the denominator of this equation is positive, it follow $\eta_x = \frac{\partial \eta}{\partial x} < 0$, and $\eta = \eta(x)$ decreases seaward forced by the river discharge and simultaneously by the ebb tide. To calculate the magnitude order of η_x, the following numeric data are used: $u_f = 0.23\ \text{m s}^{-1}$, $N_z = 1.0 \times 10^{-3}\ \text{m}^2\ \text{s}^{-1}$, $C_B = 1.0 \times 10^{-3}\ \text{m s}^{-1}$ and $gH_0^2 = 10^3\ \text{m}^3\ \text{s}^{-2}$, which gives $\eta_x = -1.0 \times 10^{-6}$. In the case of maximum friction at the bottom $u(x, H_0) = 0$, $C_B \to \infty$ and the solution is reduced to

$$\eta_X = -\frac{3N_z}{gH_0^2} u_f. \quad (10.89c)$$

With the same data values the free surface slope is $\eta_x = -0.7 \times 10^{-6}$, indicating a decrease of 30%. An estimate of the gradient barotropic pressure forced due to ± 2.0 m tidal oscillation with a wave length of 500 km = 5.0×10^5 m will give a higher value which is estimated as $\eta_x = \pm 4.0 \times 10^{-6}$. It should be observed that the surface slope expressed by this equation has the same analytical expression as the Eq. (10.78), calculated with no wind forcing ($\tau_W = 0$) and estuaries with weak longitudinal density gradient.

Returning to Eq. (10.89a) with the following simplifications: fresh water discharge equal zero ($u_f = 0$), and using the relationships $N_z = 2.5 \times 10^{-3} U_T H_0$, and $C_B = (4/\pi) \times 2.5 \times 10^{-3} U_T$, the free surface slope forced by the wind stress (τ_W) is the same as the one presented by Prandle (2009):

$$\eta_W = \frac{\tau_W\left(\frac{1}{2} - \frac{N_z}{C_B H_0}\right)}{\rho g H_0\left(\frac{1}{3} - \frac{N_z}{C_B H_0}\right)} = 1.15\frac{\tau_W}{\rho g H_0}. \quad (10.90a)$$

For maximum bottom friction ($C_B \to \infty$) this equation simplifies to the following expression:

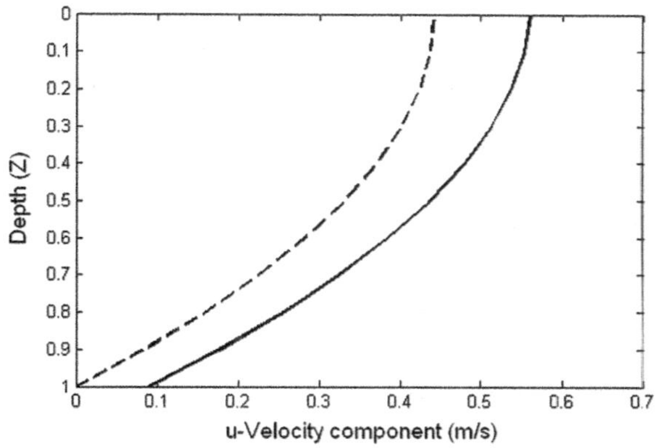

Fig. 10.12 Vertical velocity profiles of a well-mixed estuary forced by the barotropic gradient pressure force, wind stress, calculated with maximum friction (*dashed line*), and with moderate bottom friction (*thin line*). The calculated slippery velocity due to the bottom roughness is $u_B = 0.09\ m\ s^{-1}$

$$\eta_W = 1.5 \frac{\tau_W}{\rho g H_0}. \tag{10.90b}$$

For further analysis of the theoretical results the Eq. (10.86b) is used to calculate the u-velocity profile presented in Fig. 10.12, with the following physical variables: $N_z = 5.0 \times 10^{-4}\ m^2\ s^{-1}$, $C_B = 3.5 \times 10^{-3}\ m\ s^{-1}$, $\tau_W = 1.0 \times 10^{-3}$ Pa, $H_0 = 15.0$ m, e $\rho = 1020.0\ kg\ m^{-3}$. This figure indicates a slippery bottom of $u_B = 0.09\ m\ s^{-1}$. As has been described by Prandle (2009) this bottom roughness velocity is the initial process of the sediment dynamics (erosion, transport and sedimentation), which is composed of sediments with different origins and sizes influenced by actual and past dynamic processes and also by morphological, biological and chemical action. The general features of sedimentary dynamics in tidal regimes are treated at length in (Postma 1967, quoted in Prandle (2009)).

Neglecting the wind stress ($\tau_W = 0$) the solution (10.86b) may be combined with the η_x expression (10.89b), and the vertical velocity profile in function of the river discharge velocity is given by,

$$u(x, Z) = \frac{u_f H_0}{2\left(\frac{1}{3} H_0 - \frac{N_z}{C_B}\right)} \left(1 - Z^2 - \frac{2N_z}{C_B H_0}\right), \tag{10.91a}$$

and the slippery velocity at the bottom (Z = 1) is

$$u(x, 1) = u_B = -\frac{u_f N_z}{\left(\frac{1}{3} C_B H_0 - N_z\right)}. \tag{10.91b}$$

Taking into account that $U_T \gg u(z)$, the kinematic eddy viscosity coefficient may be approximate by $N_z = \kappa U_T H_0$, and Eq. (10.91a) may be rewritten as:

$$u(x, Z) = \frac{u_f}{2\left(\frac{1}{3} - \frac{\kappa U_T}{C_B}\right)}\left(1 - Z^2 - \frac{2\kappa U_T}{C_B}\right). \qquad (10.92)$$

10.6.5 Barotropic and Baroclinic Pressure Gradient, River Discharge, Wind Stress and Bottom Friction Proportional to the Square of the Velocity

The equation system of (10.3) and (10.4) has been solved by Officer (1976, 1977) for a bottom boundary condition expressed by the following quadratic relationship:

$$\rho N_z \frac{\partial u}{\partial z}\Big|_{z=H_0} = -\rho k|u(z)|u(z), \qquad (10.93)$$

with k denoting a non-dimensional bottom friction coefficient. In this case, instead the solution (10.22) deduced for a maximum friction at the bottom, the theoretical vertical velocity profile is (Officer 1977):

$$\begin{aligned} u(x, Z) = {} & u(x, 0)(1 - 9Z^2 + 8Z^3) + u(x, 1)(1 - 3Z^2 + 4Z^3) \\ & + u_f(12Z^2 - 12Z^3). \end{aligned} \qquad (10.94)$$

where Z is the non-dimensional depth. In this solution the surface and bottom velocities are denoted by u(x, 0) and u(x, 1), respectively, and are calculated by:

$$u(x, 0) = u_f + \frac{gH_0^2}{6N_z}\left(\eta_x - \frac{1}{4}\frac{\rho_x}{\rho}H_0\right), \qquad (10.95a)$$

$$u(x, 1) = u_f - \frac{gH_0^2}{3N_z}\left(\eta_x - \frac{3}{8}\frac{\rho_x}{\rho}H_0\right), \qquad (10.95b)$$

where ρ_x and η_x are the longitudinal density gradient and the free surface slope, respectively.

The steady-state vertical velocity profile of a well-mixed estuary calculated with Eq. (10.94) with the parameters: depth 8.0 m, $u_f = 0.1$ m s^{-1}, $N_z = 3.5 \times 10^{-1}$ m^2 s^{-1}, $\eta_x = 2.0 \times 10^{-4}$, $\rho_x/\rho \approx 1$, and the longitudinal density gradient 1.7×10^{-3} kg m^{-4}, is presented in Fig. 10.13.

The comparison of this result with the simulation presented in Fig. 10.12, a lower slippery bottom velocity (0.025 m s^{-1}) was obtained.

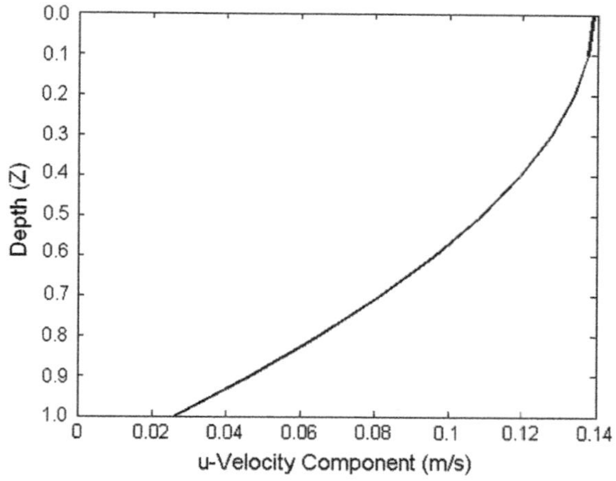

Fig. 10.13 Steady-state vertical velocity profile of a well-mixed estuary forced by the barotropic and baroclinic gradient pressure force, river discharge and wind stress, calculated with bottom roughness proportional to the square of the bottom velocity

10.6.6 Vertical Salinity Profile

Another analytic solution for the steady-state vertical salinity profile, S = S(Z), of a well-mixed estuary is presented. For the longitudinal salt balance, as indicated in (Eq. 7.98 and Chap. 7) the essential steady-state balance may be approximate by the advective circulation and the vertical salt-flux diffusion (under the assumption that the depth-mean longitudinal salinity gradient and the kinematic vertical diffusion coefficient are known). Thus, the differential equation to be solved is:

$$\frac{\partial^2 S}{\partial Z^2} = \frac{\frac{\partial \overline{S}}{\partial x} H_0^2}{K_z} u(x, Z), \tag{10.96a}$$

and the general solution is formulated by,

$$S(x, Z) = \frac{H_0^2}{K_z} \frac{\partial \overline{S}}{\partial x} \int [\int u(x, Z) dZ] dZ + C_1 Z + C_2. \tag{10.96b}$$

C_1 and C_2 are non-dimensional integration constants which may be obtained applying the following boundary conditions:

- The salt flux across the bottom is zero.

$$\frac{\rho K_z}{H_0} \left(\frac{\partial S}{\partial Z}\right)|_{Z=1} = 0. \rightarrow \left(\frac{\partial S}{\partial Z}\right)|_{Z=1} = 0. \tag{10.97a}$$

- C_2 is assumed to be equal to the mean-depth vertical salinity, $\overline{S}$,

$$C_2 = \overline{S} = \int_0^1 S(x, Z)dZ, \tag{10.97b}$$

which is to be specified.

Imposing the first boundary condition,

$$\left(\frac{\partial S}{\partial Z}\right)|_{Z=1} = -\frac{H_0^2\overline{S_x}}{K_z}\int_0^1 u(x, Z)dZ + C_1 = 0, \tag{10.98a}$$

where, $\overline{S_x} = \frac{\partial \overline{S}}{\partial x}$, and

$$C_1 = \frac{H_0^2\overline{S_x}}{K_z}[u(x, 0) - u(x, 1)], \tag{10.98b}$$

where the velocities u(x, 0) and u(x, 1) are given in the expressions (10.95a), (10.95b).

Introducing into the integrand of (10.96b) the analytical expression of u = u(x, Z) presented in the solution (10.94), solving the double integration and combining with the integration constants C_1 and C_2, the steady-state solution of vertical salinity profile is:

$$\begin{aligned} S(x, Z) - \overline{S} = \frac{H_0^2}{K_z}\frac{\partial \overline{S}}{\partial x}\{&u(x, 0)[0.4xZ^5 - 0.75 \times Z^4 - 0.5 \times Z^2 - 8.3 \times 10^{-2}] \\ &+ u(x,1)[0.2 \times Z^5 - 0.25 \times Z^4 + 1.66 \times 10^{-3}] \\ &+ u_f[-0.6 \times Z^5 + 1.0 \times Z^4 - 0.5 \times Z^2 + 6.66 \times 10^{-2}]\}. \end{aligned} \tag{10.99}$$

As an example of application of this result, the mean vertical salinity profile (Fig. 10.14) was calculated for the following numerical values: $N_z = 3.5 \times 10^{-1}\ m^2\ s^{-1}$, $K_z = 6.2 \times 10^{-7}\ m^2\ s^{-1}$, $u_f = 0.1\ m\ s^{-1}$, $H_0 = 8$ m, $\eta_x = 2.0 \times 10^{-4}$, $\rho_x/\rho \approx 1.0$; $\overline{S} = 20.0‰$ and longitudinal density gradient $\partial\rho/\partial x = 1.7 \times 10^{-3}\ kg\ m^{-4}$.

This solution (Fig. 10.14) indicates the weak halocline with the mean salinity profile 20‰ characterizing the well-mixed estuary, which is associated with the vertical velocity profile presented in Fig. 10.13.

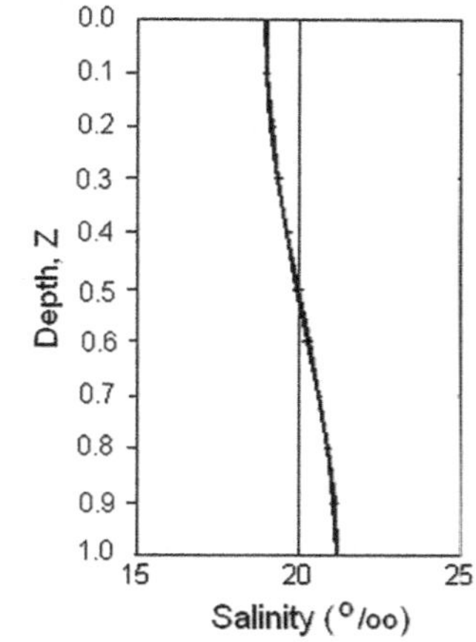

Fig. 10.14 Steady-state vertical salinity profile of a well-mixed estuary forced by the barotropic and baroclinic gradient pressure force, river discharge and wind stress, calculated with bottom roughness proportional to the square of the bottom velocity. The vertical line is the depth-mean vertical salinity profile

References

Andutta, F. P.; Miranda, L.B.; Schettini, C.A.F.; Siegle, E.; Silva, M.P.; Izumi, V.M. & Chagas, F. M. 2013. Temporal variations of temperature,salinity and circulation in the Peruípe river estuary (NovaViçosa, BA). Cont. Shelf Res., 70:36–45.

Arons, A. B. & Stommel, H. 1951. "A Mixing-Lenght Theory of Tidal Flushing". Trans. Am. Geophys.Un., 32(3):419–421.

Bowden, K. F. 1953. Note on Wind Drift in a Channel in the Presence of Tidal Currents. Proc. R. Soc. Lond., A219, pp. 426–446.

Bowden, K. F. & Fairbairn, L. A. 1952. A Determination of the Frictional Forces in a Tidal Current. Proc. R. Soc. Lond. A, 214. pp. 371–392.

Fontes, R. F. C.; Miranda, L.B. & Andutta, F.P. 2015. Estuarine Circulation. Encyclopedia of Earth Sciences Series: Estuaries. Ed. Kennish. M.J., Springer-Verlag, pp. 247–258.

Geyer, W.R. 2010. Estuarine salinity structure and circulation. In: ed. Valle-Levinson A. Contemporary Issues in Estuarine Physics. Cambridge University Press, pp. 12–26.

Hansen, D. V. & Rattray Jr., M. 1966. New Dimensions in Estuary Classification. Limnol. Oceanogr., 11(3):319–325.

Miranda, L. B.; Mesquita, R. A. & França, C. A. S. 1995. Estudo da Circulação e dos Processos de Mistura no Extremo Sul do Mar de Cananeia. Bolm Inst. Oceanogr., São Paulo, 43(2):153-154.

Miranda, L. B. & Castro, B. M. de. 1996. On the Salt Transport in the Cananeia Sea during a Spring Tide Experiment. Rev. Bras. Oceanogr., 44(2):123-133.

Maximon, L. C. & Morgan, G. W. 1955. A Theory of Tidal Mixing in a Vertically Homogeneous Estuary. J. Mar. Res., 14:157–175.

Officer, C. B. 1976. Physical Oceanography of Estuaries (and Associated Coastal Waters). New York, Wiley. 465 p.

Officer, C. B. 1977. Longitudinal Circulation and Mixing Relations in Estuaries. Estuaries, Geophysics, and the Environment. Washington, D. C., National Academy of Sciences, pp. 13–21.

Prandle, D. 1985. On Salinity Regimes and the Vertical Structure of Residual Flows in Narrow Tidal Estuaries. Estuar. Coast. Shelf Sci., 20:615–635.

Pritchard, D. W. 1955. Estuarine Circulation Patterns. Proc. Am. Soc. Civ. Eng., 81:717:1–11.

Prandle, D. 2009. Estuaries: Dynamics, Mixing, Sedimentation and Morphology. Cambridge University Press, N.Y., 236 p.

Postma, H. 1967. Sediment transport and sedimentation in the estuarine environment. In: Lauff G, H. (ed.) Estuaries, Publication no. 83. American Association for the Advancement of Science, Washington, DC., pp. 158–179 (quoted in Prandle (2009), p. 175).

Stommel, H. 1951. Recent Development in the Study of Tidal Estuaries. Tech. Rept., Massachusetts, Woods Hole Oceanographic Institution, n. 51–33. 18 p.

Chapter 11
Circulation and Mixing in Steady-State Models: Partially Mixed Estuary

The first steady-state analytical model for determining time mean longitudinal velocities in a coastal plain estuary was developed by Pritchard and Kent (1956) using the lateral and longitudinal components of the equation of motion, the tidal velocity amplitude, and the relationship between the vertical and lateral eddy stress. In this article the relationship between the vertical and transverse eddy diffusion coefficients were demonstrated using the vertical velocity profile near the bottom. The method was applied and validated with data from stations sampled in the James river estuary (Virginia, USA) during several tidal cycles in the summer (June and July, 1950) in a water column with mean depth of 8 m. The theoretical velocity profiles agreed well with the observational data, showing typical velocity profiles of partially mixed estuary, with seaward and landward motions in the upper and lower layers, respectively, and no motion at mid depths. Pritchard and Kent's paper was also a pioneering article showing the importance of comparing theoretical results with observational data.

Complementing the results of this pioneering study. This chapter presents analytical investigations of relatively narrows estuaries, assumed to be bi-dimensional systems in the Oxz plane, with vertical salinity stratification, and thus classified as partially mixed estuaries (types 2 or B). Stable salinity stratification reduces the intensity and scale turbulence in open channel flow, thereby reducing the rate of vertical mixing. Theoretically, partially mixed estuaries are adequately represented by lateral averages of the equations of mass and salt conservation and motion, as presented in Chaps. 7 and 8 which have as unknowns the density, salinity the velocity components (u, w) and the slope of free surface as a function of the independent variables (x, z, t). Laboratory investigations were conducted by Sumer and Fischer (1977) to investigate whether the rate of transverse mixing is similarly reduced in this type of estuary. However, to determine relatively simple analytical solutions, steady-state conditions must be assumed and, for validation, the theoretical solutions must be compared with time mean velocity and salinity values calculated from observational data for one or more tidal cycles.

L. Bruner de Miranda et al., *Fundamentals of Estuarine Physical Oceanography*, Ocean Engineering & Oceanography 8, DOI 10.1007/978-981-10-3041-3_11

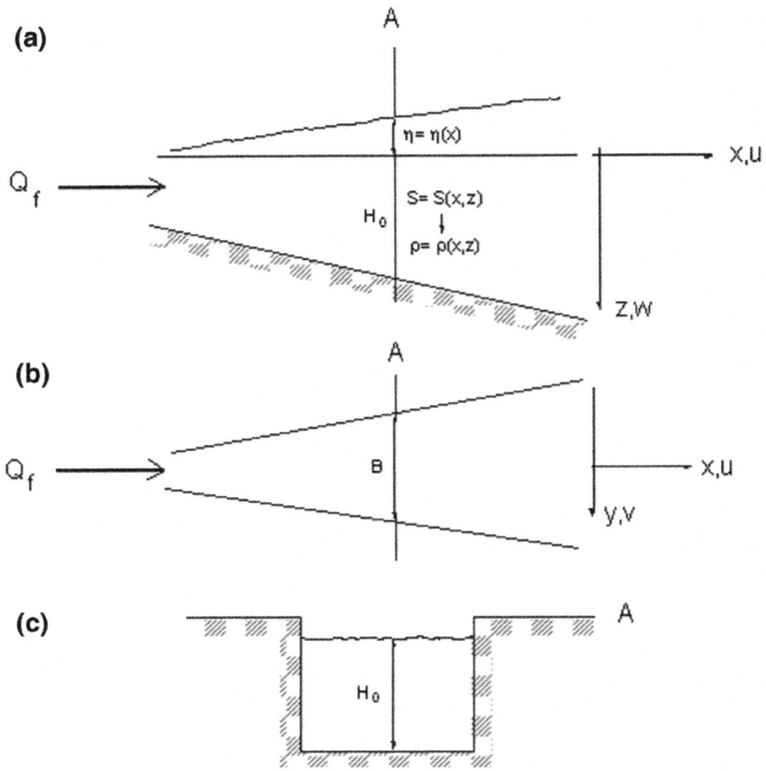

Fig. 11.1 Longitudinal (**a**, **b**) and transverse (**c**) sections of a bi-dimensional model with the adopted referential system (Oxz). H_0 is the depth in relation to a level surface, and $H_o + \eta$ is the local depth, and $B = B(x)$ is the estuary width. The Oz axis is positively oriented in the direction of the gravity acceleration

The influences of topography, transverse salinity (density) gradient and free surface elevation, may cause lateral non-homogeneity and circulation in estuaries, and a simple longitudinal and transverse section will be used (Fig. 11.1). The estuary width, $B = B(x)$, will be considered only as a function of the longitudinal distance, and its depth dependence will not be taken into account.

Neglecting topographic effects makes the mathematics considerably more tractable; however, the features of the depth dependent circulation are still basic, even though there will be modifications due to transverse effects which are not accounted for by the laterally averaged equations (Hamilton and Rattray 1978). With these simplifications, the dynamic influence of the Earth's rotation may be disregarded, and the Coriolis acceleration (f_0) does not need to be included in the longitudinal component (Ox) of the equation of motion.

Pioneering investigations by Pritchard (1952a, 1954, 1956) demonstrate that for coastal plain estuaries, the dynamic balance of the mean motion is predominantly based on the barotropic and baroclinic pressure gradients, the dissipative friction forces and the tidal non-steady state circulation. The salt balance is mainly maintained by the advective and diffusive longitudinal and vertical fluxes. With these simplifications, for practical purposes, the steady-state equation of motion has its linearity granted, disregarding the advective acceleration. As indicated by Kjerfve et al. (1991), the inertial terms may be disregarded in shallow estuarine channels, in which the bottom friction becomes more important.

11.1 Physical-Mathematical Formulation

The simplified equations of mass conservation (Eq. 7.73a) and motion (Eq. 8.57a), which are necessary for analytical and numerical treatment of problems related to steady-state bi-dimensional estuaries, have been presented in Chaps. 7 and 8, and are reproduced below,

$$\frac{\partial(uB)}{\partial x}+\frac{\partial(wB)}{\partial z}=0, \tag{11.1}$$

and

$$-\frac{1}{\rho}\frac{\partial p}{\partial x}+\frac{1}{B}\left[\frac{\partial}{\partial z}\left(BN_z\frac{\partial u}{\partial z}\right)\right]=0, \tag{11.2}$$

where B is the estuary width. The gradient pressure force decomposed in the barotropic and baroclinic pressure gradients are expressed as Eq. (2.10a, b and Chap. 2), but with the Oz axis oriented in the direction of gravity acceleration ($\vec{g}$),

$$-\frac{1}{\rho}\frac{\partial p}{\partial x}=g\frac{\partial \eta}{\partial x}-\frac{g}{\rho}\int_{\eta}^{z}\frac{\partial \rho}{\partial x}dz. \tag{11.3}$$

For hydrodynamic closure, the inclusion of the steady-state salt conservation (Eq. 7.77 and Chap. 7) is necessary

$$u\frac{\partial S}{\partial x}+w\frac{\partial S}{\partial z}=\frac{1}{B}\left[\frac{\partial}{\partial x}\left(BK_x\frac{\partial S}{\partial x}\right)+\frac{\partial}{\partial z}\left(BK_z\frac{\partial S}{\partial z}\right)\right], \tag{11.4}$$

as well as the linear equation of state of sea water (Eq. 4.11 and Chap. 4),

$$\rho(S)=\rho_0(1+\beta S). \tag{11.5}$$

The analytical solution of the equation system (11.1) to (11.5) is dependent on the boundary and integral boundary conditions used for the determination of the unknowns u, w, η, S and ρ, under some simplifications. This system has been solved by Hansen and Rattray (1965), Fisher et al. (1972), Hamilton and Rattray (1978) among others. In these articles, the main results of which will be presented in this chapter, the basic foundations for the theoretical determinations of the following were established: the steady-state vertical velocity and salinity profiles generated by the river discharge, the longitudinal gradient pressure forces due to the longitudinal density (salinity) gradients caused by the wind stress and mixing processes (advection and diffusion).

The value of the kinematic eddy viscosity coefficient (N_z), and kinematic diffusion coefficients (K_x and K_z) of the conservation equations are usually unknown, however, if the solution for the above set of equations can be shown to agree with or be validated by observational data by proper fitting of these coefficients, one must assume that either all the neglected terms are zero, or more correctly, that the neglected terms have been absorbed into these coefficients.

In the analytical simulation of the u-velocity component and salinity profiles, the estuary mixing zone (MZ) will be approximated by a simple geometry (Fig. 11.1). For this solution, it will also be necessary to formulate the following simplifying assumptions:

- The longitudinal salinity (density) gradient is independent of the depth $\partial/\partial z(\partial S/\partial x) = 0$ or $\partial/\partial z(\partial \rho/\partial x) = 0$.
- The longitudinal term of the turbulent salinity diffusion is disregarded [first term on the right hand side of Eq. (11.4)].
- The eddy kinematic coefficients of viscosity (N_z) and diffusion (K_z) are independent of the depth.

The first assumption may be justified, taking into account the results of Pritchard (1954, 1956) whose observational data demonstrated that in partially mixed estuaries the longitudinal salinity (density) gradient does not vary appreciably with depth, and the dynamical influence of this term is small in the central region of the MZ. Therefore, without great alterations in the physical aspects of the results, the longitudinal salinity (density) gradient, $\partial S/\partial x$ ($\partial \rho/\partial x$), will be substituted by the longitudinal gradients of a depth average salinity (density) $\partial \bar{S}/\partial x$ ($\partial \bar{\rho}/\partial x$), which will be calculated by the steady-state mean salinity (density) value in the water column, which will be denoted by ($\bar{S}$) or ($\bar{\rho}$), and calculated by,

$$\bar{S} = \frac{1}{(H_0 + \eta)} \int_{\eta}^{H_0} \langle S(x,z) \rangle dz \approx \frac{1}{H_0} \int_{0}^{H_0} \langle S(x,z) \rangle dz, \qquad (11.6)$$

and a similar expression for the density, $\bar{\rho}$.

The second simplifying assumption may also be justified, taking into account the observational results of Pritchard (1956) in the partially mixed James river estuary (Virginia, USA). In this estuary, the non-advective longitudinal term of the salt conservation equation (Eq. 11.4) (first term on the right hand side) has a relatively low contribution (about 1%) in comparison to the last term of this equation (vertical salt diffusion), and may be disregarded. Finally, the third assumption cannot be easily justified, and these coefficients (kinematic eddy viscosity and diffusion) will be considered as constants and representative of their mean value in the water column in order to simplify the mathematics and make the integration of equations easier.

With these assumptions, combining Eq. (11.3) with the linear equation of state of seawater (11.5) and substituting the result into the equation of motion (11.2), it follows that:

$$g\frac{\partial\eta}{\partial x}-\frac{g}{\rho}\frac{\partial\bar{\rho}}{\partial x}(z-\eta)+N_z\frac{\partial^2 u}{\partial z^2}=0, \tag{11.7a}$$

or in terms of the longitudinal salinity gradient,

$$g\frac{\partial\eta}{\partial x}-g\beta\frac{\partial\bar{S}}{\partial x}(z-\eta)+N_z\frac{\partial^2 u}{\partial z^2}=0, \tag{11.7b}$$

where according to the Boussinesq approximation, $\rho_0/\rho \approx 1$, and $\bar{\rho}(\bar{S})$ are vertical mean values of density (salinity) in the water column.

As the continuity Eq. (11.1) assures the non-divergence of the volume transport per unit width of the cross-section, it is possible to introduce the *stream function* $\psi = \psi(x, z)$, with the definitions:

$$\frac{\partial\psi}{\partial x}=wB, \text{ and } \frac{\partial\psi}{\partial z}=-uB. \tag{11.8}$$

These equalities identically satisfy the continuity equation, and it should be pointed out that the stream function, $\psi = \psi(x, z)$, with dimension $[L^3T^{-1}]$, obeys the mathematical rules which state that its mixed derivatives should be equal. The introduction of this function is very convenient because the velocity components, u and w, may be calculated by a simple derivation.

Performing the derivative of Eq. (11.7b) in relation to z and taking into account that the barotropic pressure gradient and the longitudinal salinity gradient are independent of the depth, we have

$$N_z\frac{\partial^3 u}{\partial z^3}-g\beta\frac{\partial\bar{S}}{\partial x}=0, \tag{11.9}$$

and introducing the stream function, $\psi = \psi(x, z)$, into this equation, it is reduced to

$$\frac{N_z}{B}\frac{\partial^4\psi}{\partial z^4}+g\beta\frac{\partial\bar{S}}{\partial x}=0. \tag{11.10}$$

The salt conservation Eq. (11.4) may also be combined with the stream function definition, and is then reduced to

$$-\left(\frac{\partial\psi}{\partial z}\right)\left(\frac{\partial\bar{S}}{\partial x}\right)+\left(\frac{\partial\psi}{\partial x}\right)\left(\frac{\partial S}{\partial z}\right)=BK_z\left(\frac{\partial^2 S}{\partial z^2}\right). \quad (11.11)$$

In this equation, only the salinity in the first term of the left hand side was substituted by a mean salinity value ($\bar{S}$), which will be considered a known quantity, allowing an analytical solution to be found. Only with this artifice the salinity will remain unknown, and will be determined with the salt balance between the steady-state longitudinal and vertical advection, and the vertical diffusivity terms (the first and second terms on the left hand side, and the term on the right hand side of this equation, respectively).

Equations (11.10) and (11.11) are formulations equivalent to the initial Eqs. (11.2) and (11.4), respectively. As the mean-depth salinity, $\bar{S}$, and the longitudinal gradient, $\partial\bar{S}/\partial x$, are given, this system of equations has two equations of fourth and second orders, respectively, with two unknowns, $\psi = \psi(x, z)$ and $S = S(z)$, which now govern the dynamics and the mixing processes in the MZ.

The boundary conditions necessary for finding a unique solution for the equation system (11.10) and (11.11) are the same as previously used in Chap. 10 (Eqs. 10.10 to 10.12), but now they will be expressed in terms of the stream function:

- Upper boundary condition

At the free surface ($z = \eta \approx 0$), the wind shear stress (τ_{Wx}) acts seaward or landward (positive or negative) and also may be disregarded, which are expressed by:

$$\frac{A_z}{B}\left(\frac{\partial^2\psi}{\partial z^2}\right)|_{z=0}=\pm\tau_{Wx},\quad \text{and}\quad \frac{A_z}{B}\left(\frac{\partial^2\psi}{\partial z^2}\right)|_{z=0}=0, \quad (11.12)$$

where $A_z = \rho N_z$, $[A_z] = [ML^{-1}T^{-1}]$, is the eddy viscosity coefficient.

- Lower boundary condition

At the bottom, three conditions may be formulated:

(a) Maximum friction (or no slip):

$$\left(\frac{\partial\psi}{\partial z}\right)_{z=H_0}=0, \quad (11.13)$$

(b) Moderate friction (or slippery):

$$\frac{\rho N_z}{B}\left(\frac{\partial^2\psi}{\partial z^2}\right)|_{z=H_0}=-\tau_{Bx}. \quad (11.14)$$

(c) Minimum friction ($\tau_{Bx} = 0$):

$$\frac{\partial^2\psi}{\partial^2 z}|_{z=H_0} = 0. \tag{11.15}$$

The upper and lower boundary conditions to be applied for the salt conservation Eq. (11.1) must specify zero salt flux at the surface and at the bottom, which are expressed as:

$$\rho K_z(\frac{\partial S}{\partial z})|_{z=\eta} = 0, \tag{11.16a}$$

and

$$\rho K_z(\frac{\partial S}{\partial z})|_{z=H_0} = 0. \tag{11.16b}$$

To close the hydrodynamic system (Eqs. 11.10 and 11.11), it is necessary to impose integral boundary conditions. The first is formulated by Eq. (11.6), which defines the mean salinity in the water column, and the second condition is a consequence of the continuity Eq. (11.1). As the solution is under steady-state conditions the mass (fresh water) conservation can be accomplished by

$$B\int_{\eta}^{H_0} u(x,z)dz = Q_f, \tag{11.17}$$

which, combined with the stream function definition (11.8), is reduced to

$$-\int_{\eta}^{H_0} \frac{\partial\psi}{\partial z}dz = -\psi(x,H_0) + \psi(x,\eta) = Q_f. \tag{11.18}$$

As the stream function has the dimension of volume transport $[L^3T^{-1}]$, which must be zero at the surface and bottom, the integral boundary condition may be expressed by: $\psi(x,z)|_{z=\eta} = \psi(x,z)|_{z=0} = 0$.

As with most fluid dynamics problems the analytical solution of the system of Eqs. (11.10 and 11.11) will be developed in a dimensionless form in order to permit generalized discussions of the results (Fisher et al. 1972). For this purpose, the following variables are defined:

$$Z = z/H_0;\ X = x/L;\ \Psi = \psi/Q_f;\ \$ = S/S_0; \tag{11.19a}$$

$$\bar{\$} = \bar{S}/S_0;\ T_W = \tau_{Wx}/\tau_{W0};\ T_B = \tau_{Bx}/\tau_{B0}. \tag{11.19b}$$

In these definitions, S_0 is the salinity at the coastal region, L is the mixing zone (MZ) length, and τ_{W0} and τ_{B0} are characteristics values of the wind shear and bottom stress, respectively. It should be observed that, as the axis Oz is oriented in the direction of the gravity acceleration, Z = 0 and Z = 1 are the dimensionless ordinates of the surface and bottom, respectively. With the introduction of the dimensionless variables, the equations of motion and salt conservation are expressed as:

$$\frac{gS_0\beta}{L}\frac{\partial\bar{\$}}{\partial X} + \frac{N_zQ_f}{BH_0^4}\frac{\partial^4\Psi}{\partial Z^4} = 0, \tag{11.20}$$

and

$$-\frac{Q_fS_0}{LH_0}\left(\frac{\partial\Psi}{\partial Z}\right)\left(\frac{\partial\bar{\$}}{\partial X}\right) + \frac{Q_fS_0}{LH_0}\left(\frac{\partial\Psi}{\partial X}\right)\left(\frac{\partial\$}{\partial Z}\right) = \frac{BK_zS_0}{H_0^2}\left(\frac{\partial^2\$}{\partial Z^2}\right), \tag{11.21}$$

where Ψ, \$, $\bar{\$}$, X and Z are all dimensionless variables. These equations may be further simplified as,

$$\frac{\partial^4\Psi}{\partial Z^4} + C_1(X)\frac{\partial\bar{\$}}{\partial X} = 0, \tag{11.22}$$

and

$$-\left(\frac{\partial\Psi}{\partial Z}\right)\left(\frac{\partial\bar{\$}}{\partial X}\right) + \left(\frac{\partial\Psi}{\partial X}\right)\left(\frac{\partial\$}{\partial Z}\right) = C_2(X)\frac{\partial^2\$}{\partial Z^2}, \tag{11.23}$$

with the coefficients, $C_1(X)$ and $C_2(X)$, expressed by:

$$C_1(X) = \frac{gBH_0^4\beta S_0}{LN_zQ_f} = \frac{\beta gH_0^3S_0}{LN_zu_f}, \tag{11.24}$$

and

$$C_2(X) = \frac{BLK_z}{H_0Q_f} = \frac{LK_z}{H_0^3u_f}. \tag{11.25}$$

The differential equations of this system (Eqs. 11.22 and 11.23) are dimensionless and at fourth and second degree, respectively, and its unknowns are: $\Psi = \Psi(X, Z)$ and $\$ = \$(X, Z)$. The quantities $C_1(X)$ and $C_2(X)$ are dimensionless, and their dependency on X is not well known and will not be taken into account.

Before being applied to the new equation system (11.22) and 11.23), the boundary conditions (11.12) to (11.18) must be altered to the following expressions:

$$\frac{\rho N_z Q_f}{BH_0^2 \tau_{W0}}\left(\frac{\partial^2 \Psi}{\partial Z^2}\right)\Big|_{Z=0} = T_W, \tag{11.26}$$

$$\left(\frac{\partial \Psi}{\partial Z}\right)\Big|_{Z=1} = 0, \tag{11.27}$$

$$\frac{\rho N_z Q_f}{BH_0^2 \tau_{B0}}\left(\frac{\partial^2 \Psi}{\partial Z^2}\right)\Big|_{Z=1} = -T_B, \tag{11.28}$$

$$\left(\frac{\partial \$}{\partial Z}\right)\Big|_{Z=0} = \left(\frac{\partial \$}{\partial Z}\right)\Big|_{Z=1} = 0, \tag{11.29}$$

$$\int_0^1 \$(X, Z)dZ = \overline{\$}, \tag{11.30}$$

and the integral boundary conditions are,

$$\Psi(X, 0) = 1; \Psi(X, 1) = 0. \tag{11.31}$$

Taking into account the relations (11.8) and the equalities $\partial z = H_0 \partial Z$, $\partial x = L\partial X$ and $\partial \psi = Q_f \partial \Psi$, it follows that,

$$u(X, Z) = -\frac{Q_f}{BH_0}\frac{\partial \Psi}{\partial Z} = -u_f \frac{\partial \Psi}{\partial Z}, \tag{11.32}$$

and

$$w(X, Z) = \frac{Q_f}{BL}\frac{\partial \Psi}{\partial X} = \frac{H_0 u_f}{L}\frac{\partial \Psi}{\partial X}. \tag{11.33}$$

11.2 Hydrodynamic Solution: Maximum Bottom Friction

Consider the solution of Eq. (11.22). As the first member of this differential equation is a function of X and the mean longitudinal salinity gradient should be known, this equation may be solved for the stream function, $\Psi = \Psi(X, Z)$. By integrating with Z four times, the general solution is:

$$\Psi(X, Z) = -\frac{C_1(X)}{24}\left(\frac{\partial\overline{\$}}{\partial X}\right)Z^4 + \frac{a_1(X)}{6}Z^3 + \frac{a_2(X)}{2}Z^2 + a_3(X)Z + a_4(X). \quad (11.34)$$

The dimensionless quantities, $a_1(X)$, $a_2(X)$, $a_3(X)$ and $a_4(X)$, are all function of X and are calculated from the application of the boundary conditions (11.26) and (11.27) and the integral boundary condition (11.31). Applying the last condition $\Psi(X, 0) = 1$, it follows immediately that:

$$a_2(X) = \frac{BH_0^2}{A_z Q_f}\tau_{WO}T_W = \frac{BH_0^2}{A_z Q_f}\tau_{Wx}, \quad (11.35)$$

where $A_z = \rho N_z$ and

$$a_4(X) = 1. \quad (11.36)$$

In the following step, with the boundary conditions (11.27) and $\Psi(X, 1) = 0$ from the integral boundary conditions (11.31), the result is an algebraic equation system with two unknowns, $a_1(X)$ and $a_3(X)$,

$$-\frac{C_1(X)}{6}\left(\frac{\partial\overline{\$}}{\partial X}\right) + \frac{a_1(X)}{2} + \frac{BH_0^2}{A_z Q_f}\tau_{Wx} + a_3(X) = 0, \quad (11.37)$$

$$-\frac{C_1(X)}{24}\left(\frac{\partial\overline{\$}}{\partial X}\right) + \frac{a_1(X)}{6} + \frac{BH_0^2}{2A_z Q_f}\tau_{Wx} + a_3(X) + 1 = 0. \quad (11.38)$$

Subtracting these equations in order to eliminate $a_3(X)$ and solving the result for $a_1(X)$, we have

$$a_1(X) = 3 + \frac{3}{8}C_1(X)\frac{\partial\overline{\$}}{\partial X} - \frac{3}{2}\frac{BH_0^2}{A_z Q_f}\tau_{Wx}. \quad (11.39)$$

Finally, substituting Eqs. (11.39) into (11.37) or (11.38), it follows that the value for $a_3(X)$ is,

$$a_3(X) = -\frac{3}{2} - \frac{C_1(X)}{48}\frac{\partial\overline{\$}}{\partial X} - \frac{BH_0^2}{4A_z Q_f}\tau_{Wx}, \quad (11.40)$$

Substituting the expressions $a_1(X)$, $a_2(X)$, $a_3(X)$ and $a_4(X)$ into the general solution (11.34) yields

$$\begin{aligned}\Psi(X, Z) = {} & \frac{C_1(X)}{48}\left(\frac{\partial\overline{\$}}{\partial X}\right)(-2Z^4 + 3Z^3 - Z) \\ & + \frac{1}{2}(Z^3 - 3Z + 2) + \frac{BH_0^2}{4A_z Q_f}\tau_{Wx}(-Z^3 + 2Z^2 - Z).\end{aligned} \quad (11.41)$$

Combining this result with the expression of $C_1(X)$ (Eq. 11.24), it follows that,

$$\begin{aligned}\Psi(X, Z) = {} & \frac{\beta g B H_0^4 S_0}{48 N_z Q_f L}\left(\frac{\partial \bar{\$}}{\partial X}\right)(-2Z^4 + 3Z^3 - Z) \\ & + \frac{1}{2}(Z^3 - 3Z + 2) + \frac{B H_0^2}{4 A_z Q_f}\tau_{Wx}(-Z^3 + 2Z^2 - Z).\end{aligned} \tag{11.42}$$

Rewriting this solution as a function of the dimensional distance (x) and the salinity (S) yields,

$$\begin{aligned}\Psi(x, Z) = {} & \frac{\beta g B H_0^4}{48 N_z Q_f}\left(\frac{\partial \bar{S}}{\partial x}\right)(-2Z^4 + 3Z^3 - Z) \\ & + \frac{1}{2}(Z^3 - 3Z + 2) + \frac{B H_0^2}{4 A_z Q_f}\tau_{Wx}(-Z^3 + 2Z^2 - Z),\end{aligned} \tag{11.43}$$

or, recalculating the numeric coefficients and expressing the result as a function of the mean value of the longitudinal density gradient and the river velocity $u_f = Q_f/A = Q_f/BH_0$,

$$\begin{aligned}\Psi(x, Z) = {} & \frac{g H_0^3}{A_z u_f}\left(\frac{\partial \bar{\rho}}{\partial x}\right)(-4.17 \times 10^{-2} Z^4 + 6.25 \times 10^{-2} Z^3 - 2.08 \times 10^{-2} Z) \\ & + (0.5Z^3 - 1.5Z + 1) + \frac{H_0}{A_z u_f}\tau_{Wx}(-0.25Z^3 + 0.5Z^2 - 0.25Z).\end{aligned} \tag{11.44}$$

From this analytical expression the dimensionless stream function, the u- and w-velocity components may be calculated by derivation, according to the relations (11.32 and 11.33), and the results are:

$$\begin{aligned}u(x, Z) = {} & \frac{g H_0^3}{A_z}\left(\frac{\partial \bar{\rho}}{\partial x}\right)(1.67 \times 10^{-1} Z^3 - 1.88 \times 10^{-1} Z^2 + 2.08 \times 10^{-2}) \\ & + 1.5u_f(-Z^2 + 1.0) + \frac{\tau_{Wx} H_0}{A_z}(0.75Z^2 - Z + 2.5 \times 10^{-1}),\end{aligned} \tag{11.45}$$

and

$$w(x, Z) = \frac{g}{BA_z}[\frac{\partial(BH_0^4)}{\partial x}(\frac{\partial\overline{\rho}}{\partial x}) + (BH_0^4)(\frac{\partial^2\overline{\rho}}{\partial x^2})]$$
$$- (4.17 \times 10^{-2}Z^4 - 6.25 \times 10^{-2}Z^3 + 2.08 \times 10^{-2}Z)$$
$$+ \frac{\tau_{Wx}}{BA_z}[\frac{\partial(BH_0^2)}{\partial x}](-2.5 \times 10^{-1}Z^3 + 5.0 \times 10^{-1}Z^2 - 2.5 \times 10^{-1}Z). \quad (11.46)$$

When the wind stress is zero ($\tau_{Wx} = 0$), these solutions are similar to the solutions deduced in the article of Fisher et al. (1972), and the u-velocity component (Eq. 11.45) is also similar to the Officer (1976) solution, but it has been improved with the introduction of bottom nonlinear tidal frictional influences.

Solutions (11.45) and (11.46) determine the motion in any longitudinal position of the mixing zone (MZ) of a partially mixed estuary. This result indicates that the steady-state velocity field is dependent on the longitudinal density (salinity) gradient, the river discharge, and the wind stress. And the first (11.45) and the solution (10.22 and Chap. 10) are formally identical, even though solution (10.22) was developed for a well-mixed estuary using a different deduction. This is justifiable because the initial basic hydrodynamic equations were similar, and in relatively homogeneous deep estuaries, the integrated influence of the baroclinic pressure gradient may increase, generating the typical gravitational circulation of partially-mixed estuaries, characterized by bidirectional circulation.

Calculating the velocity at the surface ($Z = 0$) from Eqs. (11.45) and (11.46), gives the following expressions:

$$u(x, 0) = 2.08 \times 10^{-2}(\frac{gH_0^3}{A_z})(\frac{\partial\overline{\rho}}{\partial x}) + 1.5u_f + 2.5 \times 10^{-1}(\frac{\tau_{Wx}H_0}{A_z}), \quad (11.47)$$

and $u(x, 1) = w(x, 0) = w(x, 1) = 0$, confirming the superior and inferior boundary conditions. A convenient expression, equivalent to the analytical profile (11.45), may be obtained combining with the surface expression, $u(x, 0)$. For this purpose, we must solve the expression (11.47) for the first term of the right hand side, which is associated with the baroclinic pressure gradient,

$$\frac{gH_0^3}{A_z}(\frac{\partial\overline{\rho}}{\partial x}) = \frac{10^2}{20.08}[u(x, 0) - 1.5u_f - 2.5 \times 10^{-1}(\frac{\tau_{Wx}H_0}{A_z})]. \quad (11.48)$$

Combining this expression with solution (11.45), it can be further simplified and yields:

$$u(x, Z) = u(x, 0)(8Z^3 - 9Z^2 + 1) + 12u_f(-Z^3 + Z^2)$$
$$+ \frac{\tau_{Wx}H_0}{A_z}(-2Z^3 + 3Z^2 - Z). \quad (11.49)$$

The power series of the dimensionless variable (Z), on the right hand side of this equation, determinates the depth variation of the surface velocity, the velocity generated by the river discharge and, in the last term, the velocity component generated by the wind shear. Under the assumption that the river discharge and the wind shear stress may be disregarded, this solution simplifies and yields the theoretical profiles obtained by Officer (1976, 1977),

$$u(x, Z) = u(x, 0)(8Z^3 - 9Z^2 + 1). \tag{11.50}$$

As the u-velocity component has been calculated (Eq. 11.45), we are able to calculate the free surface slope ($\partial\eta/\partial x$). In order to achieve this, the equation of motion (11.7b) must be applied to the free surface ($z = \eta$) and solved for, $\partial\eta/\partial x$, and in terms of the non-dimensional depth ($Z = z/H_0$) the result is,

$$(\frac{\partial\eta}{\partial x})|_{z=\eta} = \eta_x = -\frac{N_z}{gH_0^2}(\frac{\partial^2 u}{\partial Z^2})|_{Z=0}. \tag{11.51}$$

The final step is to introduce the second derivative of $u = u(x, Z)$ at the surface, $(\partial^2 u/\partial Z^2)|_{Z=0}$, into this equation and further simplified to the following expression:

$$\eta_x = 0.375\frac{H_0}{\rho_0}\frac{\partial\bar{\rho}}{\partial x} + 3\frac{N_z}{gH_0^2}u_f - 1.5\frac{\tau_{Wx}}{\rho g H_0}. \tag{11.52a}$$

This equation is equal Eq. (10.19 and Chap. 10), which has been obtained for a well-mixed estuary; however, for a partially-mixed estuary, the baroclinic pressure gradient predominates. For example, let us assume the following numeric values: $H_0 = 10.0$ m, $g = 10\ m\ s^{-2}$, $\rho = \rho_0 = 10^3\ kg\ m^{-3}$, $\partial\rho/\partial x \approx \Delta\rho/\Delta x = 3.0 \times 10^{-3}\ kg\ m^{-4}$, $u_f = 0.1\ m\ s^{-1}$, $N_z = 10^{-2}\ m^2\ s^{-1}$ and $\tau_{Wx} = 0.2\ kg\ m^{-1}\ s^{-2}$. Then, it follows that $\partial\eta/\partial x > 0$ and the first term of Eq. (11.52a) is 10 times greater than the other terms (10^{-5} compared to 10^{-6}). Only stronger landward winds may invert the free surface slope ($\partial\eta/\partial x < 0$). To calculate the analytical expression of the free surface, $\eta = \eta(x)$, the Eq. (11.52a) must be integrated, with the result being a linear variation from $\eta(x)|_{x=0} = 0$ to $\eta = \eta(x)$,

$$\eta(x) = (0.375\frac{H_0}{\rho_0}\frac{\partial\bar{\rho}}{\partial x} + 3\frac{N_z}{gH_0^2}u_f - 1.5\frac{\tau_{Wx}}{\rho g H_0})x. \tag{11.52b}$$

11.3 Hydrodynamic Solution: Moderate Bottom Friction

Following the same development as in Topic 10.3 (Chap. 10), let us now adopt the bottom boundary condition (11.28) expressed by the semi-empirical relation $T_B = \tau_{Bx}/\tau_{B0}$, and expressed by the semi-empirical boundary condition

$\tau_{Bx} = \tau_{B0}T_B = \rho(4/\pi)kU_Tu|_{z=H_0} = \rho(4/\pi)kU_Tu_{Z=1}$, indicating a moderate bottom friction (slippery condition). As previously indicated, this condition is applied when the tidal velocity amplitude is $U_T \gg u$ (Bowden 1953). Let us also assume, according to Prandle (1985), that the kinematic eddy viscosity coefficient may be empirically simulated by $N_z = kU_TH_0$, where the numeric coefficient $k = 2.5 \times 10^{-3}$ is dimensionless. Applying the upper boundary condition (11.26) and the integral boundary condition (11.31), expressed by $\Psi(x, 0) = 1$, to the general solution (11.34) yields:

$$a_2(X) = \frac{BH_0^2}{A_zQ_f}\tau_{W0}T_W = \frac{BH_0^2}{A_zQ_f}\tau_{Wx}, \tag{11.53}$$

and

$$a_4(X) = 1. \tag{11.54}$$

Therefore, with the bottom boundary condition (11.28) applied and if $a_2(X)$ is known, we have the following expression for $a_1(X)$:

$$a_1(X) = \frac{C_1(X)}{2}\left(\frac{\partial \overline{\$}}{\partial X}\right) - \frac{BH_0^2}{A_zQ_f}(\tau_{Wx} + \tau_{Bx}), \tag{11.55}$$

and, applying the second integral boundary condition (11.31), that is, $\Psi(X, 1) = 0$, yields the integration function $a_3(X)$,

$$a_3(X) = -\frac{C_1(X)}{24}\left(\frac{\partial \overline{\$}}{\partial X}\right) + \frac{1}{6}\frac{BH_0^2}{A_zQ_f}(-2\tau_{Wx} + \tau_{Bx}) - 1. \tag{11.56}$$

Substituting the integration functions $a_1(X)$, $a_2(X)$, $a_3(X)$ and $a_4(X)$, into the general solution (11.34) and further simplifying to the simplest expression yields the following expression for the stream function:

$$\begin{aligned}\Psi(X,Z) = {} & \frac{C_1(X)}{28}\left(\frac{\partial \overline{\$}}{\partial X}\right)(-Z^4 + 2Z^3 - Z) + (-Z + 1) \\ & + \frac{BH_0^2}{6A_zQ_f}[\tau_{Wx}(-Z^3 + 3Z^2 - 2Z) + \tau_{Bx}(-Z^3 + Z).\end{aligned} \tag{11.57}$$

Substituting the expression of $C_1(X)$ (11.24) into (11.57), expressing them in terms of the dimensional longitudinal distance (x) and the mean salinity $(\overline{S})$, it follows that:

$$\Psi(x, Z) = \frac{\beta g H_0^3}{24 N_z u_f}\left(\frac{\partial \overline{S}}{\partial x}\right)(-Z^4 + 2Z^3 - Z) + (-Z + 1) + \frac{H_0}{6 A_z u_f}[\tau_{Wx}(-Z^3 + 3Z^2 - 2Z) + \tau_{Bx}(-Z^3 + Z), \tag{11.58}$$

which is directly proportional to the depth and the longitudinal salinity gradient and inversely proportional to the dynamic (kinematic) eddy viscosity coefficient. In function of the longitudinal density gradient and taking into account that $N_z = kU_T H_0$, another expression for the current function is:

$$\begin{aligned}\Psi(x, Z) = {} & \frac{g H_0^2}{kU_T u_f}\left(\frac{1}{\rho_0}\frac{\partial \overline{\rho}}{\partial x}\right)(-4.17 \times 10^{-2}Z^4 + 8.3 \times 10^{-2}Z^3 - 4.17 \times 10^{-2}Z) \\ & + (-Z + 1) + \frac{1}{\overline{\rho} kU_T u_f}[\tau_{Wx}(-1.67 \times 10^{-1}Z^3 + 5 \times 10^{-1}Z^2 - 3.33 \times 10^{-1}Z) \\ & + \tau_{Bx}(-1.67 \times 10^{-1}Z^3 + 1.67 \times 10^{-1}Z)].\end{aligned} \tag{11.59}$$

According to the equalities (11.32) and (11.33), which define the u- and w-velocity components as derivatives of the stream function, the analytical expression (11.59) is used to calculate these velocity components as:

$$\begin{aligned}u(x, Z) = {} & u_f + \frac{g H_0^2}{kU_T}\frac{1}{\rho_0}\frac{\partial \overline{\rho}}{\partial x}(1.67 \times 10^{-1}Z^3 - 2.5 \times 10^{-1}Z^2 + 4.17 \times 10^{-2}) \\ & + \frac{1}{\rho k U_T}[\tau_{Wx}(5.0 \times 10^{-1}Z^2 - 1.0Z + 3.33 \times 10^{-1}) + \tau_{Bx}(5.0 \times 10^{-1}Z^2 - 1.67 \times 10^{-1})].\end{aligned} \tag{11.60}$$

and

$$\begin{aligned}w(x, Z) = {} & \left[\frac{g H_0}{kU_T}\frac{\partial (H_0^2)}{\partial x}\frac{1}{\rho_0}\left(\frac{\partial \overline{\rho}}{\partial x}\right) + \frac{g H_0^3}{kU_T}\frac{1}{\rho_0}\left(\frac{\partial^2 \overline{\rho}}{\partial x^2}\right)\right] \\ & .(-4.17 \times 10^{-2}Z^4 + 8.3 \times 10^{-2}Z^3 - 4.17 \times 10^{-2}Z).\end{aligned} \tag{11.61}$$

These solutions indicate that under normal conditions the u-velocity component is forced directly by the baroclinic pressure gradient, the river discharge and the wind stress, but the w-velocity component is dependent only on the density gradient and its second derivative.

Let us now calculate the u-velocity component (Eq. 11.60) at the bottom ($Z = 1$), in order to calculate the bottom stress. In doing so, and after simplifications, τ_{Bx} is determined by,

$$\tau_{Bx} = -9.2 \times 10^{-2}\rho g H_0^2 \frac{1}{\rho_0}\frac{\partial \bar{\rho}}{\partial x} + 2.212 u_f k U_T \rho - 3.69 \times 10^{-1}\tau_{wx}. \quad (11.62)$$

Calculating the magnitude of these terms, it is possible to see that the second term is of higher magnitude than the others terms, and positive values of the bottom friction ($\tau_{Bx} > 0$) are generally found in natural estuarine environment. Combining this result with Eq. (11.60) and simplifying the resulting expression to a more convenient solution for practical applications gives,

$$u(x, Z) = \frac{gH_0^2}{kU_T}\frac{1}{\rho_0}\frac{\partial \bar{\rho}}{\partial x}(1.67 \times 10^{-1}Z^3 - 2.96 \times 10^{-1}Z^2 - 5.8 \times 10^{-2}) + u_f(1.106Z^2 + 6.3 \times 10^{-1}) + \frac{\tau_{Wx}}{\rho k U_T}(3.16 \times 10^{-1}Z^2 - Z + 3.95 \times 10^{-1}). \quad (11.63)$$

This solution for the u-velocity component for a partially mixed estuary has the same formalism as (Eq. 10.48 and Chap. 10) for a well-mixed estuary (type 1 or C). Calculating this component at the surface (Z = 0) yields the following expression:

$$u(x, 0) = 5.8 \times 10^{-2}\frac{gH_0^2}{kU_T}\frac{1}{\rho_0}\frac{\partial \bar{\rho}}{\partial x} + 6.3 \times 10^{-1}u_f + 3.95 \times 10^{-1}\frac{\tau_{Wx}}{\rho k U_T}, \quad (11.64)$$

and, with a similar development to that used in the deduction of Eq. (11.49), under maximum friction at the bottom (non-slippery bottom), the equation to calculating the u-velocity component (11.63) may be rewritten as,

$$u(x, Z) = u(x, 0)(2.879Z^3 - 5.103Z^2 + 1) + u_f(-1.814Z^3 + 4.321Z^2) + \frac{\tau_{Wx}}{\rho k U_T}(-1.137Z^3 + 2.016Z^2 - Z). \quad (11.65)$$

This solution is similar to Eq. (10.48), which was calculated with maximum bottom friction. Comparing these equations, we may observe an increase in the importance of the baroclinic pressure gradient and the wind stress in driving the motion, and a decrease in the river discharge forcing. It is also possible to apply the equality (11.51) to this solution to calculate the steady-state free surface slope ($\partial\eta/\partial x$),

$$\left(\frac{\partial \eta}{\partial x}\right)\Big|_{z=\eta} = 5.92 \times 10^{-1}\frac{H_0}{\rho_0}\frac{\partial \bar{\rho}}{\partial x} - 2.212\frac{kU_T}{gH_0}u_f - 6.31 \times 10^{-1}\frac{\tau_{Wx}}{\rho g H_0}. \quad (11.66)$$

In comparing this result with Eq. (11.52a) we may observe an accentuated variation in the river discharge coefficient (second term in the right hand side); its numeric coefficient changes from 3.0 to −2.21. Taking into account the same orders of magnitude in these equations, the free surface slope is positive ($\partial\eta/\partial x > 0$) in both equations, being slightly higher under the first boundary condition.

The development of solutions using zero friction at the bottom as a boundary condition are easy to be demonstrated, and derived in other books, such as Officer (1976).

11.4 Theoretical Vertical Salinity Profile

We will now proceed with the solution of the second order partial differential Eq. (11.23), complemented with its coefficients $C_1(X)$ (11.24) and $C_2(X)$ (11.25), to calculate the salinity field; in the first moment the dimensionless $\$ = \$(X, Z)$ will be calculated, and further, its transformation to the solution $S = S(x, Z)$ will be obtained. This solution is dependent on the stream function, $\Psi = \Psi(x, Z)$, which has already been calculated for distinct boundary conditions (11.43) or (11.44) and (11.58) or (11.59). Of course, these solutions will be dependent on the upper and lower boundary conditions (11.29) and (11.30), respectively, and the integral boundary condition (11.31).

Let us introduce, according to Fisher et al. (1972), an auxiliary (dummy) continuous function $f = f(X, Z)$, defined as $f(X, Z) = \partial\$/\partial Z$, to the solutions of these differential equations. As its second derivative is $\partial f(X, Z)/\partial^2\$/\partial Z^2$, substituting these quantities into the Eq. (11.23) yields the following first order non-homogeneous partial differential equation with variable coefficients:

$$\frac{\partial f}{\partial Z} - \frac{B}{C_2(X, Z)} f(X, Z) = \frac{A(X, Z)}{C_2(X, Z)}. \tag{11.67}$$

where $C_2 = C_2(X)$ has previously been defined in (11.25), and the quantities A(X, Z) and B(X, Z) are expressed by,

$$A(X, Z) = -\frac{\partial\Psi(X, Z)}{\partial Z}\frac{\partial\bar{\$}}{\partial X}, \tag{11.68}$$

and

$$B(X, Z) = \frac{\partial\Psi(X, Z)}{\partial X}. \tag{11.69}$$

These quantities are in function of the stream function, $\Psi = \Psi(X, Z)$ (Eq. 11.57), and the longitudinal salinity gradient $(\frac{\partial \bar{\$}}{\partial X})$ in the A(X, Z) expression is assumed to be known.

With the definition of f = f(X, Z) yielding the differential Eq. (11.67), the boundary conditions (11.29) must be applied separately and are given by,

$$f(X, Z)|_{Z=0} = (\frac{\partial \$}{\partial Z})|_{Z=0} = \frac{1}{S_0}(\frac{\partial S}{\partial Z})|_{Z=0} = 0. \quad (11.70)$$

and

$$f(X, Z)|_{Z=1} = (\frac{\partial \$}{\partial Z})|_{Z=1} = \frac{1}{S_0}(\frac{\partial S}{\partial Z})|_{Z=1} = 0. \quad (11.71)$$

Therefore, in order for the salt flux (or salt transport) through the free surface and bottom to be zero, the f = f(X, Z) function must satisfy the conditions f(X, 0) = f(X, 1) = 0, respectively.

The general solution of Eq. (11.67) may be found in Wylie (1960) and Fisher et al. (1972) and is given by:

$$f(X, Z) = \exp\{[\int (\frac{B}{C_2})dZ] \int (\frac{A}{C_2})[\exp[-\int (\frac{B}{C_2})dZ\}dZ + b_1(X)\exp[\int (\frac{B}{C_2})dZ]. \quad (11.72)$$

The function $b_1(X)$ in the last term of the right hand side of this equation may be calculated using one of the boundary conditions, (11.70) or (11.71); however, it is convenient to adopt the latter condition because it equals zero, and then the solution is reduced to

$$f(X, Z) = \exp\{[\int (\frac{B}{C_2})dZ] \int (\frac{A}{C_2})[\exp[-\int (\frac{B}{C_2})dZ\}dZ. \quad (11.73)$$

As the quantities $C_2(X)$, A(X, Z) and B(X, Z) are given by (11.25), (11.68) and (11.69), respectively, and taking into account Eqs. (11.32) and (11.33), the integrand ratios, A/C_2 and B/C_2, are transformed in,

$$\frac{A(X, Z)}{C_2(X)} = -\frac{H_0 Q_f}{BK_z L}(\frac{\partial \bar{\$}}{\partial X})(\frac{\partial \Psi}{\partial Z}) = \frac{uH_0^2}{K_z L}(\frac{\partial \bar{\$}}{\partial X}) = \frac{uH_0^2}{K_z}(\frac{\partial \bar{\$}}{\partial x}), \quad (11.74)$$

and

$$\frac{B(X, Z)}{C_2(X)} = \frac{H_0 Q_f}{BK_z L}\left(\frac{\partial \Psi}{\partial X}\right) = \frac{wH_0}{K_z}, \tag{11.75}$$

where Ψ, u and w are known functions of x and Z, and its analytical expressions are dependent on the boundary conditions. Then, although the function f = f(X, Z) has a complicated expression (11.73), it may be numerically calculated without as many difficulties in terms of the stream function and the velocity components. Using the velocity components yields the following expression:

$$\begin{aligned} f(x, Z) = {} & \frac{H_0^2}{K_z}\left(\frac{\partial \overline{\$}}{\partial x}\right)\exp\left[\frac{H_0}{K_z}\int w(x, Z)dZ\right] \\ & \cdot \int u(x, Z)\exp\left[-\frac{H_0}{K_z}\int w(x, Z)dZ\right]dZ. \end{aligned} \tag{11.76}$$

With the analytical expression of the function f(x, Z) known, the steady-state vertical salinity profile may be calculated by:

$$S(x, Z) = S_0 \int (f(x, Z)dZ + b_2(x), \tag{11.77}$$

where $b_2(x)$ is the integration function, which is calculated by the integral boundary condition, and S_0 is the constant salinity value at the coastal region, as previously defined. This condition may be expressed by Eq. (11.30) or its equivalent mean salinity value at the water column,

$$\overline{S} = \int_0^1 S(x, Z)dZ, \tag{11.78}$$

yielding,

$$\overline{S} = S_0 \int_0^1 \left[\int f(x, Z)dZ\right]dZ + \int_0^1 b_2(x)dZ, \tag{11.79}$$

and the integration function, $b_2(x)$, is calculated by

$$b_2(x) = \overline{S} - \int_0^1 \left[S_0 \int f(x, Z)dZ\right]dZ. \tag{11.80}$$

Substituting (11.80) into the partial solution (11.77) yields an analytical expression for calculating the steady-state vertical salinity profile:

$$S(x, Z) = \overline{S} + S_0 \int f(x, Z)dZ - \int_0^1 [S_0 \int f(x, Z)dZ]dZ. \quad (11.81)$$

Although this solution is apparently complicated, when rewritten in terms of the stream function or the velocity components, it may be calculated by numerical integration. Combining Eq. (11.76) with the solution (11.81), the result is the following expression for calculating S = S(x, Z) as function of the velocity components (Fisher et al. 1972):

$$\begin{aligned} S(x, Z) = \overline{S} + (\frac{H_0^2}{K_z}\frac{\partial \overline{S}}{\partial x}) \int \exp[(\frac{H_0}{K_z}) \int w(x, Z)dZ]dZ \\ .\int u(x, Z)\exp[(-\frac{H_0}{K_z}) \int w(x, Z)dZ]dZ \\ -(\frac{H_0^2}{K_z}\frac{\partial \overline{S}}{\partial x}) \int_0^1 \int (\exp[(\frac{H_0}{K_z}) \int w(x, Z)dZ]dZ \\ .\int u(x, Z)[\exp(-\frac{H_0}{K_z}) \int w(x, Z)dZ]dZ. \end{aligned} \quad (11.82)$$

The second term on the right hand side of this equation is an indefinite integral, and its result is an expression with Z as an independent variable; the third term is a definite integral calculated in the closed interval [0 − 1], and its final result is a numeric value.

It should be noted that the theoretical steady-state velocity and salinity profiles deduced by Fisher et al. (1972) were evaluated with laboratory experimental data from the Vicksburg and the Delft Hydraulic Laboratory (Delft, Holland) salinity flume and observation data from the James River estuary (Virginia, USA). The combined dataset covered a wide range of flow conditions and degrees of salinity stratification, some of which may be partially invalidate the model assumptions, but these studies helped to define the limits of the analytical model application.

As the intensity of the u-velocity component is several orders of magnitude higher than the vertical component (w), several authors, for example, Officer (1976) and Hamilton and Wilson (1980), had neglected the vertical salt advection. With this assumption, the theoretical vertical salinity profile is established by the balance of the longitudinal advection and the vertical eddy diffusion, according to the simplified expression of Eq. (11.11):

$$-(\frac{\partial \psi}{\partial z})(\frac{\partial \overline{S}}{\partial x}) + u\frac{\partial \overline{S}}{\partial x} = K_z\frac{\partial^2 S}{\partial z^2}, \quad (11.83)$$

and the simplest solution of which may be obtained from Eq. (11.82) with the simplification w(x, Z) = 0 is:

$$S(x, Z) = \overline{S} + (\frac{H_0^2}{K_z}\frac{\partial \overline{S}}{\partial x})[\iint u(x, Z)dZ]dZ - \int_0^1 [\iint u(x, Z)dZdZ]dZ. \quad (11.84)$$

This solution may also be obtained directly by integrating the differential Eq. (11.83). In doing so, rewriting this equation in terms of the dimensionless depth (Z) and separating the variables yields (Officer (1976, 1978):

$$\frac{\partial S}{\partial Z} = (\frac{H_0^2}{K_z}\frac{\partial \overline{S}}{\partial x}) \int u(x, Z)dZ + b_3(x). \quad (11.85)$$

where the quantity, $b_3(x)$, is a dimensionless variable of integration. Taking into account the assumption that at the upper boundary condition there is no salt flux, $\frac{\rho K_z}{H_0}\frac{\partial S}{\partial Z}|_{Z=0} = 0$, it follows that $b_3(x) = 0$. Then, with a new integration,

$$S(x, Z) = (\frac{H_0^2}{K_z}\frac{\partial \overline{S}}{\partial x}) \int u(x, Z)dZ + b_4(x). \quad (11.86)$$

To calculate this second dimensionless variable of integration, $b_4(x)$, the integral boundary condition (11.78) must be applied, and its value is given by

$$b_4(x) = \overline{S} - (\frac{H_0^2}{K_z}\frac{\partial \overline{S}}{\partial x})\{\int_0^1 [\iint u(x, Z)dZdZ]dZ\}. \quad (11.87)$$

Then, substituting $b_4(x)$ into solution (11.86), the result is the vertical analytical salinity profile, S = S(x, Z), which is the same as expression (11.84).

The dependence of the u-velocity and salinity vertical profiles on the N_z (or its dynamic value, A_z), and on the kinematic eddy diffusion coefficient (K_z), which makes the comparison between experimental and theoretical results more difficult. However, as we will be seen later in this chapter, the best numerical values for these coefficients may be estimated, when the validation methodology is applied to improve the comparison of experimental data and theoretical results.

11.5 Theoretical and Experimental Velocity and Salinity Profiles

To exemplify the analytical solution of steady-state vertical profiles of the u- and w-velocity components and the salinity, let us consider an estuary with a transverse section with width $B = 10^3$ m and a depth of 12 m, forced by a river discharge of $Q_f = 20\ m^3\ s^{-1}$, where the wind stress is disregarded ($\tau_{Wx} = 0$).

11.5.1 Longitudinal and Vertical Velocity Profiles

The analytical expressions that will be used to calculate the u-velocity component, u = u(x, Z) are Eqs. (11.49) and (11.65), respectively, for maximum and moderate bottom friction, respectively. To calculate the vertical velocity profile, w = w(x, Z), the corresponding simplified Eqs. (11.46) and (11.61) will be used with the same bottom friction characteristics.

As the transverse area at a longitudinal position, x, is $BH_0 = 12 \times 10^3\ m^2$ the velocity generate by the river discharge is $u_f \approx 0.017\ m\ s^{-1}$. Let us adopt for the kinematic eddy viscosity coefficient $N_z = kU_TH_0 = 1.2 \times 10^{-2}\ m^2\ s^{-1}$, and $k = 2.5 \times 10^{-3}$, considering the tidal amplitude velocity $U_T = 0.4\ m\ s^{-1}$. Under the assumption that the mixing zone (MZ) has a length of 10^4 m (10 km) and the salinity at the mouth is 30‰, the mean longitudinal salinity gradient has an order of magnitude of $3.0 \times 10^{-6}\ m^{-1}$ and its second derivative, $\partial^2S/\partial x^2$, is estimated in $2.5 \times 10^{-8}\ m^{-2}$. These values may be converted in the corresponding values of the mass field using the linear equation of state of seawater (Eq. 11.5) with the saline contraction coefficient, $\beta = 7.0 \times 10^{-4}$ and $\rho_o = 1.0 \times 10^3\ kg\ m^{-3}$, and the following estimates are obtained: $\rho(30) = 1021.0\ kg\ m^{-3}$, $\partial\overline{\rho}/\partial x = 2.1 \times 10^{-3}$ kg m^{-4}, and $\partial^2\overline{\rho}/\partial x^2 = 4.0 \times 10^{-8}\ kg\ m^{-5}$.

In the u-velocity profile, u = u(x, Z), shown in Fig. 11.2 upper (a), we may observe gravitational circulation that is typical of partially mixed estuaries (types 2 or B), which is symmetric to the velocity generated by the river discharge ($\approx 0.017\ m\ s^{-1}$). In the moderate bottom friction condition (Fig. 11.2 upper b), the motion has higher velocity in comparison to the first condition, u(x, 1) = 0, to compensate due to the moderate bottom friction.

Values of the vertical velocity component, w = w(x, Z), can be various orders of magnitude lower than the u-velocity component, and its intensity is higher for a moderate bottom friction condition (Fig. 11.2 lower a, b). The negative value indicates the occurrence of upward motions (note that the Oz axis is oriented in the direction of the gravity acceleration), closing the continuity of the longitudinal motion, and the maximum value occurs at the middle of the water column.

11.5.2 Vertical Salinity Profile

To calculate the vertical salinity profile, S = S(x, Z), it is necessary that the mean salinity value in the water column is known, and let us adopt the value $\overline{S} = 20$‰. As the salinity is dependent on mixing processes (advection and diffusion), the advective process will be simulated by the u-velocity profile given by the solution (11.49) under the assumption that $\tau_{wx} = 0$, and, for the diffusive process, the kinematic eddy diffusion coefficient will be taken as: $K_z = 1.0 \times 10^{-6}\ m^2\ s^{-1}$.

Combining the simplified solution of the vertical salinity profile (Eq. 11.84) with the analytical equation u = u(x, Z) indicated above, which satisfies the bottom

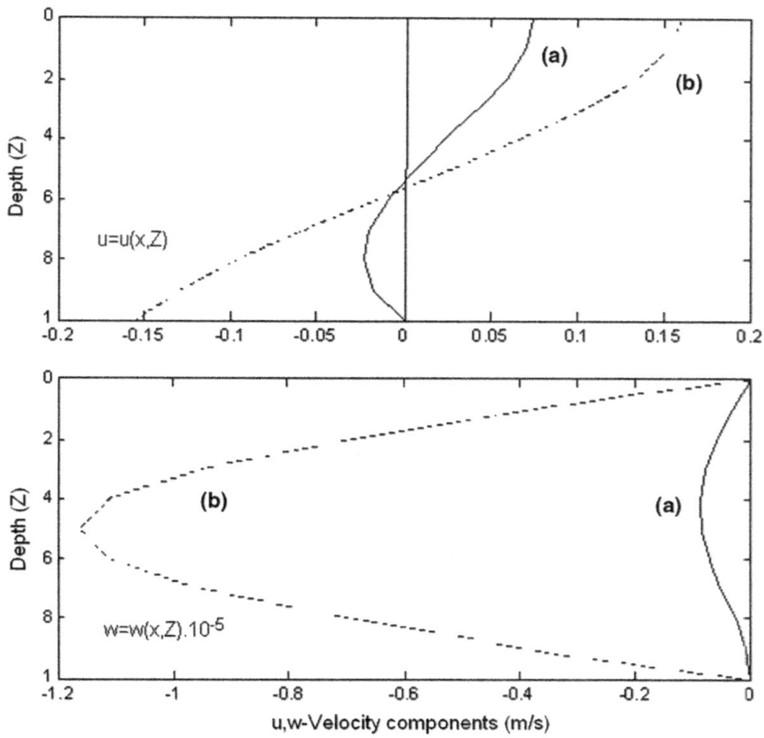

Fig. 11.2 Vertical velocity profiles, u = u(x, Z) and w = w(x, Z), calculated with Eqs. (11.49) and (11.65), and (11.46) and (11.61), respectively, with the following surface and bottom boundary conditions: zero wind stress ($\tau_{Wx} = 0$), maximum friction at the bottom, u(x, 1) = 0, (*bold line*), and a moderate bottom friction u(x, 1) $\neq$ 0 (*dashed line*)

boundary condition, u(x, 1) = 0, the steady-state vertical salinity profile is calculated by:

$$S(x, Z) = \overline{S} + (\frac{H_o^2}{K_z}\frac{\partial \overline{S}}{\partial x}u(x, 0)(8Z^3 - 9Z^2 + 1) + u_f(-12Z^3 + 12Z^2) \quad (11.88)$$

where the u-velocity component at the surface, u = u(x, 0), must be calculated by Eq. (11.64), and its solution is presented in Fig. 11.3a. Using the u-velocity component with the moderately bottom boundary condition (Eq. 11.65), we have the following expression for the vertical salinity profile:

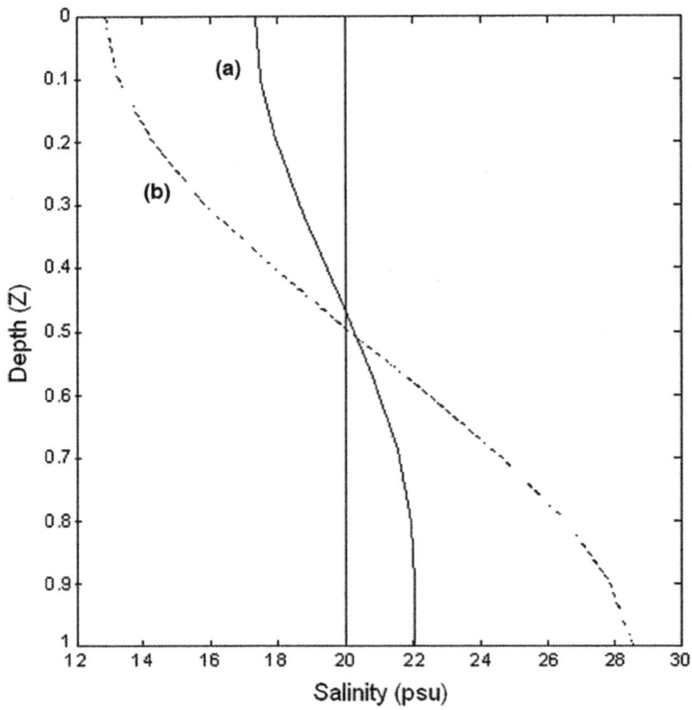

Fig. 11.3 Steady-state vertical salinity profiles calculated with Eqs. (11.88) and (11.89), under the assumption of the following surface and bottom boundary conditions: $\tau_{Wx} = 0$ and $u(x, 1) = 0$ (*bold line*), and (**b**) $\tau_{Wx} = 0$ and $u(x, 1) \neq 0$ (*dashed line*), respectively

$$S(x, Z) = \overline{S} + (\frac{H_0^2}{K_z}\frac{\partial \overline{S}}{\partial x})u(x, 0)[1.44 \times 10^{-1}Z^5 - 4.25 \times 10^{-1}Z^4 + 5.0 \times 10^{-1}Z^2 - 1.05 \times 10^{-1}]$$
$$+ u_f(-9.1 \times 10^{-2}Z^5 + 3.6 \times 10^{-1}Z^4 - 5.7 \times 10^{-2}], \quad (11.89)$$

where the current velocity at the surface (Z = 0) must be calculated by Eq. (11.64). The steady-state vertical salinity profiles under these boundary conditions are presented in Fig. 11.3.

In the Fig. 11.3a we may observe that under maximum friction bottom boundary condition, $u(x, 1) = 0$, the stratification parameter $(\delta S/\overline{S})$ is equal to 0.23. However, with a moderate bottom friction (Fig. 11.3b), there is an increase in the stratification parameter which is equal to 0.78. This increase is due to a higher influence of the advection in the vertical salinity distribution as the u-velocity component is higher under this bottom boundary condition (Fig. 11.2b-upper). Due to these changes in the vertical stratification, the circulation parameter increases from $u_s/u_f = 4.4$ to $u_s/u_f = 9.6$, and the images of these parameters on the Stratification-Circulation

Diagram (Fig. 3.11, Chap. 3) are located in the semi-plane of partially mixed estuaries and highly stratified (type 2b), because $\delta S/\overline{S} > 0.1$.

11.5.3 Validation of Experimental Velocity and Salinity Vertical Profiles

Practical examples on the validation of nearly steady-state observational u-velocity components and salinity vertical profiles with the solutions using Eqs. 11.45 and 11.84 are shown in Figs. 11.4 (upper and lower), according to the investigations of Bernardes (2001) and Bernardes and Miranda (2001). The hydrographic and current velocity were sampled in a mooring station located in the southern region of Cananéia Estuarine System (Fig. 1.5 and Chap. 1), and good agreement between theoretical and experimental data may be observed.

11.6 Hansen and Rattray's Similarity Solution

Hansen and Rattray (1965) theory is a classical theoretical development using the *similarity method* to obtain the solution of a coupled set of partial differential Eqs. (11.1) to (11.4) and associated boundary conditions, in order to describe the circulation and the salt-flux steady-state processes for coastal plain and laterally homogeneous estuaries, where turbulent mixing is primarily forced by tidal currents.

The longitudinal salinity distribution in many coastal plain estuaries takes the general form of the hyperbolic tangent function, with the maximum gradient in the estuarine region named *central regime* and tailing off asymptotically to terminal values towards the mouth and the estuary head. In the central regime, the vertical salinity stratification is nearly independent of the longitudinal position, while in the outer and inner regimes, it is proportional to the departure of the sectional mean salinities from their asymptotic values.

The salinity stratification characteristic in the central regime makes it possible for a theoretical treatment to describe the bi-dimensional velocity and salinity fields generated by external (river discharge and wind), and internal (gradient pressure and friction) forces. As noted by (Hansen and Rattray, op. cit.), analysis of the estuarine regime, therefore, constitutes a problem of both forced and free convection, with the latter influenced by density gradients on the velocity distribution. Thus, the basic non-tidal circulation associated with, and active in, maintaining the salinity distribution in estuaries consists of a seaward flow of river water and a system of currents induced by the density difference between freshwater and seawater.

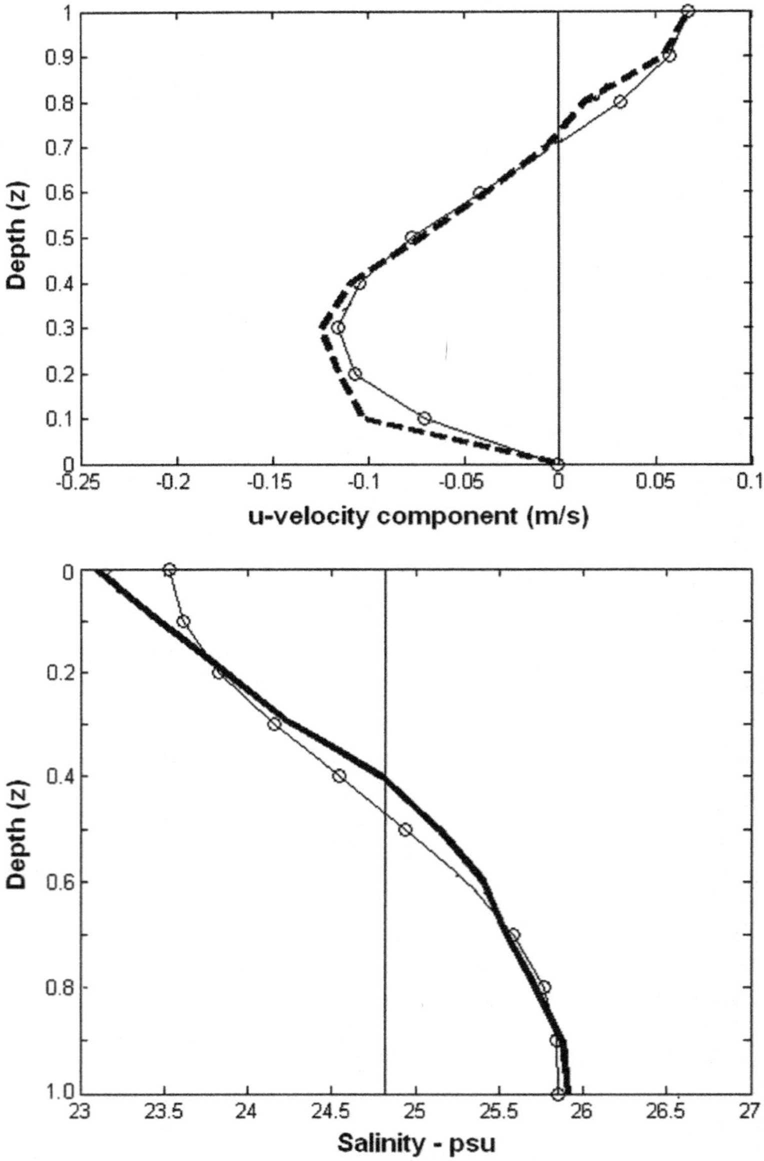

Fig. 11.4 Comparison of the steady-state experimental u-velocity (*upper*) and salinity (*lower*) vertical profiles (*thin lines*) and the corresponding theoretical results (*dashed* and *bold lines*), calculated with Eqs. (11.45) and (11.84) [adapted from Bernardes (2001) and Bernardes and Miranda (2001)]

In the central regime, the following assumptions are made: the estuary has a laterally homogeneous geometry (width and depth), and the river discharge is constant and there is a known salinity (S_0) at the estuary mouth. The basic partial differential equations, which formulate the physical-mathematical problem in relation to the Oxz referential system (Fig. 11.1), in terms of the stream function, $\psi = \psi(x, z)$, and the linear equation of state of seawater, the equation of motion and the salt conservation equations are:

$$g\beta\frac{\partial S}{\partial x} + \frac{1}{B}\frac{\partial}{\partial z}(BN_z\frac{\partial^3\psi}{\partial z^3}) = 0, \tag{11.90}$$

and

$$-(\frac{\partial\psi}{\partial z})(\frac{\partial S}{\partial x}) + (\frac{\partial\psi}{\partial x})(\frac{\partial S}{\partial z}) = B\frac{\partial}{\partial x}(K_x\frac{\partial S}{\partial x}) + B\frac{\partial}{\partial z}(K_z\frac{\partial S}{\partial z}). \tag{11.91}$$

The salt conservation equation presents the following differences in relation to the former formulation (Eq. 11.11): all terms of this equation have the salinity as an unknown, and the term that formulates the longitudinal salt diffusion (first on the right hand side) is included.

The boundary conditions that guarantee a unique solution to Eq. (11.90) are: the wind stress acting on the free surface and maximum bottom friction (non slip bottom), which are formulated by (11.12) and (11.13), respectively. The net volume transport is equal to the river discharge (11.18), due to the steady-state hypothesis. As in the salt conservation Eq. (11.91), the salt fluxes due to advection and turbulent diffusion are included, and the salt balance at the estuary mouth must be null. With the exception of the surface and bottom boundary conditions which annul the salt fluxes through these surfaces, it is necessary to impose the following integral boundary condition:

$$\rho\int_0^{H_0(x)}(-BSu + BK_x\frac{\partial S}{\partial x})dz = 0, \tag{11.92a}$$

or

$$\int_0^{H_0(x)}(S\frac{\partial\Psi}{\partial z} + BK_x\frac{\partial S}{\partial x})dz = 0. \tag{11.92b}$$

In the second integral, it was taken into account that by the current function definition (11.8), $u = -(1/B)\partial\psi/\partial z$. Then, according the steady-state condition, the resulting salt transport T_S, $[T_S] = [MT^{-1}]$, due to the advection and diffusion, must be null. The sought solutions will portrait the transition from river ($S = 0$) to the oceanic conditions, i.e., for $\partial S/\partial x > 0$, and in the classical article of Hansen and

Rattray (1965), three types of similarity solutions with this property were developed. The particular conditions required for these solutions indicate relationships among the external parameters which may be expected to result in particular velocity and salinity distributions. However, for mathematical simplicity, only the central regime of an idealized estuary, which has a rectangular cross-section and the exchange coefficients independent of the depth, will be presented. Further results on the outer and inner regimes may be found in the Hansen and Rattray's article.

In the similarity method, solutions for the stream function and salinity fields are investigated, with the following separation of variables:

$$\psi(x, z) = \psi(Z) = Q_f\Psi(Z), \tag{11.93}$$

and

$$S(\xi, Z) = S_0[\xi\nu + S(Z)], \tag{11.94}$$

where ν is a dimensionless (mixing parameter), $Z = z/H_0$ is the non-dimensional depth, $\xi = \xi(x) = Q_f x/BH_0K_{x0} = u_f x/K_{x0}$ is the non-dimensional longitudinal distance, K_{x0} is the longitudinal kinematic eddy diffusion coefficient at the estuary mouth and S_0 its mean salinity. As the estuary is laterally homogeneous, S_0 is the mean value in the water column, located at the estuary mouth. Taking into account these definitions, the stream function and the salinity are now functions of the dimensional coordinates (z) and (x, z), respectively, because $\Psi[Z(z)] = \Psi(z)$ and $S[\xi(x), Z(z)] = S(x, z)$.

As the river discharge, Q_f, is taken as constant, the stream function (11.93) is independent of the longitudinal distance (x). Then, the w-velocity component, according to its definition in terms of the stream function (Eq. 11.8), is not resolved by this analytical model, and its influence on the salt conservation Eq. (11.91) is null.

The linear longitudinal salinity variation in the central regime is assured by the linear dependence of $\xi = \xi(x)$,

$$\frac{\partial S(x, z)}{\partial x} = \frac{\partial S}{\partial \xi}\frac{d\xi}{dx} = \frac{\nu S_0 Q_f}{BH_0K_{x0}} = \nu\frac{S_0 u_f}{K_{x0}}. \tag{11.95a}$$

or

$$(K_{x0}\frac{\partial S(x, z)}{\partial x}) = \nu S_0 u_f, \tag{11.95b}$$

and the diffusive upstream salt flux (per density unit) at estuary mouth (x = 0) is the fraction of the mixing parameter (ν) of the advective the salt flux advected seaward by the river flow and is the product of the mean cross-section salinity, S_0, and the river velocity. Solving this equation for the dimensionless mixing parameter (ν), we have:

$$\nu = \frac{K_{x0}\frac{\partial S}{\partial X}}{u_f S_0} = \frac{\rho K_{x0}\frac{\partial S}{\partial X}}{\rho u_f S_0} = \frac{\Phi_{dif}}{\Phi_{adv}}. \tag{11.96a}$$

This result indicates that the mixing parameter, ν, is determined by the following salt flux ratio: the landward salt transport by eddy diffusion to the advective seaward salt transport by the river discharge. To close the salt balance in the central regime, an advective term related to the up-estuary salt flux due to the gravitational circulation (Φ_{adv}) must be included, and the mixing parameter is defined as:

$$\nu = \frac{\Phi_{dif}}{\Phi_{dif} + \Phi_{adv}}. \tag{11.96b}$$

From this expression of the mixing parameter, it follows that $0 < \nu \le 1$, and when ν = 1, there is no gravitational circulation ($\Phi_{adv} \to 0$) and the salt flux ratio (11.96b) is in balance; otherwise, if $\nu \to 0$ the salt transport by diffusion is less important ($\Phi_{dif} \ll \Phi_{adv}$), and the salt flux is mainly due to advection (river discharge and gravitational circulation), and the tidal mixing is very low and may be disregarded (Hansen and Rattray 1966; Hamilton and Rattray 1978). As we have seen in the Stratification-circulation Diagram (Chap. 3), for ν = 1 and $\nu \to 0$ corresponds to estuaries classified as well-mixed and partially mixed, respectively.

The similarity condition in the central regime also needs to satisfy the following hypothesis: the kinematic eddy viscosity (N_z) and diffusion (K_z) coefficients are constant, as is the case of the Fisher et al. (1972) analytical model; however, the kinematic eddy diffusivity, K_x, increases seaward at a rate equivalent to the river discharge (Hansen and Rattray 1966),

$$\frac{d(K_x)}{dx} = \frac{Q_f}{BH_0} = u_f. \tag{11.97}$$

Introducing the new formulations of the stream function (11.93) and salinity (11.94) into the Eqs. (11.90) and (11.91), respectively, and taking into account the last equality (11.97), yields the following dimensionless differential equations:

$$\frac{d^4\Psi(Z)}{dZ^4} + \nu Ra = 0, \tag{11.98a}$$

and

$$\frac{d^2S(Z)}{dZ^2} + \frac{\nu}{M}\left(\frac{d\Psi}{dZ} + 1\right) = 0. \tag{11.98b}$$

In these equations, Ra and M, are the dimensionless *Rayleigh estuarine number*[1] and the *mixing tidal parameter*,[2] respectively, which are defined by:

$$Ra = \frac{\beta g S_0 H_0^3}{N_z K_{x0}}, \text{ and } M = \frac{K_z K_{x0} B^2}{Q_f^2} = \frac{K_z K_{x0}}{H_0^2 u_f^2}. \tag{11.99}$$

The R_a number is a measure of how efficiently the salinity (density) generates gravitational circulation, and M represents the ratio of the tidal mixing to the river discharge.

The system of differential Eqs. (11.98a) and (11.98b) must be solved in order to satisfy the following boundary conditions, which may be obtained from the corresponding expressions (11.12), (11.13), (11.15), (11.16a, b) and (11.18):

$$\Psi(Z)|_{z=1} = \Psi(1) = 0; \text{ and } \frac{d\Psi}{dZ}|_{Z=1} = 0, \tag{11.100a}$$

$$\Psi(Z)|_{z=0} = \Psi(0) = 1; \text{ and } \frac{d^2\Psi}{dZ^2}|_{Z=0} = T_W, \tag{11.100b}$$

and

$$\frac{d(S(Z))}{dZ}|_{Z=0} = \frac{d(S(Z))}{dZ}|_{Z=1} = 0. \tag{11.100c}$$

In the boundary condition (11.100b), the wind stress, T_w is the third dimensionless parameter and is given by: $T_w = BH_0^2\tau_{Wx}/K_z\rho Q_f = H_0\tau_{Wx}/A_z u_f$.

To complete the boundary conditions of the salt conservation Eq. (11.98b), it is necessary to use the integral boundary condition (11.92b) in the dimensionless formulation, taking into account the similarity relations (11.93) and (11.94), and the expression of the longitudinal salinity gradient, $\partial S/\partial x = (\nu S_0 Q_f)/(BH_0 K_{x0})$,

---

[1]This number is an analog of the Rayleigh number, which is used to forecast the convection of compressible fluids in between plates with different temperatures. This number is proportional to the cubic power of the distances between the plates and a dimensionless combination of physical properties such as: density, gravity acceleration, thermal expansion coefficient, viscosity, specific heat and thermal diffusion.

[2]M is a non-dimensional number analogue to the ratio G/J defined by Ippen and Harleman (1961), which introduced the first number used in the estuary classification (see Chap. 3).

$$\int_0^{H_0} (S\frac{\partial \Psi}{\partial z} + BK_x\frac{\partial S}{\partial x})dz$$

$$= H_0 \int_0^1 \{S_0[\xi\nu + S(Z)]\frac{Q_f}{H_0}\frac{d\Psi}{dZ} + \frac{BK_x \nu S_0 Q_f}{BH_0 K_{x0}}\}dZ \tag{11.101}$$

$$= \overline{S}Q_f \int_0^1 [\xi\nu\frac{d\Psi}{dZ} + S(Z)\frac{d\Psi}{dZ} + \frac{\nu K_x}{K_{x0}}]dZ = 0.$$

To satisfy the salt conservation, the net salt transport at the estuary mouth ($x = \xi = 0$) must be null, and the equality $K_x = K_{x0}$, holds for this position. Then, the integral boundary condition (11.92b) in terms of the non-dimensional depth is simplified to:

$$\nu + \int_0^1 [S(Z)\frac{d\Psi}{dZ}]dZ = 0. \tag{11.102}$$

The equation of motion (11.98a) may be solved with the same procedure as used in the non-dimensional Eq. (11.22). Then, by successive integrations, we find the solution which is equivalent to (11.43). By applying the boundary conditions (11.100a) and (11.100b), the integration functions $a_3(X)$ and $a_4(X)$ will be obtained, yielding the following expression for the stream function (Hansen and Rattray 1965):

$$\begin{aligned}\Psi(Z) = &-\frac{\nu Ra}{48}(2Z^4 - 3Z^2 + Z)\\ &+\frac{1}{2}(Z^3 - 3Z + 2) - \frac{T_W}{4}(Z^3 - 2Z^2 + Z).\end{aligned} \tag{11.103}$$

With this analytical expression, which is equivalent to solution (11.43), we can easily calculate the u-velocity component in the *central regime* using the relationship $u(Z) = -u_f(d\Psi/dZ)$,

$$\begin{aligned}u(Z) = &u_f\nu Ra(0.167Z^3 - 0.188Z^2 + 0.0208)\\ &+ 1.5u_f(-1.0Z^2 + 1) + u_f T_W(0.75Z^2 - Z + 0.25),\end{aligned} \tag{11.104}$$

where T_W is the non-dimensional wind stress ($T_W = \tau_{Wx}/\tau_{W0}$). This result is similar to solution (11.45). Comparing these solutions, we may observe that the dimensionless coefficients νRa and $(gH_0^3/N_z u_f \rho_0)(\partial\overline{\rho}/\partial x)$, are equivalent, performing the same dynamical function (baroclinic pressure gradient) in the gravitational circulation.

Equation (11.104) expresses the steady-state circulation (u-velocity component) as the sum of three modes: (i) the gravitational-convection mode, associated with the Rayleigh (Ra) estuarine number; (ii) the river discharge mode; and, (iii) the

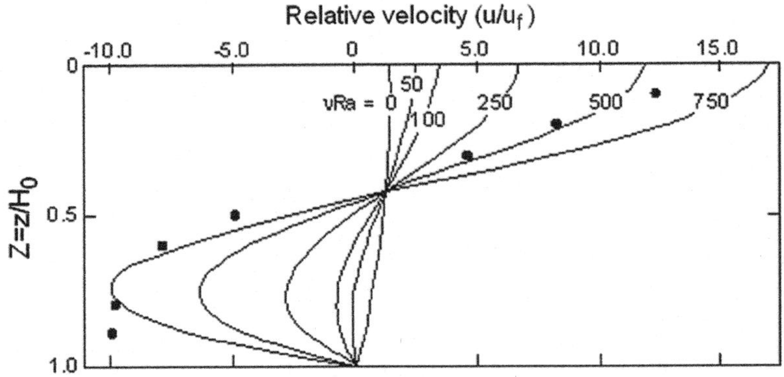

Fig. 11.5 Relative horizontal velocity profiles (u/u_f) with $\tau_{Wx} = 0$, parameterized by the Rayleigh number multiplied by the mixing parameter (νRa). Observed values (*solid dots*) are for the River James estuary (St. 17) (from Hansen and Rattray 1965)

wind-stress mode. If, for example, Ra and T_W are null, the u-velocity profile assumes a parabolic form, which is characteristic of uniform motion and has a constant eddy viscosity. As νRa increases, the baroclinic pressure gradient associated with the density (salinity) field increases and the motion becomes bidirectional for Ra > 30, as illustrated in Fig. 11.5.

The parabolic profile obtained from Eq. (11.104) when νRa = 0 and $T_W = 0$, shown in Fig. 11.5, has almost the same analytic expression of that obtained from Eq. (8.86, Chap. 8).

As the stream function, $\Psi = \Psi(Z)$, as a power series of the dimensionless depth, has already been determined (Eq. 11.103), the salt Eq. (11.98b) only has the salinity, S = S(Z), as an unknown. Integrating this equation and applying the boundary condition, $(dS/dZ)|_{Z=0}=0$, yields the following expression for the vertical salinity gradient:

$$\frac{dS}{dZ} = -\frac{\nu}{M}\int_0^Z \left(\frac{d\Psi}{dZ}\right)dZ - \frac{\nu}{M}\int_0^Z dZ + b_5, \tag{11.105}$$

where b_5 is an integration constant. The integration of the first term of the right hand side of this equation may be completed, and the variable of integration changes from Z to Ψ and the inferior integration limit becomes 1. Progressing further with this integration and applying the boundary condition (11.100a), which states that for $Z = 1 \rightarrow \Psi(1) = 0$, gives:

$$\frac{dS}{dZ} = -\frac{\nu}{M}\Psi(Z) - \frac{\nu}{M}Z + b_5. \tag{11.106}$$

Applying the boundary conditions (11.100a, b, c), it follows that $\Psi(1) = 0$, $\Psi(0) = 1$ and $dS/dZ|_{Z=1}=0$, and we find $b_5 = \nu/M$. Substituting this constant into expression (11.106) and integrating the result, we find the following solution for the vertical salinity profile, $S = S(Z)$:

$$S(Z) = S(0) - \frac{\nu}{M}\int_0^Z \Psi(Z)dZ + \frac{\nu}{M}(Z - \frac{1}{2}Z^2) + b_6, \tag{11.107}$$

where $S(0) = S_s$ is the salinity at the surface, and b_6 is a new dimensionless integration constant, which will be calculated with the boundary condition (11.102), resulting in:

$$\nu + \int_0^1 [S(Z)\frac{d\Psi}{dZ}]dZ = \nu + \int_0^1 S(0)d\Psi + \frac{\nu}{M}\int_0^1 [\int_0^Z \Psi(Z)dZ]dZ$$
$$- \frac{\nu}{M}\int_0^1 (Z - \frac{1}{2}Z^2)dZ + b_6\int_0^1 d\Psi = 0. \tag{11.108}$$

Completing the integrations and solving to the constant, b_6, yields,

$$b_6 = \nu - S(0) + \frac{\nu}{M}\int_0^1 [\int_0^Z \Psi(Z)dZ]dZ + \frac{1}{2}\frac{\nu}{M}(-1 + \frac{1}{3}). \tag{11.109}$$

Substituting this expression of the integration constant, b_6, into the partial solution (11.108), we find the solution for the steady-state vertical salinity profile,

$$S(Z) = \nu + \frac{\nu}{M}(Z - \frac{1}{2}) - \frac{1}{2}(Z^2 - \frac{1}{3})$$
$$- \int_0^Z \Psi(Z)dZ + \int_0^1 [\int_0^Z \Psi(Z)dZ]dZ, \tag{11.110}$$

or, according to the expression (11.94), for $S(x, Z) = S[\xi(x), Z]$, the final analytical expression to calculate the steady-state vertical salinity profile, obtained by Hansen and Rattray (1965), is:

$$S(x, Z) = S_0\{1 + \nu\xi + \frac{\nu}{M}[(Z - \frac{1}{2}) - \frac{1}{2}(Z^2 - \frac{1}{3})$$
$$- \int_0^Z \Psi(Z)dZ + \int_0^1 (\int_0^Z \Psi(Z)dZ)dZ]\}, \tag{11.111}$$

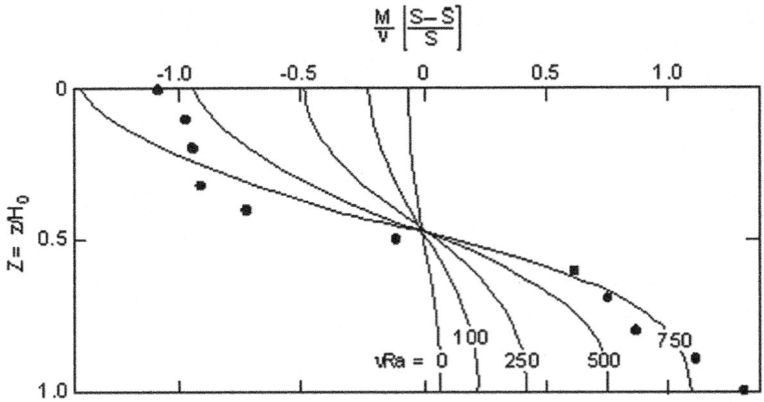

Fig. 11.6 Vertical relative salinity profiles $(M/\nu)[(S - S_O)/S_O]$ at $\xi = 0$ and $\tau_{Wx} = 0$, parameterized by the Rayleigh number multiplied by the mixing parameter (νR_a). Observed values (*solid dots*) for the James river estuary (St. 17), with $\nu R_a = 750$ (from Hansen and Rattray 1965)

where the integral in the last term on the right hand side of this solution is a constant value. Analysis of this solution indicates that the vertical salinity profile depends explicitly on the tidal mixing parameter (M). However, as the stream function $\Psi = \Psi(Z)$, is dependent on the Rayleigh estuarine number (Ra), this profile depends simultaneously on these two dimensionless parameters.

The relative salinity profiles, in relation to the salinity at the estuary mouth (S_O) multiplied by the dimensionless ratio M/ν, at $\xi = 0$ with no wind stress, and parameterized in the dimensionless product, νRa, are illustrated in Fig. 11.6. The relative stratification, like the gravitational convection, increases with νRa, but is also proportional to M/ν. Observational steady-state salinity profiles in the James river estuary (Virginia, USA) indicate good correspondence with the theoretical profiles for $\nu Ra = 750$.

The wind forcing influences on the u-velocity component and salinity profiles were also investigated in the classical articles of Rattray and Hansen (1962) and Hansen and Rattray (1965). This theoretical study was expanded by Officer (1976, 1977), Prandle (1985), among others, imposing moderate bottom friction, and the tidal currents are predominantly responsible for the eddy diffusion, but without other influence in the steady-state circulation.

11.7 Estuary Classification: Stratification-Circulation Diagram

Applying the integral boundary condition (Eq. 11.102), the mixing parameter (ν) which measures the relative importance of diffusion and advection to the salt fluxes (Eq. 11.96b), may be correlated with the dimensionless parameters M, Ra and the wind stress T_W. This correlation may be achieved by finding the positive square root of the following second degree algebraic equation (Hansen and Rattray 1965):

$$1680M(1-\nu) = (32 + 10T_W + T_W^2)\nu + (76 + 14T_W)\frac{Ra}{48}\nu^2 + \frac{152}{3}\left(\frac{Ra}{48}\right)^2\nu^3. \tag{11.112}$$

In the subsequent article, Hansen and Rattray (1966) used this equation as the starting point to analytically determine a quantitative method for use in estuary classification. This method was named the Stratification-circulation Diagram, and its practical application was presented in Chap. 3. For this purpose, Eq. (11.112) is simplified, disregarding the wind stress ($T_w = 0$). As an artifact, the first member of the equation is multiplied and divided by ν, and rearranging its terms yields the following incomplete second grade equation for the mixing parameter, ν:

$$1680\left(\frac{M}{\nu}\right)\nu^2 + \left[32 - 1680\left(\frac{M}{\nu}\right) + 76\left(\frac{\nu Ra}{48}\right) + \frac{152}{3}\left(\frac{\nu Ra}{48}\right)^2\right]\nu = 0. \tag{11.113}$$

For practical purposes, considering the parameter ν as unknown, this equation may be expressed as a function of the following dimensionless parameters: the ratio of the u-velocity at the surface (u_s) to the fresh water velocity (u_s/u_f), and the ratio of the salinity at the bottom (S_b) minus the salinity at the surface (S_s), divided by the mean salinity value in the water column ($\overline{S}$), yielding $(S_b - S_s)/\overline{S}$. As previously presented in Chap. 3, these parameters are the definitions of the circulation and stratification parameters, respectively. Then, calculating the solution of the u-velocity component (Eq. 11.104) at the surface (Z = 0), we have the following results:

$$\frac{u(0)}{u_f} = \frac{u_s}{u_f} = 1.5 + 2.08 \times 10^{-2}\nu Ra = 1.5 + \frac{\nu Ra}{48}, \tag{11.114}$$

and

$$\frac{\nu Ra}{48} = \left(\frac{u_s}{u_f} - \frac{3}{2}\right). \tag{11.115}$$

In the following step, the vertical salinity profile presented in the Eq. (11.111) will be solved at the surface (Z = 0) and bottom (Z = 1), and the last two terms on the right hand side will be integrated in the closed interval [0 – 1], and the results are:

$$\int_0^0 \Psi(Z)dZ = 0, \tag{11.116a}$$

$$\int_0^1 \Psi(Z)dZ = \frac{3}{8} - \frac{\nu Ra}{320}, \tag{11.116b}$$

and

$$\int_0^1 [\int_0^Z \Psi(Z)dZ]dZ = \frac{11}{40} - \frac{\nu Ra}{576}. \quad (11.116c)$$

Substituting these results into the Eq. (11.111), yields the following values of the salinity at the surface (Z = 0) and bottom (Z = 1):

$$\frac{S(x,0)}{\overline{S}} = \frac{S_s}{\overline{S}} = 1 + \nu\xi - \frac{\nu}{M}(\frac{7}{120} + \frac{\nu Ra}{576}), \quad (11.117)$$

and

$$\frac{S(x,1)}{\overline{S}} = \frac{S_b}{\overline{S}} = 1 + \nu\xi - \frac{\nu}{M}(\frac{1}{15} + \frac{\nu Ra}{720}). \quad (11.118)$$

By subtraction of Eqs. (11.117) and (11.118), it follows that the stratification parameter may be calculated by,

$$\frac{S_b - S_s}{\overline{S}} = \frac{\delta S}{\overline{S}} = \frac{\nu}{M}(\frac{1}{8} + 3.125 \times 10^{-3}\nu Ra), \quad (11.119)$$

or

$$\frac{M}{\nu} = (\frac{\delta S}{\overline{S}})^{-1}(0.125 + 3.125 \times 10^{-3}\nu Ra). \quad (11.120)$$

Finally, substituting expressions (11.116a), (11.116b), (11.116c) and (11.120) into Eq. (11.113) the unknown (ν) of this equation may be calculated as a function of the stratification, $\delta S/\overline{S}$, and circulation, u_s/u_f, parameters. As previously seen, these parameters may be determined in the estuary region where the central regime predominates. Although this equation has already been presented and used in the estuaries classification (Chap. 3), it is presented bellow as a complementary equation for this topic,

$$\begin{aligned}&(\frac{\delta S}{\overline{S}})^{-1}[210 + 252(\frac{u_s}{u_f} - \frac{3}{2})]\nu^2 \\ &+ [32 - (\frac{\delta S}{\overline{S}})^{-1}(210 + 252(\frac{u_s}{u_f} - \frac{3}{2})) \\ &+ 76(\frac{u_s}{u_f} - \frac{3}{2}) + \frac{152}{3}(\frac{u_s}{u_f} - \frac{3}{2})^2]\nu = 0.\end{aligned} \quad (11.121a)$$

This equation indicates the functional relation for the unknown, ν,

$$\nu = \nu(\frac{\delta S}{\overline{S}}, \frac{u_s}{u_f}), \qquad (11.121b)$$

which has $\nu = 0$ as a trivial solution. However, its solution in the real numeric field is only possible if the constant, 32, in the third term on the left hand side of the Eq. (11.121a) is disregarded. With this simplification, for $\nu = 1$, when the turbulent eddy diffusion is predominant to the landward salt transport, the solution is $u_s/u_f = 1.5$, and for $\nu \to 0$ the advective process is predominant to the seaward salt flux. Using this solution, Hansen and Rattray (1966) were able to classify estuaries with correlation of the parameters, $(\delta S/\overline{S})$ and (u_s/u_f), in the Stratification-circulation Diagram with ν $(0 < \nu \leq 1)$ as parameter. The graphical solution of this equation, forming the base of an analytical method of estuary classification, has already been presented in figures of the Chap. 3, the defined parametric values enabling four estuary types to be identified, which were closely checked with observational data of natural estuaries.

11.8 Hansen and Rattray's Velocity and Salinity Vertical Profiles: Results and Validation

For practical applications of the vertical u-velocity and salinity profile solutions of Hansen and Ratttray (1965) (Eqs. 11.104 and 11.111), describing the dynamical steady-state of the central regime of the mixing zone of estuaries due to the river discharge, the baroclinic pressure gradient and wind stress are obtained from derivations of the stream current function, $\Psi = \Psi(Z)$ Eq. (11.103). However, it should be observed that the theoretical solution, $u = u(Z)$, is only function of the vertical coordinate (z, or Z); however, the salinity solution, $S = S(x, Z)$ or $S = S(x, z)$, is also a function of the longitudinal coordinate, x, due to its dependence on the dimensionless longitudinal coordinate, $\xi = \xi(x)$. In these applications, because the local depth is dependent on the longitudinal position, $h = h(x)$, and the river discharge velocity (u_f) must usually be substituted by the vertical mean velocity in water column (u_a) at the transverse section in the longitudinal position x, the theoretical velocity becomes indirectly dependent on the longitudinal position. Thus, $u = u(x, Z)$ or $u = u(x, z)$.

In relation to the salinity, the theoretical mean value ($\overline{S}$) used in the calculation the stratification parameter, must be substituted by the corresponding value (S_a), i.e., the time-mean value at the transverse section. Furthermore, as the velocity is also dependent on the advective influence of the river velocity, it must be changed to the corresponding mean value at the section (u_a). Then, due to these simplifications, in practical applications, the analytical expressions of the theoretical velocity and salinity profiles will be denoted by $u_c = u_c(x, Z)$ and $S_c = S_c(x, Z)$, respectively, and their analytical expressions are:

$$u_c(x, Z) = \left(\frac{\nu}{48}\right)\frac{gh^3\frac{\partial\rho}{\partial x}}{\rho N_z}(1 - 9Z^2 - 8Z^3) - \left(\frac{3}{2}\right)u_f(-1 + Z^2) + \frac{1}{4}\frac{\tau_W h}{\rho N_z}(1 + 4Z + 3Z^2), \tag{11.122}$$

and

$$S_c(\mathbf{x}, Z) = S_o\left[\begin{array}{l} 1 + \upsilon\xi + \frac{\upsilon}{\mathbf{M}}[(-Z - \frac{1}{2}) - \frac{1}{2}(Z^2 - \frac{1}{3}) - \frac{1}{2}(-2Z - \frac{3}{2}Z^2 + \frac{1}{4}Z^4)\ldots \\ + (\frac{\upsilon}{48})\frac{gh^3\frac{\partial\rho}{\partial y}}{\rho \mathbf{N}_z\mathbf{u_f}}(\frac{1}{2}Z^2 - \frac{3}{4}Z^4 - \frac{2}{5}Z^5) - \frac{1}{4}\frac{\tau_W h}{\rho \mathbf{N}_z\mathbf{u_f}}(\frac{1}{2}Z^2 + \frac{2}{3}Z^3 + \frac{1}{4}Z^4)\ldots \\ + (\frac{11}{40}) - (\frac{\upsilon}{576})(\frac{gh^3\frac{\partial\rho}{\partial y}}{\rho \mathbf{N}_z\mathbf{u_f}}) - (\frac{1}{80})\frac{\tau_W h}{\rho \mathbf{N}_z\mathbf{u_f}}] \end{array}\right] \tag{11.123}$$

In these analytical formulae, the vertical Oz axis is oriented in the opposite direction of the acceleration of gravity, and the dimensionless depth varies from $Z = 0$ and $Z = -1$ at the surface and bottom, respectively. To obtain a detailed depth discretization, intervals of $|\Delta Z| = 0.1$ are adequate.

These solutions indicate that others geometric and physical quantities which must be known are: longitudinal distance, x, the estuary depth, h, the longitudinal density gradient, $\partial\rho/\partial x \approx \Delta\rho/\Delta x$, salinities at the the estuary head, S_{head}, and mouth, S_{mouth}, the wind stress, τ_{Wx}, and the mixing parameter, ν, previously determined by the Stratification-circulation diagram. Taking into account the hypothesis of Hansen and Rattry's theory the eddy coefficients N_z, K_z and K_{x0}, used in the definitions of the dimensionless quantities $\xi = \xi(x)$ and M, and the wind stress (τ_{Wx}) are considered free parameters, i.e., they must be conveniently adjusted to validate theoretical profiles in comparison to those from observational data. It is known that validation of analytical and numerical models for observational conditions requires a data set of sufficient length to cover variations in tidal cycles, river discharge and wind conditions.

There are several methods that can be used to establish the relative agreement between theoretical and experimental results. One of these is the validation method of the Relative Mean Absolute Error-RMAE (Walstra et al. 2001) and the Skill proposed by Wilmott (1981) which was further improved by Warner et al. (2005).

The method of the Relative Mean Absolute Error-RMAE is formulated by:

$$\text{RMAE} = \frac{\left\langle\left|\vec{V}_m - \vec{V}_c\right|\right\rangle}{\left\langle\left|\vec{V}_m\right|\right\rangle}, \tag{11.124}$$

where $\vec{V}_m$ and $\vec{V}_c$ are the field measured and the computed velocity vectors, respectively, and the symbol ⟨ ⟩ indicate time-mean values. This definition has been particularly applied for comparison of current velocities, but it may also be used to scalar properties. A limited and preliminary qualification of the RMAE ranges of this method indicate a variation between excellent (RMAE < 0.2) and bad (RMAE > 1.0) validation results.

The Skill method is defined by the following relationship of observed data (X_{Obs}), its time (or space) averaged value ($\overline{X}_{Obs}$), and the corresponding theoretical results (X_{Model}):

$$\text{Skill} = 1 - \frac{\sum |X_{Model} - X_{Obs}|^2}{\sum \left(|X_{Model} - \overline{X}_{Obs}| + |X_{Obs} - \overline{X}_{Obs}|\right)^2}. \tag{11.125}$$

According to the definition, the Skill parameter varies between one (1) and zero (0), indicating a perfect adjustment between calculated and observed values, or a complete disagreement, respectively. The validation of theroretical results with this parameter was applied by Andutta et al. (2006), using observational data series over two tidal cycles to validate the u-velocity component and salinity profiles calculated with a tridimensional numerical model applied to the Curimatú river estuary (Rio Grande do Norte, Brazil).

To illustrate a practical exercise to validate the analytical simulation of the u-velocity component and salinity profiles (Eqs. 11.122 and 11.123), the following physical quantities, which were calculated from hourly observational data measured in the Piaçaguera estuarine channel during three semi-diurnal tidal cycles (northern region of the Santos-São Vicente Estuary, São Paulo, Brazil, Fig. 1.5), whose time-mean values represent nearly-steady values are listed bellow:

(i) Mean values of velocity ($u_a \approx u_f$). salinity (S_a), and depth (h).
(ii) The mixing parameter, ν, obtained from the Stratification-circulation Diagram.
(iii) Mean salinities at the mouth and head (S_{mouth}, S_{head}).
(iv) Longitudinal density gradient $\partial\rho/\partial x \approx \Delta\rho/\Delta x$, adjusted to the best validated theoretical result.

Table 11.1 Free parameters N_z, K_z and K_{H0} and that obtained from observational data (*) used in the theoretical simulation the steady-state vertical salinity, $S_c = S_c(x, Z)$, and the u-velocity component, $u_c = u_c(x, Z)$, in the Piaçaguera channel (Santos-São Vicente Estuary, São Paulo)

Free and experimental parameters	Numerical values
$^*u_f = u_a$	$0.009\ \text{m s}^{-1}$
$^*h$	11.0 m
$^*\nu$	0.85
$^*\overline{S} = S_a$	26.5‰
$^*S_{mouth}$	33.0‰
$^*S_{head}$	1.0‰
$^*L_x$	20×10^3 m
$^*x$	17×10^3 m
N_z	$4.0 \times 10^{-3}\ \text{m}^2\ \text{s}^{-1}$
K_z	$1.5 \times 10^{-4}\ \text{m}^2\ \text{s}^{-1}$
K_{H0}	$1.0 \times 10^3\ \text{m}^2\ \text{s}^{-1}$
τ_{Wx}	2.0×10^{-2} Pa

(According to Miranda et al. 2012)

Others physical quantities used were the mean depth, estuary length, distance of the data sampling position to the estuary mouth, wind stress and the free parameters N_z, K_z and K_{H0}. The numerical values of these quantities are shown in Table 11.1.

Using the time-mean vertical profiles of salinity andthe u-velocity obtained with the experimental data during three tidal cycles (tick profiles of Fig. 11.7), the calculate stratification and circulation parameters were $S_P = 0.07$ and $C_P = 11.4$,

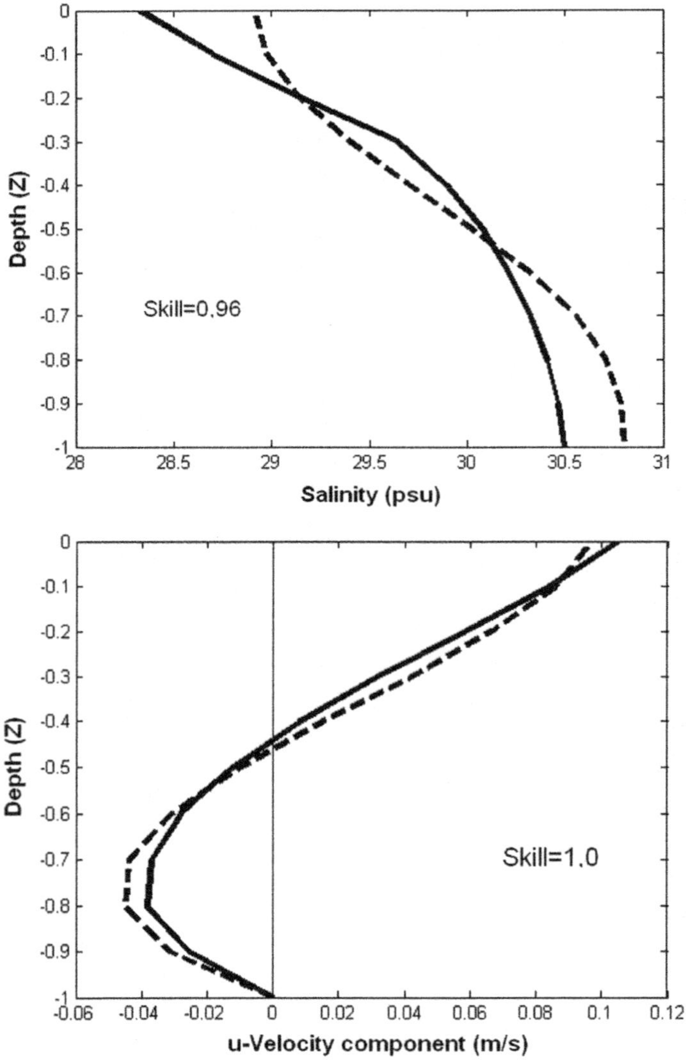

Fig. 11.7 Theoretical (*dashed line*) and observational (*tick line*) profiles of salinity (*upper*), and u-velocity vertical (*lower*) validated with observational data with the Skill parameter. Measurements made during three semi-diurnal neap tidal cycles in the Piaçaguera Channel (Santos-São Vicente Estuary, São Paulo) (according to Miranda et al. 2012)

respectively, and the estuary was classified as partially mixed and low stratified (type 2a). The mixing parameter, $\nu = 0.85$, associated with these parameters indicate that the diffusion and advection processes were responsible for 85 and 15 % to the mixing, respectively.

The theoretical vertical profiles of the salinity and the u-velocity component in comparison to the observational data are presented in the Fig. 11.7. In both profiles the mean Skill value is 0.96 and 1.0, respectively.

The nearly steady-state salinity stratification and the vertical velocity shear-stress, observed in these results were forced by the oscillatory motion generated by the tidal currents during the neap tidal cycle and a small contribution of the river discharge ($\approx$0.01 m s^{-1}); according to Miranda et al. (2012), these currents were almost the same intensity as those observed during the spring tidal period. The depth of no-motion at $Z = -0.45$ (≈ -5 m) corresponds to a mean value observed during three semi-diurnal neap-tidal cycles.

11.9 Salinity Intrusion

A steady-state theory on the salinity intrusion length (X_C) in salt wedge estuaries was presented in Chap. 9 (Eq. 9.72), based in classical theories and confirmed by experimental results. It was shown that this length is directly proportional to the reduced gravity times the square of the depth at the estuary mouth, and inversely proportional to the square of the velocity generated by the river discharge.

Due to the great importance on salt intrusion investigations into estuaries, experiments on these phenomena have been performed since the 1960 decade, in laboratory experiments at Waterways Experiment Station (WES), Vicksburg, Miss. (USA), and in the Delft Hydraulics Laboratory (DHL), Delft (Holland). The laboratory results of the maximum and minimum salt intrusions forced by fresh water discharge, tidal amplitude, mean water depth, roughness, and densities were further compared with *in situ* measurements and published in the articles of Ippen and Harleman (1961) and Rigter (1973).

In the Rigter's article, experiments in a tidal salinity flume channel (101.5 m long) are described, taking into account the tidal amplitude induced by sinusoidal tides, mean-water depth, water input discharge and bottom roughness; the flume has a vertical scale (1:64) to approximate physical characteristics of Rotterdam Waterway (Holland). The intrusion length (L_i) investigations in these experiments were supposed to be dependent on eleven physical quantities. Several experiments were investigated during which some properties were taken as constant, and others were submitted to controlled variations. A detailed analysis of these experiments, using the dimensional analysis approach may be found in the Rigter's original article.

Further investigations indicated that the following functional equation to calculate the minimum saline intrusion length (L_i) in a partially mixed estuary may be used (Prandle 1985, 2004):

$$L_i = \delta\frac{H}{k}\{\frac{[(\frac{\Delta\rho}{\rho})gH]^{1/2}}{\hat{u}}\}\{\frac{[(\frac{\Delta\rho}{\rho})gH]^{1/2}}{\bar{u}}\}, \qquad (11.126)$$

where δ and k are dimensionless coefficients, H is the channel depth, $\hat{u}$ is the tidal velocity and $\bar{u}$ its mean-depth value. The validity of this equation has been compared with observational data enabling its simplification and the following expression was suggested:

$$L_i = \delta_1\frac{\Delta\rho}{\rho}\frac{gH^2}{k\hat{u}\,\bar{u}} + \delta_2\lambda. \qquad (11.127)$$

In this equation δ_1 and δ_2 are non-dimensional numeric values, $\lambda = (gH)^{1/2}T_P$ (tidal wave velocity times the tidal period T_P), and the second term in the right hand side ($\delta_2\lambda$) was introduced to allow for variations in the degree of mixing at the estuary mouth. A least square fitting procedure was used to determine the constants δ_1 and δ_2 values; in the Rotterdam Water Way the corresponding values were 0.187 and −0.006, respectively. Similar calculations in the WES tank produced values of $\delta_1 = 0.134$ and $\delta_2 = 0.026$. Comparisons of observed and computed values for intrusions lengths indicated excellent agreement between the in situ results and the laboratory experiments.

Investigation on the saline intrusion length in an estuary located in the Southern Brazilian coast (Santa Catarina, Brazil) was published by Döbereiner (1985, quoted in Schettini (2002)). In this article, the hydraulic and sediment behavior of the Itajaí-açu river estuary was investigated during low and high river discharges. A synthesis on the Döbereiner's results is summarized as follow: for low river discharge (300 $m^3\ s^{-1}$) the saline intrusion length (L_i) was located at approximately 18 km landward from the estuary mouth, however, for river's discharges higher than 1000 $m^3\ s^{-1}$ the salt water was completely removed to the nearshore turbidity zone (NTZ).

Further studies by Schettini and Truccolo (1999) and Schettini (2002), based on seasonal observational data of saline intrusion lengths (L_i), and the associated river discharges (Q_f), were correlated by the following exponential relationship:

$$L_i = -10.72 + 32.69e^{-2.17\times10^{-3}Q_f},$$

with the root mean square error estimated in 0.7.

11.10 Secondary Circulation

The hypothesis of a longitudinal circulation laterally uniform should not be generalized to most natural estuaries, because as a tridimensional system its water masses are also driven by the flow that is normal to the main along channel flow, usually known as secondary circulation. Thus, the composition of the longitudinal and secondary flow generates along the estuarine channel a complex tri-dimensional motion similar to a helical lateral flow.

In the decomposition of velocity measurements, it is usually possible to observe the presence of v-velocity components, although with low intensity. For example, in the velocity decomposition presented in Table 5.1 (Chap. 5), it is possible to observe the presence of the secondary flows (v-velocity components).

The occurrence of secondary circulation and the associated transverse mixing due to turbulent diffusion and advection in estuaries may be generated by the interaction of the following influences (Pritchard 1956; Dyer 1977; Sumer and Fischer 1977; Chant 2010):

- Topographic deflection due to natural curvatures along the channel and irregularities at the bottom and margins;
- Non-uniform lateral and vertical salinity (density) stratification generated by mixing processes;
- Barotropic and baroclinic pressure gradients;
- Coriolis and centrifuge accelerations.

Formation of fronts may also be observed in estuarine channels, having been generated by longitudinal and transverse motions. An analysis of these fronts has been made in terms of density forced motions in laboratory and in situ experiments by Nunes and Simpson (1985). The occurrence of this phenomenon may be visually observed because convergence lines frontal zones acts as a filter, gathering organic and inorganic detritus and debris.

The dynamical balance of the secondary circulation and the relative importance of its terms in the equation of motion, have been investigated by Pritchard (1956), Dyer (1973, 1977), Sumer and Fischer, 1977) and Ong et al. (1994) with in situ and laboratory experiments. Sumer and Fischer (op. cit.), Nunes and Simpson (op. cit.), Chant (2010) have provided evidence on the secondary circulation in laterally stratified partially stratified and well-mixed estuaries. Following the theoretical formulation of Nunes and Simpson (op. cit.) for an analytical solution of the secondary circulation, it is necessary to formulate some simplifying hypotheses:

- Steady-state bi-dimensional motion in the Oyz plane.
- Cross channel baroclinic pressure gradient force, $\partial S/\partial y = f(y)$ and $\partial \rho/\partial y = g(y)$.
- Transverse density gradient is independent of the depth, $\partial/\partial z(\partial \rho/\partial y) = 0$; and
- Uniform longitudinal and transverse sections (centrifugal accelerations are disregarded).

As in the analytical solutions for calculating steady-state longitudinal circulation, let us consider an estuary which, by hypothesis, has a constant depth (H_0), as indicated schematically in Fig. 11.8, and a known salinity (density) field. This figure indicates the reference (Oyz) with the Oz axis origin at the free surface and oriented in the acceleration gravity ($\vec{g}$) direction, and schematically indicates the

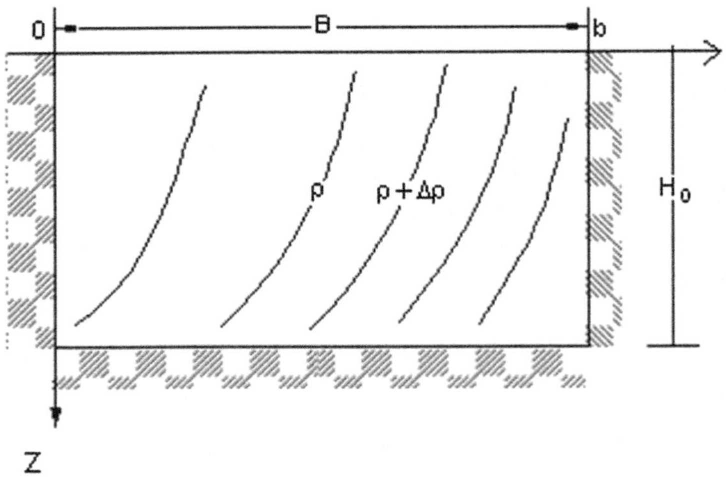

Fig. 11.8 Transversal section and reference system for an analytical study of the steady-state secondary circulation in a laterally non-homogeneous estuary

laterally non-homogeneous density field. Hence, according to the simplifying hypotheses, the cross-channel baroclinic pressure gradient force is formulated by:

$$-\frac{1}{\rho}\frac{\partial p}{\partial y} = g\frac{\partial \eta}{\partial y} - \frac{g}{\rho}\frac{\partial \rho}{\partial y}z. \quad (11.128)$$

As the transverse density gradient is independent of depth the baroclinic term on the right hand side of Eq. (11.128) increases linearly with depth.

As the longitudinal motion will not be taken into account in this simple model, the system of equations to be solved, associated with the linear equation of state of seawater, $\rho = \rho_0(1 + \beta S)$, are expressed by:

$$g\frac{\partial \eta}{\partial y} - \frac{g}{\rho}\frac{\partial \rho}{\partial y}z + N_z\frac{\partial^2 v}{\partial z^2} = 0, \quad (11.129a)$$

$$\frac{\partial v}{\partial y} + \frac{\partial w}{\partial z} = 0. \quad (11.129b)$$

The dynamical balance expressed by Eq. (11.129a) does not take into account the dynamical balance between friction and the Coriolis acceleration, which has been investigated by Chant (2010). However, according to the latitudinal estuary position and intensity of the longitudinal and transverse motions, this effect may or may not be disregarded. Examples of estuaries in which these conditions may be found are presented in the articles of Dyer (1977) and Ong et al. (1994).

As the geometry of the cross-section, the salinity field and the turbulent kinematic eddy viscosity are known, this is a closed hydrodynamic system and the boundary and integral boundary conditions warrant a unique solution. In the continuity equation v- and w-velocity components may be expressed in terms of the *current function*, $\Psi = \Psi(y, z)$, $[\Psi] = [L^2T^{-1}]$, and the velocity components may be expressed by: $v(y, z) = -\frac{\partial \psi}{\partial z}$ and $w(y, z) = \frac{\partial \psi}{\partial y}$. Manipulating and rewriting Eq. (11.129a) in terms of the barotropic and baroclinic pressure gradients, and using the linear equation of state of seawater, the general solution for the transverse velocity is:

$$v(y, z) = -\frac{\partial \psi}{\partial z} = \frac{\bar{c}S_y}{6}z^3 + \frac{\bar{a}\eta_y}{2}z^2 + C_1 z + C_2, \tag{11.130}$$

where the notations $S_y = \partial S/\partial y$, $\eta_y = \partial \eta/\partial y$, and the $\bar{c}$ and $\bar{a}$ coefficients are expressed by $\bar{c} = \beta g/N_z$, and $\bar{a} = -g/N_z$, and the integration constants C_1 and C_2, with dimensions $[C_1] = [T^{-1}]$ and $[C_2] = [LT^{-1}]$, must be determined with the following surface and bottom boundary conditions: wind stress (τ_{Wy}) and the maximum friction at the bottom:

$$\rho N_z \frac{\partial v}{\partial z}\Big|_{z=0} = \tau_{Wy}, \tag{11.131a}$$

and

$$v(y, z)\big|_{z=H_0} = 0. \tag{11.131b}$$

Others solutions may be obtained simulating different bottom conditions as, for example, a moderate (slippery) bottom friction:

$$\rho N_z \frac{\partial v}{\partial z}\Big|_{z=H_0} = \tau_{By}. \tag{11.132}$$

Applying the surface and boundary conditions (11.131a), (11.131b) the following values of the integration constants are obtained:

$$C_1 = \frac{\tau_{Wx}}{\rho N_z}, \tag{11.133a}$$

and

$$C_2 = -\frac{1}{6}\bar{c}S_y H_0^3 + \frac{1}{2}\bar{a}\eta_y H_0^2. \tag{11.133b}$$

Combining these constants with the general solution (11.130) and simplifying the resulting expression, the solution for the v-velocity component is:

$$v(y, z) = \frac{1}{6}\bar{c}S_y(z^3 - H_0^3) + \frac{1}{2}\bar{a}\eta_y(z^2 - H_0^2) + \frac{\tau_{Wy}}{\rho N_z}z, \quad (11.134a)$$

or in function of the non-dimensional depth Z,

$$v(y, Z) = \frac{H_0^3}{6}\bar{c}S_y(Z^3 - 1) + \frac{H_0^2}{2}\bar{a}\eta_y(Z^2 - 1) + \frac{\tau_{Wy}H_0}{\rho N_z}Z. \quad (11.134b)$$

This solution isn't in the most convenient formulation for practical applications, because it contains the free surface slope (η_y) as an unknown, which may be calculated with the imposition of an integral boundary condition, that turns the transversal volume transport to zero:

$$\iint_A v(y, z)dydz = 0, \quad (11.135a)$$

or

$$\int_0^b [\int_0^{H_0} v(y, z)dz]dy = B\int_0^{H_0} v(y, z)dz = 0, \quad (11.135b)$$

As the integrand v = v(y, z) is already known (Eq. 11.134a) the integration may be concluded,

$$-\frac{1}{8}\bar{c}S_yH_0^3 - \frac{1}{3}\bar{a}\eta_yH_0^3 + \frac{\tau_{Wy}}{2N_z\rho}H_0 = 0. \quad (11.136)$$

Solving this result for the unknown, η_y, and using the expressions $\bar{a} = -g/N_z$, and $\bar{c} = \beta g/N_z$, we have:

$$\eta_y = \frac{\partial\eta}{\partial y} = -\frac{3}{8}\beta S_yH_0 - \frac{3\tau_{Wy}}{\rho g H_0}, \quad (11.137a)$$

or, neglecting the wind stress ($\tau_{Wy} = 0$),

$$\eta_y = \frac{\partial\eta}{\partial y} = -\frac{3}{8}\frac{H_0}{\rho_0}\frac{\partial\rho}{\partial y} = -0.375\frac{H_0}{\rho_0}\frac{\partial\rho}{\partial y}. \quad (11.137b)$$

This result is similar to the longitudinal component of the free surface slope of Eq. 10.19 (Chap. 10) with $u_f = 0$ and $\tau_{Wx} = 0$, and the transverse slope of the free surface is directly proportional to the corresponding density gradient, but with the opposite signal (<0), because $H_o > 0$ and $\partial\rho/\partial y > 0$. Thus, if η_y is known it may be substitute into Eq. (11.134a). Simplifying the result and writing the solution in terms of the non-dimensional depth and the transverse density gradient, it follows that the equation to calculate the v-velocity component according to (Nunes and Simpson 1985) is:

$$v(y, Z) = \frac{gH_0^3}{A_z}\frac{\partial \rho}{\partial y}(0.167Z^3 - 0.188Z^2 + 0.021). \quad (11.138)$$

where $\rho_o N_z = A_z$. This result indicates that the direction and intensity v-velocity component is directly dependent on the local depth (H_0) and the transverse density gradient ($\partial\rho/\partial y$), and is inversely proportional to the vertical dynamical eddy viscosity coefficient. Therefore, transverse intensity variations or orientation changes in the density gradient may generate convergence or divergence of the velocity field. When $\partial\rho/\partial y > 0$, the steady-state free surface slope (Eq. 11.137b) is negative ($\partial\eta/\partial y < 0$), and velocity of the secondary circulation in the surface is positive and oriented in the direction of increasing density. If there is a change in the density gradient ($\partial\rho/\partial y < 0$) at a given depth along the transverse section, the secondary circulation has its direction inverted.

At the free surface ($Z = 0$), the solution (11.138) for the v-velocity component is reduced to:

$$v(y, 0) = 2.1 \times 10^{-2}\frac{gH_0^3}{A_z}\frac{\partial \rho}{\partial y}. \quad (11.139)$$

Changes in the transverse density gradient, from $\partial\rho/\partial y > 0$ to $\partial\rho/\partial y < 0$ in the well-mixed Conway estuary (North Wales, Scotch) were successfully used by Nunes and Simpson (1985) to theoretically explain the visible accumulation of organic and inorganic matter and debris along axial convergence lines.

The vertical velocity profile of the v-velocity component, calculated with Eq. (11.138) is presented in Fig. 11.9a to exemplify the bilateral divergence of the velocity field, and was used to calculate the ascending vertical velocity component ($w < 0$). These profiles were calculated for different values of the transverse density gradient, $\partial\rho/\partial y = 2.5 \times 10^{-2}$ kg m^{-4} and $\partial\rho/\partial y = 1.5 \times 10^{-2}$ kg m^{-4}, in water columns separated by a distance of 200 m. For the others quantities, the following values were used: $H_0 = 10.0$ m, $N_z = 1.0 \times 10^{-2}$ m^2 s^{-1}, $\rho_0 = 1005.0$ kg m^{-3}, and $A_z \approx 10$ kg m^{-1}s^{-1}.

Once the transverse vertical velocity profile has been calculated, the profile of the vertical velocity component $w = w(y, Z)$, generated by the convergence (divergence) of the v-velocity field, may be calculated using the continuity equation solved by finite increments,

$$w(y, Z) = \frac{\partial \psi}{\partial y} = -H_0 \int_0^1 \frac{\Delta v}{\Delta y} dZ, \quad (11.140)$$

with the following boundary conditions: $w(y, 0) = w(y, 1) = 0$. The vertical velocity profile, $w = w(y, Z)$, calculated by finite increments is presented in

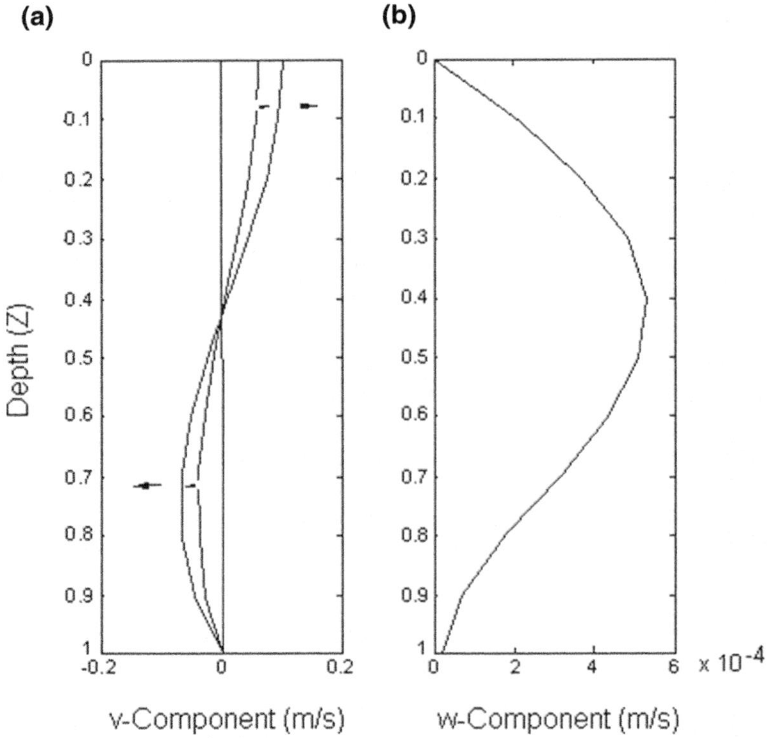

Fig. 11.9 Steady-state vertical v-velocity profiles (**a**) and w-velocity profiles (**b**) calculated with Eqs. (11.138) and (11.140), respectively, to demonstrate the velocity field divergence (indicated by the *arrows*) generating descending motion (w > 0)

Figure 11.9b, indicating a flow towards the bottom (w > 0). Comparing the magnitudes of the v- and w-velocity components, we may observe that $|w| \ll |v|$ and that the highest value of w (5.5×10^{-4} m s^{-1}) is at $Z \approx 0.4$.

Let us now apply the general solution (11.130) imposing the moderate bottom boundary condition (Eq. 11.132) and the no-wind stress ($\tau_{Wx} = 0$) will remain for the upper boundary condition. Let us assume, according to Prandle (1985), that the vertical kinematic eddy viscosity coefficient is given by the relation $N_z = kU_TH_o$, with the non-dimensional coefficient k equal to 2.5×10^{-3}. With these boundary conditions, $C_1 = 0$ and

$$C_2 = \frac{1}{3}\bar{c}S_yH_0^3 + \frac{1}{2}\bar{a}\eta_yH_0^2. \tag{11.141}$$

Substituting these values of the integration constants into the general solution (11.130) and reducing it to the simplest analytical expression, we have the following solution, which is similar to (11.134a):

$$v(y, z) = \frac{1}{6}\bar{c}S_y(z^3 - 2H_0^3) + \frac{1}{2}\bar{a}\eta_y(z^2 - H_0^2), \tag{11.142a}$$

or in function of the non-dimensional depth Z,

$$v(y, Z) = \frac{H_0^3}{6}\bar{c}S_y(Z^3 - 2) + \frac{H_0^2}{2}\bar{a}\eta_y(Z^2 - 1). \tag{11.142b}$$

As noted previously, this solution isn't in the most convenient formulation for practical applications, because it contains the free surface slope (η_y) as an unknown, which may be calculated with the imposition of the integral boundary condition (11.135a), and the result is

$$\frac{\partial \eta}{\partial y} = \eta_y = \frac{9}{16}\frac{H_0}{\rho_0}\frac{\partial \rho}{\partial y} = 0.562\frac{H_0}{\rho_0}\frac{\partial \rho}{\partial y}, \tag{11.143}$$

which is similar to the solution (11.137b), but with a different numeric coefficient. Substituting this result into Eq. (11.142b), simplifying the result and introducing the non-dimensional depth ($Z = z/H_0$), an alternative expression to calculate the transverse vertical velocity profile as a function of the density gradient is

$$v(y, Z) = \frac{gH_0^3}{A_z}\frac{\partial \rho}{\partial y}(0.167Z^3 - 0.281Z^2 + 0.052), \tag{11.144a}$$

or using the relation $N_z = A_z/\rho_0 = kU_TH_0$ ($k = 2.5 \times 10^{-3}$)

$$v(y, Z) = \frac{gH_0^2}{kU_T}\frac{1}{\rho_0}\frac{\partial \rho}{\partial y}(0.167Z^3 - 0.281Z^2 + 0.052). \tag{11.144b}$$

To close the circulation field the w-velocity component generated by the convergence or divergence of the $v = v(y, Z)$ component may be calculated.

Experimental and theoretical investigations of the convergence and divergence of secondary circulation in well-mixed and laterally homogeneous estuaries may be found article of Nunes and Simpson (1985).

The simple steady-state analytic model just described, don't take into account transverse bottom variations ($\partial H_0/\partial y \neq 0$ or $\partial h/\partial y \neq 0$), and has the lateral density gradients as the main forcing mechanism to drive the secondary circulation and the related axial convergences. Studies of the non-steady-state analytic model of lateral flow convergences, arising from the interaction of tidal flow with the estuary bathymetry are described by Li and Valle-Levinson (1999) and Valle-Levinson et al. (2000). The models are based on the solution of the depth-averaged, first order equations for momentum balance, forced by a single-frequency semidiurnal tide at the estuary mouth. In this model, the energy dissipation is simulated by a linear friction coefficient, κ, with dimension of velocity $[\kappa] = [LT^{-1}]$.

Considering a channel with straight parallel boundaries, with the Ox axis extended along the right estuary margin looking into the estuary, and the Oy axis along the open boundary, and with a laterally variable depth distribution, the depth-averaged first-order equation for momentum balance is:

$$\frac{\partial u}{\partial t} = -g\frac{\partial \eta}{\partial x} - \frac{\kappa}{h}u. \quad (11.145a)$$

The second term on the right hand side was introduced to simulate the energy dissipation by assuming a linear frictional dependence, which is directly and inversely proportional to the velocity u and depth, h, respectively. To the hydrodynamic system closure, the continuity Eq. (7.66a, Chap. 7) is

$$h\frac{\partial u}{\partial x} + \frac{\partial (vh)}{\partial y} + \frac{\partial \eta}{\partial t} = 0. \quad (11.145b)$$

The Equations (11.145a), (11.145b) are formally derived from the Navier-Stokes equations by integrating these equations over a rectangular cross section, and are similar to the one-dimensional formulation used by Blumberg (1975) in studies of the wave propagation in a uniform channel.

For a single-frequency co-oscillating tide the solution for this equation system can be expressed in terms of the complex exponential number, expressed by (Li and Valle-Levinson 1999, and Valle-Levinson et al. 2000) as:

$$u = Ue^{i\sigma t}, \quad v = Ve^{i\sigma t} \quad \text{and} \quad \eta = Ae^{i\sigma t}. \quad (11.146)$$

In these formulations U and V are the complex amplitudes of the longitudinal and the secondary circulation velocity (m s^{-1}), and A is the amplitude (m) of the complex tidal elevation, σ is the tidal angular frequency (s^{-1}), and i is the imaginary number unit ($\sqrt{-i}$). Substituting (11.146) into (11.145a), (11.145b) yields:

$$i\sigma U = -g\frac{\partial A}{\partial x} - \frac{\kappa}{h}U, \quad (11.147a)$$

and

$$i\sigma A + h\frac{\partial U}{\partial x} + \frac{\partial (uV)}{\partial y} = 0. \quad (11.147b)$$

Under the assumption that the co-oscillating tidal amplitude is known, and imposing the following boundary conditions: (i) at the estuary head (x = L) the velocity is U; and (ii) at the lateral side boundaries (y = 0, D), the velocity (V) vanishes. Thus, applying these conditions to Eq. (11.147a) yields the solution for the U-velocity component.

$$U = -\frac{g}{i\sigma + \frac{\kappa}{h}}(\frac{\partial A}{\partial x}). \tag{11.148}$$

In this equation, the tidal elevation amplitude (A) may be treated as independent of the transverse direction (y), as demonstrated by Li (1996); Li and O'Donnell (1997) and Li et al. (1998). This assumption led to a dramatic simplification of the problem, and the solution of A may be expressed by (Li and Valle-Levinson 1999):

$$A = \eta_0 \frac{\cos[\omega(x - L)]}{\cos(\omega L)}, \tag{11.149}$$

where η_0 is the tidal amplitude. The longitudinal (U) and transverse (V) velocities components, and the angular frequency (ω) are calculated by,

$$U = \frac{g}{i\sigma + \kappa/h} \frac{\eta_0 \omega}{\cos([\omega(x - L)]} \sin[\omega(x - L)], \tag{11.150a}$$

$$V = -\frac{A}{h}\{i\sigma y + \int_0^y [\frac{gh\omega^2}{(i\sigma + \beta/h)}]dy\}, \tag{11.150b}$$

and

$$\omega^2 = \frac{i\sigma B}{-\int_0^B (\frac{gh}{i\sigma + \beta/h})dy}. \tag{11.150c}$$

As pointed out by Li and Valle-Levinson (1999), the transverse velocity (V) is insensitive to the transverse momentum balance, and may be obtained from the continuity Eq. (11.145b).

This non-steady-state analytic solution was applied by Li and Valle-Levinson (1999) and Valle-Levinson et al. (op. cit.) in two transects of the James river estuary (Virginia, USA), and were compared with observational ADCP measurements. The analytical model reproduced well the timing and location of the convergences in agreement with the experimental results. The mechanisms which generates convergences due to the laterally bottom variation, during flood and ebb conditions are schematically presented in Fig. 11.10.

In addition to the cross-channel the baroclinic pressure gradient force and the interaction of tidal flow with the estuary bathymetry, a detailed analysis of the following mechanisms that can drive secondary flows have been presented by Chant (2010): (i) Ekman forcing characterized by the balance between friction and the Coriolis acceleration; (ii) flow curvature, which has long been recognized to drive a helical flow normal to the stream-wise flow; and (iii) diffusive boundary layers.

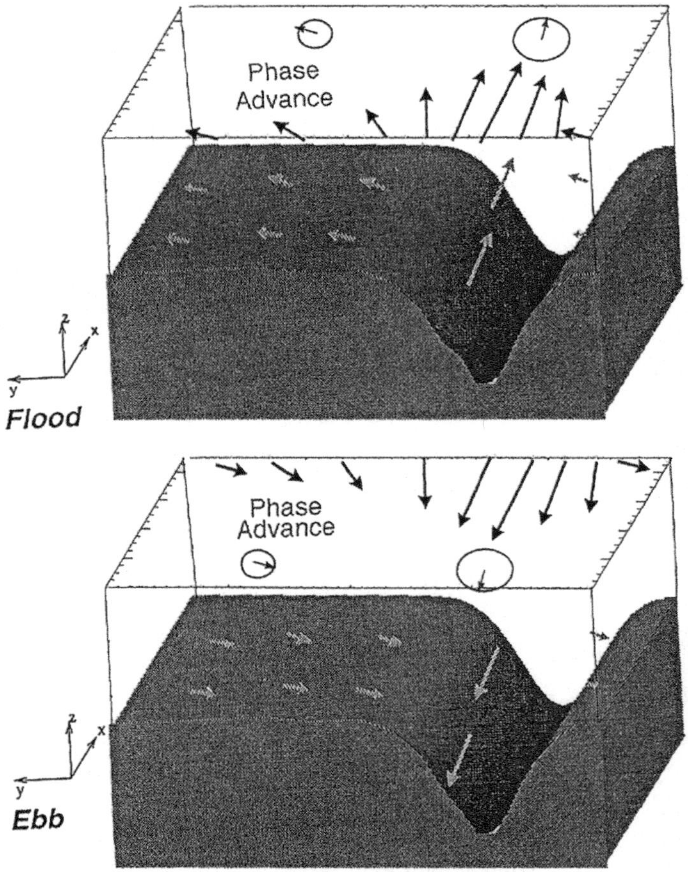

Fig. 11.10 Mechanisms generating bottom induced convergences for flood and ebb conditions are schematically represented. The *upper* and *lower* diagrams are looking into the estuary and towards its mouth, respectively. The ellipticity of the tidal motion is greatly exaggerated to illustrate the transverse flows. According to Valle-Levinson et al. (2000)

References

Andutta, F. P.; Miranda, L. B., Castro, B. M. & Fontes, R. F. C. 2006. Numerical Simulation of the Hydrodynamic in the Curimataú Estuary, RN Brazil. Oceanography and Global Changes, SP-Brazil, pp. 545–558.

Bernardes, M.E.C. 2001. Circulação Estacionária e Estratificação de Sal em Canais Estuarinos parcialmente Misturados. Dissertação de Mestrado. São Paulo, Instituto Oceanográfico, Universidade de São Paulo. 202 p.

Bernardes, M.E.C. & Miranda, L.B. 2001. Circulação Estacionária e Estratificação de Sal em Canais Estuarinos: Simulação com Modelos Analíticos. Rev. bras. oceanogr., 49(1/2):115–132. (DOC 39).

Blumberg, A. F. 1975. A Numerical Investigation into the Dynamics of Estuarine Circulation. Tech. Rept. Chesapeake Bay Institute, The Jonhs Hopkins University. n. 91. 110 p. + Apêndices.

Bowden, K. F. 1953. Note on Wind Drift in a Channel in the Presence of Tidal Currents. Proc. R. Soc. Lond., A219, pp. 426–446.

Chant, R.J. 2010. Estuary secondary circulation. In: ed. Valle-Levinson A. Contemporary Issues in Estuarine Physics. Cambridge University Press, pp. 100–124.

Döbereiner, C.E. 1985. Comportamento hidráulico e sedimentológico do estuário do rio Itajaí, SC. Rio de Janeiro, Instituto Nacional de Pesquisas Hidroviárias (INPH), Relatório 700/03, 34 p. (quoted in Schettini (2002), p. 132).

Dyer, K. R. 1973. Estuaries: A Physical Introduction. London, Wiley. 140 p.

Dyer, K. R. 1977. Lateral Circulation Effects in Estuaries. Estuaries, Geophysics and the Environment. Washington, D. C., National Academy of Sciences, pp. 22–29.

Fisher, J. S.; Ditmars, J. D. & Ippen, A. T. 1972. Mathematical Simulation of Tidal Time Averages of Salinity and Velocity Profiles in Estuaries. Massachusetts Institute of Technology, Mass., Rept. MITSG 72–11, 157 p.

Hamilton, P. & Rattray Jr., M. 1978. Theoretical Aspects of Estuarine Circulation. In: Kjerfve B. (ed.). Estuarine Transport Processes. Columbia, Univesity of South Carolina, pp. 37–73. (Belle W. Baruch Library in Marine Science, 7).

Hamilton, A. D. & Wilson, R. E. 1980. Nontidal Circulation and Mixing Processes in the Lower Potomac Estuary. Estuaries, 3(1):11–19.

Hansen, D. V. & Rattray Jr., M. 1965. Gravitational Circulation in Sraits and Estuaries. J. Mar. Res., 23(1):102–122.

Hansen, D. V. & Rattray Jr., M. 1966. New Dimensions in Estuary Classification. Limnol. Oceanogr., 11(3):319–325.

Ippen, A. T. & Harleman, D. R. F. 1961. One-Dimensional Analysis of Salinity Intrusion in Estuaries. Committee on Tidal Hydraulics. Tech. Bull. Corps of Engineers U. S. Army, n. 5. 120 p.

Kjerfve, B.; Miranda, L. B. & Wolanski, E. 1991. Modelling Water Circulation in an Estuary and Intertidal Salt Marsh System. Neth. J. Sea Res., 28(3):141–147.

Li, C.; O'Donnell, J.; Valle-Levinson, A.; Li, H.; Wong, K-C & Lwiza, K.M.M., 1998. In: Tide induced mass-flux in shallow estuaries. in Ocean Waves Measurement and Analysis, (eds.) B. L. Edge and J.M. Hemsley,. Am. Soc. Civ. Eng., Reston, VA. of, v. 2, pp. 1510–1524.

Li, C. 1996. Tidally induced residual circulation with cross shelf bathymetry. Ph D. Dissertation, 242 pp. Univ. of Conn Storrs.

Li, C. & Valle-Levinson A. 1999. A two-dimensional analytical model for a narrow estuary of arbitrary lateral depth variation. The intra-tidal motion. J. Geophys. Res. 104, pp. 23,525–23,543.

Li, C. & O'Donnell, J. 1997. Tidally driven residual circulation in shallow estuaries with lateral depth variations. J. Geophys. Res. V. 102, pp. 27,915–27,1929.

Miranda, L. B.; Dalle Olle, E.; Bérgamo, A.L.; Silva, L.S. & Andutta, F.P. 2012. Circulation and salt intrusion in the Piaçaguera Channel, Santos (SP). Braz. J. Oceanography, 60(1):11–23.

Nunes, R. A. & Simpson, J. H. 1985. Axial Convergence in a Well-Mixed Estuary. Estuar. Coast. Shelf Sci., 20:637–649.

Ong, J. -E.; Gong, W. -K. & Uncles, R. J. 1994. Transverse Structure of Semi-diurnal Currents Over a Cross-section of the Merbok Estuary, Malaysia. Estuar. Coast. Shelf Sci., 38:283–290.

Officer, C. B. 1976. Physical Oceanography of Estuaries (and Associated Coastal Waters). New York, Wiley. 465 p.

Officer, C. B. 1977. Longitudinal Circulation and Mixing Relations in Estuaries. Estuaries, Geophysics, and the Environment. Washington, D. C., National Academy of Sciences, pp. 13–21.

Officer, C. B. 1978. Some Simplified Tidal Mixing and Circulation Flux Effects in Estuaries. In: Kjerfve, B. (ed.). Estuarine Transport Processes. Columbia, University of South Carolina Press, pp. 75–93. (The Belle W. Baruch Library in Marine Science, 7).

Prandle, D. 1985. On Salinity Regimes and the Vertical Structure of Residual Flows in Narrow Tidal Estuaries. Estuar. Coast. Shelf Sci., 20:615–635.

Prandle, D. 2004. Saline Intrusion in Partially Mixed Estuaries. Est. Coast. Shelf Sci. (59):385–397.

Pritchard, D. W. 1952a. Salinity Distribution and Circulation in the Chesapeake Bay Estuarine System. J. Mar. Res., 11(1):106–123.

Pritchard, D. W. 1954. A Study of Salt Balance in a Coastal Plain Estuary. J. Mar. Res., 13 (1):133 144.

Pritchard, D. W. 1956. The Dynamic Structure of a Coastal Plain Estuary. J. Mar. Res., 15(1):33–42.

Rattray Jr., M. & Hansen, D. V. 1962. A Similarity Solution for Circulation in an Estuary. J. Mar. Res., 20(2):121–133.

Rigter, B.P. 1973. Minimum Length of Salt Intrusion in Estuaries. Proceedings of the America Society of Civil Engineers. Journal of Hydraulics Division, 99 (HY9):1475–1496.

Schettini, C. A. F. & Truccolo. E. C. 1999. Dinâmica da Intrusão Salina no Estuário do Rio Itajaí-açu. In: Congresso Latino Americano de Ciências do Mar, 8, Trujillo, Peru, Resumenes ampliados, Tomo II, UNT/ALICMAR, p. 639–640.

Schettini, C. A. F. 2002. Caracterização Física do Estuário do Rio Itajaí-açu, SC. Revista Brasileira Recursos Hídricos, 7(1):123–142.

Sumer, S.M & Fischer, H.B. 1977. Transverse Mixing in Partially Stratified Flow. Proceedings of the American Society of Civil Engineers. Vol 103, No. HY6, pp. 587–600.

Valle-Levinson, A.; Li, C.; Wong, K-C & Lwiza, K.M.M. 2000. Convergence of lateral flow along a coastal plain estuary. J. Geophys. Res., v. 105, NO C7: pp. 17045–17061.

Walstra, D., Sutherland, J., Hall, L., Blogg, H., and van Ormondt, M. (2001) Verification and Comparison of Two Hydrodynamic Area Models for an Inlet System. Coastal Dynamics' 01: pp. 433–442. doi:10.1061/40566(260)44.

Wilmott, C. J. 1981. On the Validation Models. Physical Geography, 2 (2), pp. 184–194.

Warner, J.C.; Geyer, W.R.; Lerczak, J.A. 2005. Numerical modeling of an estuary: A comprehensive skill assessment. J. Geophys. Res. V.110(CO5001), p. 1–13.

Wylie, C. R. 1960. Advanced Engineering Mathematics. New York, McGraw-Hill. 696 p.

Chapter 12
Numerical Hydrodynamic Modelling

As estuaries are three dimensional and time dependent, numerical models have been developed to overcome the simplifications inherent to the already studied analytical models (simple geometry, steady-state) and calculate estuarine circulation and salinity distributions. These models can be numerically integrated at selected grid points spatially distributed in the system domain; the governing partial differential equations use methods of finite-difference or finite-elements in curvilinear horizontal coordinates or sigma vertical coordinates, respectively.

Numerical models have been developed and published since the end of the 1960s. This method involves replacing the differential partial equations with equivalent finite difference algebraic equations, which are solved numerically. Applications of the solutions of time-dependent numerical equations, allowing the probable distribution of pollutants in coastal regions and estuaries to be determined, have become increasingly more important with the increase in speed and memory capacity of computers.

A technical review and critical appraisal of various aspects of numerical modeling techniques of estuaries were made by a selection of eminent scientists and engineers in the field, and these essays were supplemented by discussions at technical conferences held during the course of the report's preparation, edited by George H. Ward Jr. and William H. Espey Jr., published in early 1971 (Ward and Espey 1971). Topics discussed included one-, two-, and three-dimensional mathematical models for estuarine hydrodynamics, water quality models of chemical constituents (nitrogen forms) and biological (phytoplankton and zooplankton), and estuarine temperature field related to modeling thermal discharges, and principles and applicability of physical models in estuarine analysis. This report also included a review of solution techniques, with a detailed discussion of finite-difference methods. Scientists who took part in these discussions were D.W. Pritchard, D.R.F. Harleman, M. Rattray Jr, D.J. O'Connor, R.V. Thomann, J.E. Edinger, A.T. Ippen, G.J., Paulik, J.J. Lendertsee, J.A. Harder.

L. Bruner de Miranda et al., *Fundamentals of Estuarine Physical Oceanography*,
Ocean Engineering & Oceanography 8, DOI 10.1007/978-981-10-3041-3_12

In the finite-difference method, the fluid domain is subdivided into arbitrary elements, for instance, a triangular grid format, enabling better adjustment to the geometric space of the fluid domain, even in complex boundary configurations. Also, as noted by Lendertsee and Critton (1971) (Lendertsee et al. (1973)), many different approximations can generally be made for each term of the partial equations to be solved, resulting in a wide selection of approximations. One approximation may be considered better than another, depending on the scope of the investigation and the processes described by the equation for a particular situation. Useful considerations in the design of numerical computation schemes may also be found in Lendertsee's article.

As with other numerical investigations of natural phenomena, using a numerical model to simulate an estuarine system only works when the modeler fully understands the model's limitations and the physical processes involved, and conducts adequate calibration and validation. The complexity of estuaries often requires a grid that will result in a scientific credible, yet computationally feasible model. The grid should provide a compromise between depicting the physical realities of the estuarine system and the computational feasibility. As estuarine channels have irregular shore-lines, islands and shipping channels, numerical models require very small grid sizes to resolve these boundaries in detail; in these environments, curvilinear grids provide a better representation (Ji 2008). Also, as previously seen, estuarine circulation is driven by tides, river discharge, baroclinic pressure gradient force and wind, forming a very complex tri-dimensional system. As such, specification of the open boundary conditions that link the estuarine water mass to the river, coastal sea, bottom and atmosphere, is also required.

12.1 Briefy Outline on Numerical Models

The Princeton Ocean Model (POM) was originally developed at Princeton University by G. Mellor and A.F. Blumberg in collaboration with Dybalysis of Princeton (H.J. Herring and R.C. Patchen). The model incorporates the Mellor-Yamada turbulence scheme developed in early 1970 by George Mellor and Ted Yamada, widely used by oceanic and atmospheric models. At the time, early computer ocean models such as the Bryan–Cox model, which was developed in the late 1960s at the Geophysical Fluid Dynamics Laboratory and later became the Modular Ocean Model (MOM), were mostly aimed at coarse-resolution simulations of the large-scale ocean circulation. Thus, there was a need for a numerical model that could handle high-resolution coastal ocean processes.

The Blumberg–Mellor model (which later became POM) included new features such as free surface to handle tides, sigma vertical coordinates (i.e., terrain-following) to handle complex topographies and shallow regions, a curvilinear grid to better handle coastlines, and a turbulence scheme to handle vertical mixing. In the early 1980s, the model was primarily used to simulate estuaries such as the Hudson–Raritan Estuary (by Leo Oey) and the Delaware Bay (Boris

Galperin). At this time, attempts had also been made to use a sigma coordinate model for basin-scale problems, with the coarse resolution model of the Gulf of Mexico (Blumberg and Mellor) and models of the Arctic Ocean (with the inclusion of ice-ocean coupling by Lakshmi Kantha and Sirpa Hakkinen).

In the late 1990s and the 2000s, many other terrain-following community ocean models were developed; some of their features can be traced back to those included in the original POM, while other features are additional numerical and parameterization improvements. Several ocean models are direct descendents of POM, for example, the commercial version of POM known as the Estuarine and Coastal Ocean Model (ECOM), the Navy Coastal Ocean Model (NCOM) and the Finite-Volume Coastal Ocean Model (FVCOM).

The Delft Hydraulics MOR module of Delft 3-D Flow fully integrates the effects of waves, currents and sediment transport for morphological development (e.g. see Nicholson et al. 1997). The module simulates the processes on the same curvilinear grid system as used in the flow module, which allows a very efficient and accurate representation of complex areas. This module is a multi-dimensional (2D or 3D) hydrodynamic and transport simulation program which calculates the non-steady state circulation and transport phenomena resulting from tides, river discharge and meteorological forces due to wind-stress.

The Danish Hydraulic Institute (DHI) developed the version MIKE21 modeling package. This advanced software employs state-of-the-art computer simulation techniques to model hurricanes and associated storm surge and waves, and hydrodynamic processes in coastal and estuarine waters, water quality, sediment transport processes and morphological changes.

The Environmental Fluid Dynamics Code (EFDC) (Hamrick 1992) is a public-domain modeling package for simulating three-dimensional (3D) flow, transport and biogeochemical processes in rivers, lakes, estuaries, reservoirs, wetlands and coastal regions. This code was originally developed at the Virginia Institute of Marine Sciences and is currently supported by the U.S. Environmental Protection Agency (EPA). The EFDC model has been extensively tested and documented in more than 100 modeling studies, and is presently being used by universities, research organizations, governmental agencies, and consulting firms. This advanced 3D time-variable model provides the capability of internally linking hydrodynamics, water quality and eutrophication, sediment transport and toxic chemical transport and fate sub-models in a single source code framework. It includes four major modules: (i) hydrodynamics; (ii) water quality; (iii) sediment transport; and, (iv) and toxics substances. The full integration of the four modules is unique and eliminates the need for complex interfacing of multiple modes to address different processes. Representative applications of the EFDC model may be found in Ji (2008).

In this chapter, only simple problems related to the discretization of solutions of the two-dimensional equation of motion will be treated, and some case studies are presented to demonstrate how hydrodynamic modeling and validation can be applied to practical problems of estuarine circulation, tide oscillations and salinity distributions using the Delft 3-D Flow numeric model.

12.2 The Finite Difference Method

The basic formulation for establishing an expression of a differential partial equation using the method of finite differences may be obtained from the Taylor series expansion, as exemplified in Fig. 12.1, with a simple bi-dimensional (2D) rectangular grid in the Oxy plane. In this example, the indexes i and j denote positions along the Ox and Oy axes, respectively, and Δx_i and Δy_j, denote finite increments along axes directions. To solve a third dimension, such as depth, the Oz axis normal to the Oxy plane must also be specified, and the index k may be used to denote the position (z_k) and finite intervals (Δz_k) along this axis. For a non-steady-state problem, the discrete time intervals may be referred by the index n, for instance, preferentially as a superscript of the symbol denoting the function or variable (f^n). Some basic principles will be presented to establish a formulation for finite difference equations for the dynamics of an estuary, following classical books and articles of Lendertsee and Criton (1971), Lendertse et al. (1973), Blumberg (1975) and Roache (1982).

As indicated in Fig. 12.1, the grid spacing in the directions i and j are indicated by $\Delta x = x_{i+1} - x_{i,j}$ and $\Delta y = y_{i,j+1} - y_{i,j}$ and, for the Oz direction, $\Delta z = z_{i,k+1} - z_{i,k}$; for convenience, the intervals Δx, Δy and Δz, which define the elemental volumes ($\Delta x \,.\, \Delta y \,.\, \Delta z$), are considered constant, unless indicated otherwise. Discrete time intervals will be indicated by $\Delta t = t_{n+1} - t_n$.

The symbol f = f(x, y, z, t) (or in two dimensions f = f(x, y, t)) will be used to denote a continuous function in space and time, with the corresponding discrete functions, f = f(i, j, k, t) or f = f(i, j, t), in the tri- and bi-dimensional space; the governing differential equations are replaced with finite difference equations that operate only on the grid positions. If L, B and H are the estuary's length, width and depth, which are subdivided into the positions n, m and k, respectively, the following relations exist according to the discrete format: $L/n = \Delta x_i \rightarrow L = \sum_{i=1}^{i=n} \Delta x_i$

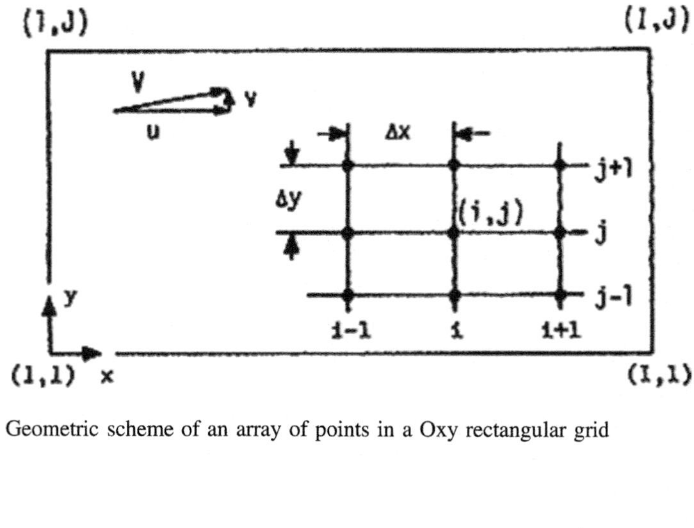

Fig. 12.1 Geometric scheme of an array of points in a Oxy rectangular grid

$B/m = \Delta y_i \rightarrow B = \sum_{i=1}^{i=m} \Delta y_i$, and $H/k = \Delta k_i \rightarrow H = \sum_{i=1}^{i=k} \Delta k_i$, respectively. In the same way, the discrete intervals of the time domain (T) are $T/k = \Delta t_j$ and $T = \sum_{j=0}^{k-1} \Delta t_j$.

Consider a bi-dimensional space and a continuous function, $f = f(x, y, t)$, with continuous higher orders derivatives. The first order derivative, $\partial f/\partial x$, may be deduced by the Taylor series expansion. Then, considering a known continuous function in the space point (i, j), we may write:

$$f_{i+1,j} = f_{i,j} + \frac{\partial f}{\partial x}\Big|_{i,j}(x_{i+1,j} - x_{i,j}) + \frac{1}{2}\frac{\partial^2 f}{\partial x^2}\Big|_{i,j}(x_{i+1,j} - x_{i,j})^2 + \cdots + \text{SOT}, \quad (12.1a)$$

or

$$f_{i+1,j} = f_{i,j} + \frac{\partial f}{\partial x}\Big|_{i,j}\Delta x + \frac{1}{2}\frac{\partial^2 f}{\partial x^2}(\Delta x)^2 + \cdots + \text{SOT}, \quad (12.1b)$$

where SOT indicates "*superior order terms*". Solving Eq. (12.1b) for the partial derivative, $\partial f/\partial x$, gives:

$$\frac{\partial f}{\partial x}\Big|_{i,j} = \frac{1}{\Delta x}(f_{i+1,j} - f_{i,j}) - \frac{1}{2}\frac{\partial^2 f}{\partial x^2}\Big|_{i,j}(\Delta x)^2 + \cdots + \text{SOT}, \quad (12.2)$$

and the last expression may be written as:

$$\frac{\partial f}{\partial x}\Big|_{i,j} = \frac{1}{\Delta x}(f_{i,+1,j} - f_{i,j}) + O(\Delta x), \quad (12.3)$$

where $O(\Delta x)$ indicates an approximation error. This approximated expression of this derivative ($\partial f/\partial x$) will be denoted by $\delta f/\delta x$, or simplified to $\delta_x f$; it is a *forward approximation*, with the i index increasing in the Ox direction:

$$\frac{\partial f}{\partial x}\Big|_{i,j} = \delta_x f\big|_{i,j} = \frac{1}{\Delta x}(f_{i+1,j} - f_{i,j}) + O(\Delta x). \quad (12.4)$$

The finite difference equations are generally classified according to the lower power of truncation. In Eq. (12.4) there is a first order error, and the equation is named a *first order equation*. Of course, second and third order approximations are better than first order approximations.

With an analogous procedure, but expanding the function $f_{i,j}$ backwards in order to obtain the expression for $f_{i-1,j}$, we have another finite difference approximation equivalent to (12.4). In this case, the first order approximation is given by:

$$\frac{\partial f}{\partial x}\Big|_{i,j} = \delta_x f\big|_{i,j} = \frac{1}{\Delta x}(f_{i,j} - f_{i-1,j}) + O(\Delta x). \quad (12.5)$$

The known *central finite difference* is obtained by subtracting the *backward* from the *forward* expansion; considering for these expressions the third order approximation which is expressed by:

$$f_{i+1,j} = f_{i,j} + \frac{\partial f}{\partial x}\Big|_{i,j}\Delta x + \frac{1}{2}\frac{\partial^2 f}{\partial x^2}\Big|_{i,j}(\Delta x)^2 + \frac{1}{6}\frac{\partial^3 f}{\partial x^3}\Big|_{i,j}(\Delta x)^3 + \cdots + \text{SOT}, \quad (12.6)$$

and

$$f_{i-1,j} = f_{i,j} - \frac{\partial f}{\partial x}\Big|_{i,j}\Delta x - \frac{1}{2}\frac{\partial^2 f}{\partial x^2}\Big|_{i,j}(\Delta x)^2 - \frac{1}{6}\frac{\partial^3 f}{\partial x^3}\Big|_{i,j}(\Delta x)^3 + \cdots + \text{SOT}. \quad (12.7)$$

Subtracting (12.7) from (12.6) yields:

$$f_{i+1,j} - f_{i-1,j} = 2\frac{\partial f}{\partial x}\Big|_{i,j}\Delta x + \frac{1}{3}\frac{\partial^3 f}{\partial x^3}\Big|_{i,j}(\Delta x)^3 + \cdots + \text{SOT}. \quad (12.8)$$

and solving the last expression for $\partial f/\partial x$,

$$\frac{\partial f}{\partial x}\Big|_{i,j} = \frac{1}{2}(f_{i+1,j} - f_{i-1,j})\Delta x - \frac{1}{6}\frac{\partial^3 f}{\partial x^3}\Big|_{i,j}(\Delta x)^3 + \cdots + \text{SOT}, \quad (12.9a)$$

and using the notation $\delta_x f$,

$$\delta_x f\big|_{i,j} = \frac{1}{2}(f_{i+1,j} - f_{i-1,j})\Delta x + O(\Delta x^2). \quad (12.9b)$$

From this last expression, it follows that the second order approximation of the partial derivative $\partial f/\partial x$ ($\delta_x f$), using the central difference approach, is calculated by:

$$\frac{\delta f}{\delta x}\Big|_{i,j} = \delta_x f\big|_{i,j} = \frac{1}{2\Delta x}(f_{i+1,j} - f_{i-1,j}). \quad (12.10)$$

Analogous expressions follow immediately for derivations in relation to the independent variables y and t. Thus, second order central finite differences for the derivations $\delta f/\delta y$ and $\delta f/\delta t$, for example, are calculated by,

$$\frac{\delta f}{\delta y}\Big|_{i,j} = \delta_y f\big|_{i,j} = \frac{1}{2\Delta y}(f_{i,j+1} - f_{i,j-1}), \quad (12.11)$$

and

$$\frac{\delta f}{\delta t}\Big|_{i,j}^{n} = \delta_t f\big|_{i,j}^{n} = \frac{1}{2\Delta t}(f_{i,j}^{n+1} - f_{i,j}^{n-1}), \quad (12.12)$$

where $\Delta t = \delta_t = (t^{n+1} - t^{n-1})$ is a constant time interval.

Let us now determine the second order derivative, $\delta^2 f/\delta x^2 = \delta_x^2 f$, by central finite differences. For this purpose, the summation of expressions (12.6) and (12.7) is

$$f_{i+1,j} + f_{i-1,j} = 2f_{i,j} + \frac{\partial^2 f}{\partial x^2}|_{i,j}(\Delta x)^2 + \frac{1}{12}\frac{\partial^4 f}{\partial x^4}|_{i,j}(\Delta x)^3 + \cdots + \text{SOT}, \qquad (12.13)$$

Solving this equation for $\delta_x^2 f$ $(\partial^2 f/\partial x^2)$

$$\delta_x^2 f|_{i,j} = \frac{(f_{i+1,j} + f_{i-1,j} - 2f_{i,j})}{(\Delta x)^2} + O[(\Delta x)^2], \qquad (12.14)$$

or

$$\delta_x^2 f|_{i,j} = \frac{(f_{i+1,j} + f_{i-1,j} - 2f_{i,j})}{(\Delta x)^2}. \qquad (12.15)$$

As an example of the property application of the operator $\delta f/\delta x$ $(\delta_x f)$, let us make the deduction of the expression (12.15), starting with an approximation of the first derivative of Eq. (12.10), and rewriting it in terms of the half interval Δx $(\Delta x/2)$,

$$\frac{\delta f}{\delta x}|_{i,j} = \frac{(f_{i+\frac{1}{2},j} - f_{i-\frac{1}{2},j})}{\Delta x}. \qquad (12.16)$$

Taking into account that $\delta^2 f/\delta^2 x = (\delta/\delta x(\delta f/\delta x))$ we have,

$$\frac{\delta^2 f}{\delta x^2}|_{i,j} = \frac{[f_{i+1,j} - f_{i,j} - (f_{i,j} - f_{i-1,j})]}{(\Delta x)^2}, \qquad (2.17a)$$

then

$$\delta_x^2 f|_{i,j} = \frac{(f_{i+1,j} + f_{i-1,j} - 2f_{i,j})}{(\Delta x)^2}, \qquad (12.17b)$$

which is equivalent to Eq. (12.15).

As another example, let us calculate the approximation by second order finite differences of the second order derivative of $f = f(x, y, t)$ with two spatial variables, i.e., $\delta^2 f/\delta x\delta y$. The expression of this derivative may be easily deduced if we observe that,

$$\frac{\delta^2 f}{\delta x \delta y} = \frac{\delta}{\delta x}\left(\frac{\delta f}{\delta y}\right). \qquad (12.18)$$

Applying a similar procedure to this expression, as used for Eq. (12.10),

$$\frac{\delta^2 f}{\delta y \delta x}\Big|_{i,j} = \frac{\delta}{\delta y}\left[\frac{(f_{i+1,j} - f_{i-1,j})}{2\Delta x}\right], \tag{12.19}$$

and

$$\frac{\delta^2 f}{\delta y \delta x}\Big|_{i,j} = \frac{[f_{i+1,j+1} - f_{i-1,j+1} - (f_{i+1,j-1} - f_{i-1,j-1})]}{2\Delta x \Delta y}, \tag{12.20a}$$

or

$$\frac{\delta^2 f}{\delta y \delta x}\Big|_{i,j} = \frac{[f_{i+1,j+1} - f_{i+1,j-1} - f_{i-1,j+1} + f_{i-1,j-1})]}{2\Delta x \Delta y}. \tag{12.20b}$$

The second order derivative (12.20b) in the x and y coordinates has a truncation error indicated generically by $O[(\Delta x)^2 + (\Delta y)^2]$. Also, it should be noted that the operator, $\delta^2 f/\delta x \delta y$, obeys the same rules as the derivation of a continuous function and, in relation to the mixed derivatives used above, holds the identity, $\delta^2 f/\delta x \delta y = \delta^2 f/\delta y \delta x$.

Finite differences of partial derivatives, such as those presented above, may be combined in order to obtain an expression of physical-mathematical laws, for example, the second-order partial differential of the Laplace equation which, in the two-dimension scalar form, $\psi = \psi(x, y)$, is given by,

$$\frac{\partial^2 \psi}{\partial x^2} + \frac{\partial^2 \psi}{\partial y^2} = 0. \tag{12.21}$$

Combining the developed expression (12.20b) with second order derivatives, we may write:

$$\frac{\delta^2 \psi}{\delta x^2} + \frac{\delta^2 \psi}{\delta y^2} = \frac{1}{(\Delta x)^2}(\psi_{i+1,j} + \psi_{i-1,j} - 2\psi_{i,j}) + \frac{1}{(\Delta y)^2}(\psi_{i,j+1} + \psi_{i,j-1} - 2\psi_{i,j}) = 0, \tag{12.22a}$$

or

$$\psi_{i+1,j} + \psi_{i-1,j} + \beta^2(\psi_{i,j+1} + \psi_{i,j-1}) - 2(1 + \beta^2)\psi_{i,j} = 0, \tag{12.22b}$$

where $\beta = \Delta x/\Delta y$ is the characteristic ratio of the grid spacing in the Ox and Oy directions, respectively. This equation is usually referred to as the five point approximation of the Laplace equation. In the condition when $\Delta x = \Delta y$, it follows that the expression for $\psi_{i,j}$ is:

$$\psi_{i,j} = \frac{1}{4}(\psi_{i+1,j} + \psi_{i-1,j} + \psi_{i,j+1} + \psi_{i,j-1}), \tag{12.23}$$

which shows that in the five point approximation for the Laplace equation the unknown, $\psi_{i,j}$ is calculated by its mean value at four neighboring points.

In the following example, let us consider a differential equation (11.24) representing a linear model which governs the space-time variation of a property defined in a one-dimensional space $\zeta = \zeta(x, t)$;

$$\frac{\partial \zeta}{\partial t} + u\frac{\partial \zeta}{\partial x} - N\frac{\partial^2 \zeta}{\partial x^2} = 0. \tag{12.24}$$

If $\zeta = \zeta(x, t)$ is the u-velocity and N the kinematic eddy viscosity coefficient $[N] = [L^2T^{-1}]$, this equation represents the one-dimensional equation of motion (or momentum equilibrium). The second order approximation of the finite difference in space (x) and time (t) is written as:

$$\frac{1}{2\Delta t}(\zeta_i^{n+1} - \zeta_i^{n-1}) + \frac{1}{2\Delta x}(u\zeta_{i+1}^n - u\,\zeta_{i-1}^n) - \frac{N}{(\Delta x)^2}(\zeta_{i+1}^n + \zeta_{i-1}^n - 2\zeta_i^n) = 0. \tag{12.25}$$

It should be noted that this equation may be explicitly solved to the unknown (ζ_i^{n+1}), taking into account the previous known values in space and time. However, for N > 0 and Δt > 0, the solution may be numerically unstable, and random solutions may be generated without any relation to the differential solution. This clearly indicates the difference between an algebraic finite difference expression, which is mathematically correct, and the desirable solution to the differential equation (Roache 1982).

If, for instance, instead of using central differences for all independent variables, the finite difference of the partial differential (12.24) is calculated by the forward finite difference scheme, first and second order numerical approximations will be obtained for time and space, respectively,

$$\frac{(\zeta_i^{n+1} - \zeta_i^n)}{\Delta t} + \frac{(u\zeta_{i+1}^n - u\,\zeta_{i-1}^n)}{2\Delta x} - \frac{N\,(\zeta_{i+1}^n + \zeta_{i-1}^n - 2\zeta_i^n)}{(\Delta x)^2} = 0. \tag{12.26}$$

According to Roche (op. cit.), at least for some conditions of the independent variables (t, x and intervals Δt, Δx), and for the dependent variables, u and N, this solution becomes stable.

In applying the finite-difference scheme to the equation of motion in the Eulerian formulation, which isn't a non-linear equation, care must be taken in the formulation. For example, if the term of the advective acceleration is formulated by the central finite differences scheme, that is,

$$u\frac{\partial u}{\partial x} = u\delta_x u = \frac{1}{2\Delta x}u_i^{n+1}(u_{i+1}^n - u_{i-1}^n), \tag{12.27}$$

this approximation is not satisfactory, because the von Neumann condition will not be satisfied for any fixed value of the ratio Δt/Δx, unless for the trivial solution u = 0. This problem may be solved with the forward and backward finite difference scheme when u < 0 or u ≥ 0, respectively, using the expression of Richtmeyer and Morton (1967),

$$(u_i^{n+1})[\frac{1}{\Delta x}(u_{i+1}^n \pm u_i^n)], \tag{12.28}$$

where $u_i^{n+1} < 0$ or $u_i^{n+1} \geq 0$, respectively. Similar expressions may be written for the remaining non-linear terms of the advective acceleration, or other terms of any non-linear equation.

Other relations may be obtained from the Taylor's expansion series, for instance, adding Eqs. (12.6) and (12.7), we have the following second order approximation:

$$2f_{i,j} = f_{i+1,j} + f_{i-1,j} + O(\Delta x)^2, \tag{12.29a}$$

or

$$f_{i,j} = \frac{(f_{i+1,j} + f_{i-1,j})}{2} + O(\Delta x)^2. \tag{12.29b}$$

If expansions are made in relation to the time variable to investigate the non-steady-state characteristics of the function f = f(x, y, t), it follows that the algebraic finite difference approximation is:

$$f_i^n = \frac{(f_i^{n+1} + f_i^{n-1})}{2} + O(\Delta t)^2. \tag{12.30}$$

First order approximations in the time domain may also be taken from the previous series expansions (12.6 and 12.7), and written as:

$$f_{i-1}^n = f_i^n + O(\Delta t), \tag{12.31a}$$

and

$$f_i^{n+1} = f_i^n + O(\Delta t). \tag{12.31b}$$

Linearization of the terms of the motion equation (12.24) may also be achieved from diagonal mean values, yielding:

$$f_i^n = \frac{1}{2}(f_{i+1}^{n-1} + f_{i-1}^{n+1}) + O(\Delta t), \quad (12.32a)$$

and

$$f_{i-2}^n = \frac{1}{2}(f_{i-1}^{n+1} + f_{i-3}^{n-1}) + O(\Delta t), \quad (12.32b)$$

where i = 2, 3, 4, ... and n = 1, 2, 3, ...

12.3 Explicit and Implicit Schemes

The numerical schemes for the analytical solution to an equation of finite differences for a partial differential equation are classified as *explicit* and *implicit*. The difference between these methods will be shown using the particular second order differential equation (12.24), which represents a one-dimensional space, simulating the spatial and temporal variations of the property $\zeta = \zeta(x, t)$. This equation has already been solved by forward and central finite differences and, without loss of generality, let us assume that to simplify the mathematical treatment, the middle term may be disregarded ($u\partial\zeta/\partial x = 0$). Then, the equation is approximated by finite differences as,

$$\frac{1}{\Delta t}(\zeta_i^{n+1} - \zeta_i^n) = \frac{N}{(\Delta x)^2}(\zeta_{i+1}^n + \zeta_{i-1}^n - 2\zeta_i^n), \quad (12.33)$$

where i = 1, 2, ... I − 1 and n = 0, 1, 2, ... I. The boundary and initial conditions for this equation may be established as,

$$\zeta_0^n = \zeta_I^n = 0, \quad \text{for } n = 0, 1, 2, \ldots I - 1, \quad (12.34a)$$

and

$$\zeta_i^0 = \varphi(i\Delta x), \quad \text{for } i = 0, 1, 2, \ldots I. \quad (12.34b)$$

Then, solving Eq. (12.33) explicitly for ζ_i^{n+1}, we have:

$$\zeta_i^{n+1} = \zeta_i^n + [\frac{N(\Delta t)}{(\Delta x)^2}](\zeta_{i+1}^n + \zeta_{i-1}^n - 2\zeta_i^n), \quad (12.35)$$

which may be solved recursively for the determination of all values of ζ_i^n for $0 \le i \le I$ and $n \ge 0$. This is named an *explicit* scheme and *one step* solution, meaning that all values of the second member are known and only one calculation is required to reach the next time step; thus, the solution progresses, and the values

ζ_{i+1}^{n+1} do not appear in the second equation member. It is also named *two-time-steps* because only two instants of time are necessary to calculate the property value, i.e., to calculate the property at the time instant n + 1, it is only necessary to know its value at t = n. As previously stated, the approximation of this equation is at first and second order for time and space, respectively $\{O[\Delta t,(\Delta x)^2]\}$; further details related to the stability of this solution may be found in Richtmyer and Morton (1967) and Roache (1982).

At this stage, we should mention that although the solution (12.25) centered in space and time has an approximation order of $\{O[(\Delta t)^2, (\Delta x)^2]\}$, it is not acceptable because it is unstable for any value of the coefficient N and for t > 0. However, if N = 0 the solution will have stable characteristics and this method is frequently known as *leapfrog*; the numerical solution of ζ_i^{n+1} under this simplification is:

$$\zeta_i^{n+1} = \zeta_i^{n-1} - u\frac{\Delta t}{\Delta x}(\zeta_{i+1}^n - \zeta_{i-1}^n). \tag{12.36}$$

This solution has second order approximations for space and time and is explicit with one step. Its solution requires three time instants to be known because values at times n and n − 1 are necessary to calculate the value of the next time step (n + 1). Under the same initial and boundary conditions as indicated in (12.34a, b), it follows from the above solution that the new value of ζ_i^{n+1} is calculated from the known value ζ_i^{n-1} minus the last term on the right hand side of Eq. (12.36), skipping over the value in the time instant n (ζ_i^n); this procedure justifies the name, leapfrog, given to this calculation scheme. At this point, we should be reminded that the numeric solution (12.36) is the solution to a partial differential equation and an advective equation when ζ(x, t) = u(x, t) is a velocity component. An initial condition of this equation may be expressed by ζ = ζ(x,0) or ζ_i^0 and its solution using the finite difference scheme was demonstrated by Roache (op. cit.).

The presented method is explicit because, as we have seen, it is only necessary to know the values of ζ = ζ(x.t) at time instants t = n, n − 1, n − 2 …, to advance the computation for the new time n + 1. However, we should note the stability criteria of Richtmyer and Morton (op. cit.), which indicates that

$$2N[\frac{\Delta t}{(\Delta x)^2}] \leq 1, \tag{12.37}$$

i.e., if the value chosen for Δx in the solution is too small, the time step Δt, will also be small, increasing the computational cycles required to satisfactorily finish the problem, due to the above relationship between (Δt) and the square $(\Delta x)^2$.

The implicit method uses values of the spatial derivatives in advanced time steps, which means that the solution needs a system with (n + 1) equations to advance the data processing to the next time step. To exemplify this method, let us start with Eq. (12.24), writing its first two terms on the left hand side with forward time steps (Δt),

$$\frac{(\zeta_i^{n+1} - \zeta_i^n)}{\Delta t} = u\frac{\delta\zeta}{\delta x}. \qquad (12.38)$$

Calculating the derivative of the right hand side by central differences, and rearranging its terms, this equation is written as:

$$\zeta_i^{n+1} = \zeta_i^n - \frac{u\Delta t}{2\Delta x}(\zeta_{i+1} - \zeta_{i-1}), \qquad (12.39)$$

and, with an analogous procedure for the non-linear term of Eq. 12.24, and solving for ζ_i^{n+1}, the result is:

$$\zeta_i^{n+1} = \zeta_i^n + \frac{N\Delta t}{(\Delta x)^2}(\zeta_{i+1} + \zeta_{i-1} - 2\zeta_i). \qquad (12.40)$$

If the spatial derivations in Eqs. (12.39) and (12.40) were calculated in the time instant, n, this method would be explicit. However, if these derivatives were calculated in the time step n + 1, the calculation scheme would be completely implicit. And, as consequence, expressions (12.39) and (12.40) are calculated by:

$$\frac{(\zeta_i^{n+1} - \zeta_i^n)}{\Delta t} = \frac{u}{2\Delta x}(\zeta_{i+1}^{n+1} - \zeta_{i-1}^{n+1}), \qquad (12.41)$$

and

$$\frac{(\zeta_i^{n+1} - \zeta_i^n)}{\Delta t} = \frac{N}{(\Delta x)^2}(\zeta_{i+1}^{n+1} + \zeta_{i-1}^{n+1} - 2\zeta_i^{n+1}), \qquad (12.42)$$

respectively. These solutions have an estimated error of the order $O[\Delta t,(\Delta x)^2]$, however, as indicated in several investigations, this method has advantage in relation to its stability. The determination of the property, ζ, in a generic time step, ζ_i^{n+1}, requires the simultaneous solution of a number of linear algebraic equations, with M indicating the net knots not specified by known boundary conditions.

For a generalization of the implicit and explicit schemes introduced above, let us introduce the following notation for a single variable function, generically defined by $f = f(x)$, to the central finite difference δf_i or $(\delta f)_i$, then:

$$\delta f_i = (\delta f)_i = f[(i + 1/2)\Delta x - f(i - 1/2)\Delta x], \qquad (12.43)$$

where the index (i) is an integer value. With this notation, the symbols $\delta^2 f_i$ or $(\delta^2 f)_i$ indicate the following expressions:

$$\delta^2 f_i = (\delta^2 f)_i = f[(i + 1)\Delta x - f(i\Delta x] - [f(i\Delta x) - f(i - 1)\Delta x)], \qquad (12.44a)$$

or

$$\delta^2 f_i = (\delta^2 f)_i = f[(i+1)\Delta x + f(i-1)\Delta x)] - 2f(i\Delta x)]. \quad (12.44b)$$

Once the above notation is introduced, lets us consider the following system:

$$\frac{(\zeta_i^{n+1} - \zeta_i^n)}{\Delta t} = \frac{N}{\Delta t(\Delta x)^2}[\theta(\delta^2\zeta)_i^{n+1} + (1-\theta)(\delta^2\zeta)_i^n], \quad (12.45)$$

where θ is a real number varying in the interval $0 \le \theta \le 1$. If $\theta = 0$ this algebraic system becomes explicit, as previously indicated (Eq. 12.33). Each equation of this system furnishes an unknown (ζ_i^{n+1}) in terms of the quantities (ζ_i^n). If $\theta \neq 0$ it is necessary to simultaneously solve a set of linear equations to calculate the unknown in the next time step (ζ_i^{n+1}) and, as previously seen, the system is implicit.

The simultaneous solution of the linear equations of an implicit system is not as easily solved as an explicit system of equations, because its solution is obtained iteratively. To illustrate the solution of an implicit system, let us present an example of the implicit system solution from the Roache (1982), starting with Eq. (12.46) under the assumption that the initial and boundary conditions are known, i.e., the n + 1 values of ζ_1 and ζ_I. Then the equation for a generic knot may be written as:

$$\zeta_{i-1}^{n+1} + a\,\zeta_i^{n+1} + c\zeta_{i+1}^{n+1} = b, \quad (12.46)$$

where $a = \frac{2\Delta X}{u\Delta t}$, c = −1 and $b = -a\,\zeta_i^n$. According to the boundary conditions, the value of ζ_1^n is known; thus, this equation may be solved to i = 2 and it is possible to calculate the value of ζ_3^{n+1} as a function of ζ_1 and ζ_2,

$$\zeta_3^{n+1} = f(\zeta_1, \zeta_2), \quad (12.47)$$

continuing to i = 3, it follows that:

$$\zeta_4^{n+1} = f(\zeta_2, \zeta_3), \quad (12.48)$$

which combined with the functional relation (12.47) yields,

$$\zeta_4^{n+1} = f(\zeta_1, \zeta_2). \quad (12.49)$$

Progressing further with this procedure, for i = I − 2, we have

$$\zeta_{I-1}^{n+1} = f(\zeta_{I-3}, \zeta_{I-2}) = f(\zeta_1, \zeta_2), \quad (12.50)$$

and finally for i = I − 1 the result is:

$$\zeta_I^{n+1} = f(\zeta_{I-2}, \zeta_{I-1}) = f(\zeta_1, \zeta_2). \qquad (12.51)$$

Thus, as the boundary conditions (ζ_1, ζ_I) are known, the last Eq. (12.51) may be solved for ζ_2^{n+1}. Subsequently, with a second calculation of Eq. (12.46), the final results may be obtained. This procedure has only been described to illustrate the sequence for obtaining the solution; however, it is subject to the influence of truncation errors, which may be overcome with the utilization of the *triangular algorithmic* to simultaneously solve a system of linear equations; this algorithmic is named as such because the matrix used to solve the system of equations,

$$[A][\zeta] = [B], \qquad (12.52)$$

which must be inverted, is a diagonal matrix, i.e., its elements are only different from zero in the principal diagonal and at the two adjacent diagonals, and the others elements are null. A diagonal matrix has an easy solution, and a FORTRAN computational subroutine is presented in Roache's book.

12.4 The Volume Method of Finite Difference

As an example of formulating a solution to a hydrodynamic system of equations using finite difference for this method, let us initially consider the one-dimensional mass conservation equation (Eq. 7.92a, Chap. 7):

$$\frac{\partial(uA)}{\partial x} + \frac{\partial A}{\partial t} = 0, \quad \text{or} \quad \frac{\partial(uA)}{\partial x} + B\frac{\partial h}{\partial t} = 0, \qquad (12.53)$$

where A is the cross section area, u is the mean u-velocity component in the transverse section A, B is the width of the estuarine channel, which is assumed to be uniform (B = cte), and $\underline{h}$ is a reference level (horizontal datum). Thus, the product uA = Q is the volume transport [uA] = [Q] = $[L^3T^{-1}]$ through the cross section area A. From this equation, it follows that the quantities Q = Q(x, t) and u = u(x, t) may be considered as unknowns if the geometric characteristics of the system are known.

Figure 12.2 schematically presents the spatial-temporal variations of the free surface height (a), the transverse section A and width B (b), and the bi-dimensional space-time (c) subdivided into Δx and Δt intervals. Let us also consider a well-mixed estuary (type 1 or C), and a volume transport (uA) crossing a control transverse section (i) generated by the tidal oscillation, and thus, forced by the barotropic pressure gradient force.

Indicating by L the estuary mixing zone (MZ) length, which is subdivided in I − 1 regular space intervals, the number of knots in the longitudinal direction is

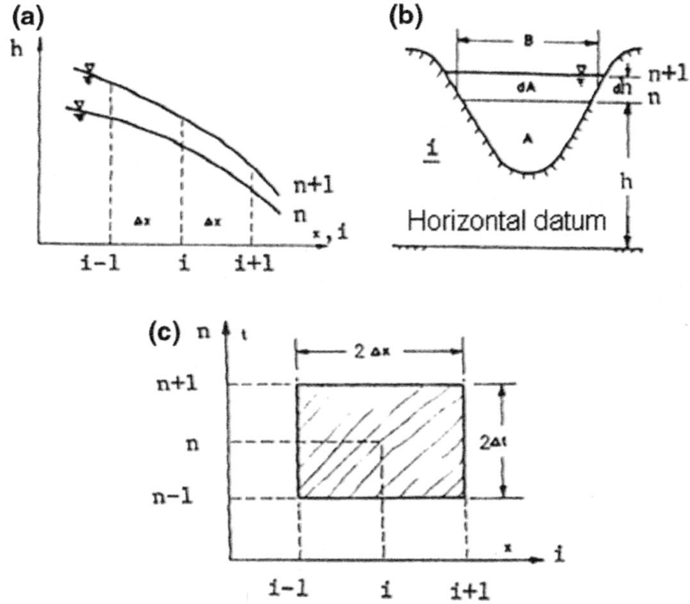

Fig. 12.2 **a** The schematic representation of the tide oscillation h = h(i). **b** The transverse section characteristics (A), and **c** The space-time showing the grid Δx versus Δt of the finite difference approximation of the continuity equation (after McDowell and O'Connor 1977)

equal to I (i = 1,2,3,…, I), the Δx length interval is equal to the ratio L/(I − 1), and the sub-volume of each cell is the cross-section area (A) times (Δx).

According to what we have already seen, the partial differential equation (12.53) may be approximated by finite differences with different orders. In this application, the first order approximation, O(Δx, Δt), will be chosen for simplicity. Then,

$$\frac{(Q^n_{i+1} - Q^n_i)}{\Delta x} + \frac{B^n_i(h^{n+1}_i - h^n_i)}{\Delta t} = 0, \tag{12.54a}$$

or

$$\frac{(Q^n_{i+1} - Q^n_i)}{\Delta x} = -\frac{B^n_i(h^{n+1}_i - h^n_i)}{\Delta t}, \tag{12.54b}$$

with i = 1, 2, 3, …, I − 1 and n = 0, 1, 2, …. As the right hand side of Eq. (12.54b) contains the temporal variable, which is a function of known data, the initial condition of the problem will be imposed naturally. Let us assume that the longitudinal coordinate, Ox, is oriented seaward, and its origin (x = 0) is located at the estuary head, which is the transitional zone of the tidal river zone (TRZ) and mixing zone (MZ). Then, we have the following boundary condition:

$$Q_i^n = Q_f, \tag{12.55}$$

where Q_f is the river discharge, which is known and constant.

Under the assumption that the stability conditions are satisfied and solving Eq. (12.54a) explicitly for the quantity Q_{i+1}^n, which is associated with the river discharge, it follows that the expression to calculate the volume transport across any transverse section is:

$$Q_{i+1}^n = Q_i^n - \frac{\Delta x}{\Delta t} B_i^n (h_i^{n+1} - h_i^n). \tag{12.56}$$

This result, with dimension $[L^3T^{-1}]$, indicates that the volume transport may be calculated at any instant of time (n) if the free-surface elevation and the estuary width are known. In the next along channel time step i = 1,

$$Q_2^n = Q_f - \frac{\Delta x}{\Delta t} B_1^n (h_1^{n+1} - h_1^n). \tag{12.57}$$

If, in this equation, $h_1^{n+1} \approx h_1^n$, i.e., for i = 1 the time variation of h may be disregarded, it follows that $Q_2^n = Q_f$, and the tidal influence may be disregarded in the sub-volume 2. Otherwise, the difference $h_1^{n+1} - h_1^n$ may be positive (>0) or negative (<0), indicating an ebb or a flood tide condition.

Continuing to the next volume (i = 2),

$$Q_3^n = Q_f - \frac{\Delta x}{\Delta t} B_2^n (h_2^{n+1} - h_2^n), \tag{12.58}$$

and the volume transport in the next sub-volume (3) may be calculated at any time. The second term on the right hand side may be positive or negative according to $h_2^{n+1} > h_2^n$ or $h_2^{n+1} < h_2^n$, respectively, indicating the ebb or flood tide and will be subtracted or added to the fresh water discharge Q_f.

As this iterative process must proceed to the last sub-volume (i = I − 1), it follows that,

$$Q_I^n = Q_f - \frac{\Delta x}{\Delta t} B_{I-1}^n (h_{I-1}^{n+1} - h_{I-1}^n). \tag{12.59}$$

Now, taking into account that $Q_{i+1}^n = (Au)_{n+1}^n$, it follows immediately from the calculated volume transport that the mean value of the u-velocity component is calculated by:

$$u_{i+1}^n = \frac{Q_{i+1}^n}{A_{i+1}^n}. \tag{12.60}$$

The finished iterative process for i = 1, 2, 3, … I − 1 corresponds to the computation along the longitudinal axis, and in turn, a similar procedure must be applied in the time domain of interest, i.e., for n = 0, 1, 2,…, over one or more tidal cycles.

This method is named the *volume method,* which is justified because the second term on the right hand side of Eq. (12.56) generates volumes per time unit. In practice, the variable $\underline{h}$ must be known at regular distances intervals (Δx). According to McDowell and O'Connors (1977), this quantity must be known with an accuracy greater than 10^{-2} m, and the numerical solution must be validated with experimental results.

12.5 A Simple Unidimensional Numeric Model

12.5.1 Explicit Solution

Under the assumption of a well-mixed estuary, let us formulate the main hydrodynamic processes that characterize a one-dimensional estuary using the explicit method of finite differences. The starting hydrodynamic equations are simplified expressions of the equations of motion and continuity, which were used in the development of a mathematical model for prediction of unsteady salinity intrusion in estuaries by Thatcher and Harleman (1972):

$$\frac{\partial u}{\partial t}+\frac{\partial (uu)}{\partial x}+g\frac{\partial h}{\partial x}+\frac{gu|u|}{C_y^2 R_H}=0, \tag{12.61a}$$

and

$$\frac{\partial (uA)}{\partial x}+B\frac{\partial h}{\partial t}=0, \tag{12.61b}$$

where C_y and $R_H \approx H_o$ are the Chézy coefficient and the hydraulic radius, respectively, as defined in Chap. 8. These equations indicate that the local and advective accelerations, plus the barotropic gradient, are in balance with the frictional force. These equations will be numerically integrated to calculate the field of motion, $u = u(x, t)$ or u_i^n, and the elevations of the free surface, $h = h(x, t)$ or h_i^n, forced by the tidal oscillation.

As in the preceding application, the plane x-t is subdivided into integration cells Δx and Δt. The longitudinal number of knots is equal to the ratio L/(I − 1), with I denoting the numbers of points in the longitudinal direction. As before, the sub-volumes are equal to A. Δx, and the schemes in Fig. 12.3 indicate the spatial-temporal grid (a), where the volume transport and velocity will be calculated, and (b) the longitudinal positions where the velocity and volume transports will be alternatively calculated.

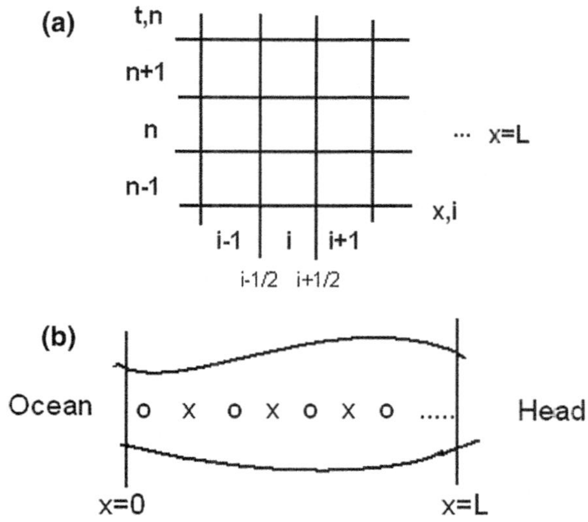

Fig. 12.3 **a** Integration cells in the x-t plane. **b** Longitudinal plane section with positions where the quantities u (in positions o) and h (in positions x) and the transport volume (uA = Q) will be calculated. Adapted from Thatcher and Harleman (1972)

In Fig. 12.3a the cell structure and the free surface heights h_i^n and h_{i+1}^n, are defined at the knots localized in the cell center, and in Fig. 12.3b the u-velocity components and the volume transport (uA) are calculated at the left and right cell's limits indicated by the longitudinal positions o and x, respectively, for example: $u_{i+1/2}^n$ and $u_{i-1/2}^n$, or, $Q_{i+1/2}^n$ $(uA_{i+1/2}^n)$, and $Q_{i-1/2}^n$ or $(uA_{i-1/2}^n)$ are calculate for $i = 0, 1, 2, \ldots I - 1$.

The formulations using the finite difference scheme of Eqs. (12.61a, b) are written as:

$$\frac{\delta u}{\delta t} + \frac{\delta(uu)}{\delta x} + g\frac{\delta h}{\delta x} + \frac{gu|u|}{C_y^2 H_o} = 0, \qquad (12.62a)$$

and

$$\frac{\delta(uA)}{\delta x} + B\frac{\delta h}{\delta t} = 0. \qquad (12.62b)$$

Using the Forward Time Central Scheme (FTCS) in an unique cell spacing yields the following finite difference expressions for local and advective accelerations:

$$\frac{\delta u}{\delta t} = \left(\frac{u_{i+1/2}^{n+1} - u_{i+1/2}^n}{\Delta t}\right), \qquad (12.63a)$$

and

$$\frac{\delta(uu)}{\delta x} = [\frac{(uu)^n_{i+1} - (uu)^n_i}{\Delta x}], \tag{12.63b}$$

respectively. However, the advective acceleration is non-linear and requires a redefinition of the u-values at the knots (i ± 1) in terms of its mean values,

$$u^n_{i+1} = \frac{1}{2}(u^n_{i+3/2} + u^n_{i+1/2}), \tag{12.64a}$$

and

$$u^n_i = \frac{1}{2}(u^n_{i+1/2} + u^n_{i-1/2}). \tag{12.64b}$$

The barotropic pressure gradient force and the friction due to viscosity are calculated as:

$$g\frac{\delta h}{\delta x} = \frac{g}{\Delta x}(h^n_{i+1} - h^n_i), \tag{12.65a}$$

and

$$\frac{g(u^n_{i+1/2}|u^n_{i+1/2}|)}{(C_y|^n_{i+1/2})^2 h^n_{i+1/2}}, \tag{12.65b}$$

respectively. The last finite difference requires the definition of $h^n_{i+1/2}$ in terms of a mean value, and

$$h^n_{i+1/2} = \frac{1}{2}(h^n_{i+1} + h^n_i). \tag{12.66}$$

In order to eliminate possible instabilities during the computation of the friction term (last term in Eq. 12.61a), this term must be delayed for one time step, i.e., it must be calculated by,

$$\frac{g(u^{n-1}_{i+1/2}|u^{n-1}_{i+1/2}|)}{(C_y|^{n-1}_{i+1/2})^2 h^{n-1}_{i+1/2}}. \tag{12.67}$$

For the continuity Eq. (12.61b), it follows that the finite difference expression is:

$$\frac{[(Au)^n_{i+1/2} - (Au)^n_{i-1/2}]}{\Delta x} + \frac{B^n_i(h^{n+1}_i - h^n_i)}{\Delta t} = 0. \tag{12.68}$$

The non-linear term of the advective acceleration (last term in Eq. 12.63b) will be transformed into a linear term, according to the following approximation:

$$\frac{[(uu)_{i+1}^{n} - (uu)_{i}^{n}]}{\Delta x} = \frac{u_i^n (u_{i+1}^n - u_{i-1}^n)}{\Delta x}. \tag{12.69}$$

The term u_{i-1}^n, on the right hand side of this equation, may be calculated by a mean equivalent of Eq. (12.64a), which is a linear expression of the advective acceleration. Another linear expression, suggested by McDowell and O'Connors (1977), may also be obtained by an artifact of the second member of Eq. (12.68) and from some approximations which have already been presented. In effect, from expression (12.30) we have:

$$u_i^n = \frac{(u_i^{n+1} + u_i^{n-1})}{2} + O(\Delta t^2), \tag{12.70a}$$

with i = 1, 2, … and n = 0, 1, 2, … I − 1, and, from the approximations (12.32a, b) we may write,

$$u_{i+1}^n = \frac{1}{2}(u_{i+2}^{n-1} + u_i^{n+1}) + O(\Delta t, \Delta x), \tag{12.70b}$$

and

$$u_{i-1}^n = \frac{1}{2}(u_i^{n+1} + u_{i-2}^{n-1}) + O(\Delta t, \Delta x). \tag{12.70c}$$

Substituting approximations (12.70a, b, c) into expression (12.69) yields,

$$\frac{[(uu)_{i+1}^{n} - (uu)_{i}^{n}]}{\Delta x} = \frac{[(u_i^{n+1} + u_i^{n-1})(u_{i+2}^{n-1} - u_{i-2}^{n-1})]}{4\Delta x}, \tag{12.71a}$$

or alternatively

$$\frac{[(uu)_{i+1}^{n} - (uu)_{i}^{n}]}{\Delta x} \approx \frac{[(u_i^{n+1})(u_{i+2}^{n-1} - u_{i-2}^{n-1})]}{2\Delta x}, \tag{12.71b}$$

and

$$\frac{[(uu)_{i+1}^{n} - (uu)_{i}^{n}]}{\Delta x} \approx \frac{[(u_i^{n-1})(u_{i+2}^{n-1} - u_{i-2}^{n-1})]}{2\Delta x}. \tag{12.71c}$$

Expressions (12.71a, b, c) are equivalent to those presented by McDowell and O'Connors (op. cit.).

Combining Eqs. (12.63a, b; 12.64a, b; 12.65a, b, 12.67, 12.71a or b, c) yields the final finite difference expressions of the equations of motion and continuity equivalent to the corresponding partial differential equations (12.61a, b):

$$u_{i+1/2}^{n+1} = u_{i+1/2}^{n} + \frac{\Delta t}{\Delta x}[-(uu)_{i+1}^{n} - (uu)_{i}^{n}] - \frac{g}{\Delta x}(h_{i+1}^{n} - h_{i}^{n}) - \frac{g}{[(C_y)_{i+1/2}^{n}]^2 h_{i+1/2}^{n}}(u_{i+1/2}^{n}|u_{i+1/2}^{n}|), \tag{12.72a}$$

and

$$h_{i}^{n+1} = h_{i}^{n} + \frac{\Delta t}{B_{i}^{n}\Delta x}[(uA)_{i+1/2}^{u} - (uA)_{i-1/2}^{n}]. \tag{12.72b}$$

Assuming that the stability of this analytical system is satisfied, that the initial and boundary conditions are known, and the geometric characteristics of the estuary are also known, Eq. (12.72b) may be used at the initial time instant (n = 0) to calculate the free surface height for i = 1, thus obtaining the first value h_1^1. In the following step, the second member of Eq. (12.72a) may also be solved. Following this, in the initial time-space step (n = 0 and i = 1), the second member of Eq. (12.72a) may be solved, and $h_{1/2}^{n}$ may be calculated as a mean value and applied to equation similar to (12.66),

$$h_{i/2}^{0} = \frac{1}{2}(h_1^0 + h_{-1}^0), \tag{12.73}$$

where the value of h_{-1}^0 is extrapolated from the initial condition. Then, with Eq. (12.72a), the unknown $h_{i+1/2}^1$ may be calculated. In the following step, for i = 2, 3, 4, …, and from the value at the initial time (n = 0), it is possible to determine the unknowns, h_i^n and $u_{i+1/2}^n$ for i = 1, 2, 3, …, I − 1, using iteratively Eqs. (12.72a, b). This process must be repeated for the other times (n > 0), enabling knowledge of the unknowns h_i^n and $h_{i+1/2}^n$, which will satisfy the imposed initial and boundary conditions.

This method may also be expanded to include in the mass conservation equation, the lateral fresh water input from tributaries and the free surface processes of precipitation-evaporation. In the equation of motion, changes in the influences on the system dynamics, due to variations in the estuary geometry, may also be included.

When the estuarine channel presents a bifurcation due to the presence of a tributary, this may also be included in the computational scheme. This may be accomplished by including a knot located in the neighboring area just before the junction, which must be the same knot used in the determination of the free-surface height, as indicated in Fig. (12.4).

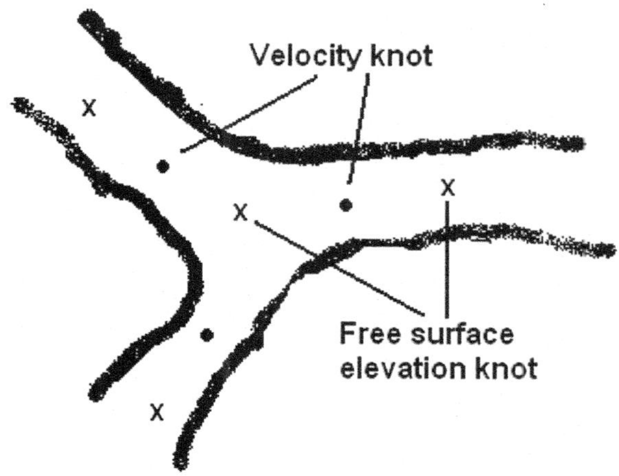

Fig. 12.4 Schematic diagram indicating a bifurcation in an estuarine channel. The symbols • and x indicate the positions of the u-velocity component and the surface height calculations, respectively

Then, for instance, for the i-knot (x in Fig. 12.4), the free surface elevation must be calculated by the following expression:

$$h_i^{n+1} = h_i^n + \frac{\Delta t}{B_i^{-n} \Delta x} [(uA)_m^n - (uA)_{i-1/2}^n], \tag{12.74a}$$

and

$$(uA)_m^n = \frac{1}{2} [(uA)_{i+1/2}^n - (uA)_{i+1/2}^{-n}]. \tag{12.74b}$$

Then, in Eq. (12.74a), the term on the right hand side indicates the volume transport at the bifurcation, and B_i^{-n}, is the channel width calculated as the mean value at positions (i − 1/2) and (1 + 1/2). Subsequently, the computation will follow independently along each one of the channels.

12.5.2 *Implicit Solution*

The same problem formulated by the differential partial equations of motion and continuity (12.61a, b) may be solved by the implicit method, and the scheme for the integration cells is similar to that presented in Fig. 12.3. In this solution, the finite differences for the equation of motion is calculated in a given time step; however, the continuity equation must be displaced forward by a space interval, Δx. Then, the equation system to be numerically integrated is composed of the following algebraic equations:

$$\frac{(u_{i+1/2}^{n+1} - u_{i+1/2}^{n})}{\Delta t} + \frac{(uu)_{i+1}^{n+1} - (uu)_{i}^{n+1}}{\Delta x} + \frac{g}{\Delta x}(h_{i+1}^{n+1} - h_{i}^{n+1})$$
$$+ \frac{g}{[(C_y)_{i+1/2}^{n+1}]^2 h_{i+1/2}^{n+1}}[u_{i+1/2}^{n+1}|u_{i+1/2}^{n+1}|] = 0, \qquad (12.75a)$$

and

$$\frac{[(uA)_{i+3/2}^{n+1} - (uA)_{i+1/2}^{n+1}]}{\Delta x} + \frac{[B_{i+1}^{n+1}(h_{i+1}^{n+1} - h_{i+1}^{n})]}{\Delta t} = 0. \qquad (12.75b)$$

The non-linear terms in Eq. (12.75a) must be written in linear format, and the advective acceleration will be given by:

$$\frac{[(uu)_{i+1}^{n+1} - (uu)_{i}^{n+1}]}{\Delta x} \approx \frac{u_i^{n+1}(u_{i+1}^{n} - u_i^{n})}{\Delta x}, \quad \text{if } u_i^{n+1} < 0, \qquad (12.76a)$$

and

$$\frac{[(uu)_{i+1}^{n+1} - (uu)_{i}^{n+1}]}{\Delta x} \approx \frac{u_i^{n+1}(u_i^{n} - u_{i-1}^{n})}{\Delta x}, \quad \text{if } u_i^{n+1} \geq 0. \qquad (12.76b)$$

The simultaneous application of Eqs. (12.75a, b) with the initial and the associated boundary conditions will generate a system of $I - 1$ equations with same quantity of unknowns, which must be solved for each time step for $n \geq 0$. This whole process is successively and iteratively repeated for each time interval of interest (one or more tidal cycles).

Thus, if the initial and boundary conditions and the estuary geometry are known, it is possible to calculate, with repeated solutions of this linear equation system, the free surface elevation, $h = h(x, t)$ or the surface elevation (tidal height), and the longitudinal velocity field, $u = u(x, t)$, during successive time intervals.

12.6 The Blumberg's Bi-dimensional Model

This classical non-steady-state bi-dimensional (Oxz) numerical model was developed by Blumberg (1975), applying the explicit method of finite differences to a system of equations similar to Eqs. (8.55–8.59, Chap. 8), which corresponds physically to a partially mixed, laterally homogenous estuary (type 2, or B). In this model, the Ox axis is landward orientated with $x = 0$ and $x = L$ indicating the mouth and head positions, and the Oz axis is oriented in the direction contrary to the acceleration of gravity, with $z = -h(x)$ indicating the estuary depth. Thus, the basic equations system which will be numerically integrated, considering for the turbulence a semi-empirical approach, are:

$$\frac{\partial(uB)}{\partial x} + \frac{\partial(wB)}{\partial z} = 0, \tag{12.77a}$$

$$\frac{\partial(uB)}{\partial t} + \frac{\partial(uuB)}{\partial x} + \frac{\partial(uwB)}{\partial z} = \frac{\partial}{\partial x}(BN_x \frac{\partial u}{\partial x}) + \frac{\partial}{\partial z}(BN_z \frac{\partial u}{\partial z})$$

$$- ku|u| \frac{\partial B}{\partial z} - gB \frac{\partial \eta}{\partial x} - \frac{gB}{\rho} \frac{\partial}{\partial x}(\int_z^0 \rho dz), \tag{12.77b}$$

$$\frac{\partial(BS)}{\partial t} + \frac{\partial(uBS)}{\partial x} + \frac{\partial(wBS)}{\partial z} = \frac{\partial}{\partial x}(BK_x \frac{\partial S}{\partial x}) + \frac{\partial}{\partial z}(BK_z \frac{\partial S}{\partial x}), \tag{12.77c}$$

In these equations B is the estuary width, $\rho(S) = \rho_0(\alpha + \beta S)$ is the density calculated by the linear equation of state of seawater, with the following numeric values: $\rho_0 = 0.99891$ g cm^{-3}, $\alpha = 1.0$ and for saline contraction coefficient, $\beta = 7.6 \times 10^{-4}$ $(‰)^{-1}$.

The solution of this system of equations is dependent on the following boundary conditions:

- Salinities at the estuary head, $S(L, z, t)|_{x=L}$, and mouth, $S(0, z, t)|_{x=0}$.
- River discharge Q_f.
- Elevation in relation to the level surface $\eta = \eta(z, t)|_{z=0}$.
- Wind (τ_W) and bottom (T_B) shear stresses, formulated by:

$$BN_z(\frac{\partial u}{\partial z})|_{z=\eta} = B_\eta \frac{\tau_W}{\rho}, \quad \text{and,} \quad BN_z(\frac{\partial u}{\partial z})|_{z=-h} = \frac{T_B}{\rho}. \tag{12.78a}$$

In practical applications these stresses are simulated by semi-empirical expressions, such as (Eqs. 8.25 and 8.31, Chap. 8):

$$\tau_W = \rho_{air} C_D U_V |U_V| \quad \text{and} \quad \tau_B = \rho k(x)|u|u, \tag{12.78b}$$

where $k = k(x)$ is a non-dimensional coefficient calculated in function of the Manning number (n) defined by:

$$k(x) = \frac{gn^2}{(8.23)^2}[h(x)]^{1/3}. \tag{12.79}$$

In this equation, the Manning number is in c.g.s. units, and a typical value for the Potomac river estuary (Washington, USA) is $n = 3.9 \times 10^{-2}$ $(cm)^{1/6}$ (Blumberg 1975).

The precipitation-evaporation balance will be disregarded at the free surface (z = 0), and at the bottom (z = −h) the salt flux is zero. These boundary conditions are expressed by:

$$\rho K_z(\frac{\partial S}{\partial z})_{z=\eta} = \rho K_z(\frac{\partial S}{\partial z})_{z=-h} = 0. \tag{12.80}$$

An additional equation used in the Blumberg's model is obtained from the vertical integration of the continuity equation (12.77a) from the depth z = −h, up to the free surface, z = η, resulting in the following expression:

$$(wB)|_{z=\eta} - (wB)|_{z=-h} + \int_{-h}^{\eta} [\frac{\partial(uB)}{\partial x}] dz = 0. \tag{12.81a}$$

Applying the Leibnitz rule to the last term of this equation yields:

$$(wB)|_{z=\eta} - (wB)|_{z=-h} + (uB)|_{z=\eta}\frac{\partial\eta}{\partial x} - (uB)|_{z=-h}\frac{\partial(-h)}{\partial x} + \frac{\partial}{\partial x}[\int_{-h}^{\eta} (uB)dz] = 0. \tag{12.81b}$$

For the kinematic boundary conditions, the vertical velocity component at the bottom (z = −h) is null, $w(x, z)|_{z=-h} = 0$, and at the surface (z = 0) it is equal to the time variation of the free surface, $w(x, z)|_{z=\eta}=\eta(x, t)$. Thus it follows that:

$$-(wB)|_{z=-h} + (uB)|_{z=-h}\frac{\partial(-h)}{\partial x} = 0, \tag{12.81c}$$

and

$$(wB)|_{z=\eta} = (uB)|_{z=\eta}\frac{\partial\eta}{\partial x} + \frac{\partial(B_\eta\eta)}{\partial t}. \tag{12.82}$$

Applying these results to Eq. (12.81b) yields:

$$\frac{\partial(B_\eta\eta)}{\partial t} + \frac{\partial}{\partial x}[\int_{-h}^{\eta} (uB)dz] = 0. \tag{12.83}$$

The initial conditions imposed on the hydrodynamics equations may be arbitrary because they are parabolic in time, and thus any initial value may be chosen for the forward time solution (t > 0), because they may quickly remove all initial influences (Blumberg 1975).

Solutions to the equation system (Eqs. 12.77a, b, c), can not be obtained analytically. In the Blumberg's technical article, the explicit finite difference method was applied, enabling an algebraic solution which conserves mass (volume), salt and motion in the presence of dissipative effects. To apply this method, the estuary volume was subdivided into a grid defining the knots of interest, containing (I − 1) . (K − 1) partial sub-volumes, where I and K indicate the total number of grid points in the Ox and Oz directions, respectively. Thus, if B is the estuary width at a given longitudinal position, this sub-volume is calculated by B . Δx . Δz.

The corresponding algebraic equations, which satisfy the conservation laws and will enable the determination of u, w, η and S as functions of (x, z, t), are defined at the grid locations shown in Fig. 12.5. This figure indicates that salinity (S) and pressure (p) are defined at the center of each sub-volume, while the vertical velocity component (w) is defined at the top and bottom of it. The grid containing the u-velocity components is staggered with respect to the basic grid as these velocities are defined at the center of the vertical sides of the sub-volume. This staggered arrangement permits easy application of the boundary conditions and evaluation of the dominant pressure gradient forces without interpolation or averaging; the articles of Bryan (1969) and Lendertse et al. (1973) have used similar grids (quoted in Blumberg 1975).

Since most partially mixed estuaries (mainly those that are highly stratified) have higher vertical velocity and salinity gradients than horizontal gradients, the vertical grid spacing must be made much smaller than the horizontal spacing to ensure an adequate resolution of the vertical dimension. The vertical thickness of each

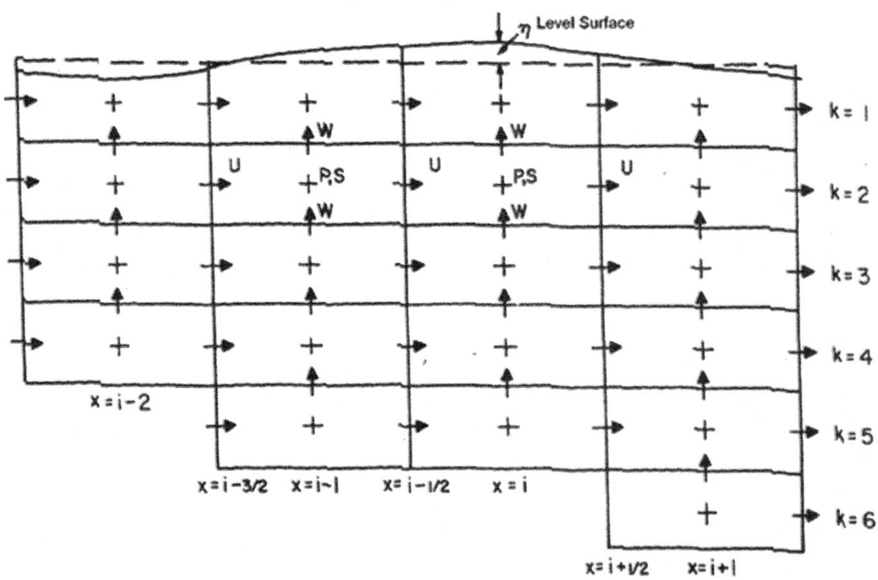

Fig. 12.5 Finite difference grid scheme (after Blumberg 1975)

sub-volume is constant, except in the upper layer where, due to free surface tidal oscillations, its thickness varies in time and space.

To derive the finite difference equations, the following sum and difference operators defined by Schuman (1962) (quoted in Blumberg 1975) were used:

$$\overline{f(x,\ z,\ t)}^{x} \equiv \frac{[f(x+\frac{1}{2}\Delta x,\ z,\ t)+f(x-\frac{1}{2}\Delta x,\ z,\ t]}{2}, \tag{12.84a}$$

$$\delta_x[(f(x,\ z,\ t)] = \frac{f(x+\frac{\Delta x}{2},\ z,\ t)-f(x-\frac{\Delta x}{2},\ z,\ t)}{\Delta x}, \tag{12.84b}$$

and

$$\delta_x\overline{f(x,z,t)}^{x} \equiv \frac{[f(x+\Delta x,z,t)+f(x-\Delta x,z,t]}{2\Delta x}. \tag{12.84c}$$

The $\overline{f(x)}^{n}$ notation is used to mean the function evaluation at a time step, n. The bar and delta operators form a commutative and distributive algebraic operation. Similar operators are defined for x ± Δx and also for the independent variables z and t. Following the method proposed by Lendertse et al. (1973) (quoted in Blumberg (1975)) for vertical integration, and applying the sum and difference operators, the partial differential Eqs. (12.77a, b, c) become:

$$\delta_x(uB)^n+\delta_z(wB)^n=0, \tag{12.85a}$$

$$\begin{aligned}&\delta_t(\overline{SB})^t+\delta_x(\overline{S}^x uB)^n+\delta_z[(wB)\overline{S}^z]^n-\delta_x[BK_x\delta_x(S)]^{n-1}\\&-\delta_z[BK_z\delta_z(S)]^{n-1}=0,\end{aligned} \tag{12.85b}$$

$$\begin{aligned}&\partial_t(\overline{uB})^t+\delta_x(\overline{u}\overline{u}^xB)^n+\delta_z[(\overline{w}^xB)\overline{u}^z]^n-\delta_x[BN_x\delta_x(u)]^{n-1}\\&-\delta_z[BN_z\delta_z(u)]^{n-1}+ku|u|\delta_z(B)^{n-1}_{1/2}\\&+Bg\delta_x\eta^n+Bg\beta\delta_x(\sum_{j=1}^{k}S_j\Delta z_j)^n=0,\end{aligned} \tag{12.85c}$$

$$\begin{aligned}&\delta_t[\overline{S_1B_1(\Delta z+\eta)}]^t+\delta_x[\overline{u_1B_1S_1(\Delta z+\eta)}^x]^n-[(wB)_{3/2}S^{-z}_{3/2}]^n\\&+\delta_x[BK_x(\Delta z+\eta)\delta_x(S)]^{n-1}+[BK_z\delta_z(S)^{n-1}_{3/2}=0,\end{aligned} \tag{12.85d}$$

and, in the top layer, the equations are obtained by vertical integration of Eqs. (12.77b, c) from z = −Δz to z = η,

$$\begin{aligned}&\delta_t[\overline{u_1 B_1(\Delta z+\eta^{-x})}]^t+\delta_x[\overline{u_1 u_1}^x B_1(\Delta z+\eta)]^n-[\overline{w}^x B)_{3/2}\overline{u}^z_{3/2}]^n\\&+\delta_x[(\Delta z+\eta)BN_x\delta_x(u)]^{n-1}+[BN_z\delta_z(u)]^{n-1}_{3/2}\\&+[ku_1|u_1|(B_{1/2}-B_{3/2})]^{n-1}-[\tau_W B_{1/2}]^{n-1}+gB(\Delta z+\overline{\eta}^x)^n\delta_x(\eta)^n\\&-g\beta B\Delta z(\Delta z+\eta^{-x})^n\delta_x(S_1)^n=0.\end{aligned} \tag{12.85e}$$

Equation (12.83) was obtained by vertical integration of the continuity equation over the entire water column depth, and its finite difference expression is written as:

$$B(\eta)\frac{\partial\eta}{\partial t}+\frac{\partial}{\partial x}\int_{-h}^{0}(uB)dz+\frac{\partial}{\partial x}\int_{0}^{\eta}(uB)dz=0, \tag{12.86}$$

and its middle term may be approximated by

$$\frac{\partial}{\partial x}\int_{-h}^{0}(uB)dz\approx\int_{-h}^{0}\frac{\partial(uB)}{\partial x}dz, \tag{12.87}$$

because $(uB)_{z=-h}\approx 0$, taking into account that by the continuity equation, the integrated function of this equation may be approximated by $-\frac{\partial(wB)}{\partial x}$, and it follows that:

$$\frac{\partial}{\partial x}(\int_{-h}^{0}(uB)dz=-\int_{-h}^{0}\frac{\partial(wB)}{\partial x}dz\approx-(wB)_{k=1/2}. \tag{12.88}$$

Under the assumption that the u-velocity component at the free surface is equal to that of the first sub-volume (k = 1) of this layer ($u_{i+1/2,1}$), it is possible to obtain the following approximation for the last term of Eq. (12.86):

$$\frac{\partial}{\partial x}\int_{0}^{\eta}(uB)dz\approx\delta_x(\overline{u_{K=1}B_{k=1/2}\eta})^x. \tag{12.89}$$

Combining the approximations (12.88) and (12.89) with Eq. (12.86), we have:

$$\delta_t(\overline{B\eta})^t-(wB)^n_{1/2}+\delta_x(u_1 B_1\overline{\eta}^x)^n=0. \tag{12.90}$$

The algebraic Eqs. (12.85a, b, c, d, e) and (12.90) constitute a finite difference system. All terms are written in central finite differences in space and time, with the exception of diffusion and friction; the diffusion terms are delayed by one time step to simplify the scheme without losing the equation's conservative property, and the friction terms are delayed one time step to maintain the stability. The full program documentation, a linear stability analysis and its application to the Potomac River estuary (Washington, DC, USA) are presented in Blumberg's technical report.

The finite difference formulation for a non-linear equation may give rise to a special type of instability. As first pointed out by Phillips (1969) (quoted Blumberg 1975), non-linear instability cannot be suppressed by using smaller values of the time step. Although no rigorous theory exists to explain the phenomena the instability, which arises when short-wave disturbances are not damped out, must be removed. In the numeric finite differences program of Blumberg, instability did not become dominant, primarily because of the lack of substantial horizontal gradients, and due to the introduction of an artificial viscosity term written as,

$$K_x = N_x = (\frac{c\Delta x}{\sqrt{2}})^2 |\frac{\partial u}{\partial x}|, \tag{12.91a}$$

where c is an adjustable constant, and Δx is the horizontal grid spacing. Starting with this coefficient, it was demonstrated that the computational procedure becomes stable if the following condition is achieved between the diffusion coefficient K_x, the time step interval (Δt) and the longitudinal grid spacing (Δx):

$$\frac{K_x \Delta t}{(\Delta x)^2} \leq \frac{1}{4}. \tag{12.91b}$$

The computational boundary condition for the velocities are that the water passing through the ocean boundary is constrained to be horizontal (w = 0), and that there is no momentum flux imparted to the estuary by the ocean. The presence of salinity requires additional boundary conditions: (i) when inflow occurs, the salinity is prescribed, and; (ii) in the outflow, hydrodynamic equations governing the interior region determine the salinity. Since the horizontal gradients of salinity near the boundary are small and the flow is horizontal, simple advection can be taken as the governing process for salinity distribution:

$$\frac{\partial S}{\partial t} + \frac{\partial (uS)}{\partial x} = 0, \tag{12.92}$$

and the boundary value is extrapolated along the characteristic solution of the centered difference analogous to this equation.

In the application of Eqs. (12.85a, b, c, d, e), semi-empirical relationships of the vertical kinematic eddy viscosity (N_z) and diffusion (K_z) coefficients were used. The closure of this system of equations is made with a semi-empirical approach using the following equations (Blumberg 1975):

$$K_z = k_1^2 z^2 (1 - \frac{z}{h})^2 |\frac{\partial u}{\partial z}| (1 - \frac{Ri}{Ri_c})^{1/2}, \tag{12.93}$$

and

$$N_z = K_z(1 + Ri), \quad \text{for } Ri < Ri_c, \tag{12.94a}$$

$$N_z = \gamma_c K_z, \quad \text{for } Ri^3 \geq Ri_c, \tag{12.94b}$$

where γ_c is a critical condition of the non-dimensional quantity γ defined by the ratio K_z/N_z.

As should be expected, the staggered grid arrangement and the finite difference increments influence the schematization and the resolution of the velocity and salinity fields. Thus, the grid spacing should be small enough to describe the estuary bathymetry and resolve the salt intrusion limit. In the Blumberg's-2D numerical model, the vertical grid spacing faced the following constraints: (i) stability arising from the finite difference method for the diffusive and viscous terms, and; (ii) the thickness of the upper layer should be larger than the gravity wave amplitude. For optimal numerical results, the restriction was $\Delta z > 4\eta_{max}$, and the equation system required the use of a semi-implicit method.

The usefulness of the numerical model was assessed with several tests, which investigated whether the governing equations were correctly formulated and properly programmed. The first test run checked for the conservation of volume and simulated a non-tidal river flow demonstrated for the following conditions: (i) volume transport through any cross-section using an equation similar to Eq. (7. 103c, Chap. 7); (ii) tidal wave propagation for a long channel with uniform transverse sections; (iii) channel with varying cross-sectional areas; (iv) non--steady-state comparison between flume measurements and computed solutions for times of high and low water, and; (v) comparison of time-averaged numeric model solutions simulated during a tidal cycle with an analytic steady-state solution.

All these tests performed in the numerical model are well documented in Blumberg's technical report. For the last condition (v), an analytical steady-state model was used, with the non-linear terms (advective acceleration) neglected, the gradient pressure force reduced to the baroclinic component and the kinematic eddy viscosity coefficient constant. With these simplifications, the longitudinal equation of motion (Eq. 11.2, combined with 11.3 and 11.5, Chap. 11) is used to calculate the u-velocity component, according to Hunter (1975, quoted in Blumberg (1975)) is given by:

$$-g\beta \frac{\partial}{\partial x} \int_{z}^{0} S dz + N_z \frac{\partial^2 u}{\partial z^2} = 0. \tag{12.95}$$

Disregarding the variation of salinity with depth, $\partial S/\partial z \approx 0$, (weakly stratified or well-mixed estuaries) yields:

$$g\beta\frac{\partial S}{\partial x}+N_z\frac{\partial^3 u}{\partial z^3}=0, \tag{12.96}$$

which will be solved with the following boundary and integral boundary conditions:

- Wind stress at the free surface:

$$\rho N_z\frac{\partial u}{\partial z}\Big|_{z=0}=\tau_W. \tag{12.97a}$$

- Null velocity at the bottom:

$$u(z)|_{z=-h}=u(-h)=0. \tag{12.97b}$$

- Fresh water (volume) conservation:

$$\int_{-h}^{0}u(z)dz=\frac{Q_f}{B},\quad \text{or,}\quad \int_{-1}^{0}u(Z)dZ=\frac{Q_f}{Bh}=u_f, \tag{12.97c}$$

  where B and h are the estuary width and depth, respectively. In Hunter's original article, instead $u(-h)=0$ the bottom boundary condition it was assumed that the bottom shear stress (τ_{Bx}) is linearly related to the velocity. The integral boundary condition (12.97c) indicates that the volume (mass) continuity is preserved.

With three successive integrations of Eq. (12.96) the vertical velocity profile is calculated by:

$$u(z)=-\frac{g\beta}{N_z}\iiint\frac{\partial S}{\partial x}dzdzdz, \tag{12.98}$$

and the general solution is:

$$u(z)=-\frac{gS_x\beta}{6N_z}z^3+\frac{C_1}{2}z^2+C_2z+C_3, \tag{12.99}$$

where $\frac{\partial S}{\partial x}=S_x$, C_1, C_2 and C_3 are integrations constants with the following dimensions $[C_1]=[L^{-1}T^{-1}]$, $[C_2]=[T^{-1}]$ and $[C_3]=[LT^{-1}]$, respectively. Applying the first boundary condition (12.97a), it follows immediately that $C_2=\frac{\tau_W}{\rho N_z}$ and, the boundary conditions (12.97b, c) yield two equations with the unknowns C_1 and C_3:

$$\frac{gS_x\beta h^3}{6N_z} + \frac{1}{2}h^2C_1 + C_3 - \frac{\tau_W h}{\rho N_z} = 0, \tag{12.100a}$$

and

$$\frac{1}{24}\frac{gS_x\beta h^3}{N_z} + \frac{1}{6}C_1h^2 - \frac{1}{2}\frac{\tau_W h}{\rho N_z} + C_3 = \frac{Q_f}{Bh}. \tag{12.100b}$$

Solving this system of equations for the unknowns, C_1 and C_3, its analytical expressions are:

$$C_1 = -\frac{3}{8}\frac{gS_x\beta h}{N_z} - \frac{3Q_f}{2Bh^3} + \frac{3}{2}\frac{\tau_W}{\rho N_z h}, \tag{12.101a}$$

and

$$C_3 = \frac{1}{48}\frac{gS_x\beta h^3}{N_z} + \frac{3}{2}\frac{Q_f}{Bh} + \frac{1}{4}\frac{\tau_W h}{\rho N_z}. \tag{12.101b}$$

Substituting the calculated values of C_1, C_2 and C_3 into the general solution, (12.99), yields the steady-state vertical velocity profile,

$$\begin{aligned} u(z) = &-\frac{1}{6}\frac{gS_x\beta z^3}{N_z} + \left(-\frac{3}{16}\frac{gS_x\beta h}{N_z} + \frac{3}{2}\frac{Q_f}{Bh^3} + \frac{3}{4}\frac{\tau_W}{\rho N_z h}\right)z^2 \\ &+ \frac{\tau_W}{\rho N_z}z + \frac{1}{48}\frac{gS_x\beta h^3}{N_z} + \frac{1}{4}\frac{\tau_W h}{\rho N_z} + \frac{3}{2}\frac{Q_f}{Bh}. \end{aligned} \tag{12.102a}$$

Factoring to reduce the solution to its simplest expression of the u-velocity component in terms of the non-dimensional depth ($Z = z/h$), gives:

$$\begin{aligned} u(Z) = &-\frac{1}{48}\frac{gS_x\beta h^3}{N_z}(-8Z^3 - 9Z^2 + 1) + \frac{3}{2}u_f(Z^2 - 1) \\ &+ \frac{1}{4}\frac{\tau_W h}{\rho N_z}(3Z^2 + 4Z + 1). \end{aligned} \tag{12.102b}$$

As would be expected, the vertical velocity profile is driven by the baroclinic pressure gradient force, river discharge and wind stress, and it is easy to demonstrate that this solution identically satisfies the surface and bottom boundary conditions and the volume (mass) continuity.

With this analytical solution, the steady-state profile of the u-velocity component is calculated using the following values: $N_z = 1.6 \times 10^{-2}$ m^2 s^{-1}, $\beta = 7.0 \times 10^{-4}$, $h = 10.5$ m, $u_f = 0.02$ m s^{-1}, $\tau_W = 0.5$ kg m^{-1} s^{-2} and $\partial S/\partial x = 2.3 \times 10^{-2}$ m^{-1}. The results are presented comparatively in Fig. 12.6, indicating a good agreement of the ebb and flood motion with the analytical and numerical profiles obtained by

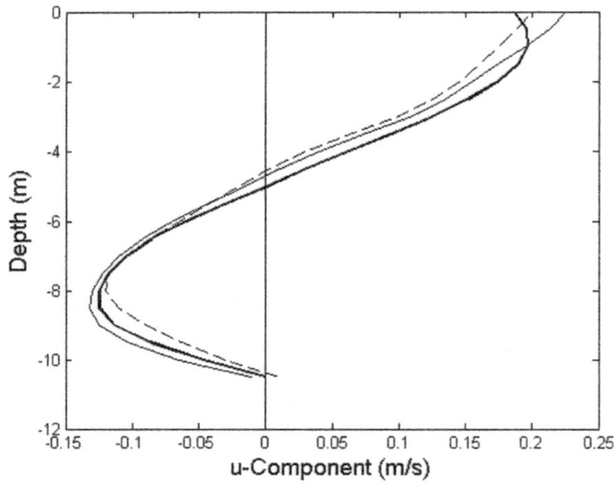

Fig. 12.6 Validation of the vertical u-velocity component profile calculated by a numeric solution (*dashed line*) by comparison with steady-state analytic solutions from the Hunter (1975) model under the conditions of moderate (*thin line*) and maximum bottom friction (*thick line*), respectively. Negative and positive values indicate ebb and flood tidal conditions, respectively (adapted from Blumberg 1975)

Hunter (1975) and Blumberg (1975), respectively. It should be remembered that Hunter's analytical profile, shown in this figure, was calculated with moderate bottom boundary condition.

12.7 Results on Numerical Modelling: Caravelas-Peruípe Rivers Estuarine System

The coastal plain estuary where the Caravelas and Peruípe rivers empty into the coastal sea in the southern Bahia State (Bahia, Brazil), is a complex transitional environment bordering on a mangrove forest and vestigial areas of South Atlantic Forest. The Caravelas-Peruipe Rivers Estuarine System (CPRES) empties its water mass almost 60 km west of the Abrolhos National Marine Park (Fig. 12.7).

This system was sampled during an interdisciplinary, inter-university thematic project "Productivity, Sustainability and Uses of the Abrolhos Banks Ecosystem", sponsored by the National Brazilian Council for Research and Technology Development (CNPq) and the Ministery of Science and Technology (MST). To accomplish the objective of the project, fortnightly estuarine field work was performed in spring and neap tidal cycles in the austral winter and summer of 2007 and 2008, respectively, providing an observational data basis for numerical modeling validation for the estuarine region (Fig. 12.7). Further investigations of this project may be found in the special edition of Continental Shelf Research (2013, v. 70, 176 p.).

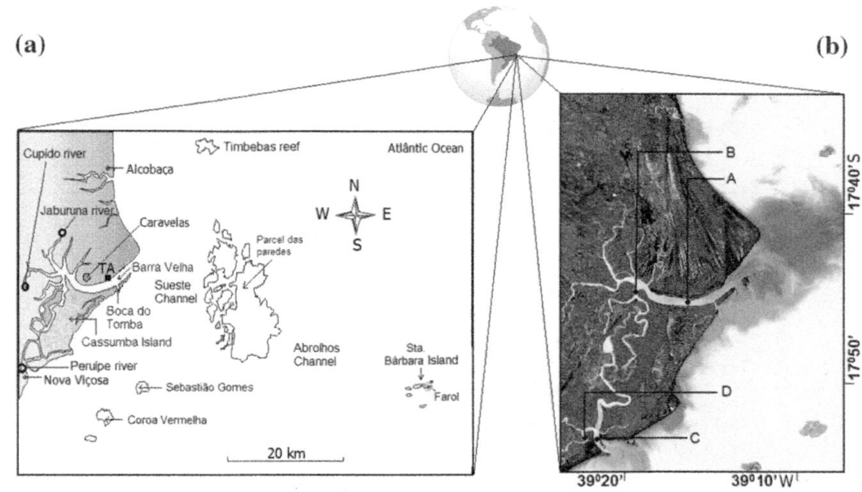

Fig. 12.7 The Caravelas-Peruípe Rivers Estuarine System, the Aracruz—TA harbor, the Sueste and Abrolhos channels, the Abrolhos National Marine Park (*left*). Location of the oceanographic stations in Caravelas (*A*, *B*) and Nova Viçosa (*C*, *E*) in the North and South, respectively (*right*) (according to Andutta 2011)

An outline of some the numerical model (Delft 3-D Flow) results of the estuarine system's spatial and temporal tidal oscillation variability, thermohaline properties, and circulation will be presented in this topic.

The numerical curvilinear grid used in the simulations is presented in Fig. 12.8. To allow better resolution, the grid spacing was locally refined in the estuarine channels to $15 \times 15\ \mathrm{m}^2$, increasing in the coastal region and reaching up to $300 \times 300\ \mathrm{m}^2$. The model results were quantitatively validated using the Skill parameter, with field measurements of tidal oscillation, currents and salinity during neap and spring tides at mooring stations. In the numerical data processing, homogeneous conditions were initially used for the fields of salinity, density and the kinematic vertical coefficients of viscosity and diffusivity; after four weeks of running simulations, these fields were saved under spatially varied conditions. These new initial conditions allowed transient time to be avoided, thus optimizing the simulations.

The model evaluation was performed using field measurements undertaken during the summer and winter austral seasons, and longitudinal measurements in the main channel of the Caravelas estuary, presented in the articles of Schettini and Miranda (2010) and Pereira et al. (2010). River discharge values were taken from the Brazilian National Water Agency (ANA), and were estimated as $\approx 20.0\ \mathrm{m}^3\ \mathrm{s}^{-1}$ with extrapolation of $\approx 4.0\ \mathrm{m}^3\ \mathrm{s}^{-1}$, for the unified Cúpido and Jaburuna rivers, which are tributaries of the Caravelas estuary.

Tidal oscillations at neap and spring tides were well simulated at the four control sites (A, B, C and D) shown in Fig. 12.7, right, which were also used to validate the spatial distribution of tidal heights, circulation and salinity. The best results, with

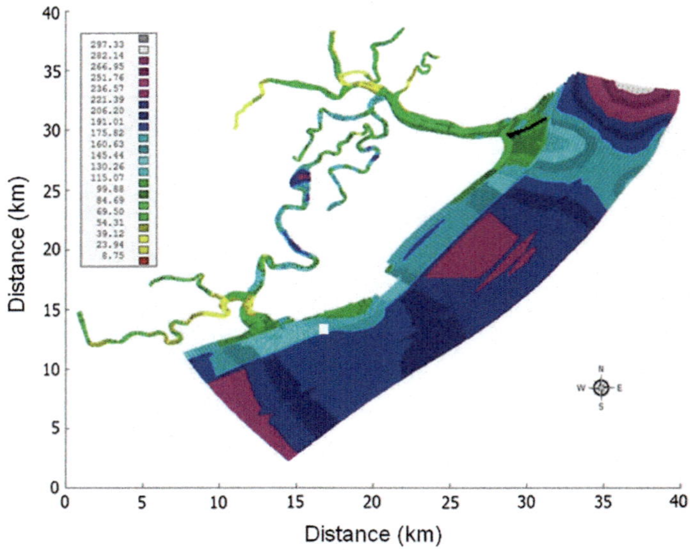

Fig. 12.8 The curvilinear numeric grid and size distributions in the investigated region (according to Andutta 2011)

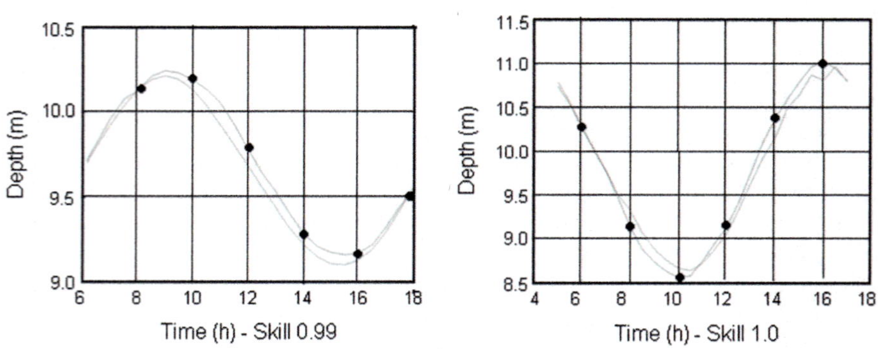

Fig. 12.9 Experimental and theoretical tidal oscillations in the Caravelas estuarine channel at neap (*left*) and spring (*right*) tide in January, 2008 (according to Andutta 2011)

mean values of over 0.9 for the Skill parameter, were obtained for tidal amplitudes between 1.3 and 2.5 m at neap and spring tides (Fig. 12.9), respectively.

Longitudinal velocity component and salinity simulations for January, 2008, at mooring station, A, during neap and spring tides are shown in Figs. 12.10 and 12.11, respectively. The numerical results of the velocities were simulated better in spring tides, with the mean skill values in the range of 0.77 and 0.93, while at neap tides this parameter was lower, with values between 0.38 and 0.65. Good results were achieved for the salinity structure at spring tide, with mean skill values over 0.83, hence comprising all the control station for validation. At neap tides, the

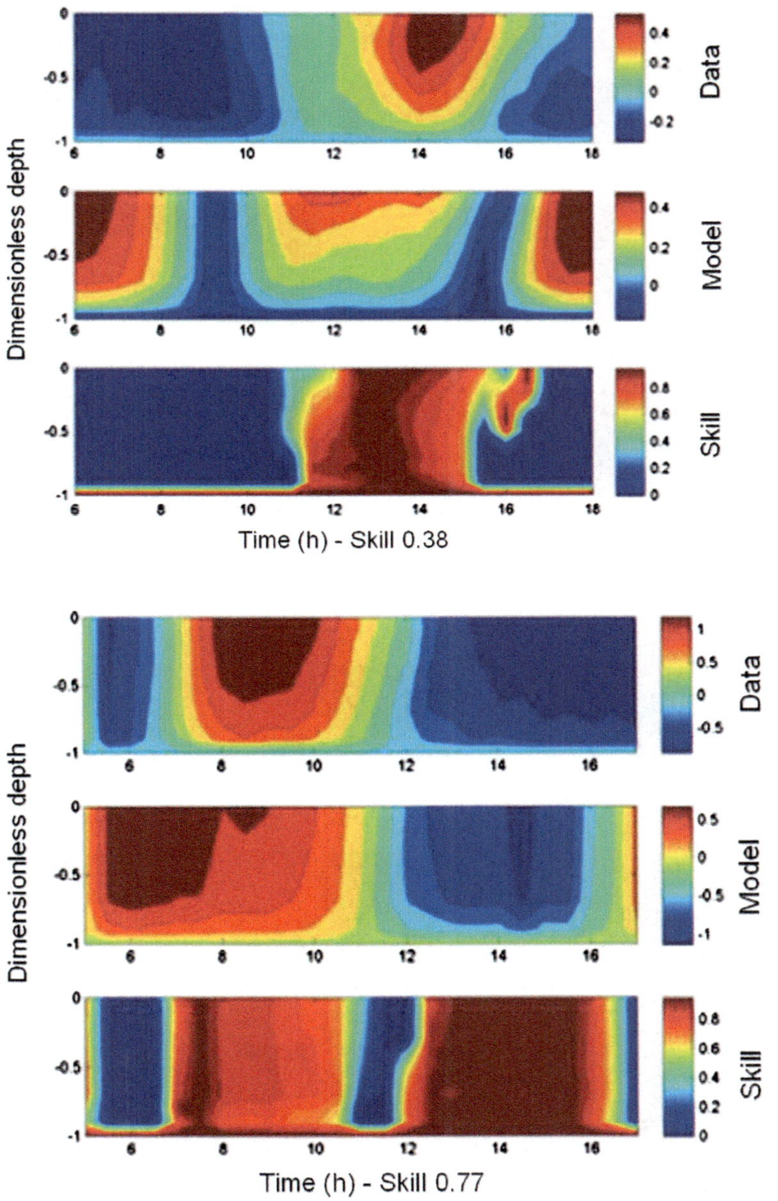

Fig. 12.10 Comparison of time variability of observational u-velocity profiles (m s^{-1}), the corresponding theoretical profiles, and the Skill parameter at station, A, during neap (*upper*) and spring (*lower*) tides. Flood and ebb motions are indicated by $u > 0$ and $u < 0$, respectively (according to Andutta 2011)

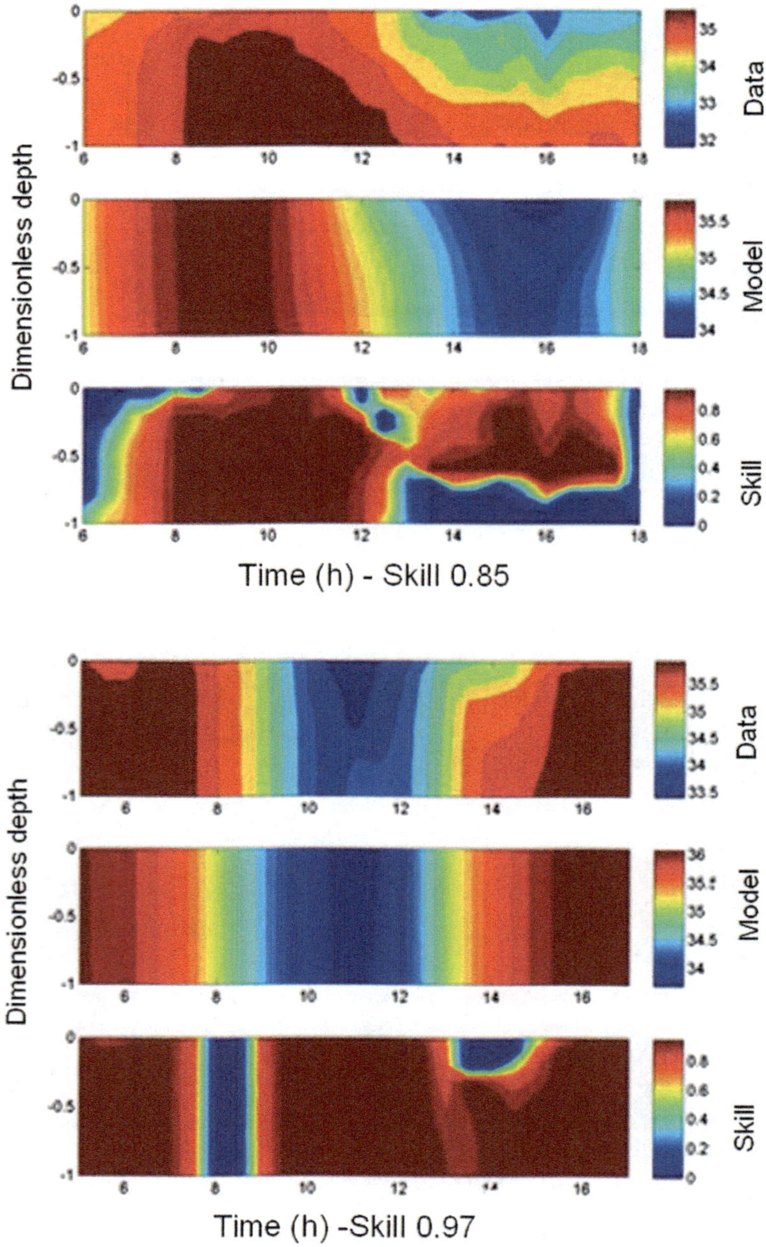

Fig. 12.11 Comparison of time variability of observational salinity (‰) profiles, the corresponding theoretical, and the Skill parameter at station A during neap (*upper*) and spring (*lower*) tidal cycles (according to Andutta 2011)

corresponding mean skill values were relatively high, varying in the range of 0.73–0.85. In addition, there were difficulties adequately simulating the highly vertical and longitudinal salinity stratification in the Nova Viçosa estuary (not shown), which was due to the stronger river inflow on the Peruípe river causing difficulties in the measurements of hydrographic properties and currents in the field.

Time variations of the u-velocity profile, u = u(Z, t), during a semi-diurnal tidal cycle calculated by the model are presented comparatively with the experimental profiles, together with the corresponding Skill parameter. At neap tide (Fig. 12.10, upper), relatively high Skill values (0.6 to 0.8) were obtained, validating theoretical results during the higher current intensities between the time period of $\approx$12 h to $\approx$16 h, but an accentuated phase difference can be observed between theoretical and observational data. However, outside of this high intensity period the Skill parameter indicated very low values (<0.2), reducing its tidal mean value to only 0.38. For the spring tidal period, there was an increase in the mean Skill value, which was double (0.77) that observed in the neap tidal period, indicating a good correspondence between observed and simulated salinity variation at the mooring station (Fig. 12.10, lower).

Variations in the simulated and observational vertical salinity profiles, S = S (z, t), during the neap and spring tidal cycles are shown in Fig. 12.11, upper. In the neap tide cycle, the observational and the theoretical salinity profiles varied in the intervals 32.0–35.8‰ and 34.0–35.8‰, respectively, and the calculated mean skill value was relatively high (0.85). However, in the time interval between 13 and 16 h, the theoretical simulation indicated only a small vertical salinity stratification compared with the observed values, and near bottom, low Skill values were observed (<0.2), and the salt water intrusion near the bottom was well simulated

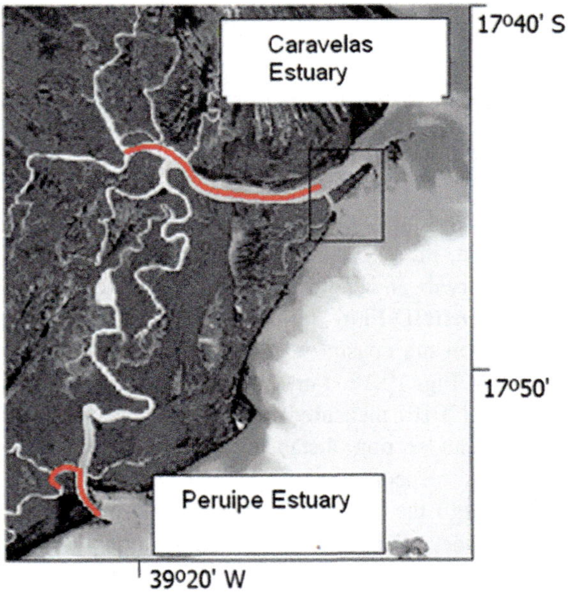

Fig. 12.12 Longitudinal section in the Caravelas estuary used in the numerical simulation of the longitudinal salinity intrusions during neap and spring tidal conditions. The longitudinal section is indicated by the *red line*

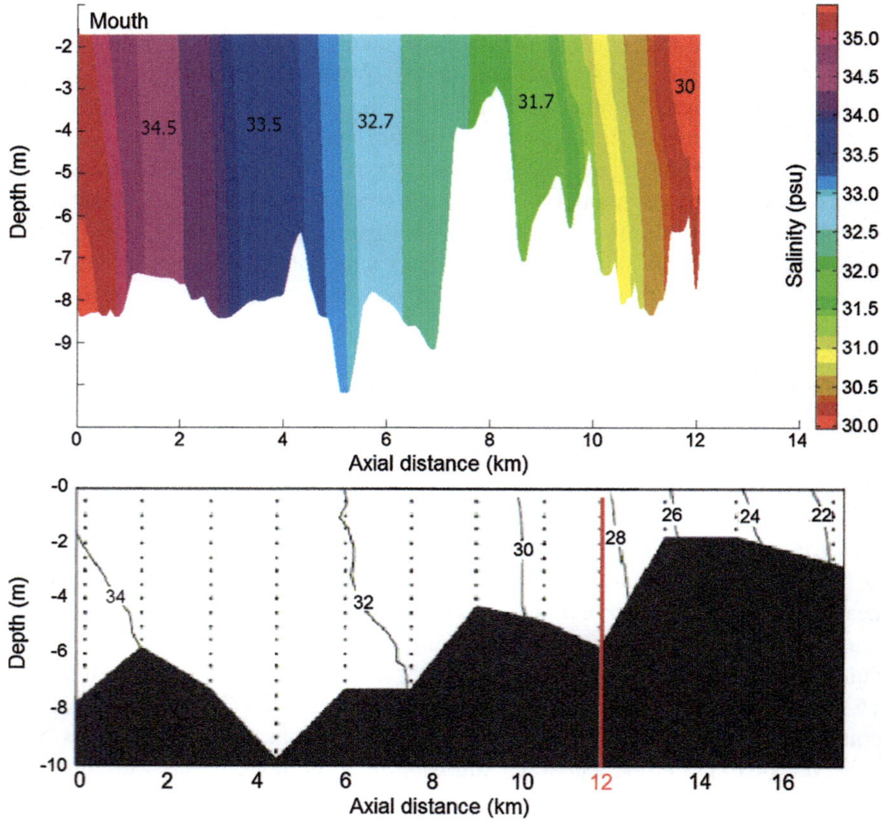

Fig. 12.13 Nearly steady-state longitudinal salinity (‰) distributions in the Caravelas estuary during the spring low tide. Simulated (*upper*) and observational data (*lower*) from Schettini and Miranda (2010) (according to Andutta 2011)

(~35.5‰). In the spring tide cycle (Fig. 12.11, lower), as observed in the velocity simulation, the experimental and simulated salinity values presented close variations intervals and the mean skill values were high (0.85 and 0.97).

The Delft3D-Flow numeric model has also been applied to a comparative analysis on the coastal water mass intrusions in the Caravelas estuarine channel, shown in Fig. 12.12. For these simulations, experimental results of Schettini and Miranda (2010) measured in April, 2001 were used in the validation, covering a longitudinal section distance of 16 and 26 km, for low and high tide salinity intrusions, respectively.

Although the simulations have been compared and validated with no simultaneous observational data and probably under different forcing conditions, the model parameters were adjusted to a higher validation Skill parameter.

During the spring low tide, the nearly steady-state salinity fields indicates a low vertical salinity stratification, with the theoretical and observational salinity varying in the intervals ≈34.5–35.0‰ and ≈34.0–34.5‰, respectively (Fig. 12.13). About 6 km from the mouth, the theoretical and observational values are very close varying from ≈32.0 to ≈32.5‰, and up to ≈12 km landward from the estuary mouth, the salinity decreases to ≈ 30.0 and ≈28.0–28.5‰ for the observational and theoretical distributions, respectively.

The longitudinal nearly-state salinity distribution in the Caravelas estuary at spring high tide is shown in Fig. 12.14. The model results (upper) indicate a vertically well-mixed estuary, which is in agreement with the observational data (lower). It should be noted that the highly saline water (>36.0‰) shows the Tropical Water mass (TW) intrusion advancing up to 6 km into the estuary. A good agreement between the numerical simulation and the experimental data is also observed up to 12 km from the

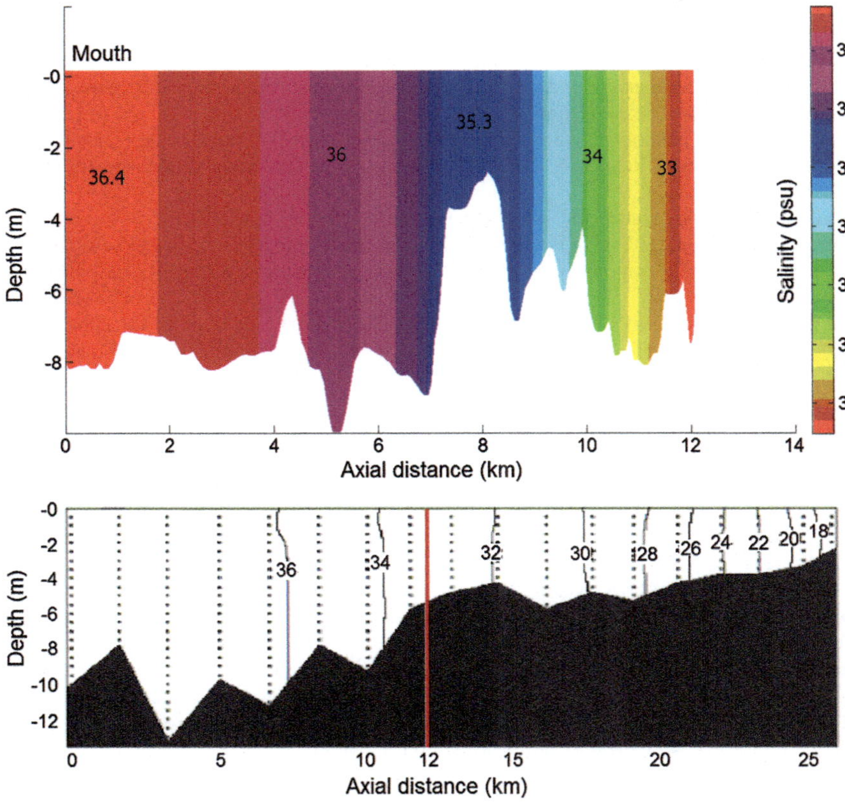

Fig. 12.14 Nearly steady-state longitudinal salinity (‰) distribution in the Caravelas estuary during the spring high tide. Simulated (*upper*), according to Andutta (2011), and the observational data (*lower*) according to Schettini and Miranda (2010)

estuary mouth, with salinities in the interval 36.0 psu —36.4 psu and ≈33 psu for the observational and theoretical longitudinal profiles, respectively. Thus, according to Andutta (2011), these results conclusively indicate that the Tropical Water (TW) mass intrusion into the Caravelas estuary was adequately simulated, including the low salinity estuarine water mass intrusion towards its head.

References

Andutta, F. P. 2011. O Sistema Estuarino dos Rios Caravelas e Peruipe (BA): Observações, Simulações, Tempo de Residência e Processos de Difusão e Advecção. Tese de Doutorado. São Paulo, Instituto Oceanográfico, Universidade de São Paulo. 121 p.

Blumberg, A. F. 1975. A Numerical Investigation into the Dynamics of Estuarine Circulation. Tech. Rept. Chesapeake Bay Institute, The Jonhs Hopkins University. n. 91. 110 p. + Apêndices.

Bryan, K. 1969. A numerical method for the study of the circulation of the world ocean. J. Computational Phys., 4(3):347–376.

Hamrick, J. M. 1992. A Three-dimensional Environmental Fluid Dynamics Computer Code: Theoretical and Computational Aspects. The College of William and Mary, Virginia Institute of Marine Science, Special Report 317, 63 pp.

Hunter, J.R. 1975. A method of velocity interpolation applicable to stratified estuaries. Chesapeake Bay Institute. Special Report 45, The Johns Hopkins University (quoted in Blumberg (1975), p. 68).

Ji, Z-G. 2008. Hydrodynamics and Water Quality. John Wiley & Sons, 676 p.

Lendertsee and Critton, E.C. (1971). A water quality simulation model of well mixed estuaries and coastal seas. Vol. 2. Computational Procedures. Rand Corporation, Report R-708-NYC, New York, 53 pp.

Lendertsee, J.J.; Alexander, R.C. & Liu, S-K. 1973. A three dimensional model for estuaries and coastal Seas, V. 1. Principles of Computation. The Rand Corporation. Santa Monica, CA. No. R-1417-OWRR, 57 p.

McDowell, D.M. & O'Connors, B.A. 1977. Hydraulic Behaviour of Estuaries. John Wiley & Sons, N.Y. 292 p.

Nicholson, J., Broker, I., Roelvink, J.A., Price, D., Tanguy, J.M., Moreno, L. (1997). Intercomparison of coastal area morphodynamic models. Coastal Engineering, 31(1–4):97–123.

Pereira, M. D.; Siegle, E.; Miranda, L. B. & Schettini, C. A. F. 2010. Hidrodinâmica e Transporte de Material Particulado em Suspensão em um Estuário Dominado pela Maré. Rev. Bras. de Geofísica (RBGf), 28(3):427–444.

Phillips, O.M. 1969. The Dynamics of the Upper Ocean. Cambridge University Press, London.

Richtmyer, R.D. & Morton, K.W. 1967. Difference Methods for Initial Value Problems. Interscience, N.Y., 463 p.

Roache, P.J. 1982. Computational Fluid Dynamics. Hermosa Publishers. Albuquerque, M.N., 446 p.

Schettini & Miranda, L. B. 2010. Circulation and Suspended Particulate Matter Transport in a Tidally Dominate Estuary: Caravelas Estuary, Bahia, Brazil. Braz. J. Oceanogr. 58(1):1–11.

Schuman, F.G. 1962. Numerical Experiments with Primitive Equations. Symposium of Numerical Weather Prediction. Met. Soc. Of Japan, Tokyo. pp. 85–107.

Thatcher, M. L. & Harleman, D. R. F. 1972. Prediction of Unsteady Salinity Intrusion in Estuaries: Mathematical Model and User's Manual. Massachusetts Institute of Technology, Mass., Rep. MITSG 72–21. 193 p.

Ward Jr., G. & Espey Jr., W. H. 1971. Estuarine modeling: An Assessment. Water Quality Office, Environmental Protection Agency. Control. Research Series. p. 225.

Printed by Printforce, the Netherlands